The Microbiological Safety of Food in Healthcare Settings

Edited by

Barbara M. Lund, Ph.D., M.Pharm., FIFST
Visiting Scientist,
Formerly, Head of the Microbiological Food Safety and Spoilage Group,
Institute of Food Research,
Norwich Research Park,
Norwich, UK

Paul R. Hunter M.B.A., M.D., F.R.C.Path.,
F.F.P.H., F.I.Biol., Dp. Bact.
Professor of Health Protection
School of Medicine, Health Policy and Practice,
University of East Anglia,
Norwich, UK

Blackwell
Publishing

Blackwell Publishing editorial offices:
Blackwell Publishing Ltd, 9600 Garsington Road, Oxford OX4 2DQ, UK
Tel: +44 (0)1865 776868
Blackwell Publishing Professional, 2121 State Avenue, Ames, Iowa 50014-8300, USA
Tel: +1 515 292 0140
Blackwell Publishing Asia Pty Ltd, 550 Swanston Street, Carlton, Victoria 3053, Australia
Tel: +61 (0)3 8359 1011

First published 2008 by Blackwell Publishing Ltd

ISBN 978-1-4051-2220-7

Library of Congress Cataloging-in-Publication Data

The microbiological safety of food in healthcare settings / edited by Barbara M. Lund, Paul R. Hunter.
p. ; cm.
Includes bibliographical references and index.
ISBN-13: 978-1-4051-2220-7 (hardback : alk. paper)
ISBN-10: 1-4051-2220-X (hardback : alk. paper)
1. Health facilities–Food service–Safety measures. 2. Health facilities–Food service–Quality control.
3. Foodborne diseases–Prevention. 4. Food–Microbiology.
I. Lund, Barbara M. II. Hunter, Paul R.
[DNLM: 1. Food Microbiology. 2. Food Poisoning–prevention & control. 3. Health Facilities–organization & administration. 4. Safety Management–methods.
WC 268 M6269 2008]

RA975.5.D5M53 2008
363.19′26–dc22
2007018633

A catalogue record for this title is available from the British Library

Set in 10/12.5 pt Times
by Aptara Inc., New Delhi, India
Printed and bound in Singapore
by Utopia Press Pte Ltd

The publisher's policy is to use permanent paper from mills that operate a sustainable forestry policy, and which has been manufactured from pulp processed using acid-free and elementary chlorine-free practices. Furthermore, the publisher ensures that the text paper and cover board used have met acceptable environmental accreditation standards.

For further information on Blackwell Publishing, visit our website:
www.blackwellpublishing.com

Contents

Contributors

David W.K. Acheson, MD, FRCP
Assistant Commissioner for Food Protection
United States Food and Drug Administration
5600 Fishers Lane
Rockville, Maryland, 20857, USA

Dinah Barrie, MB, BS, FRCPath
Consultant Microbiologist
Mayday Hospital
Mayday Road
Thornton Heath
Surrey CR7 7YE, UK

Dianne L. Benjamin, MS, RD, CDN, CFSP
Consumer Safety Officer, Retail Food Protection Team
FDA/CFSAN/Office of Food Safety
5100 Paint Branch Parkway
College Park
Maryland 20740, USA

John M. Cowden, MB, ChB, FFPH
Consultant Epidemiologist
Health Protection Scotland
Clifton House, Clifton Place
Glasgow. G3 7LN, UK

Paul R. Hunter, MBA, MD, FRCPath., FFPH., F.I.Biol., Dp. Bact.
Professor of Health Protection
School of Medicine, Health Policy and Practice
University of East Anglia,
Norwich NR4 7TJ, UK

Lisa F. Lubin MS, RD
United States Food and Drug Administration
5600 Fishers Lane
Rockville
Maryland, 20740, USA

Barbara M. Lund, PhD, MPharm, FIFST
Visiting Scientist
Institute of Food Research
Norwich Research Park
Colney
Norwich, NR4 7UA, UK

Sarah J. O'Brien, MB, BS, FRCP, FFPH, DTM&H
Professor of Health Sciences and Epidemiology
University of Manchester
Oxford Road
Manchester, M13 9PT, UK

Jacqui Reilly, PhD, BA (Hons), PgC, LTHE, LPE, RGN
Consultant Epidemiologist/Head of HAI Group
Health Protection Scotland
1 Cadogan Square
Glasgow, G2 7HF, UK

Robert V. Tauxe, MD, MPH
Division of Foodborne, Bacterial and Mycotic Diseases
National Center for Zoonotic, Vectorborne and Enteric Diseases
Centers for Disease Control and Prevention
MS C-09
Atlanta, Georgia 30306, USA

Patrick G. Wall, MVB, MB, BAO, BCh, MSc, MBA, MRCVS, FFPHM, MFPHMI
Professor of Public Health
School of Public Health and Population Science
University College Dublin
Belfield, Dublin 4, Ireland

Jon-Mikel Woody, MS
Policy Analyst, Food Defense Oversight Team
Office of Food Defense, Communication, and Emergency Response
Center for Food Safety and Applied Nutrition
Food and Drug Administration
5100 Paint Branch Parkway
College Park
Maryland 20740, USA

Foreword

The challenge of ensuring safe food in the healthcare setting occurs on the interface between the worlds of curative medicine and public health. One point of intersection is the patient with an infection that might be foodborne. As a centre for diagnosis and treatment, the hospital is a listening post for problems affecting the community at large, and for problems that may be specific to that institution. That infection may have been acquired in the community, and reports of clusters of such infections from astute hospital-based clinicians and microbiologists have often been the first indicator of outbreaks that affect the entire community. A single infection or a cluster of infections may have been acquired at the hospital itself. Conducting surveillance for infections that might have an origin within the hospital depends on rigorous monitoring of illnesses and on setting a low threshold for obtaining diagnostic specimens from patients with gastroenteritis when an outbreak is suspected. Because patients continuously move out of the hospital into the community, infections that may be related to hospital food service may also be identified among persons who are recently discharged. Thus, detecting infections that may be related to hospital exposures also depends on regular communication between the public health authorities conducting surveillance in the community and infection control groups conducting surveillance in hospitals or other healthcare settings.

Because the populations in hospitals and long-term care facilities include many of the most vulnerable individuals, those who become ill there may be at particularly high risk for serious and sometimes fatal consequences. Many kinds of infections may have a foodborne source, though it is rarely possible to definitely determine that source for a single isolated case. Typically, the specific food source of a foodborne infection is only determined in the course of an outbreak investigation. However, sporadic cases of foodborne infections are far more common than outbreaks in community-acquired foodborne infections, and the same may be true for hospital-acquired infections. Because a sporadic individual case may be the herald event that presages an outbreak, it is important to heighten surveillance efforts after a single case is identified, and to consider the food sources within the hospital itself when the likely exposure period includes the time for which the person was in the hospital.

When an outbreak occurs as a result of contaminated food served in a hospital, the persons affected can include patients, staff, visitors and others in the community. The investigation and control of outbreaks of foodborne illness in the healthcare setting can be challenging, because the population turns over rapidly, the patients have a variety of other illnesses, a multitude of possible food and non-food sources need to be considered, and there is a natural reluctance to disrupt routines of care and treatment. In addition to the efforts of the institution's epidemiologist and infection control staff, close coordination with the local health authorities is vital to help clarify links the outbreak may have to events beyond the hospital. Engagement of the local food safety authorities will also help bring their expertise in the evaluation of food service premises, a skill that few hospital infection control staff would be expected to have. The successful investigation may well require the combined efforts of community public health authorities with experience and skill in food safety issues, as well as the hospital and local epidemiological and microbiological staff.

The investigations reveal that most of these outbreaks can be prevented by policies and procedures within the institutions. These same measures are likely to also prevent many of the sporadic cases that may be occurring. It is important to remember that the population served in many healthcare settings includes many persons at higher risk for poor outcomes or even death, because of underlying diseases, or immunocompromising treatments. Therefore, the safety standards for foods provided to such populations need to be set higher than those used for ordinary restaurants or other foods service establishments. In the United States, for example, the Model Food Code, a set of draft regulations recommended by the Food and Drug Administration for adoption by state or local governments, has a special section that covers establishments serving high-risk populations. This section has additional requirements that are beyond what is required of restaurants. For example, it requires the use of pasteurized eggs in many recipes, in place of unpasteurized shell eggs, and has other provisions that reduce the risk for the most vulnerable. Hospital authorities can consider and implement a menu of prevention practices. They could require that kitchen staff receive training and certification in the principles of food safety. They could routinely provide paid sick leave to dietary staff. They could ensure that the layout of the kitchen minimizes cross-contamination and maximizes handwashing, and could provide incentives to make regular handwashing happen. They could put food safety requirements in their food purchase contracts, just as some major restaurant and grocery store chains do. They could routinely use nothing but pasteurized eggs, either as bulk liquid product in a carton or as in-shell pasteurized eggs purchased by the dozen. They could use irradiated meat and poultry products. It is possible that the lower risk could even translate into lower costs of illness, and that economic considerations favour prevention.

In addition, because healthcare facilities serve persons who are most vulnerable to food-borne infections, they can also be educated about food safety, and how they can reduce their risk after they leave the hospital. Consulting dietitians can be a source of preventive information, providing food safety advice to all patients in addition to the advice specific to their condition, as well as consulting with high-risk persons on food safety before they leave hospital and return home.

The dietary service should be seen as an integral part of the general effort to prevent healthcare-associated infections. The daily routines of the dietary service are a central bulwark against disaster. It is unfortunate that this is often ignored by those with direct oversight of hospitals. Routine hospital inspection for accreditation, and audits of the food service consider many aspects of patient safety in some detail. However, they often leave the safety of the foods that are produced and served to hospital patients to the attention of local public health authorities, who may lack resources and the perceived authority to inspect and enforce the higher standards that are needed for the healthcare setting. The recent trend towards outsourcing hospital food to external food establishments should be used as opportunity to require higher standards.

At the same time, the challenges of food safety should remind us that hospitals do not exist in isolation, and that foodborne disease prevention extends all the way back to the farm. Food is one of the primary routes by which microbes and microbial genes flow from farms into communities and hospitals. Outbreaks can occur as the result of contamination in remote pastures, feedlots, henhouses and produce fields as well as because of specific food handling errors in the kitchen itself. Foodborne illnesses show how closely the clinical world can be connected to the ways that food animals and plants are raised and processed. Egg-associated outbreaks of salmonellosis are a recurrent reminder of this. Locally produced foods have been the source of many infections, and need to be chosen with regard to both their quality and safety. At the same time, some infections are imported with foodstuffs from other countries. The rapid shipment of foods across international boundaries, and the rapid transmission of illnesses around the world mean that the state of the world's food supply matters to hospitals as well as to travel clinics.

This text brings together the most up-to-date information on all these dimensions from a panel of experts on both sides of the Atlantic, with a focus on the experience in the United Kingdom. It covers in one volume the salient features of each of the major foodborne pathogens, the intricacies of outbreak investigations, the size and complexity of the at-risk population, the recent regulatory experiences and practical approaches to making sure that the food and water supplies in healthcare facilities are managed for safety. It should be a standard reference work for professionals and students alike, for those managing the safety of food in healthcare institutions, and those providing for the treatment, safety and well-being of the patients that come to them.

Robert V. Tauxe, MD, MPH
Acting Deputy Director
Division of Foodborne, Bacterial and Mycotic Diseases
National Center for Zoonotic, Vectorborne and Enteric Diseases
Centers for Disease Control and Prevention
Atlanta, Georgia 30333, USA

Preface

Healthcare settings include hospitals, nursing homes, institutions and homes for the elderly and for the disabled, nurseries, organizations supporting sick or elderly persons in their homes, and relief for malnourished populations. In many cases individuals in healthcare settings are particularly vulnerable to infections, including foodborne diseases, because of their illness, drug treatment, impaired immune response, or age. Thus, standards for the microbiological safety of foods in healthcare settings need to be more stringent than those for the general population. In addition, advice on safe foods needs to be given to persons in vulnerable groups who live in the community.

The aim of this book is to describe the ascertainment and risks of foodborne disease and its incidence in healthcare settings, and to highlight important features of the provision of safe food in these settings.

This book is intended to be suitable for physicians, doctors and nurses responsible for the control of infection, clinicians, those responsible for catering management, microbiologists, environmental health officers, food scientists and food technologists. The aim is to assist all who have a responsibility for the supply of safe food for persons in healthcare settings and for dietary advice given to vulnerable individuals.

Barbara M. Lund
Paul R. Hunter

Acknowledgement

The editors wish to express their thanks to Dr Tony C. Baird-Parker, for his support for this book and his helpful advice throughout the work, to Professor J.G. Collee for his support and encouragement, and to the staff at Blackwell Publishing for their help in the production of this book.

BML wishes to thank particularly the following scientists who kindly reviewed sections of Chapter 2:

Dr T.C. Baird-Parker and Professor G.W. Gould, formerly of Unilever, Bedford, UK;
Dr G.C. Barker, Professor J. Hinton and Professor M.W. Peck, Institute of Food Research, Norwich, UK;
Dr J.E. Coia, Glasgow Royal Infirmary, Department of Bacteriology, Glasgow, Scotland;
Professor M.P. Doyle, Center for Food Safety, University of Georgia, USA;
Professor S. Fanning, Centre for Food Safety, University College Dublin, Ireland;
Dr C. Gallimore and Dr J. McLauchlin, Centre for Infections, Health Protection Agency, London, UK;
Dr D.N. Lees, Centre for Environment, Fisheries and Aquaculture, Weymouth, UK;
Professor S.J. O'Brien, University of Manchester, Manchester, UK;
Dr R.B. Tompkin, Food Safety Consultant, LaGrange, IL, USA.

She also wishes to thank Mr Paul Pople, Institute of Food Research, Norwich, for assistance with production of figures.

Abbreviations used in the text

AIDS	Acquired immune deficiency syndrome
BSE	Bovine spongiform encephalopathy
CCP	Critical control point
CDC	Centers for Disease Control and Prevention (USA)
CFSAN	Center for Food Safety and Applied Nutrition (USA)
CFU	Colony-forming unit
DNA	Deoxyribonucleic acid
DH	Department of Health (UK)
EFSA	European Food Safety Authority
EHO	Environmental Health Officer
EU	European Union
FDA	Food and Drug Administration (USA)
FSA	Food Standards Agency (UK)
HACCP	Hazard Analysis Critical Control Point
HAI	Healthcare-associated infection
HPA	Health Protection Agency (UK)
HTST	High temperature short time
HUS	Haemolytic uraemic syndrome
ICT	Infection control team
kDa	Kilodaltons
NHS	National Health Service (UK)
PCR	Polymerase chain reaction
PFGE	Pulsed field gel electrophoresis
PHF	Potentially hazardous food
RNA	Ribonucleic acid
RTE	Ready-to-eat
STEC	Shiga toxin-producing *Escherichia coli* (Vero cytotoxin-producing *E. coli*)
TCS	Time/temperature control for safety food
UK	United Kingdom
US	United States of America
USDA	United States Department of Agriculture
WHO	World Health Organization

Interconversion of degrees Centigrade and degrees Fahrenheit

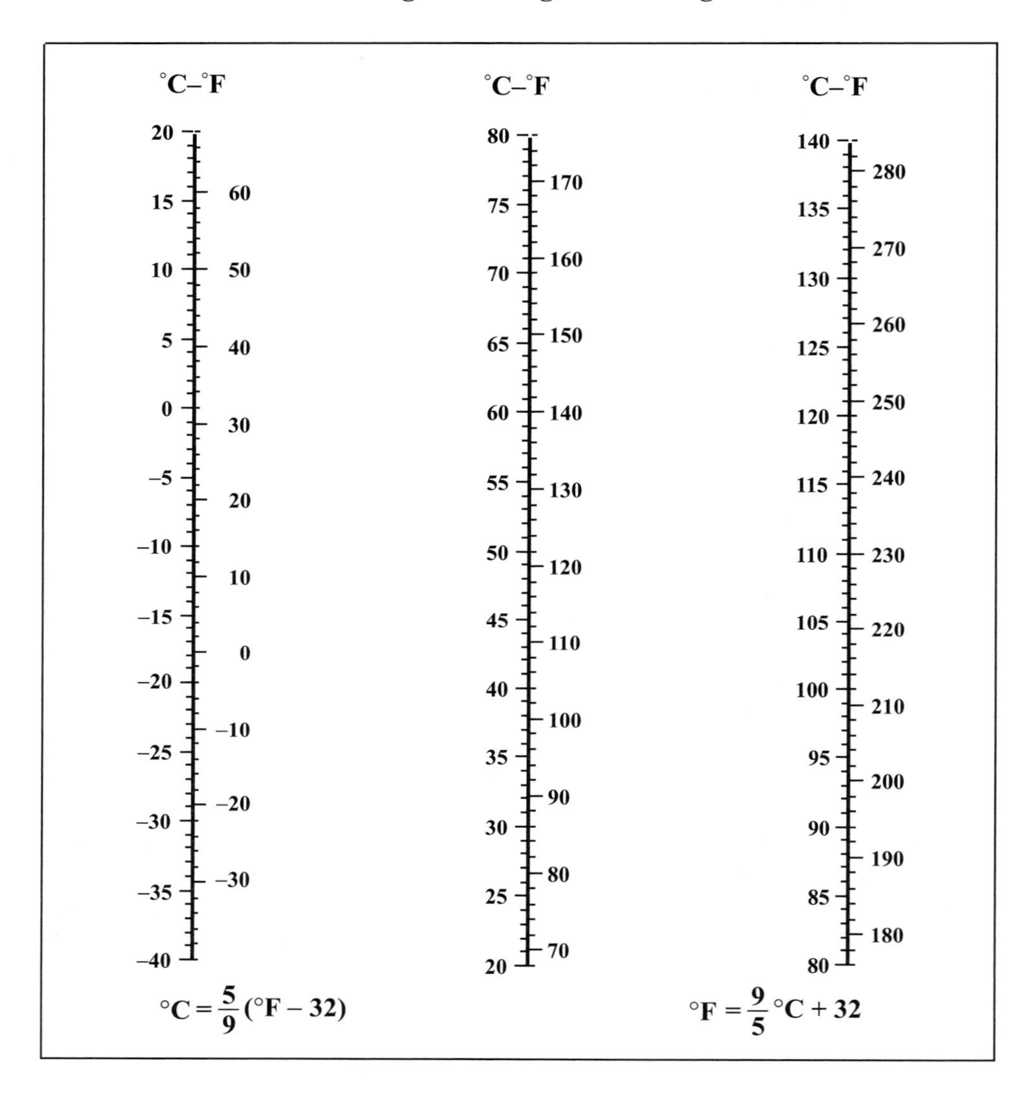

$$°C = \frac{5}{9}(°F - 32)$$

$$°F = \frac{9}{5}°C + 32$$

1 Overview

Patrick G. Wall

1.1 Food safety for vulnerable groups

A series of food-related problems, culminating in BSE, have placed food safety high on the agenda of policy makers and local and international media. The confidence of consumers at large in the safety of food, in the food industry's commitment to produce safe food, and in the ability of enforcement agencies to police the food chain, has been damaged. In the context of healthcare institutions, however, it is not new variant CJD or dioxin contamination that is contributing to morbidity and mortality but a range of bacteria, viruses and protozoa (Chapter 2), causing illnesses that are easily preventable (Rocourt *et al.*, 2003).

The position of catering services within healthcare institutions is often given low priority compared to high-profile medical services. However, catering is pivotal to the operation of the institution. For any patient to recover they must be in an anabolic state, and appetizing, nutritious food is the foundation for achieving this. The patient's diet is as important as therapeutic and surgical interventions. Unfortunately, this is rarely recognized and while there are often fund-raising initiatives for a new MRI scanner or coronary care unit, and politicians love the high media coverage associated with announcing the opening of these facilities, the need for a new kitchen or a blast chiller does not generate the same level of interest.

Food poisoning may not cause major morbidity in robust adults but can be life-threatening in small infants, the frail elderly, or people suffering from concurrent morbidity (Kendall *et al.*, 2003). The frail elderly (Kendall *et al.*, 2006) and the ill (Trevejo *et al.*, 2005) often have lower immunity than healthy adults and the infectious dose to precipitate an infection is lower (Chapter 5). For example, *Enterobacter sakazakii* can cause sepsis, meningitis or necrotizing enterocolitis in neonates and the case-fatality rate has been reported to be as high as 33%, however this pathogen rarely causes disease in adults (Lai, 2001). Similarly, persons over 65 years old are about 7.5 times more susceptible to *Listeria monocytogenes* than persons under 65 with no other condition, and persons with a range of conditions from alcoholism through diabetes, cancer, AIDS and transplant patients show an increased relative susceptibility ranging from 18 to 2584 (WHO, 2004). The problem of low-level contamination of a food

Box 1.1 Food safety advice from the American Cancer Society (2005).

During the time when your white blood cell count is low:

Food safety advice

- Avoid raw milk or milk products; any milk or milk product that has not been pasteurized, including cheese and yoghurt made from unpasteurized milk
- Avoid raw or undercooked meat, fish, chicken, eggs, tofu
- Do not eat cold smoked fish
- Do not eat hot dogs, deli meats, processed meats (unless they have been cooked again just before eating)
- Avoid any food that contains mold (for example blue cheese, including that in salad dressings)
- Avoid uncooked vegetables and fruits
- Avoid uncooked grain products
- Avoid unwashed salad greens
- Do not eat vegetable sprouts (alfalfa, bean and others)
- Avoid fruit and vegetable juices that have not been pasteurized
- Avoid raw honey (honey that has not been pasteurized)
- Do not eat raw nuts or nuts roasted in their shells
- Do not drink beer that has not been pasteurized (home brewed and some microbrewery beers); also avoid brewer's yeast
- Avoid any outdated food
- Do not eat any cooked food left at room temperature for 2 hours or more
- Avoid any food that has been handled or prepared with unwashed hands
- Talk to your doctor about any dietary concerns you may have, or ask to talk with a registered dietician

and resulting outbreak of listeriosis in a vulnerable group is illustrated well by the report by Maijala *et al.* (2001) (Chapter 4). Food safety, hygienic practices and an effective food safety management system are, therefore, crucial in healthcare institutions and other healthcare settings.

From one viewpoint, food safety issues in healthcare environments are different from food safety issues in other commercial catering establishments. Certain high-risk foods should be excluded from all hospital menus. Unpasteurized milk, unpasteurized milk products and undercooked foods that could contain pathogens have no place on the menu in a healthcare institution. For immunocompromised and other high-risk patients, in particular those with haematological malignancies, a low microbial diet (LMD) can be used. LMDs are not sterile, but are diets in which specific foods are excluded in an effort to reduce the risk of infection (UCSF, 2006). An LMD has been advised by the American Cancer Society (2005) for patients when their white blood cell count is low (Box 1.1). Some US hospitals have developed lists of foods that are prohibited and lists of foods that are allowed while on LMD (Roswell Park Cancer Institute, 2005; Kansas City Cancer Center, 2006; San Francisco Children's Hospital, UCSF, 2006). A study carried out in the UK and Ireland in 1991 by Bibbington *et al.* (1993) on bone marrow transplant patients reported that there were wide variations in clinical practice and limited general agreement regarding LMDs. Some UK hospitals have nutritional protocols which include a list of foods that patients should avoid during treatment. It makes little sense to compromise patient well-being, in units where patients are receiving expensive healthcare and costly interventions, by exposing them to food that could contain *Listeria monocytogenes* and other pathogens; therefore, advice similar to that of the American Cancer Society should be followed more universally.

Irradiated foods are considered unacceptable by many consumers in the European Union (EU), but vulnerable patients are likely to hold a different view. The Centers for Disease Control and Prevention estimates that if food irradiation were used for half of the meat and poultry consumed in the US, there would be at least 900,000 fewer cases of foodborne illnesses annually and at least 352 fewer deaths due to foodborne illnesses (Tauxe, 2001). Following US Food and Drug Administration approval of irradiation of raw meat and meat products in February 2000, some healthcare institutions use irradiation-pasteurized chicken (Frenzen *et al.*, 2000). Although less popular, other hospitals and long-term care facilities use irradiation-sterilized foods on a limited basis, to provide immunocompromised patients with microbiologically safe foods that are more varied and higher in quality than meals prepared with the use of thermal sterilization alone (Osterholm & Norgan, 2004).

Some patients who cannot take food orally need to be fed directly into the gastrointestinal tract. The enteral formulae used for these patients are excellent growth media for bacteria; a UK report found that 30% of enteral tube feeds (ETF) were contaminated with a variety of microorganisms (NICE, 2003), and tube feeding may be associated with transmission of infection (Bliss *et al.*, 1998). The European Society for Clinical Nutrition and Metabolism has produced guidelines on enteral nutrition (Lochs *et al.*, 2006) and the Department of Health (UK) (2006) has produced advice aimed at reducing the risk of infection. Some reports suggest that in addition to hygienic preparation, storage and administration, ETF should also be pasteurized before use (Oliveira *et al.*, 2001). In the UK, clinical guidelines state that 'wherever possible pre-packaged, ready-to-use feeds should be used in preference to feeds requiring decanting, reconstitution or dilution' (NICE, 2003).

At the end of the spectrum from acute hospitals are smaller units, residential homes and meal-delivery services. In small units, food may be prepared on-site and staff engaged in patient care may also be involved in food preparation. In situations such as these, the health-care provider/food preparer must pay particular attention to avoiding cross-contamination. Outbreaks of foodborne intestinal disease in residential institutions are common (Ryan *et al.*, 1997). In some residential homes, elderly residents help with the preparation and distribution of meals; if this is occurring, adequate supervision is required to ensure that safe practices are adhered to. In terms of meal delivery services, such as Meals-on-Wheels, volunteers should receive training that addresses safe preparation, cooking and storage of food prior to delivery as well as temperature control and maximum hot/cold holding times for meals both during and after delivery.

1.2 Food procurement

Many large hospitals have outsourced their catering and food provision to contractors who are believed to provide a better service, flexibility and value for money. In this situation, the staff in the kitchens are often employees of the contractor rather than the healthcare institution, and responsibility for staff health and staff training rests with the contractor. Meals may arrive precooked and chilled or frozen, and minimal handling takes place in the healthcare institution. Furthermore, large health institutions often have shops, cafes and canteens on site that supply a range of foods, and these need to be included in the overall food safety programme. These food suppliers are often operated by outside retailers and caterers, and food from these outlets must be safe for both staff and patients.

Food is now traded on the global market and food products and ingredients are sourced from all over the world. To trade in the Single Market, products are required to meet EU standards, and if free trade is to be safe trade then all countries must pay the same attention to enforcement, to ensure full compliance with the legal requirements. The relentless price competition favours countries with economies of scale and cheaper labour and is leading to food travelling greater and greater distances. Increasing liberalization of trade under the recent World Trade Organisation round means that this trend will continue. In sourcing food, traceability is important – if the source of the food is unknown, it is not possible to know whether it has been produced to the highest standards. A requirement for full traceability should be included in all purchasing specifications. In terms of foodborne illness, the health of the residents of institutions is only as safe as the standards of the weakest supplier. Article 18 of Regulation (EC) No. 178/2002 (EC, 2002) spells out that all food business operators shall be able to identify any person from whom they have been supplied with food. There are no exemptions from this requirement for healthcare institutions.

Lessons could be learnt from the policy of the major supermarket chains that only purchase from approved suppliers. To protect their reputations and brand names they include strict food safety requirements in their purchasing specifications. These specifications use accredited industry standards such as the British Retail Consortium (BRC) Global Standard, Food (BRC, 2005), to provide reassurance of compliance. In the UK, the NHS Supply Chain management (NHS Supply Chain, 2007) organizes central tendering for contracts to supply food to many hospitals within the NHS. Hospitals participating in these arrangements and using one of the NHS Supply Chain nominated suppliers will be covered by a due diligence defence. The NHS Supply Chain specifies that successful suppliers must be approved by NHS Supply Chain appointed Hygiene Auditors in compliance with the NHS Code of Practice (NHS, 2001) or with the BRC Global Standard, Food (BRC, 2005) prior to supplying food to a Health Authority; these standards specify that the basis of the company's food safety system shall be a Hazard Analysis and Critical Control Points (HACCP) plan. It is important to source the safest food possible and products from suppliers with aggressive pathogen-reduction programmes should be given preference, e.g. sourcing eggs from flocks that have been vaccinated against *Salmonella enteritidis*. The outbreaks of *Salmonella enteritidis* in British hospitals in 2002 associated with imported, contaminated eggs (Chapter 4) illustrated that cheaper products can mean inferior products. In 1988, the Chief Medical Officer in the UK advised that in institutions with high-risk groups, raw shell eggs should be replaced with pasteurized egg in recipes for products that would not be cooked or would only be lightly cooked. This advice was reiterated by O'Brien and Ward (2002) and by the Food Standards Agency (FSA, 2002).

A procurement policy that focuses on price alone is a recipe for disaster, and a purchasing manager who follows such a policy may be deemed negligent in the advent of any adverse health effects in patients or residents occurring as a result of the consumption of contaminated food. Procedures must be in place for agreeing the contract specification, evaluating contractors who tender for the contract, monitoring the performance of the service delivered and the nutritional quality and safety of the food provided. Healthcare professionals with an expertise in infection control and food safety need to be involved in the development of procurement policies as part of an institution's food safety management system.

In EU legislation, foodstuffs for particular nutritional uses (PARNUTS) are defined as 'foodstuffs which, owing to their special composition or manufacturing process are clearly distinguishable from foodstuffs for normal consumption, which are suitable for their claimed nutritional purposes and which are marketed in such a way as to indicate such suitability' (EEC, 1989). It is essential that these foods are labelled correctly. It should be documented in the procurement policy that any changes in formulation of PARNUTS are discussed with the institutions' healthcare professional. PARNUTS such as high-calorie or high-protein drinks should contain information on the label to advise patients that the products should not be left at room temperature for extended periods of time.

1.3 Factors leading to foodborne disease

Regardless of the size of institutions, the same preventable faults contribute to outbreaks of foodborne disease and often several occur simultaneously (Fitzgerald *et al.*, 2001). The risk factors associated with food preparation procedures and employee behaviours in most need of improvement in hospitals were (FDA/CFSAN, 2004):

1. Improper holding/time and temperature (40.3%)
2. Contaminated equipment/protection from contamination (18.9%)
3. Poor personal hygiene (17.5%)
4. Chemical contamination (13.4%)
5. Inadequate cooking (6.3%)
6. Food from unsafe sources (0.5%)

Temperature control is the most frequently occurring out-of-compliance food safety risk in hospitals. The range of equipment that is required in kitchens will vary depending on whether food is prepared from a raw, cook-chill or ready-to-eat state. Sufficient refrigeration is necessary to ensure that no food that should be stored chilled or frozen is left at ambient temperature. Cookers and microwave ovens should be well-maintained and designed to cope with the throughput of food. Attention has to be given to ensure that food is transported appropriately from the kitchens to the wards within institutions. Cold food has to be kept cold and hot food has to be kept hot to prevent pathogens from multiplying in food. Ward kitchens are a potential source of problems and the staff utilizing them may be different from the catering staff in the main kitchens, and they also require a competence in food safety and hygienic practices (Chapter 6).

Almost 20% (18.9%) of non-compliances with food safety guidelines in hospitals were due to contaminated equipment and insufficient protection from contamination

(FDA/CFSAN, 2004). Often institutions are victims of their own success and as their reputation builds up throughput of patients increases, but food safety standards must be maintained. If necessary, the establishment and work practices should be re-engineered and the kitchens should be of a size appropriate to the throughput. In healthcare institutions, the situation may arise where an individual suffering from a foodborne illness is admitted for treatment, but must be barrier-nursed with rigorous attention to hygiene in order to prevent an infectious agent from spreading within the hospital to both staff and other patients (Barrie, 1996).

The person-to-person route can transmit many of the food- and water-borne pathogens, and poor personal hygiene contributes 17.5% of out-of-compliances in healthcare institutions. Person-to-person spread of bacteria can be a particular risk in healthcare institutions if hygiene standards are low, or if patients have poor personal hygiene as a result of mental or physical incapacity. The index case and initial cases in many outbreaks in institutions may be foodborne but frequently outbreaks are amplified by person-to-person spread (Wall *et al.*, 1996). An added risk in healthcare institutions is that patients may contaminate their own meals or those of other patients, therefore appropriate assistance should be given to patients at meal times and this final step in the process should be adequately supervised. Furthermore, healthcare workers' hands must be decontaminated before every episode of care that involves direct contact with patients' food (Pratt *et al.*, 2001).

It is important to have good occupational health policies to ensure that staff are excluded from work when they are infected with a communicable disease that could be transmitted directly or via food to patients and other staff members. Strategies to reduce the risk of cross-contamination from infected workers are not standardized internationally (NDSC, 2004). Guidelines for management of food handlers who are affected by a gastrointestinal infection have been produced in the UK by the Department of Health (1995, 1996) and by a Working Group of the former PHLS (Salmon, 2004), in Ireland by the National Disease Surveillance Centre (NDSC, 2004) and in the US by the FDA (FDA/CFSAN, 2005). It is recognized that in the case of workers who supply food to highly susceptible groups of the population, the risks and consequences of transmission of infection are greater than in the case of workers who do not supply food to such groups. Furthermore, for certain pathogens that cause severe morbidity and have a low infectious dose, consideration may be given to exclusion from work until stool samples are clear or confining the individual to low-risk activities (NDSC, 2004). For people at increased risk of spreading infection, such as those who work in clinical and social care who have direct contact with highly susceptible persons, it is particularly important to assess them before they return to work (Salmon, 2004).

Food from unsafe sources accounts for 0.5% of out-of-compliances in healthcare institutions and even good food safety systems will be overwhelmed by grossly contaminated product entering the kitchens (Cowden *et al.*, 1995). The water supply, including coolers supplying mineral water and ice dispensers to the institution should be from a potable source complying with all the legal requirements (Chapter 7). The issue of visitors bringing food into the institution for patients needs to be addressed. Currently, most cancer hospitals require that visitors get approval before bringing any food to patients, but one maternity hospital in Ireland, for example, allows fruit, biscuits, sweets or crisps but not cooked food to be brought into the hospital by visitors (National Maternity Hospital, Ireland, 2006). To reduce the likelihood of contaminated food being brought in for patients, the menu should have sufficient choice to cater for all dietary requirements, such that there is no need for food to be brought in for patients. If an institution is not catering for patients' dietary needs/tastes, healthcare workers in that institution should be informed. If any food is brought in for patients it should be items such as biscuits and fresh fruit that will be well washed and preferably peeled. During certain outbreaks of infection, particularly due to norovirus, it may be necessary to ask visitors not to bring in certain foods, including fruit, for patients.

1.4 Food safety management

Ensuring the safety of food is a challenge in all healthcare institutions from small residential units and food-delivery services at homes of senior citizens to large acute hospitals, long-stay psychiatric hospitals and nursing homes. The food safety challenges can vary between and within institutions, from neonatal intensive care units to geriatric wards. However, the basic requirements for good hygienic practices and effective food safety management systems are the same, and in terms of legal requirements healthcare institutions are regarded similarly to any other food business.

The identification of risks and the introduction of risk-management approaches will reduce the likelihood of out-of-compliances occurring. The food safety legislation in the EU has been consolidated and simplified. Article 5 of the EU Regulation on Hygiene of Foodstuffs states that food business operators, including those in healthcare institutions, 'shall' put in place, implement and maintain a permanent procedure, or procedures, based on the principles of (HACCP) (Regulation (EC), 2004) to ensure safe food (Chapter 8). They are required to:

1 identify any hazards that must be prevented, eliminated or reduced to acceptable levels
2 identify the critical control points (CCPs) at which control is essential to prevent or eliminate a hazard or reduce it to acceptable levels
3 establish critical limits at CCPs, which separate acceptability from unacceptability for the prevention, elimination or reduction of identified hazards
4 establish and implement effective monitoring procedures at CCPs
5 establish corrective actions when monitoring indicates that a CCP is not under control
6 establish procedures to verify that the measures outlined in 1–5 above are working effectively
7 document and record measures in 1–6

High-standard facilities appropriate to the volume of food being produced and designed to easily accommodate the operation of HACCP are required (EC, 2004). Another prerequisite to an effective HACCP plan is recruitment and retention of good staff (WHO, 1999). Insufficient staff to run a well-equipped facility with a good food safety management system can contribute to problems. It is a legal requirement that staff preparing and serving food should be supervised adequately or receive training commensurate with the staff members' responsibility (Regulation (EC), 2004). If there is a high turnover of staff, meeting this requirement can present a challenge. Staff are an institution's greatest asset, but untrained staff can be an institution's greatest liability. Despite the legislative requirement, training in itself is often not sufficient to deliver the behavioural changes necessary to work practices. The objective is to achieve competent staff. Good hygienic practices have to be as much a part of the culture of the healthcare institution's kitchens as they are of an operating theatre. Standard operating procedures, facilities and work-flow should be such that there is only one way to prepare the food and that is the hygienic way.

In terms of controlling foodborne disease in healthcare institutions, it is essential that cases and/or outbreaks are detected early (Chapter 3). It can be difficult to differentiate the various causes of diarrhoea, such as antibiotic-associated diarrhoea caused by *Clostridium difficile*, irritable bowel syndrome, or person-to-person spread of norovirus, from foodborne disease. Therefore, all cases of diarrhoea must be documented, faecal samples should be sent for analysis, vigilance should be maintained for clusters of illness or any change in the normal pattern of bowel habit in individual patients or groups of patients. All institutions should arrange to have access to a communicable disease epidemiologist and seek their assistance as soon as there is any suspicion of an outbreak.

Foodborne illness-prevention strategies should also include provision of food safety education for select populations such as immunocompromised persons (e.g. HIV-infected patients, cancer chemotherapy recipients and persons receiving long-term oral steroids or immunosuppressive agents) who are more susceptible to infection than the general population (Chapter 5). Specific information on high-risk foods that should be avoided as well as strategies to reduce their risk of foodborne infection, such as thorough cooking, avoidance of cross-contamination, and short-term refrigerated storage of cooked perishable foods, should be provided. These patients often have repeated short admissions to hospital, and because of their vulnerability it is particularly important that they do not suffer from foodborne illness either in the healthcare institution or at home. Some hospitals offer nutritional education material and classes to patients (e.g. Coborn Cancer Center, U.S., 2000) in order to prevent foodborne infections.

Because of the litany of food scares and food-related illnesses, all EU member states have reviewed the way in which they regulate the food chain and each country now has a central food safety body. These bodies do not all have the identical remit, but risk assessment, risk management and risk communication are key responsibilities of some of them. The European Food Safety Authority was created in 2002 as an agency independent of the EU Commission to undertake risk assessment and risk communication. Responsibility for risk management and legislation remains with the Commission and the Member States. The competent authorities in member states have been encouraged to develop sector-specific guidance on what the hygiene legislation requires and to develop and disseminate codes to good hygiene practice. Healthcare institutions, because of their vulnerable populations, are a priority area for these documents in all member states.

The management of healthcare institutions should familiarize themselves with the legal requirements and codes of good practice. Their core competencies may be in healthcare rather than in food safety, therefore they should seek assistance, if required, to ensure that they are in full compliance. The manager of food services within the healthcare institution has the responsibility to ensure that systems are in place to produce safe food 7 days per week,

365 days per year. Food safety is non-negotiable and an effective food safety management system is a requirement for the protection of the health of the residents and for due diligence defence.

References

American Cancer Society (2005) *Infections in People with Cancer*. Available from http://www.cancer.org/docroot/ETO/content/ETO_1_2X_Infections_in_people_with_cancer.asp. Accessed 5 March 2007.

Barrie, D. (1996) The provision of food and catering services in hospital. *Journal of Hospital Infection* **33**, 13–33.

Bibbington, A., Wilson, P. & Jones, J.M. (1993) Audit of nutritional advice given to bone marrow transplant patients in the United Kingdom and Eire. *Clinical Nutrition* **12**, 230–235.

Bliss, D.Z., Johnson, S., Savik, K. *et al.* (1998) Acquisition of *Clostridium difficile* and *Clostridium difficile*-associated diarrhoea in hospitalised patients receiving tube feeding. *Annals of Internal Medicine* **129**, 1012–1019.

British Retail Consortium (BRC) (2005) *BRC Global Standard, Food*. Issue 4. Available from http://www.brc.org.uk. Accessed 5 March 2007.

Coborn Cancer Center, U.S. (2000) *Session #4: Food Safety 101*. Available from http://www.centracare.com/sch/centers/cancer/cancer_educ.html. Accessed 5 March 2007.

Cowden, J.M., Wall, P.G., Adak, G. *et al.* (1995) Outbreaks of foodborne infectious intestinal disease in England and Wales 1992 and 1993. *Communicable Disease Report (CDR) Review* **5**(8), R109–R117. Available from http://www.hpa.org.uk/cdr/archives/CDRreview/1995/cdrr0895.pdf. Accessed 5 March 2007.

Department of Health, UK (1995) *Food Handlers. Fitness to Work. Guidelines for Food Businesses, Enforcement Officers and Health Professionals*. Department of Health, London. Available from the Food Standards Agency, UK.

Department of Health, UK (1996) *Food Handlers. Fitness to Work. Guidelines for Food Business Managers*. Department of Health, London. Available from the Food Standards Agency, UK.

Department of Health, UK (2006) *Essential Steps to Safe, Clean Care. Enteral Feeding*. Available from http://www.dh.gov.uk/assetRoot/04/13/62/76/04136276.pdf. Accessed 5 March 2007.

EC (2002) Regulation (EC) No. 178/2002 of the European Parliament and of the Council of 28 January 2002 laying down the general principles and requirements of food law, establishing the European Food Safety Authority and laying down procedures in matters of food safety. *Official Journal of the European Union* **L31**, 1 February 2002. Available from http://europa.eu.int/eur-lex/pri/en/oj/dat/2002/l_031/l_03120020201en00010024.pdf. Accessed 5 March 2007.

EC (2004) Regulation (EC) No. 852/2004 of the European Parliament and of the Council of 29 April 2004 on the hygiene of foodstuffs. *Official Journal of the European Union* **L226/3**, 25 June 2004. Available from http://europa.eu.int/eur-lex/pri/en/oj/dat/ 2004/l_139/l_13920040430en00010054.pdf. Accessed 5 March 2007.

EEC (1989) Council Directive 89/398/EEC of 3 May 1989 on the approximation of the laws of the Member States relating to foodstuffs intended for particular nutritional uses. *Official Journal* **L186**, 30 June 1989, pp. 0027–0032.Available from http://eur-lex.europa.eu/LexUriServ/LexUriServ.do?uri=CELEX:31989L0398:EN:HTML. Accessed 5 March 2007.

FDA/CFSAN (2004) *FDA Report on the Occurrence of Foodborne Illness Risk Factors in Selected Institutional Foodservice, Restaurant, and Retail Food Store Facility Types*. Available from http://www.cfsan.fda.gov/~acrobat/retrsk2.pdf. Accessed 5 March 2007.

FDA/CFSAN (2005) *Food Code 2005*. Available from http://www.cfsan.fda.gov. Accessed 5 March 2007.

Fitzgerald, M., Bonner, C., Foley, B. & Wall, P.G. (2001) Analysis of outbreaks of infectious intestinal disease in Ireland: 1998 and 1999. *Irish Medical Journal* **94**, 140 and 142–144.

Food Standards Agency (UK) (2002) *Agency Re-Emphasises Advice on Use and Handling of All Eggs and Issues Guidance on Use of Spanish Eggs Ref R507–28*. Available from http://www.food.gov.uk/news/pressreleases/2002/oct/reemphasiseeggadvice. Accessed 5 March 2007.

Frenzen, P., Majchrowicz, A., Buzby, B., Imhoff, B. & the FoodNet Working Group (2000) *Consumer Acceptance of Irradiated Meat and Poultry Products*. Issues in Food Safety Economics. USDA Economic Research Service Agriculture Information Bulletin No. 757.

Kansas City Cancer Center, US (2006) *Patient Guide: Low Microbial Diet*. Available from http://www.kccancercenters.com/content.aspx?section=search&id=32693. Accessed 5 March 2007.

Kendall, P., Medeiros, L.C., Hillers, V. *et al.* (2003) Food handling behaviors of special importance for pregnant women, infants and young children, the elderly, and immune-compromised people. *Journal of the American Dietetic Association* **103**, 1646–1649.

Kendall, P.A., Hillers, V.V. & Medeiros, L.C. (2006) Food safety guidance for older adults. *Clinical Infectious Diseases* **42**, 1298–1304.

Lai, K.K. (2001) *Enterobacter sakazakii* infections among neonates, infants, children, and adults. *Medicine* **80**, 113–122.

Lochs, H., Allison, S.P., Meier, R. *et al.* (2006) Introductory to the ESPEN guidelines on enteral nutrition: terminology, definitions and general topics. *Clinical Nutrition* **25**, 180–186.

Maijala, R., Lyytikainen, O., Johansson, T. *et al.* (2001) Exposure to *Listeria monocytogenes* within an epidemic caused by butter in Finland. *International Journal of Food Microbiology* **70**, 97–109.

National Disease Surveillance Centre (NDSC) (2004) *Preventing Foodborne Disease: A Focus on the Infected Food Handler*. Report of the Food Handlers with Potentially Foodborne Diseases Subcommittee of the NDSC's Scientific Advisory Committee. Available from http://www.ndsc.ie/hpsc/A-Z/Gastroenteric/FoodborneIllness/Publications/File,871,en.pdf. Accessed 5 March 2007.

National Health Service (NHS) (2001) *Code of Practice for the Manufacture, Distribution and Supply of Food Ingredients and Food-Related Products*. Available from http://www.pasa.doh.gov.uk/food/docs/code_of_practice_2001.pdf. Accessed 8 March 2007.

National Health Service (NHS) Supply Chain (2007) Available from http://www.supplychain.nhs.uk. Accessed 22 February 2007.

National Institute for Health and Clinical Excellence (NICE) (2003) *Prevention of Healthcare-Associated Infections in Primary and Community Care*. Available from http://www.nice.org.uk/pdf/Infection_control_fullguideline.pdf. Accessed 12 February 2007.

National Maternity Hospital, Holles Street, Ireland (2006) *Visitor Information Services*. Available from http://www.nmh.ie/Internet/index.php?page=FrontPageFormat/VisitorInfo/VisitorInfo.html. Accessed 12 February 2007.

O'Brien, S. & Ward, L. (2002) Nosocomial outbreak of *Salmonella enteritidis* PT6a (Nx, CpL) in the United Kingdom. *Eurosurveillance Weekly* **6**(43), 021024. Available from http://www.eurosurveillance.org/ew/2002/021024.asp#20. Accessed 12 February 2007.

Oliveira, M.R., Batista, C.R.V. & Aidoo, K.E. (2001) Application of hazard analysis critical control points system to enteral tube feeding in hospital. *Journal of Human Nutrition and Dietetics* **14**, 397–403.

Osterholm, M.T. & Norgan, A.P. (2004) The role of irradiation in food safety. *New England Journal of Medicine* **350**, 1898–1901.

Pratt, R.J., Pellowe, C., Loveday, H.P. *et al.* (2001) The epic project: developing national evidence-based guidelines for preventing healthcare associated infections. *Journal of Hospital Infection* **47**(Suppl 1), S3–S4.

Rocourt, J., Moy, G., Vierk, K. & Schlundt, J. (2003)*The Present State of Foodborne Disease in OECD Countries* [on-line].Food Safety Department, WHO, Geneva, Switzerland. Available from http://www.who.int/foodsafety/publications/foodborne_disease/oecd_fbd.pdf. Accessed 12 February 2007.

Roswell Park Cancer Institute, US (2005) *In-Patient Low Microbial Dieti* [on-line]. Available from http://www.roswellpark.org/files/1_2_1/BMT/Low%20Microbial%20.pdf. Accessed 12 February 2007.

Ryan, M.J., Wall, P.G., Adak, G.K. *et al.* (1997) Outbreaks of infectious intestinal disease in residential institutions in England and Wales 1992–1994. *Journal of Infection* **34**, 49–54.

Salmon, R. & Working Group (2004) Preventing person-to-person spread following gastrointestinal infections: guidelines for public health physicians and environmental health officers. Prepared by a Working Group of the former PHLS Advisory Committee on Gastrointestinal Infections. *Communicable Disease and Public Health* **7**, 362–384.

Tauxe, R.V. (2001) Food safety and irradiation: protecting the public from foodborne infections. *Emerging Infectious Disease* **7**(Suppl 3), 516–521.

Trevejo, R.T., Barr, M.C. & Robinson, R.A. (2005) Important emerging bacterial zoonotic infections affecting the immunocompromised. *Veterinary Research* **36**, 493–506.

University of California, San Francisco Children's Hospital (UCSF) (2006) *Bone Marrow Transplant. Dietary Concerns During Bone Barrow Transplant.* Available from http://www.ucsfhealth.org/childrens/medical_services/cancer/bmt/diet.html. Accessed 12 February 2007.

Wall, P.G., Ryan, M.J., Ward, L.R. & Rowe, B. (1996) Outbreaks of salmonellosis in hospitals in England and Wales: 1992–1994. *Journal of Hospital Infection* **33**, 181–190.

World Health Organization (WHO) (1999) *Strategies for Implementing HACCP in Small and/or Less Developed Businesses*, 16–19 June, The Hague, Geneva. Available from http://www.who.int/foodsafety/publications. Accessed 5 March 2007.

World Health Organization (WHO) (2004) *Risk Assessment of Listeria Monocytogenes in Ready-to-Eat Food.* MRA Series 4. Interpretative Summary. Available from http://www.who.int/foodsafety/publications/micro/mra_listeria/en/. Accessed 5 March 2007.

2 Properties of Microorganisms that Cause Foodborne Disease

Barbara M. Lund

Introduction

The aim of this chapter is to outline the importance and effect of the main foodborne pathogens covered. Links to detailed information on the mechanism of pathogenicity of the organisms and on methods of detection can be found in the references cited at the end of each section.

An understanding of the factors leading to foodborne disease has developed largely from investigations of outbreaks, which have demonstrated transmission routes, food vehicles and mechanisms of contamination. Tables listing reported foodborne outbreaks have been included in the chapter to demonstrate factors that have led to these outbreaks. In the case of many foodborne pathogens, however, recognized and reported outbreaks account for only a minority of cases of illness, and many cases appear to be sporadic.

In the best surveillance systems, the causative agent and the food vehicle of transmission in a high proportion of outbreaks are not determined. The extent to which outbreaks are reported, investigated and details are published in peer-reviewed papers, depends on the resources and expertise given to surveillance in the relevant country. Countries with the best surveillance tend to publish the most complete and detailed reports on foodborne disease; in countries with more limited surveillance a lower proportion of outbreaks is reported and detailed investigations are more restricted.

Some outbreaks may be large but dispersed over a very wide area, these outbreaks may be very difficult to detect, and cases are liable to be reported as sporadic. Increased use of molecular subtyping of strains in recent years, together with the establishment of associated surveillance networks, has greatly improved recognition of outbreaks, including those that occur over wide areas (Tauxe, 2006).

In the Summaries at the end of the chapter, Table 2A lists groups of foods and the main human pathogenic microorganisms associated with those foods; Table 2B gives a summary of the main microorganisms associated with foodborne disease, the foods with which they are associated and the main controls needed to prevent foodborne illness.

Reference

Tauxe, R.V. (2006) Molecular subtyping and the transformation of public health. *Foodborne Pathogens and Disease* **3**, 4–8.

2.1 *Bacillus cereus*

2.1.1 *Importance as a cause of foodborne disease*

Foodborne disease caused by *Bacillus cereus* is considered usually to be relatively mild and of short duration, and cases are likely, therefore, to be greatly underreported. In England and Wales, one outbreak was reported in 1999 and none in 2000, whereas in the Netherlands and in Norway in 1999 and 2000 *B. cereus* was stated as the most important causative agent identified in reported outbreaks (WHO, 2003).

It was estimated that in 2000, *Bacillus* spp. caused ~1.8% of cases (11,144 cases) of foodborne illness in England and Wales (Adak *et al.*, 2002), and *B. cereus* caused 0.2% of cases (27,360 cases) attributed to known foodborne pathogens in the US (Mead *et al.*, 1999). Other species of *Bacillus* including *B. subtilis* and *B. licheniformis*, have been associated less frequently than *B. cereus* with foodborne disease (Kramer & Gilbert, 1989; Granum & Baird-Parker, 2000); little is known about the pathogenicity of these species and the distinguishing features of strains able to cause foodborne disease (EFSA, 2004). Measures to prevent foodborne illness caused by *B. cereus* are expected to prevent that caused by other *Bacillus* spp.

Bacillus cereus causes two types of foodborne illness: (1) a 'diarrhoeal syndrome', following consumption of food containing high numbers of the vegetative bacteria, which form enterotoxins in the intestine and (2) an 'emetic syndrome' resulting from consumption of food containing emetic toxin formed by growth of bacteria in the food. The bacterium was recognized first as the cause of diarrhoeal illness in 1948, while the emetic syndrome was identified first in the UK in the 1970s.

In addition to foodborne disease, *B. cereus* is an important cause of non-gastrointestinal disease, particularly in vulnerable groups including neonates, intravenous drug users, immunocompromised persons and those with artificial prostheses (Drobniewski, 1993; Logan & Turnbull, 2003).

2.1.2 *Characteristics of the organisms*

Bacillus spp. are Gram-positive, spore-forming bacteria that can grow in aerobic or anaerobic conditions. Spores of *B. cereus* occur widely in the environment and are present in low numbers in most raw foods (Kramer & Gilbert, 1989), but do not cause illness unless conditions allow germination of the spores and multiplication of the vegetative bacteria. A serotyping scheme based on flagellar, H, antigens has been used to characterize strains; it does not identify food-poisoning strains, but certain serotypes are involved most commonly in emetic or diarrhoeal syndromes. Phage typing, plasmid analysis and molecular techniques are used also to characterize isolates.

Some strains of *B. cereus* are mesophilic, with a temperature range for growth of 10–55°C. Other strains that can grow at lower temperatures are termed psychrotrophic, those able to grow at 4–6°C have been grouped as *B. weihenstephanensis* (EFSA, 2004), some strains that grow at 7°C also grow at 42°C (Borge *et al.*, 2001). Psychrotrophic strains are of concern in milk and in chilled foods, growth at 7°C can be sufficiently rapid to give over a 1000-fold increase in numbers in 10 days (Dufrenne *et al.*, 1995), which is within the shelf-life of many chilled foods.

The spores of *B. cereus* are heat-resistant, with decimal reduction times (D values) for mesophilic strains at 121°C in phosphate buffer of between 0.3 and 2.35 minutes (ICMSF, 1996), and for spores of psychrotrophic strains D values at 90°C are between 2.2 and >100 minutes (Dufrenne *et al.*, 1995). Heat resistance may be greater than this in some foods, and these spores will survive normal cooking processes.

Strains that cause the diarrhoeal syndrome form three enterotoxins that have been shown to be involved in food poisoning (Choma & Granum, 2002) while those that cause the emetic syndrome form a low-molecular weight, cyclic peptide (cereulide) toxin. Diarrhoeal strains show great diversity in phenotypic and genotypic properties whereas emetic strains show very low diversity (Ehling-Schulz *et al.*, 2006). Both mesophilic and psychrotrophic strains can form enterotoxins, but psychrotrophic strains may be less able to cause diarrhoeal syndrome than mesophilic strains. While many diarrhoeal strains can grow at 4°C and/or 7°C, very few or no emetic toxin-producing strains have been found that grow at temperatures below 10°C (EFSA, 2004; Carlin *et al.*, 2006).

The enterotoxin(s) are heat-sensitive and are inactivated at 56°C for 5 minutes, but the emetic toxin is highly heat-resistant, withstanding 121°C for 90 minutes (Granum & Baird-Parker, 2000).

Commercial methods are available for detection of enterotoxin, but are not yet available for detection of the emetic toxin; the fact that tests for emetic toxin are not yet made routinely is likely to result in underestimation of food poisoning by emetic strains. A PCR method for detection of emetic strains, based on part of the cereulide synthetase genes, has been developed (Fricker *et al.*, 2007).

2.1.3 *Disease in humans*

The diarrhoeal syndrome is associated with the consumption of proteinaceous foods, vegetables, sauces and puddings, the emetic syndrome is associated particularly with consumption of cooked rice.

The diarrhoeal syndrome is usually caused by foods containing at least 10^5 *B. cereus*/g or mL of food (Granum & Baird-Parker, 2000). The bacteria reach the small intestine where they form enterotoxins. Symptoms occur 8–16 hours, or occasionally >24 hours, after ingestion and last for 12–24 hours, occasionally several days and include abdominal pain, watery diarrhoea and rectal tenesmus, sometimes with nausea.

The emetic syndrome occurs when food is consumed in which *B. cereus* has multiplied to give 10^5–10^8 bacteria/g of food and formed the emetic toxin. Symptoms occur between 30 minutes and 5 hours after ingestion and take the form of nausea, vomiting and stomach cramps, sometimes followed by diarrhoea, possibly due to additional production of enterotoxin.

Usually the symptoms of both types of syndrome are relatively mild, but severe symptoms including bloody diarrhoea, and deaths occurred in an outbreak associated with a diarrhoeagenic strain in a nursing home for the elderly (Lund *et al.*, 2000) and at least two reports have been published of severe illness, liver failure and death caused by emetic strains (Mahler *et al.*, 1997; Dierick *et al.*, 2005).

2.1.4 *Source and transmission*

Spores of *B. cereus* are present in soil and on vegetation and occur, usually in low numbers, in a range of foods including dairy products, meats, spices and dried cereals, particularly rice. The external surface of cereal grains is contaminated heavily with bacteria (Sarrias *et al.*, 2002). During drying about 99% of these bacteria are killed but spores, including those of *B. cereus* and some moulds, remain viable. In spices the number present is usually low but in some cases may be as high as 10^5 CFU/g (Farkas, 2000). Low numbers of *B. cereus* are often found in herbal teas, but the numbers may be high in some products (Michels, 2000). Dried milk and dehydrated potato (King *et al.*, 2007) are commonly contaminated with spores of *B. cereus*.

Spores of *B. cereus* can adhere to, and persist on, the surface of food-processing equipment, which may become a reservoir for the spores (EFSA, 2004).

Usually the spores survive cooking, and if cooked foods are then maintained at temperatures between about 10 and 55°C germination can occur followed by growth of vegetative bacteria. Most outbreaks of foodborne illness due to this organism have been caused by holding cooked food at temperatures that allowed growth of the bacterium, or by slow cooling of large quantities of food. The occurrence of outbreaks of the emetic syndrome associated with cooked (particularly fried) rice from Chinese restaurants and take-away outlets has been linked to the practice of saving portions of boiled rice from bulk cooking, allowing the rice to 'dry off' at ambient temperature after which it may be stored overnight or longer before it is fried quickly with beaten egg. This practice results in survival of spores of *B. cereus* present originally in the rice, followed by germination, growth of vegetative bacteria and toxin formation while the rice is maintained at ambient temperature. In a survey of cooked rice from restaurants and take-away premises in the UK, pre-cooked, stored samples contained the highest numbers of *Bacillus* spp., 6% contained $>10^2$ CFU *B. cereus*/g and 1.1% contained $>10^5$ CFU/g, which was considered potentially hazardous (Nichols *et al.*, 1999).

Psychrotrophic strains of *B. cereus* limit the keeping quality of pasteurized milk, causing 'sweet curdling'. The spores are present in silo tanks of raw milk (Svensson *et al.*, 2004) and survive pasteurization. Emetic toxin-forming strains have been found rarely in milk and in the farm environment, but under some circumstances they occurred frequently at a low level (Svensson *et al.*, 2006); if milk is kept under refrigeration these strains are probably of little concern, but they may be of concern if the milk is used to produce powder.

Dried milk products, infant formulae and other infant foods, and special dietetic foods are often contaminated with spores of *B. cereus*, usually fewer than 100 CFU/g (Becker *et al.*, 1994), this level is most unlikely to cause illness. When naturally contaminated infant food was reconstituted and maintained at room temperature (27°C) for 7–9 hours the number of *B. cereus* increased to more than 10^5 CFU/mL, a level that is hazardous for infants (Becker *et al.*, 1994). It is important, therefore, that foods of this type are used promptly after preparation (see also Section 2.6.). In dried milk products used in a school feeding program 46% contained *B. cereus* spores at levels up to 5×10^3 CFU/g; about 30% of strains were enterotoxigenic and could pose a risk if reconstituted products were held unrefrigerated (Reyes *et al.*, 2007).

A recent report described an association between invasive *B. cereus* infection in children with cancer and the consumption of tea preparations made from the addition of *Camellia sinensis* or other leaves to hot water (El-Saleeby *et al.*, 2004). Further information on this possible risk is needed.

Spores of *Bacillus* spp. survive the pasteurization process used in production of cooked chilled foods ('sous vide' foods, Refrigerated Foods of Extended Durability [REPFEDS]) and can be the main aerobic bacteria present immediately after production (Carlin *et al.*, 2000). Some of these *Bacillus* spp. multiply at 4–10°C, but in vegetable pureés *B. cereus* and other food-poisoning strains were isolated mainly from products that had been stored at higher (abuse) temperatures. Despite the survival of *Bacillus* spp. there do not appear to be reports of foodborne disease associated with these products.

In certain fermented meals used in food seasoning, for example Nigerian fermented vegetable proteins, *B. cereus* is able to multiply resulting in high numbers in the final product (Oguntoyinbo & Oni, 2004). The consumption of foods containing such products may pose a risk to vulnerable groups.

2.1.5 *Examples of outbreaks*

The first well-documented description of outbreaks of the diarrhoeal disease was provided by Hauge in the 1950s following four outbreaks in Norway that involved about 600 cases (Kramer & Gilbert, 1989). Because these outbreaks were similar, only one was described in detail, this involved patients and staff of a hospital. The suspect meal was a lunch that included chocolate pudding and vanilla sauce as a dessert, these items had been prepared on the previous day and stored overnight at room temperature. Vanilla sauce prepared from similar ingredients on the day before use was also the cause of other outbreaks. Samples of sauce that contained between 2.5×10^7 and 1.1×10^8 *B. cereus*/mL showed little change in the odour, taste or consistency; up to 10^4 spores of *B. cereus*/g were found in the cornstarch used to prepare the sauce.

Outbreaks of the emetic disease were described first in the UK in the early 1970s. Between 1971 and 1984, 192 episodes, associated with cooked (usually fried) rice from Chinese restaurants and 'take-away' outlets, were recorded in the UK involving more than 1000 cases (Kramer & Gilbert, 1989).

Several strains of *B. cereus* may be present in a single food associated with illness; from a home-prepared meal that caused symptoms of emesis and diarrhoea in two adults, 68% of 122 isolates of *B. cereus* formed emetic toxin and 26% formed diarrhoeal toxin (Pirhonen *et al.*, 2005).

Examples of more recent outbreaks, caused by *B. cereus*, that have been described in the literature are shown in Table 2.1.1.

2.1.6 *Main methods of prevention and control*

Cleaning and disinfection of equipment used for food production is important to avoid a build-up of spores (EFSA, 2004).

Spores can be destroyed by cooking under pressure or by irradiation but they will survive normal cooking. Cooked foods should be consumed promptly or kept hot (above about

Table 2.1.1 Examples of outbreaks of foodborne disease caused by *Bacillus cereus*.

Place, date	Cases	Deaths	Food implicated	Evidence[a]	Where food was provided	Factors leading to outbreak	Reference
UK, 1976	49[b]	1	Roast chicken, stuffing, potato, green beans, gravy	M (*C. perfringens* may have been involved also)	Meals-on-wheels	Cooked chicken stored overnight at room temperature. (Next day meat portioned, sliced, heated for 30 min at 232°C). Stuffing prepared from dried mix, gravy from chicken and gravy mix; dehydrated potato and green beans re-constituted. Meals dispatched at 80°C but temperature had fallen to 46–50°C within 2 h	Jephcott *et al.* (1977)
US, 1985	160[b]	0	Rice and chicken items	S	Hospital cafeteria	Cafeteria supplied by outside caterer. Deficiencies found in preparation and storage of food items	Baddour *et al.* (1986)
US, 1985	46[b]	0	Beef stew	S	Nursing home (38 patients and 8 staff affected)	After cooking the stew was cooled on a stove at room temperature for 3–5 h. After overnight storage in a refrigerator the stew was warmed but not re-boiled before serving to patients and staff	DeBuono *et al.* (1988)
Netherlands, 1986–89	280[c]	0	Pasteurized milk	M	nr	nr	van Netten *et al.* (1990)
US, 1989	55[b]	0	Game hens	S	Wedding reception with food supplied by caterer	Cross-contamination possible during cooking, maintenance of cooked food at unsafe temperature during delivery in an unrefrigerated van	Slaten *et al.* (1992)
US, 1991	139[b]	0	Barbecued pork	S	University-sponsored barbecue	Pork was left unrefrigerated for >18 h after primary cooking; no method of checking that pork was adequately reheated before consumption	Luby *et al.* (1993)
Norway, 1992	17[b]	0	Meat, vegetable stew	nr	nr	nr	Granum & Baird-Parker (2000)

UK, nr	300[c]	0	Cooked rice	nr; 10^6 *B. cereus*/g of cooked rice	Wedding reception	nr	Ripabelli *et al.* (2000)
US, 1993	14[c] (12 children, 2 staff)	0	Fried rice	S and M	Two jointly owned child day-care centres	Chicken-fried rice prepared in local restaurant. Cooked rice cooled at room temperature before refrigeration. Pan-fried in oil next day, delivered to day-care centres and held without refrigeration for ~1.5 h before serving	CDC (1994)
Norway, 1995	152[b]	nr	Meat, vegetable stew	M; 2×10^5 *B. cereus* detected per gram of food	Meal supplied by a hotel kitchen	nr	Granum & Baird-Parker (2000) (Granum pers. comm)
France, 1998	44[b]	3	Vegetable puree	M	Nursing home for elderly people	nr	Lund *et al.* (2000)
US, 1998	7[c]	0	Rice, handled	nr; 5.6×10^5 *B. cereus*/g of rice	Church day school	6 children and one adult had handled hydrated rice before consuming a meal	Briley *et al.* (2001)
Italy, 2000	173[d]	0	Cakes prepared by confectioner	M	2 banquets	Rolling board in confectioner's shop suspected as a source of contamination	Ghelardi *et al.* (2002)
Netherlands, 2000	~ 100[c]	nr	Vegetarian rice dish	M; emetic toxin detected in food; calculated 10^5–10^8 *B. cereus*/g food	nr	nr	EFSA (2004)
Belgium, 2003	5[c]	1	Pasta salad	M	Family meal	Pasta salad prepared on Friday, taken to a picnic on Saturday, remainder stored in a refrigerator (at 14°C) until served on the Monday evening. This temperature is optimum for formation of emetic toxin	Dierick *et al.* (2005)

[a]M (microbiological): identification of an organism of the same type from cases and in the suspect vehicle or vehicle ingredient(s), or detection of toxin in faeces or food; S (statistical): a significant statistical association between consumption of the suspect vehicle and being a case.
[b]= Diarrhoeal syndrome;
[c]= Emetic syndrome;
[d]Symptoms reported as those of intoxication (nausea, watery diarrhoea), isolates reported to form enterotoxin but not emetic toxin.
nr = not reported

63°C) until served, or cooled rapidly and kept below 7–8°C (ideally below 4°C) preferably for a short time, e.g. a few days (EFSA, 2004). The procedures specified for cooling in relation to *Clostridium perfringens* (Section 2.5.6) will also control *B. cereus*.

In relation to cooking of rice and fried rice the following recommendations have been made (Kramer & Gilbert, 1989; Doyle, 2004):

1 prepare quantities of rice as needed
2 keep prepared rice hot (55–63°C) or cool quickly and transfer to a refrigerator within 2 hours of cooking
3 reheat cooked rice thoroughly before serving

Dried milk products, infant formulae and other infant foods, and special dietetic foods should be used promptly after preparation, not maintained at room temperature for several hours.

Summary of measures advised in healthcare settings

1 Cooked foods (including rice) should be consumed promptly, or kept above 63°C (145°F) for a short time, or cooled rapidly and kept below 7–8°C (ideally below 4°C).
2 Dried milk products and similar powdered foods should be used promptly after reconstitution, not held at room temperature for several hours.
3 Equipment should be kept cleaned and disinfected to avoid build-up of spores.

References

Adak, G.K., Long, S.M. & O'Brien, S.J. (2002) Trends in indigenous foodborne disease and deaths, England and Wales: 1992 to 2000. *Gut* **51**, 832–841.

Baddour, L.M., Gaia, S.M., Griffin, R. & Hudson, R. (1986) A hospital cafeteria-related food-borne outbreak due to *Bacillus cereus*: unique features. *Infection Control and Hospital Epidemiology* **7**, 462–465.

Becker, H., Shaller, G., von Wiese, W. & Terplan, G. (1994) *Bacillus cereus* in infant foods and dried milk powders. *International Journal of Food Microbiology* **23**, 1–15.

Borge, G.I.A., Skeie, M., Sørhaug, T. *et al.* (2001) Growth and toxin profiles of *Bacillus cereus* isolated from different food sources. *International Journal of Food Microbiology* **69**, 237–246.

Briley, R.T., Teel, J.H. & Fowler, J.P. (2001) Nontypical *Bacillus cereus* outbreak in a child care center. *Journal of Environmental Health* **63**, 9–11.

Carlin, F., Fricker, M., Pielaat, A. *et al.* (2006) Emetic toxin-producing strains of *Bacillus cereus* show distinct characteristics within the *Bacillus cereus* group. *International Journal of Food Microbiology* **109**, 132–138.

Carlin, F., Guinebretiere, M.-H., Choma, C. *et al.* (2000) Spore-forming bacteria in commercial cooked, pasteurised and chilled vegetable pureés. *Food Microbiology* **17**, 153–165.

Centers for Disease Control and Prevention (CDC) (1994) Epidemiologic notes and reports *Bacillus cereus* food poisoning associated with fried rice at two child day care centers – Virginia, 1993. *Morbidity and Mortality Weekly Report* **43**(10), 177–178.

Choma, C. & Granum, P.E. (2002) The enterotoxin T (BcET) from *Bacillus cereus* can probably not contribute to food poisoning. *FEMS Microbiology Letters* **217**, 115–119.

DeBuono, B.A., Brondum, J., Kramer, J.M. *et al.* (1988) Plasmid, serotypic, and enterotoxin analysis of *Bacillus cereus* in an outbreak setting. *Journal of Clinical Microbiology* **26**, 1571–1574.

Dierick, K., Van Coillie, E., Swiecicka, I. *et al.* (2005) Fatal family outbreak of *Bacillus cereus*-associated food poisoning. *Journal of Clinical Microbiology* **43**, 4277–4279.

Doyle, M.P. (2004) *Bacillus cereus*. In: *Bacteria Associated with Foodborne Diseases*. Institute of Food Technologists. Available from http://www.foodprocessing.com/whitepapers/2004/4.html. Accessed 5 March 2007.

Drobniewski, F.A. (1993) *Bacillus cereus* and related species. *Clinical Microbiology Reviews* **10**, 324–338.

Dufrenne, J., Bijwaard, M., te Giffel, M. *et al.* (1995) Characteristics of some psychrotrophic *Bacillus cereus* isolates. *International Journal of Food Microbiology* **27**, 175–183.

Ehling-Schulz, M., Svensson, B., Guinebretiere, M.-H. *et al.* (2006) Emetic toxin formation of *Bacillus cereus* is restricted to a single evolutionary linkage of closely related strains. *Microbiology* **151**, 183–197.

El-Saleeby, C.M., Howard, S.C., Hayden, R.T. & McCullers, J.A. (2004) Association between tea ingestion and invasive *Bacillus cereus* infection among children with cancer. *Clinical Infectious Diseases* **39**, 1536–1539.

European Food Safety Authority (EFSA) (2004) Opinion of the scientific panel of biological hazards on "*Bacillus cereus* and other *Bacillus* spp. in foodstuffs". *The EFSA Journal* **175**, 1–49. Available from http://www.efsa.eu.int/science/biohaz/biohaz_opinions/839_en.html. Accessed 5 March 2007.

Farkas, J. (2000) Spices and herbs. In: Lund, B.M., Baird-Parker, T.C. & Gould, G.W. (eds). *The Microbiological Safety and Quality of Food*. Aspen Publishers, Gaithersburg, MD, pp. 897–918.

Fricker, M., Messelhäußer, U., Busch, U. *et al.* (2007) Diagnostic real-time assays for the detection of emetic *Bacillus cereus* strains in foods and recent food-borne outbreaks. *Applied and Environmental Microbiology* **73**, 1892–1898.

Ghelardi, E., Celandroni, F., Salvetti, S. *et al.* (2002) Identification and characterization of toxigenic *Bacillus cereus* isolates responsible for two food-poisoning outbreaks. *FEMS Microbiology Letters* **208**, 129–134.

Granum, P.E. & Baird-Parker, T.C. (2000) *Bacillus* species. In: Lund, B.M., Baird-Parker, T.C. & Gould, G.W. (eds). *The Microbiological Safety and Quality of Foo*d. Aspen Publishers, Gaithersburg, MD, pp. 1029–1039.

International Commission on Microbiological Specifications for Foods (ICMSF) (1996) *Bacillus cereus*. In: *Microorganisms in Foods 5. Microbiological Specifications of Food Pathogens*. Blackie Academic and Professional, London, pp. 20–35.

Jephcott, A.E., Barton, B.W., Gilbert, R.J. & Sheater, C.W. (1977) An unusual outbreak of food-poisoning associated with meals-on-wheels. *Lancet* **310** (Part 8029), 129–130.

King, N.J., Whyte, R. & Hudson, J.A. (2007) Presence and significance of *Bacillus cereus* in dehydrated potato products. *Journal of Food Protection* **70**, 514–520.

Kramer, J.M. & Gilbert, R.J. (1989) *Bacillus cereus* and other *Bacillus* species. In: Doyle, M.P. (ed). *Foodborne Bacterial Pathogens*. Marcel Dekker, New York and Basel, pp. 21–70.

Logan, N.A. & Turnbull, P.C.B. (2003) *Bacillus* and other aerobic endospore-forming bacteria. In: Murray, P.R., Baron, E.J., Jorgensen, J.H., Pfaller, M.A. & Yolken, R.H. (eds). *Manual of Clinical Microbiology*, 8th edn. ASM Press, Washington, DC, pp. 445–461.

Luby, S., Jones, J., Dowda, H. *et al.* (1993) A large outbreak of gastroenteritis caused by diarrheal toxin-producing *Bacillus cereus*. *Journal of Infectious Diseases* **167**, 1452–1455.

Lund, T., De Buyser, M.-L. & Granum, P.E. (2000) A new cytotoxin from *Bacillus cereus* that may cause necrotic enteritis. *Molecular Microbiology* **38**, 254–261.

Mahler, H., Pasi, A., Kramer, J.M. *et al.* (1997) Fulminant liver failure in association with the emetic toxin of *Bacillus cereus*. *New England Journal of Medicine* **336**, 1142–1148.

Mead, P.S., Slutsker, L., Dietz, V. *et al.* (1999) Food-related illness and death in the United States. *Emerging Infectious Diseases* **5**, 607–622.

Michels, M.J.M. (2000) Teas, herbal teas and coffee. In: Lund, B.M., Baird-Parker, T.C. & Gould, G.W. (eds) *The Microbiological Safety and Quality of Food*. Aspen Publishers, Gaithersburg, MD, pp. 960–972.

Nichols, G.L., Little, C.L., Mithani, V. & de Louvois, J. (1999) The microbiological quality of cooked rice from restaurants and take-away premises in the United Kingdom. *Journal of Food Protection* **62**, 877–882.

Oguntoyinbo, F.A. & Oni, O.M. (2004) Incidence and characterization of *Bacillus cereus* isolated from traditional fermented meals in Nigeria. *Journal of Food Protection* **67**, 2805–2808.

Pirhonen, T.I., Andersson, M.A., Jääskeläinen, E.L. *et al.* (2005) Biochemical and toxic diversity of *Bacillus cereus* in a pasta and meat dish associated with a food-poisoning case. *Food Microbiology* **22**, 87–91.

Reyes, J.E., Bastías, J.M., Gutiérrez, M.R. & Rodriguez, M. de la O. (2007) Prevalence of *Bacillus cereus* in dried milk products used by Chilean School Feeding program. *Food Microbiology* **24**, 1–6.

Ripabelli, G., McLauchlin, J., Mithani, V. & Threlfall, E.J. (2000) Epidemiological typing of *Bacillus cereus* by amplified fragment length polymorphism. *Letters in Applied Microbiology* **30**, 358–363.

Sarrias, J.A., Valero, M. & Salmeron, M.C. (2002) Enumeration, isolation and characterization of *Bacillus cereus* strains from raw rice. *Food Microbiology* **19**, 589–595.

Slaten, D.D.S., Oropeza, R.I. & Werner, S.B. (1992) An outbreak of *Bacillus cereus* food poisoning – are caterers supervised sufficiently? *Public Health Reports* **107**, 477–480.

Svensson, B., Ekelund, K., Ogura, H. & Christiansson, A. (2004) Characterization of *Bacillus cereus* isolated from milk silo tanks at eight different dairy plants. *International Dairy Journal* **14**, 17–27.

Svensson, B., Monthán, A., Shaheen, R. *et al.* (2006) Occurrence of emetic toxin producing *Bacillus cereus* in the dairy production chain. *International Dairy Journal* **16**, 740–749.

van Netten, P., van de Moosdijk, A., van Hoensel, P. *et al.* (1990) Psychrotrophic strains of *Bacillus cereus* producing enterotoxin. *Journal of Applied Bacteriology* **69**, 73–79.

World Health Organization (WHO) (2003) *WHO Surveillance Programme for Control of Foodborne Infections and Intoxications in Europe*, 8th report, 1999–2000. Available from http://www.bfr.bund.de/internet/8threport/8threp_fr.htm. Accessed 5 March 2007.

2.2 *Brucella* spp.

2.2.1 *Importance as a cause of foodborne disease*

Brucellosis is a cause of disease in animals and of foodborne disease in humans in many areas of the world. Infection is endemic in animals in Portugal, Spain, Southern France, Italy, Greece, Kuwait, Turkey, North Africa, South and Central America, Eastern Europe, Asia, Africa, the Caribbean and the Middle East (CDC, 2005). Infection is an occupational risk among people working with livestock in countries where the disease is endemic and may occur through contamination of skin wounds or by airborne transmission and inhalation. Foodborne disease is common in areas where the disease is endemic in animals and is transmitted through unpasteurized or recontaminated pasteurized milk and milk products such as cheese or ice cream (Kupulu & Sarimehmetoglu, 2004; CDC, 2005).In many developed countries the disease in animals has been brought under control, resulting in a marked decrease in cases in humans. Human brucellosis is a notifiable disease in many countries. Comparisons of data from different countries on human brucellosis suffer from limitations of differences in reporting systems and in efficiency of reporting. Examples of the reported incidence of brucellosis per 100,000 persons in European countries in 2000 were: Turkey, 15.9; Greece, 5.2; Portugal, 5.07; Italy, 1.8; Switzerland, 0.17; England and Wales, 0.01 (WHO, 2003); in middle Eastern countries such as Kuwait and Saudi Arabia, the incidence may be up to >70 per 100,000 persons (Cutler *et al.*, 2005).

2.2.2 *Characteristics of the organisms*

Brucella spp. are aerobic, non-sporing, non-motile, Gram-negative cocco-bacilli. The four main species associated with disease in humans are *Brucella melitensis*, *B. abortus*, *B. suis* and *B. canis*. *Brucella abortus* is a primary pathogen for cattle, *B. melitensis* for goats and sheep, *B. suis* for swine and *B. canis* for dogs (Moreno *et al.*, 2002). *Brucella melitensis* is the most pathogenic for humans, followed by *B. suis*, *B. abortus* and *B. canis*. Most strains of *Brucella* grow slowly on laboratory media; growth can be improved by the addition to the medium of blood, serum or tissue extracts, but incubation for up to 4 weeks may be required for colonies to form. Many strains of *B. abortus* require 5–10% carbon dioxide for growth. The use of PCR and of fluorescence in situ hybridization can reduce the time required for detection in blood cultures (Wellinghausen *et al.*, 2006). Characteristics of species and biogroups are outlined by Corbel (1998). Differentiation of strains of *Brucella* for epidemiological purposes has been based on biotyping, on patterns of resistance to antibiotics, and more recently on molecular typing methods (Whatmore *et al.*, 2006).

Strains of *Brucella* have been isolated from sea mammals since 1994 (Godfroid, 2002); in regions where these mammals are an important part of the human diet it is possible that this may be a means of transmission of *Brucella* spp. to humans.

The temperature range for growth of *Brucella* spp. is 6–42°C, with an optimum at 37°C (ICMSF, 1996). Pasteurization of milk will inactivate high numbers of brucellas. The organisms can survive for several days in milk and milk products, survival being aided by low storage temperature (ICMSF, 1996), strains surviving for >140 days in cream at 2–4°C, >100 days in yoghurt at 5–10°C and >800 days in raw milk at −40°C.

Brucella strains are classified as dangerous pathogens (Corbel, 1998). They have frequently caused laboratory-acquired infections, therefore cultures and materials likely to be contaminated should only be handled under adequate containment conditions.

2.2.3 *Disease in humans*

Brucella spp. are intracellular pathogens. The dose needed to cause illness is unknown. The incubation period following infection has been estimated as 14 ± 6 days for *B. melitensis* but may be up to 8 weeks, and for *B. abortus* and *B. suis* it may be several weeks or months (ICMSF, 1996). The onset of symptoms can be insidious. Typical symptoms include fever, fatigue, malaise, chills, sweats, myalgia, weight loss and enlargement of liver and spleen (Fosgate *et al.*, 2002). The acute phase includes an undulant fever ('Malta fever', 'Mediterranean fever'). Complications include arthritis, spondylitis, sacroiliitis, endocarditis and neurological disorders (Pappas *et al.*, 2005). The disease is classified as a severe hazard for the general population, with life-threatening or substantial chronic sequelae or long duration (ICMSF, 2002).

2.2.4 *Source and transmission*

Brucella abortus, *B. melitensis*, *B. suis* and *B. canis* cause diseases in domesticated and non-domesticated animals, particularly affecting the reproductive system and resulting in infertility, abortion and birth of weak offspring. While these species are primary pathogens for particular animals, they are not specific to these hosts and *B. melitensis* and *B. suis* have emerged as bovine pathogens in intensive dairy farms. Wild animals can act as reservoirs of infection, for example bison and elk in certain National Parks in the US and Canada are liable to carry *Brucella* sp. and to infect livestock herds (Godfroid, 2002; USDA, 2005) and in several countries in Europe there is a large reservoir of *B. suis* in wildlife, particularly wild boars, and a risk of infection of outdoor-reared herds of pigs (Godfroid & Käsbohrer, 2002). In domestic ruminants, brucellas become localized in the uterus of pregnant animals and may cause abortion. The mammary gland is involved frequently and brucellas are shed in the milk (Chomel *et al.*, 1994).

Consumption of unpasteurized dairy products, especially raw milk, soft cheese, butter and ice cream, is the most common means of foodborne transmission (Pappas *et al.*, 2005). In the US, the epidemiology of brucellosis in humans changed between 1973 and 1992. Before the mid-1970s in the US, cattle and pigs were reported as the most important sources of *Brucella* infecting humans, and the majority of cases of infection were by *B. abortus* and *B. suis* in men exposed to livestock or to contaminated carcasses in packing and rendering plants (Chomel *et al.*, 1994). Following the implementation of measures to eradicate the disease in animals, and general pasteurization of milk, brucellosis in humans has continued to be a problem in California, particularly, where it is now mainly foodborne. In California, it is caused now mainly by *B. melitensis*, and the incidence of the disease is highest in people of Hispanic origin, due probably to the consumption of Mexican soft cheeses, often made with unpasteurized milk from goats or sheep, and which is liable to be imported illegally (Thapar & Young, 1986; Chomel *et al.*, 1994; Fosgate *et al.*, 2002).

Several Northern European countries have attained brucellosis-free status by elimination of the disease in livestock (Godfroid & Käsbohrer, 2002). In Great Britain, *B. melitensis* and *B. suis* have never been reported in livestock and officially brucellosis-free status of cattle herds was attained in December 1996; this status is maintained by means of a programme of surveillance and testing of livestock (DEFRA, 2005). During 2003, some herds of cattle in Scotland were infected with *B. abortus* as a result of the import of infected cattle, but control measures prevented the spread of infection and Great Britain has retained its official brucellosis free (OBF) status. As a result of infection of animals in Northern Ireland, many herds had their OBF status suspended and *B. abortus* continues to be present in cattle. *Brucella abortus* has been practically eliminated from cattle in the US (Ragan, 2002).

2.2.5 *Examples of outbreaks of foodborne brucellosis*

Foodborne outbreaks are seldom described in detail, this is due in part to the long incubation of the disease, the slow growth of the bacterium in culture and the limited availability of methods for typing strains. An outbreak in Malta in 1985 that affected 135 people and caused one death was associated with consumption of soft cheese made with unpasteurized milk of sheep and goats (CDSC, 1995), and an outbreak in Spain in 2002 that affected 11 people was attributed to consumption of cheese made with unpasteurized goats' milk supplied from a farmhouse and by a street-seller (Mendez-Martinez *et al.*, 2003). Since January 2003, brief reports have appeared of foodborne outbreaks in Mexico, Spain, Thailand, Algeria, Lebanon, Saudi Arabia and Uzbekistan, several of which were associated with consumption of unpasteurized goats' milk or cheese made from such milk (Promed-mail, 2004–5).

2.2.6 *Main methods of control*

The incidence of brucellosis in humans can be reduced markedly by reduction or elimination of the disease in animals, using vaccination and slaughter (Cutler *et al.*, 2005). In countries that claim to be free of brucellosis this depends on a large investment in surveillance. The reservoir of brucellas in wild animals constitutes a continuing risk of infection of domestic animals.

In many regions, brucellosis causes a large burden of disease, losses in livestock and severe and debilitating disease in humans (Cutler *et al.*, 2005). In countries where brucellosis occurs in animals, foodborne disease should be prevented by avoiding the consumption of unpasteurized milk and milk products.

Summary of measures advised in healthcare settings

Especially in regions where brucellosis occurs in animals, avoid consumption of unpasteurized milk and milk products.

References

Centers for Disease Control and Prevention (CDC) (2005) *Disease Information. Brucellosis*. Available from http://www.cdc.gov/ncidod/dbmd/diseaseinfo/brucellosis_g.htm. Accessed 5 March 2007.

Chomel, B.B., DeBess, E.E., Mangiamele, D.M. *et al.* (1994) Changing trends in the epidemiology of human brucellosis in California from 1973 to 1992: a shift toward foodborne transmission. *Journal of Infectious Diseases* **170**, 1216–1223.

Communicable Disease Surveillance Centre (CDSC) (1995) *Brucellosis Associated with Unpasteurised Milk Products Abroad. Communicable Disease Report CDR Weekly* **5**(32). Available from http://www.hpa.org.uk/cdr/archives/1995/cdr3295.pdf. Accessed 5 March 2007.

Corbel, M.J. (1998) *Brucella*. In: Collier, L., Balows, A. & Sussman, M. (eds). *Topley and Wilson's Microbiology and Microbial Infections*, 9th edn. *Vol. 2: Systematic Bacteriology* (Volume editors Balows, A. & Duerden, B.I.). Arnold, London, Chapter 35, pp. 829–852.

Cutler, S.J., Whatmore, A.M. & Commander, N.J. (2005) Brucellosis – new aspects of an old disease. *Journal of Applied Microbiology* **98**, 1270–1281.

Department for Environment and Rural affairs (DEFRA) (2005) *Zoonoses Report, United Kingdom* 2004, pp. 40–42.

Fosgate, G.T., Carpenter, T.E., Chomel, B.B. *et al.* (2002) Time–space clustering of human brucellosis, California, 1973–1992. *Emerging Infectious Diseases* **8**, 672–678.

Godfroid, J. (2002) Brucellosis in wildlife. *Revue Scientifique et Technique de l'Office International des Epizooties* **21**, 277–286.

Godfroid, J. & Käsbohrer, A. (2002) Brucellosis in the European Union and Norway at the turn of the twenty-first century. *Veterinary Microbiology* **90**, 135–145.

International Commission on Microbiological Specifications for Foods (ICMSF) (1996) *Brucella*. In: *Microorganisms in Foods. Microbiological Specifications of Food Pathogens.* Blackie Academic and Professional, London, pp. 36–44.

International Commission on Microbiological Specifications for Foods (ICMSF) (2002) Ranking of foodborne pathogens or toxins into hazard groups. In: *Microorganisms in Foods. Microbiological Testing in Food Safety Management.* Kluwer Academic/Plenum Publishers, New York, pp. 166–172.

Kupulu, O. & Sarimehmetoglu, B. (2004) Isolation and identification of *Brucella*. spp. in ice cream. *Food Control* **15**, 511–514.

Mendez-Martinez, C., Paez Jimenez, A., Cortes Blanco, M. *et al.* (2003) Brucellosis outbreak due to unpasteurized raw goat cheese in Andalucia (Spain), January–March, 2002. *Euro Surveillance* **8**(7–8), 164–168. Available from http://www.eurosurveillance.org/em/v08n07/v08n07.pdf. Accessed 5 March 2007.

Moreno, E., Cloeckaert, A. & Moriyón, I. (2002) *Brucella* evolution and taxonomy. *Veterinary Microbiology* **90**, 209–227.

Pappas, G., Akritidis, N., Bosilkovski, M. & Tsianos, E. (2005) Brucellosis. *New England Journal of Medicine* **352**, 2325–2336.

Promed-mail (2004–5) Available from http://www.promedmail.org.

Ragan, V.E. (2002) The Animal and Plant Health Inspection Service (APHIS) brucellosis eradication program in the United States. *Veterinary Microbiology* **90**, 11–18.

Thapar, M.K. & Young, E.J. (1986) Urban outbreak of goat cheese brucellosis. *Pediatric Infectious Diseases* **5**, 640–643.

USDA (2005) *Brucellosis. Animal and Plant Inspection Services*. Available from http://www.aphis.usda.gov/vs/nahps/brucellosis. Accessed 5 March 2007.

Wellinghausen, N., Nöckler, K., Sigge, A. *et al.* (2006) Rapid detection of *Brucella* spp. in blood cultures by fluorescence in situ hybridization. *Journal of Clinical Microbiology* **44**, 1828–1830.

Whatmore, A.M., Shankster, S.J., Perrett, L.L. *et al.* (2006) Identification and characterization of variable-number tandem-repeat markers for typing of *Brucella* spp. *Journal of Clinical Microbiology* **44**, 1982–1993.

World Health Organization (WHO) (2003) *WHO Surveillance Programme for Control of Foodborne Infections and Intoxications in Europe*, 8th report, 1999–2000. Available from http://www.bfr.bund.de/internet/8threport/8threp_fr.htm. Accessed 5 March 2007.

2.3 *Campylobacter* spp.

2.3.1 *Importance as a cause of foodborne disease in humans*

Campylobacter jejuni subsp. *jejuni* (referred to subsequently as *C. jejuni*) and to a lesser extent *C. coli* are now the most commonly reported bacterial cause of gastrointestinal disease in humans in many industrialized countries. The majority of cases of infection are attributed to transmission in food or water. Most cases are reported as sporadic and not as part of an outbreak, but many cases reported as sporadic may be part of unrecognized outbreaks (Campylobacter Sentinel Surveillance Scheme Collaborators, 2003). Campylobacter infections are greatly under-reported (WHO, 2003), but in many European countries the number of reported cases is increasing (Takkinen *et al.*, 2003).

2.3.2 *Characteristics of the organisms*

The 'thermophilic' campylobacters, which include *C. jejuni*, *C. coli* and *C. lari*, are Gram-negative, non-sporing, curved, S-shaped or spiral-shaped bacteria with a single polar flagellum on one or both ends resulting in a corkscrew-like motility, which may facilitate their penetration through mucous layers. In old cultures and in adverse conditions the bacteria can become coccoid. They require complex growth media and usually are inhibited by atmospheric oxygen, requiring an atmosphere containing 5–7% oxygen and 10% carbon dioxide. Optimum growth temperatures are 37–42°C and little or no growth occurs at 30°C or below, thus the bacteria rarely if ever multiply in/on food in temperate or cold climates unless it is held at 30–45°C.

Campylobacter fetus subsp. *fetus* (referred to subsequently as *C. fetus*) differs from the thermophilic strains in being rod-shaped and in growing at 25°C, whereas only a proportion of strains grow at 42°C (Nachamkin & Skirrow, 1999). Cultural methods for isolation of *C. jejuni* and *C. coli* commonly use incubation at 42°C, which would prevent growth of many strains of *C. fetus*.

Injury and death of campylobacters occurs at about 45°C and higher. The bacteria are, in general, more sensitive to heat treatment, freezing, drying and antimicrobials than are other foodborne pathogenic bacteria. Although the numbers present in foods are reduced by freezing, a proportion of the bacteria can survive in frozen food, and campylobacters have been detected in foods such as frozen meat and poultry.

Most campylobacteriosis in humans is caused by *C. jejuni*. In England and Wales in 2000–2002, 91.76% of cases were caused by *C. jejuni*, 8.05% by *C. coli* and <0.2% by *C. lari*, *C. fetus* and other *Campylobacter* spp. (HPA, 2003); national data from Germany in 2002–2003 showed that 12–13% of cases were caused by *C. coli* (Gürtler *et al.*, 2005). For surveillance on a broad scale, serotyping, based mainly on either heat-stable antigens (the Penner scheme) or on heat-labile antigens (the Lior scheme), has been used widely, supplemented by biotyping and phage typing and increasingly by molecular methods particularly pulsed-field gel electrophoresis (PFGE) and multilocus sequence typing (MLST) (Sails *et al.*, 2003; ACMSF, 2005).

2.3.3 *Disease in humans*

In developing countries frequent infection with campylobacters is common in early childhood and immunity develops (Skirrow & Blaser, 2000). In developed countries immunity is not widespread in adults. Ingestion of ~500 *C. jejuni* may be sufficient to cause illness and the incubation period before development of symptoms is ~3.2 days (range 1–7 days). Symptoms include abdominal pain (which may mimic appendicitis), watery and/or bloody diarrhoea, malaise, fever, myalgia and sometimes vomiting. In previously healthy adults illness usually lasts less than 1 week, but patients continue to excrete campylobacters in their faeces for several weeks after they have recovered clinically, unless the infection has been treated with antibiotics. Bacteraemia is rare in previously healthy persons, it is most common in patients >65 years old, particularly in immunocompromised persons. In persons with AIDS the incidence of illness due to *C. jejuni* is 40–100 times greater than in the general population (Wassenaar & Blaser, 1999; Coker *et al.*, 2002). Rare complications of campylobacter infection are described by Skirrow and Blaser (2000).

Co-infection with multiple strains of *Campylobacter* species was reported in 5–10% of cases of acute human enteritis in England and Wales (Kramer *et al.*, 2000). *Campylobacter jejuni* can colonize the jejunum and ileum; probably because of their spiral shape and their motility the bacteria are able to move through the mucin layer covering the surface of the intestinal epithelium, enter into the deep crypts that line the epithelial cells, adhere to these epithelial cells and cause secretory diarrhoea (Van Vliet & Ketley, 2001). Invasion and proliferation in epithelial cells leads to damage, inflammation and inflammatory diarrhoea; a proportion of the bacteria can cross the intestinal mucosa and migrate via the lymphatic system to various extraintestinal sites.

In a low proportion of cases, infection can result in serious, long-term disease including reactive arthritis and Guillain–Barré syndrome (GBS). The reactive arthritis is similar to that following salmonella or other bacterial intestinal infections, symptoms can be incapacitating and last for weeks or months, but are followed usually by a full recovery (Skirrow & Blaser, 2000). The frequency of reactive arthritis following campylobacter infection depends in part on the incidence of the HLA B27 tissue antigen in the population. Estimated frequencies of reactive arthritis from 0 to 1.7% have been cited for some community outbreaks but 5% was calculated in hospital patients in Scandinavia, where a relatively high percentage of the population (14%) is HLA B27 positive (Skirrow & Blaser, 2000).

GBS is a demyelinating disorder that can result in acute neuromuscular paralysis; the proportion of patients with GBS who have evidence of previous campylobacter infection has been estimated as 30–40% (Nachamkin, 2002) and as 80% (Schmidt-Ott *et al.*, 2006). Serotypes associated particularly with GBS include heat-stable (HS) serotypes HS:19 and HS:41 (Penner typing system). The risk of developing GBS following infection with campylobacters is estimated as approximately 0.1% and following infection with serotype HS:19 about 0.5% (Nachamkin, 2002).

A sentinel surveillance scheme for England and Wales, started on 1 May 2000, showed the highest *risk of infection* in infants under 1 year old, and a decreasing risk of infection in age groups over 70 years (Fig. 2.3.1) (HPA, 2003). In contrast, an earlier study in England and Wales, showed the lowest incidence of campylobacter *bacteraemia* in children aged 1–4 years and the highest incidence in patients aged 65 years or more (Skirrow & Blaser, 2000) (Fig. 2.3.2).

In developing countries high exposure to campylobacters and high rates of infection with significant mortality occur early in life, inducing a high level of immunity (Wassenaar & Blaser, 1999; Oberhelman & Taylor, 2000). The rates of campylobacter diarrhoea in children less than 5 years old in developing countries are over 100-fold greater than those in

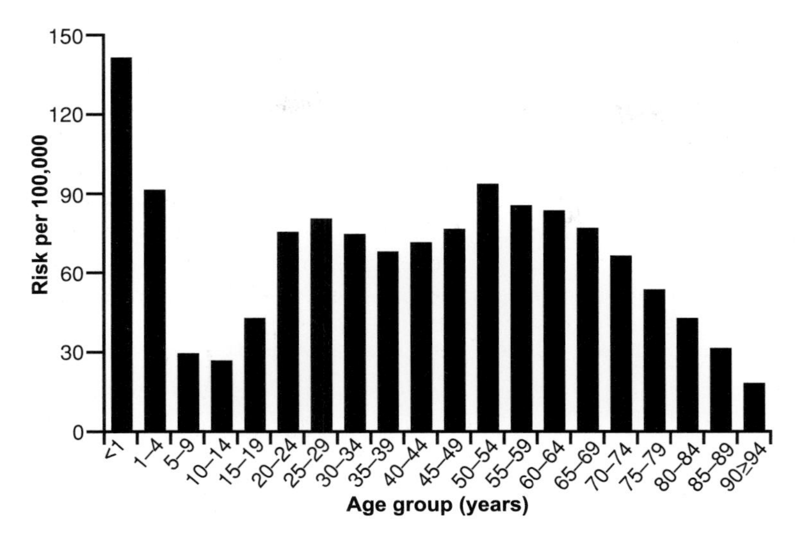

Fig. 2.3.1 The risk of campylobacter infection by age group among the Campylobacter Sentinel Surveillance Scheme population (HPA, 2003).

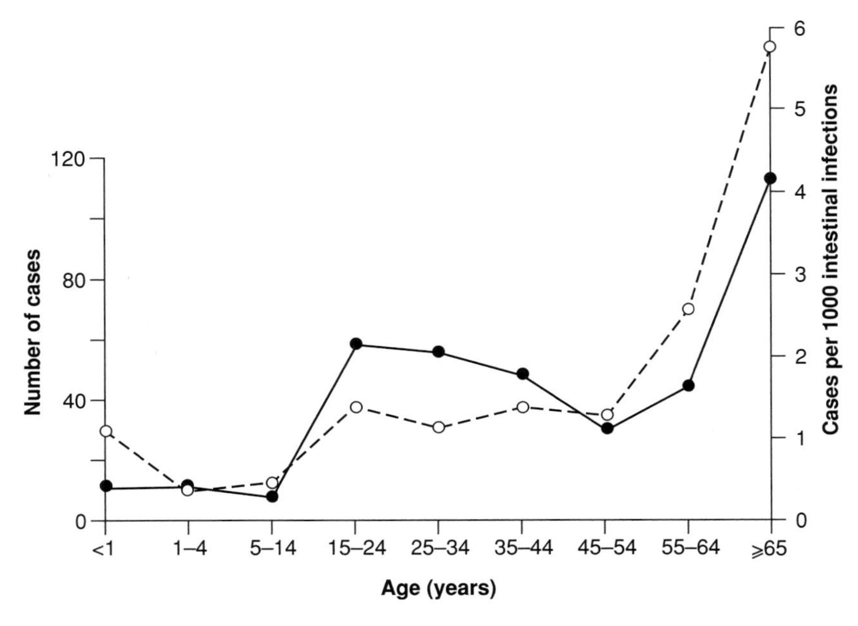

Fig. 2.3.2 Distribution of campylobacter bacteraemia cases by age, England and Wales, 1981–91. Solid line number of cases ($n = 374$); dashed line cases per 1000 intestinal infections. (Reprinted with permission from Skirrow *et al.*, 1993. Copyright Cambridge University Press)

Table 2.3.1 Comparison of the features of *C. jejuni* infections in developed with those in developing countries.

Feature	Developed countries	Developing countries
Isolation rate among people with diarrhoea (%)	5	15–40
Isolation rate among well persons (%)	<1	15–40
Average number of infections/lifetime	0–1	>5
Widespread immunity among adults	Absent	Present
Affected age groups (yr)	<5, 20–30	<2
Principal manifestation of illness	Inflammatory diarrhoea	Simple diarrhoea
Illness severity	Severe	Mild
Bloody diarrhoea (%)	50	15
Epidemics reported	Yes	No
Seasonal increase in incidence	Yes	No
Duration of excretion (days)	16	8

Reprinted with permission from Oberhelman & Taylor (2000).

England, but in older children exposure can result in asymptomatic infection and a transient carrier state. The rates of symptomatic illness in school-age children and adults in developing countries are lower than in England. Campylobacter infection is an important cause of 'Traveller's diarrhoea' in people travelling to developing countries from industrialized countries. A comparison of the features of *C. jejuni* infections in developed and in developing countries is shown in Table 2.3.1.

Antibodies to campylobacter antigens are formed during illness (Cawthraw *et al.*, 2002), and confer short-term immunity to the homologous strain. Repeated infections appear to result in immunity despite the heterogeneity of *C. jejuni* strains. Of students in the US who consumed raw milk on a farm for the first time, 22 of 25 (85%) developed illness due to *C. jejuni*, whereas ten persons who consumed raw milk regularly were not ill and their serum samples showed elevated antibody levels to *C. jejuni* (Blaser *et al.*, 1987). Occupational exposure to *C. jejuni*, for example in poultry abbatoir workers, appears to result in protective immunity and resistance to disease.

Campylobacter fetus has been reported as a cause of cellulitis and bacteraemia in patients suffering from pre-existing disease and taking cimetidine, which would decrease gastric acidity (Ichiyama *et al.*, 1998); the infection was associated with the consumption of raw beef, liver or improperly cooked pork. An outbreak of bacteremia caused by an atypical *C. fetus* subsp. *fetus* strain in nursing home residents who ate commercial cottage cheese was referred to by Friedman *et al.* (2000). An increasing proportion of *C. fetus* infections is occurring in patients with AIDS (Blaser, 1998), thus the possible risk to immunosuppressed people and those with underlying disease needs to be recognized.

2.3.4 *Source and transmission*

Animals such as dogs, cats and hamsters, used as pets, and animals used for food production, including poultry, cattle, sheep and pigs, are often reservoirs of campylobacters.

The organisms may cause diarrhoea in these animals, but usually colonization is harmless. Contact with pets is a means of transmission of infection to humans; this risk should be recognized particularly in the case of vulnerable groups of the population.

Campylobacter jejuni is found in poultry, cattle and sheep. A study in Denmark in 1998–2000 showed that organic and free-range broiler flocks of poultry had a higher level of contamination with campylobacters than conventional flocks (Heuer *et al.*, 2001); all of 22 (100%) organic broiler flocks, 29 of 50 (49.2%) of extensive indoor flocks and 29 of 79 (36.7%) of conventional flocks were positive for campylobacters mainly *C. jejuni*. Measures to reduce contamination with *Campylobacter* in conventionally produced poultry include an emphasis on bio-security and hygiene (ACMSF, 2005).

Campylobacter jejuni rarely causes disease in poultry, but carcasses become contaminated and a high proportion of raw poultry may be contaminated with the bacterium. Infection in pigs is usually with *C. coli*, rather than *C. jejuni*, and this infection is associated sometimes with enteric disease in the pigs. The carcasses of pigs are reported to be contaminated more frequently with campylobacters than are carcasses of sheep and cattle (Stern & Line, 2000). Of a range of raw meats on retail sale in the UK in 1998, 83% (of 198) chicken portions were positive for campylobacters, and the incidence in liver from lambs, ox and pigs was 72.9%, 54.2% and 71.7%, respectively (Kramer *et al.*, 2000). Almost 30% of the samples yielded more than one strain of campylobacter. Of retail raw meat samples from supermarkets in the US, 70% (of 184) samples of chicken carcasses, 14% (of 172) samples of turkey breasts, 1.7% (of 181) pork chops and 0.5% (of 182) beef steaks were positive for campylobacters, mainly *C. jejuni* or *C. coli* (Zhao *et al.*, 2001).

Infection in cattle, sheep and goats can result in contamination of milk, either as a result of faecal contamination, or as a result of mastitis. Direct udder excretion of *C. jejuni* by an asymptomatic dairy cow has been linked with human cases of gastroenteritis (Hutchinson *et al.*, 1985; Orr *et al.*, 1995). Human infection has resulted also from consumption of inadequately pasteurized milk (Birkhead *et al.*, 1988) and milk that has been contaminated after pasteurization (Pebody *et al.*, 1997; Stuart *et al.*, 1997).

Private drinking water supplies have been associated with many outbreaks of disease; of the 25 reported outbreaks of infection associated with private supplies in England and Wales between 1970 and 2002, campylobacters were a cause in 13 (52%) (Said *et al.*, 2003). Waterborne outbreaks of campylobacter infection are fairly common in Sweden, Norway and Finland where, in sparsely populated regions, groundwater is often used without disinfection (Hanninen *et al.*, 2003).

The majority of cases of foodborne campylobacteriosis are reported as sporadic, with consumption of chicken or of unpasteurized milk implicated as major causes (Friedman *et al.*, 2000; Frost *et al.*, 2002). Infection with *C. jejuni* and with *C. coli* may be associated with different foods, patients infected with *C. coli* were more likely to report consumption of bottled water or of paté than those infected with *C. jejuni* (Gillespie *et al.*, 2002). Until 1997, the investigation of possible outbreaks in the UK was hampered by the lack of widely available reference typing.

2.3.5 *Examples of outbreaks of foodborne campylobacteriosis*

Between 1 January 1995 and 31 December 1999, 50 campylobacter outbreaks were reported in England and Wales, representing 2% of the general outbreaks of infectious intestinal

Table 2.3.2 Examples of outbreaks of foodborne disease caused by *Campylobacter* spp.

Place, date	Cases	Number hospitalized	Deaths	Food implicated	Evidence[a]	Where food was provided	Factors leading to outbreak	Reference
US, 1979–81	10[b] (*C. fetus* subsp. *fetus*)	9 (patients in clinics)	1	Raw calf's liver	D	Clinics in Mexico	Underlying disease and consumption of raw liver	CDC (1981)
US, 1982	15[b] (*C. jejuni* and *C. fetus*)	2	0	Unpasteurized milk	S	Banquet at a farm	Consumption of unpasteurized milk	Klein *et al.* (1986)
UK, 1983	75	nr	0	Unpasteurized milk	M	Community	Excretion of *Campylobacter* by dairy cows and consumption of unpasteurized milk	Hutchinson *et al.* (1985)
UK, 1984–86	64–112	nr	nr	Chicken	M	Catering college and associated restaurants	Poor practices on farms, deficiencies in practice in kitchens	Pearson *et al.* (2000)
US, 1985	23	nr	nr	Raw milk	S	Dairy tour	Consumption of raw milk	CDC (1986)
US, 1986	35	nr	0	Inadequately pasteurized milk	M	Boarding school	Milk supplied by dairy farm affiliated to school. Pasteurization unit set at lower temperature than specified, and lacked proper thermometers	Birkhead *et al.* (1988)
UK, 1992	>72	nr	0	Unpasteurized milk	S	Music Festival	Temporary licence granted to 2 local farms to sell unpasteurized milk at the festival	Morgan *et al.* (1994)
UK, 1992	>110	nr	nr	Inadequately pasteurized milk	S	Community	Milk failed phosphatase test, indicating inadequate heat treatment	Fahey *et al.* (1995)
UK, 1994	23	0	nr	Unpasteurized milk	S	Farm visit by nursery school children	Consumption of unpasteurized milk	Evans *et al.* (1996)

UK, nr	12	8 needed medical attention	0	Chicken pieces	S	Departmental party for hospital staff	Staff member thought the chicken did not require further cooking, but label on the pack advised cooking	Murphy *et al.* (1995)
Belgium, 1995	>24 (*C. coli*)	0	0	Mixed salad containing feta cheese, gouda cheese ham, bacon	S	Church school for adults	Poor practices that would allow cross-contamination	Ronveaux *et al.* (2000)
US, 1995	76	39 (camp infirmary)	0	Tuna salad (tuna + mayonnaise)	S	Summer camp	Deficiencies in food preparation; food handlers not trained	Roels *et al.* (1998)
Northern Ireland, 1997	12	0	0	Seasonal leaves and tomato salad	S	Conference lunch	Probable cross-contamination in kitchen	Moore, *et al.* (2000)
US, 1997	16	3	1	Hawaiian meal	S	Senior Centre	Cross contamination in the kitchen, from raw chicken or/beef	Winquist *et al.* (2001)
US, 1997	119 (*C. jejuni* and *Salmonella heidelberg*	31	1	Chopped liver salad	S	Nursing home	Cooked chicken livers placed in a bowl containing raw chicken-liver juice. Failure of refrigerator used to store liver salad	Layton *et al.* (1997)
UK, 1997	12 (*Campylobacter* sp.)	1	0	Stir-fried chicken pieces	S	Restaurant	Undercooking of chicken	Evans *et al.* (1998)
Austria, 1998	38	0	0	Unpasteurized milk	M	Youth Centre	Outbreak strain isolated from cows that supplied the milk	Lehner *et al.* (2000)
US, 1998	27	nr	nr	Gravy, pineapple	M	Luncheon at elementary school	Food handler with *C. jejuni* diarrhoea	Olsen *et al.* (2001)

(*cont.*)

Table 2.3.2 *(continued)*

Place, date	Cases	Number of hospitalized	Deaths	Food implicated	Evidence[a]	Where food was provided	Factors leading to outbreak	Reference
Germany, 1999	11	2	0	Chocolate drink made with unpasteurized milk	M	Farm visit	Consumption of unpasteurized milk	Lieftucht (1999)
US, 2001	75	0	0	Unpasteurized cows' milk	M	Farm supplying unpasteurized milk at community events and directly to consumers	Consumption of unpasteurized milk	CDC (2002)
UK, 2001	30	nr	nr	nr	nr	Microbiology conference	nr	CDSC (2001)
Finland, 2002–3	6	1	0	Unpasteurised cows' milk	M	Farming family	Consumption of unpasteurised milk	Schildt *et al.* (2006)
Spain, 2003	81	nr	nr	Custard made with UHT milk	S	School	Suspected contamination from raw chicken. Food preparation areas for uncooked meats and ready-to-eat foods were not separated	Jiménez *et al.* (2005)
Denmark, 2005	79	1	nr	Chicken salad	S	7 company canteens supplied by same caterer	Suspected contamination from raw chicken, which was stored in refrigerator directly above fried chicken used in salad	Mazick *et al.* (2006)

[a] M (microbiological): identification of an organism of the same type from cases and in the suspect vehicle or vehicle ingredient(s), or detection of toxin in faeces or food; D (descriptive): other evidence, usually descriptive, reported by local investigators as indicating the suspect vehicle or food; S (statistical): a significant statistical association between consumption of the suspect vehicle and being a case. [b] Unless stated otherwise, outbreaks were due to *C. jejuni*.

nr = not reported.

disease for which a causative agent was identified (Frost *et al.*, 2002). Foodborne transmission was identified in 35 (70%) of the campylobacter outbreaks, the majority of which occurred in commercial catering establishments; waterborne transmission (not linked to municipal water supplies) was identified in four (8%). Poultry products were the food implicated most frequently; cross-contamination was the fault reported most commonly, with inadequate heat treatment and inappropriate storage also involved.

Examples of outbreaks that have been described in detail are given in Table 2.3.2. These illustrate factors that have led to outbreaks and that are likely to contribute to sporadic cases. The fourth of these outbreaks occurred over a period of 3 years and was attributed to a continued supply of contaminated poultry from a single farm to a restaurant, with undercooking of the poultry. An outbreak in the UK in 1997 was also associated with undercooking of poultry.

Ten of the outbreaks were caused by the use of unpasteurized or inadequately pasteurized milk. Six of the outbreaks were linked with poor practices during food preparation that would have allowed cross-contamination from other foods, and one outbreak was attributed to an infected food handler.

2.3.6 *Main methods of control and prevention*

Measures to minimize foodborne infection with campylobacters include procedures to limit the carriage of the bacteria by animals and birds, particularly poultry, on the farm, and to limit contamination during processing, distribution and sale. It is important also to prevent cross-contamination in the kitchen and to ensure thorough cooking.

Live poultry are often colonized by large numbers of campylobacters, with numbers in the small intestine and caeca ranging from 10^5 to $>10^9$ CFU/g (Jacobs-Reitsma, 2000). Efforts to reduce colonization and contamination of broiler chickens emphasize the use of proper sanitation, control of farm hygiene and of water supplies, biosecurity and treatment of litter (ACMSF, 2005). In the case of free-range chickens it may be difficult to prevent exposure of the birds to campylobacters.

Some poultry flocks are free of detectable *C. jejuni* but others are contaminated. Birds from contaminated flocks enter the slaughterhouse with high numbers of campylobacters on their feathers and skin as well as in the intestine. During defeathering and evisceration leaking of contents of the intestine contributes to contamination of the carcasses. Washing and chilling tend to reduce the numbers of campylobacters on carcasses, but final numbers can be between 10^2 and 10^5 campylobacters per fresh carcass (Jacobs-Reitsma, 2000). Freezing reduces the level of contamination but campylobacters can survive on frozen poultry for several months. Many measures to reduce contamination are used in the production of poultry meat; heat treatment or freezing of the final product are among the most effective measures that are allowed (Havelaar, 2005). Hygiene on the farm and control during slaughter can reduce greatly the numbers of campylobacter on poultry and thus reduce the public health risk (Havelaar, 2005). For the foreseeable future, however, a proportion of poultry meat is liable to be contaminated with campylobacters, and prevention of disease in humans will depend on avoidance of cross-contamination from raw poultry to other foods in kitchens, and thorough cooking of poultry products.

Laying hens often excrete campylobacters, which can contaminate the eggs externally. The bacterium is sensitive to drying, and survival on the shell appears to be poor. Penetration of the organism through the shell into eggs is poor, and table eggs are unlikely to be involved in transmission of campylobacter to humans.

Campylobacters can be detected in a proportion of samples of raw milk but are killed by properly controlled pasteurization. Infection in humans can be prevented by avoiding the consumption of unpasteurized or inadequately pasteurized milk. Dairy products such as cheese and yoghurt made from raw milk do not appear usually to present a risk of infection with campylobacters, probably because of the sensitivity of the bacteria to low water activity and to organic acids. The bacteria are liable to be present in surface water, thus the avoidance of consumption of untreated water supplies, and prevention of contamination of municipal water supplies are necessary.

Despite the fact that they are more susceptible to adverse conditions than are other foodborne pathogenic bacteria, campylobacters can survive in food products for long periods and can survive in kitchen environments and result in cross-contamination of foods.

The majority of foodborne infections with campylobacters can be prevented by avoiding consumption of uncooked, undercooked or unpasteurized foods of animal origin, particularly poultry and milk, and of untreated water supplies, avoidance of cross-contamination from raw foods of animal origin to ready-to-eat foods in the kitchen, and avoidance of contamination by infected food handlers.

Summary of measures advised in healthcare settings

1. Poultry should be obtained from sources where measures have been taken to minimize contamination with campylobacters.
2. Care must be taken to avoid cross-contamination of other foods from poultry.
3. Foods should be cooked thoroughly.
4. Raw or unpasteurized milk should not be served.
5. Catering personnel should not handle food while suffering from symptoms of infection, or handle cooked food or ready-to-eat food with bare hands; good hand washing is essential.

References

Advisory Committee on the Microbiological Safety of Food (ACMSF) (2005) *Second Report on Campylobacter.* Food Standards Agency, London. Available from http://www.food.gov.uk/multimedia/pdfs/acmsfcampylobacter.pdf. Accessed 6 March 2007.

Birkhead, G.R., Vogt, R.L., Heun, E. *et al.* (1988) A multiple strain outbreak of *Campylobacter* enteritis due to consumption of inadequately pasteurized milk. *Journal of Infectious Diseases* **157**, 1095–1097.

Blaser, M.J. (1998) Editorial response: *Campylobacter fetus* – emerging infection and model system for bacterial pathogenesis at mucosal surfaces. *Clinical Infectious Diseases* **27**, 256–258.

Blaser, M.J., Sazie, E. & Williams, L.P., Jr. (1987) The influence of immunity on raw milk-associated *Campylobacter* infection. *Journal of the American Medical Association* **257**, 43–46.

Campylobacter Sentinel Surveillance Scheme Collaborators (2003) Point source outbreaks of *Campylobacter jejuni* infection – are they more common than we think and what might cause them? *Epidemiology and Infection* **130**, 367–375.

Cawthraw, S.A., Feldman, R.A., Sayers, A.R. & Newell, D.G. (2002) Long-term antibody responses following human infection with *Campylobacter jejuni*. *Clinical and Experimental Immunology* **130**, 101–106.

CDSC (2001) *Outbreak of Campylobacter Infection Following Microbiology Conference. Communicable Disease Report CDR Weekly* **11**(11). Available from http://www.hpa.org.uk/cdr/archives/2001/cdr1101.pdf. Accessed 6 March 2007.

Centers for Disease Control and Prevention (CDC) (1981) *Campylobacter* sepsis associated with 'nutritional therapy' – California. *Morbidity and Mortality Weekly Report* **30**, 294–295.

Centers for Disease Control and Prevention (CDC) (1986) Epidemiologic notes and reports *Campylobacter* outbreak associated with raw milk provided on a dairy tour – California. *Morbidity and Mortality Weekly Report* **35**(19), 311–312.

Centers for Disease Control and Prevention (CDC) (2002) Outbreak of *Campylobacter jejuni* infections associated with drinking unpasteurized milk procured through a cow-leasing program – Wisconsin, 2001. *Morbidity and Mortality Weekly Report* **51**(25), 548–549.

Coker, A.O., Isokpehi, R.D., Thomas, B.N. *et al.* (2002) Human campylobacteriosis in developing countries. *Emerging Infectious Diseases* **8**, 237–243.

Evans, M.R., Lane, W., Frost, J.A. & Nylen, G. (1998) A campylobacter outbreak associated with stir-fried food. *Epidemiology and Infection* **121**, 275–279.

Evans, M.R., Roberts, R.J., Ribiero, C.D. *et al.* (1996) A milk-borne campylobacter outbreak following an educational farm visit. *Epidemiology and Infection* **117**, 457–462.

Fahey, T., Morgan, D., Gunneburg, C. *et al.* (1995) An outbreak of *Campylobacter jejuni* enteritis associated with failed milk pasteurization. *Journal of Infection* **31**, 137–143.

Friedman, C.R., Neimann, J., Wegener, H.C. & Tauxe, R.V. (2000) Epidemiology of *Campylobacter jejuni* infections in the United States and other industrialized nations. In: Nachamkin, I. & Blaser, M.J. (eds). *Campylobacter*, 2nd edn. ASM Press, Washington, DC, pp. 121–138.

Frost, J.A., Gillespie, J.A. & O'Brien, S.J. (2002) Public health implications of campylobacter outbreaks in England and Wales, 1995–9: epidemiological and microbiological investigations. *Epidemiology and Infection* **128**, 111–118.

Gillespie, I.A., O'Brien, S.J., Frost, J.A. *et al.* (2002) A case–case comparison of *Campylobacter coli* and *Campylobacter jejuni* infection: a tool for generating hypotheses. *Emerging Infectious Diseases* **8**, 937–942.

Gürtler, M., Alter, T., Kasimir, S. & Felhaber, K. (2005) The importance of *Campylobacter coli* in human campylobacteriosis: relevance and genetic characterization. *Epidemiology and Infection* **133**, 1081–1087.

Hänninen, M.-L, Haajanen, H., Pummi, T. *et al.* (2003) Detection and typing of *Campylobacter jejuni* and *Campylobacter coli* and analysis of indicator organisms in three waterborne outbreaks in Finland. *Applied and Environmental Microbiology* **69**, 1391–1396.

Havelaar, A.H. (2005) CARMA: a multidisciplinary project to evaluate the costs and benefits of controlling *Campylobacter* in the Netherlands. *Conference on Interventions against Campylobacter.* Cambridge, UK.

Health Protection Agency (HPA) (2003) *The Campylobacter Sentinel Surveillance Scheme – Data from the First Two Years of the Study. Communicable Disease Report CDR Weekly* **13**(10): enteric. Available from http://www.hpa.org.uk/cdr/archives/2003/cdr1903.pdf. Accessed 6 March 2007.

Heuer, O.E., Pederson, K., Anderson, J.S. & Madsen, M. (2001) Prevalence and antimicrobial susceptibility of thermophilic *Campylobacte*r in organic and conventional flocks. *Letters in Applied Microbiology* **33**, 269–274.

Hutchinson, D.N., Bolton, F.J., Hinchliffe, P.M. *et al.* (1985) Evidence of udder excretion of *Campylobacter jejuni* as the cause of milkborne *Campylobacter* outbreak. *Journal of Hygiene* **94**, 205–215.

Ichiyama, S., Hirai, S., Minami, T. *et al.* (1998) *Campylobacter fetus* subspecies *fetus* cellulitis associated with bacteremia in debilitated hosts. *Clinical Infectious Diseases* **27**, 252–255.

Jacobs-Reitsma, W. (2000) *Campylobacter* in the food supply. In: Nachamkin, I. & Blaser, M.J. (eds). *Campylobacter*, 2nd edn. ASM Press, Washington, DC, pp. 467–481.

Jiménez, M., Soler, P., Venanzi, J.D. *et al.* (2005) An outbreak of *Campylobacter jejuni* enteritis in a school of Madrid, Spain. *Euro Surveillance* **10**(4), 118–121.

Klein, B.S., Vergeront, J.M., Blaser, M.J. *et al.* (1986) *Campylobacter* infection associated with raw milk. *Journal of the American Medical Association* **255**, 361–364.

Kramer, J.M., Frost, J.A., Bolton, F.J. & Waring, D.R.A. (2000) *Campylobacter* contamination of raw meat and poultry at retail sale: identification of multiple types and comparison with isolates from human infection. *Journal of Food Protection* **63**, 1654–1659.

Layton, M.C., Callists, S.G., Gomez, T.M. *et al.* (1997) A mixed foodborne outbreak with *Salmonella Heidelberg* and *Campylobacter jejuni* in a nursing home. *Infection Control and Hospital Epidemiology* **18**, 115–121.

Lehner, A., Schneck, C., Feierl, G. *et al.* (2000) Epidemiologic application of pulsed-field gel electrophoresis to an outbreak of *Campylobacter jejuni* in an Austrian youth centre. *Epidemiology and Infection* **125**, 13–16.

Lieftucht, A. (1999) An outbreak of campylobacter infection associated with a farm in Germany. *Eurosurveillance Weekly* 25 November, **11**(48), 991125. Available from http://www.eurosurveillance.org/ew/1999/991125.asp#1. Accessed 6 March 2007.

Mazick, A., Ethelberg, S., Møller Nielsen, E. *et al.* (2006) An outbreak of *Campylobacter jejuni* associated with consumption of chicken, Copenhagen, 2005. *Euro Surveillance* **11**(5), 137–139.

Moore, J.E., Stanley, T., Smithson, R. *et al.* (2000) Outbreak of campylobacter food-poisoning in Northern Ireland. *Clinical Microbiology and Infectious Diseases* **6**, 397–398.

Morgan, D., Gunnenberg, C., Gunnell, D. *et al.* (1994) An outbreak of *Campylobacter* infection associated with the consumption of unpasteurised milk at a large festival in England. *European Journal of Epidemiology* **10**, 581–585.

Murphy, O., Gray, J., Gordon, S. & Bint, A.J. (1995) An outbreak of campylobacter food poisoning in a health care setting. *Journal of Hospital Infection* **30**, 225–228.

Nachamkin, I. (2002) Chronic effects of *Campylobacter* infection. *Microbes and Infection* **4**, 399–403.

Nachamkin, I. & Skirrow, M.B. (1999) *Campylobacter*, *Arcobacter* and *Helicobacter*. In: Collier, L., Balows, A. & Sussman, M. (eds). *Topley and Wilson's Microbiology and Microbial Infections*, 9th edn. *Vol. 2: Systematic Bacteriology* (Volume editors Balows, A. & Duerden, B.I.). Arnold, London, Chapter 54, pp. 1237–1256.

Oberhelman, R.A. & Taylor, D.N. (2000) Campylobacter infections in developing countries. In: Nachamkin, I. & Blaser, M.J. (eds). *Campylobacter*, 2nd edn. ASM Press, Washington, DC, pp. 139–153.

Olsen, S.J., Hansen, G.R., Bartlett, L. *et al.* (2001) An outbreak of *Campylobacter jejuni* infections associated with food handler contamination: the use of pulsed-field gel electrophoresis. *Journal of Infectious Diseases* **183**, 164–167.

Orr, K.E., Lightfoot, N.F., Sisson, P.R. *et al.* (1995) Direct milk excretion of *Campylobacter jejuni* in a dairy cow causing cases of human enteritis. *Epidemiology and Infection* **114**, 15–24.

Pearson, A.D., Greenwood, M.H., Donaldson, J. *et al.* (2000) Continuous source of campylobacteriosis traced to chicken. *Journal of Food Protection* **63**, 309–314.

Pebody, R.G., Ryan, M.J. & Wall, P.G. (1997) Outbreaks of campylobacter infection: rare events for a common pathogen. *Communicable Disease Report. CDR Review* **7**(3), R33–R37. Available from http://www.hpa.org.uk/cdr/archives/CDRreview/1997/cdrr0397.pdf. Accessed 6 March 2007.

Roels, T.H., Wickus, B., Bostrom, H.H. *et al.* (1998) A foodborne outbreak of *Campylobacter jejuni* (O:33) infection associated with tuna salad: a rare strain in an unusual vehicle. *Epidemiology and Infection* **121**, 281–287.

Ronveaux, O., Quoilin, S., Van Loock, F. *et al.* (2000) A *Campylobacter coli* foodborne outbreak in Belgium. *Acta Clinica Belgica* **55**, 307–311.

Said, B., Wright, F., Nichols, G.L. *et al.* (2003) Outbreaks of infectious disease associated with private drinking water supplies in England and Wales 1970-2000. *Epidemiology and Infection* **130**, 469–479.

Sails, A.D., Swaminathan, B. & Fields, P.I. (2003) Utility of multilocus sequence typing as an epidemiological tool for investigation of outbreaks of gastroenteritis caused by *Campylobacter jejuni*. *Journal of Clinical Microbiology* **41**, 4733–4739.

Schildt, M., Savolainen, S. & Hänninen, M.-L. (2006) Long-lasting *Campylobacter jejuni* contamination of milk associated with gastrointestinal illness in a farming family. *Epidemiology and Infection* **134**, 401–405.

Schmidt-Ott, R., Schmidt, H., Feldman, S. *et al.* (2006) Improved serological diagnosis stresses the major role of *Campylobacter jejuni* in triggering Guillain–Barré syndrome. *Clinical and Vaccine Immunology* **13**, 779–783.

Skirrow, M.B., Jones, D.M., Sutcliffe, E. & Benjamin, J. (1993) Campylobacter bacteraemia in England and Wales, 1981–91. *Epidemiology and Infection* **110**, 567–573.

Skirrow, M.B. & Blaser, M.J. (2000) Clinical aspects of infection. In: Nachamkin, I. & Blaser, M.J. (eds). *Campylobacter*, 2nd edn. ASM Press, Washington, DC, pp. 69–88.

Stern, N.J. & Line, J.E. (2000) *Campylobacter*. In: Lund, B.M., Baird-Parker, T.C. & Gould, G.W. (eds). *The Microbiological Safety and Quality of Food*. Aspen Publishers, Gaithersburg, MD, pp. 1040–1056.

Stuart, J., Sufi, F., McNulty, C. & Park, P. (1997) Outbreak of campylobacter enteritis in a boarding school associated with bird pecked bottle tops. *Communicable Disease Report. CDR Review* **7**(3), R38–R40. Available from http://www.hpa.org.uk/cdr/archives/CDRreview/1997/cdrr0397.pdf. Accessed 6 March 2007.

Takkinen, J., Ammon, A., Robstad, O. *et al.* (2003) European survey on Campylobacter surveillance and diagnosis 2001. *Euro Surveillance* **8**(11), 207–213. Available from http://www.eurosurveillance.org/em/v08n11/v08n11.pdf. Accessed 6 March 2007.

Van Vliet, A.H.M. & Ketley, J.M. (2001) Pathogenesis of enteric *Campylobacter* infection. *Journal of Applied Microbiology* **90**, 45S–56S.

Wassenaar, T.M. & Blaser, M.J. (1999) Pathophysiology of *Campylobacter jejuni* infection in humans. *Microbes and Infection* **4**, 399–403.

Winquist, A.G., Roome, A., Mshar, R. *et al.* (2001) Outbreak of Campylobacteriosis at a senior center. *Journal of the American Geriatrics Society* **49**, 304–307.

World Health Organization (WHO) (2003) *WHO Surveillance Programme for Control of Foodborne Infections and Intoxications in Europe*, 8th report, 1999–2000. Available from http://www.bfr.bund.de/internet/8threport/8threp_fr.htm. Accessed 6 March 2007.

Zhao, C.W., Ge, B.L., De Villena, J. *et al.* (2001) Prevalence of *Campylobacter* spp., *Escherichia coli*, and *Salmonella* serovars in retail chicken, turkey, pork and beef from the Greater Washington, DC area. *Applied and Environmental Microbiology* **67**, 5431–5436.

2.4 *Clostridium botulinum*

2.4.1 *Importance as a cause of foodborne disease*

Outbreaks and sporadic cases of foodborne illness caused by *Clostridium botulinum* are uncommon in most developed countries, as a result of controls used in the food industry. The consequences of foodborne botulism can be drastic, because the bacteria form highly lethal neurotoxins, and unless treatment is given promptly the illness is liable to be fatal. Even if death does not occur, the victim may require intensive care, and several months for recovery.

In England and Wales, only six cases of foodborne botulism were reported between 1990 and 2005 and in countries such as the Netherlands, Norway and Sweden few or no cases were reported in 1999–2000 (WHO, 2003). The numbers of reported cases in other countries in 1999 and 2000 were: Germany 19, 11; France 32, 28; Poland 97, 72; Turkey 96, 18; Uzbekistan 44, 87; the Russian Federation 478, 409 equivalent to 0.33–0.28 cases per 100,000 persons (WHO, 2003). In the US in 2003, 20 cases of foodborne botulism were reported (CDC, 2005).

Infant botulism, which is sometimes caused by foodborne transmission of the bacterium, has been described in many countries.

2.4.2 *Characteristics of the organisms*

Clostridium botulinum is an anaerobic, Gram-positive, spore-forming bacterium (Fig. 2.4.1); it was isolated first in 1897 from raw, salted ham that caused an outbreak of disease in Belgium. The name covers four groups of bacteria that have been differentiated on the basis of their physiological properties and genetic relatedness, and that differ also in the detailed structure of the toxin formed (Table 2.4.1). The majority of strains appear to produce toxin of a single antigenic type, but some strains form two types of toxin, usually a major amount of one and a minor amount of another. Such strains have been designated Af (indicating the major and minor toxin, respectively), Bf, Ab and Ba. Botulism in humans is almost always caused by toxin types A, B or E and occasionally by toxin type F. On rare

Table 2.4.1 *Clostridium botulinum* groups and neurotoxins formed.

Group	Neurotoxin formed	Main species affected
I (proteolytic)	A	Humans
	B	Humans; cattle, horses
	F	Humans
II (non-proteolytic)	B	Humans
	E	Humans; some water birds
	F	Humans
III	C	Birds, particularly water birds; farmed chicken and pheasants; minks, ferrets, foxes, cattle, horses
	D	Cattle, sheep, horses
IV	G	Not known

Modified from Lund & Peck, 2000.

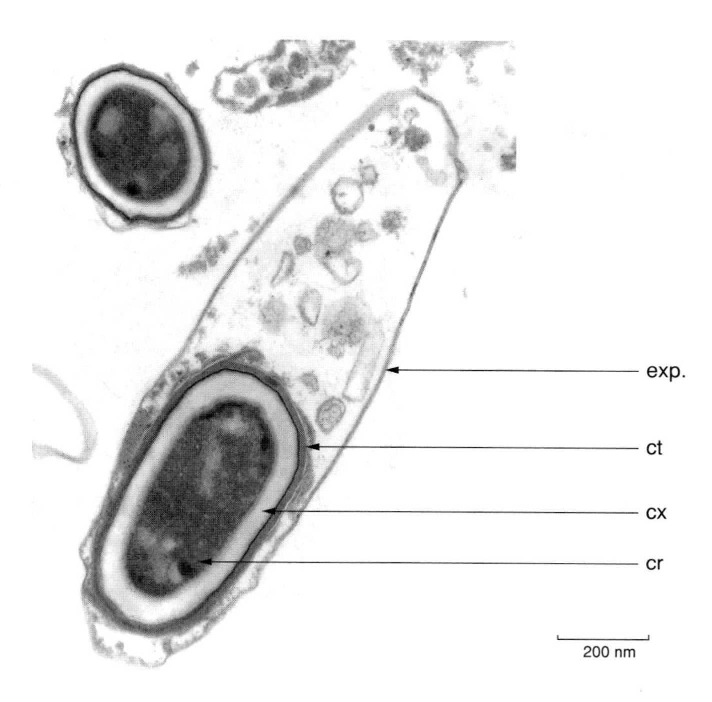

Fig. 2.4.1 Transmission electron micrograph of a spore of *Clostridium botulinum*, non-proteolytic, type B. exp, exosporium; ct, spore coat; cx, spore cortex; cr, spore core. (Electron micrograph courtesy of Drs Sandra Stringer, Martin Webb and Mary Parker, Institute of Food Research, Norwich, UK)

occasions strains of *C. butyricum* and of *C. baratii* have been isolated that form type E and F botulinum-like neurotoxin, respectively.

Groups I and II *C. botulinum* cause botulism in humans and differ markedly in their properties (Table 2.4.2), further discussion will relate mainly to these groups. Spores of *C. botulinum* can survive for long periods in air and in the environment and can germinate in the presence of oxygen, but the vegetative cells die gradually on exposure to oxygen. Spores of Group I strains are highly heat-resistant; they will survive normal cooking, and heating under pressure is required for inactivation. Spores of Group II strains are less heat-resistant but are liable to survive many cooking processes. While Group I strains will not grow at temperatures below 10°C, Group II strains will grow at temperatures as low as 3°C and are of concern in refrigerated foods.

The neurotoxin molecule consists of two polypeptide chains, of about 100 kDa and 50 kDa, joined by a disulphide bond. After consumption of toxic food, the toxin crosses the intestinal epithelium (Maksymowych & Simpson, 2004), passes into the blood and reaches cholinergic synapses, where it blocks the release of the neurotransmitter, acetylcholine. The lethal dose in the bloodstream of man is estimated as about 1 nanogram (10^{-6} mg) /kg body weight. Toxin is inactivated by heat treatment; in many foods exposure to 85°C for 5 minutes results in inactivation by a factor of about 10^5 (Lund & Peck, 2000).

Table 2.4.2 Some properties of Group I and Group II *C. botulinum*, the main causes of botulism in humans.

	C. botulinum Group I (proteolytic)	*C. botulinum* Group II (non-proteolytic)
Toxin types	A, B, F	B, E, F
Optimum growth temperature, °C	30–40	25–37
Minimum growth temperature, °C	10–12	3
Minimum pH for growth	4.6	5.0
Minimum NaCl to prevent growth (%w/v)	10	5
Maximum heat resistance of spores at 100°C ($D_{100°C}$ value)[a]	>15 min	<0.1 min[b]
Max. $D_{82.2°C}$ value	~ 2000 min	2.4 min[b]

Data from Lund & Peck, 2000; Peck & Stringer, 2005.
[a] *D* value is the time required, at the temperature specified, to reduce the number of viable spores tenfold. The *D* values given here are for spores heated in phosphate buffer, 1/15 M, pH 7.0.
[b] This *D* value is found if the heated spores of non-proteolytic *C. botulinum* are recovered in the absence of the enzyme lysozyme. The heat treatment causes sublethal damage to the spores, a proportion of which can germinate if lysozyme is present, and give vegetative growth. Lysozyme is a heat-stable enzyme that can be present in some raw foods and can greatly increase the measured heat-resistance of the spores of non-proteolytic strains to give a Max. $D_{82.2°C}$ value of 231 min.

2.4.3 *Disease in humans*

Four types of botulism have been distinguished.

1. *Foodborne botulism* results from consumption of food in which *C. botulinum* has grown and formed toxin.
2. *Infant botulism* is caused by the ingestion of spores of *C. botulinum*, which germinate in the intestine of young infants, and the resulting vegetative bacteria become established and form neurotoxin.
3. *Adult infectious botulism* is reported rarely; normally the spores do not become established in the intestine, but this can occur in adults who have a history of abdominal surgery, gastrointestinal tract abnormalities, gastric achlorhydria, or recent treatment with antibiotics (Shapiro *et al.*, 1998; Fenicia *et al.*, 1999).
4. *Wound botulism* results when anaerobic conditions occur within a wound and *C. botulinum* spores contaminate the wound, germinate, the vegetative bacteria multiply and form toxin which is absorbed and circulates in the body. The condition can follow traumatic injuries and can be associated with drug abuse (Shapiro *et al.*, 1998; Akbulut *et al.*, 2005).

In the US, where infant botulism has been studied most intensively, the incidence is much higher than that of foodborne botulism and wound botulism (Fig. 2.4.2).

In foodborne botulism the major symptoms usually occur within 12–36 hours of ingestion of toxin, but may occur within 6 hours or after 10 days. The initial symptoms are often nausea and vomiting, which are caused, probably, not by the neurotoxin but by other products of metabolism of *C. botulinum*. The first effect of the toxin is often on neuromuscular junctions in the head and neck, causing symptoms such as double vision, inability to focus,

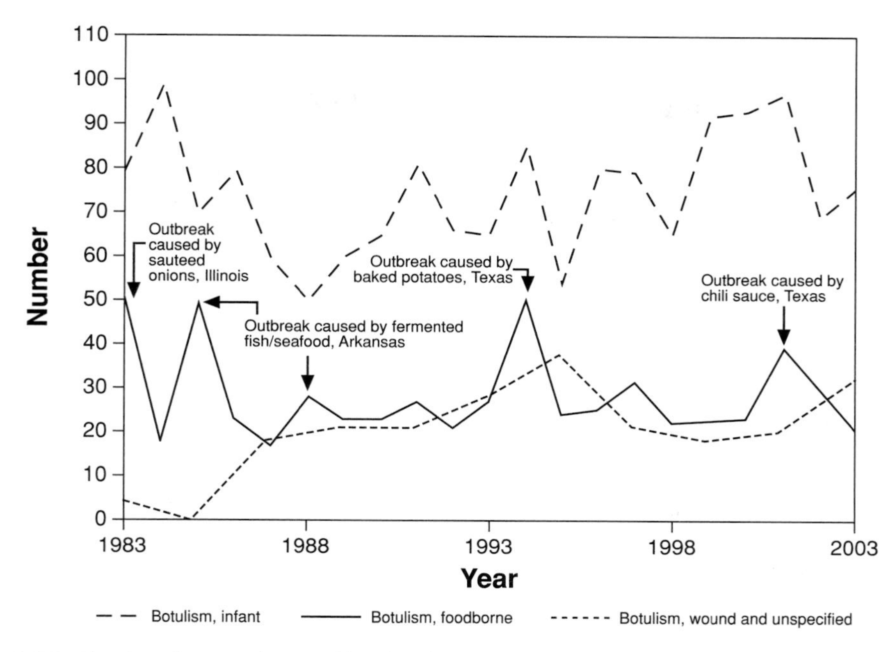

Fig. 2.4.2 Number of reported cases of botulism by year – US, 1983–2003. (Modified from CDC, 2005)

drooping eyelids (ptosis), dry mouth, difficulty in speaking clearly (dysphonia) and inability to swallow (dysphagia). If the illness is untreated muscle paralysis may progress to the arms, legs and trunk and to respiratory and heart muscles and result in death. The symptoms can appear similar to those of Guillain–Barré syndrome, stroke and myasthenia gravis. For severe botulism, supportive treatment with mechanical ventilation is required. Equine antitoxin to toxin types A, B and E is licenced and available and is effective if given early in the illness, but carries the risk of a hypersensitive reaction (Shapiro *et al.*, 1998). In severe cases ventilatory support may be required for weeks or months, clinical recovery generally requires a similar period.

The ingestion of low numbers of *C. botulinum* spores is not harmful to healthy adults, but in some conditions the spores can cause infant botulism in children under about 1 year old, whose intestines are susceptible to colonization by the bacterium before the establishment of competing intestinal microorganisms. The dose required to cause infection is estimated as 10–100 spores. The descriptions of *infant botulism* are based mainly on hospitalized patients (Arnon, 2004). Some outpatient cases that have been detected by physicians familiar with the 'classic' symptoms of infant botulism have shown only a few days of lethargy, poor feeding and some decrease in bowel movement frequency. In contrast, a few cases have occurred in which the history and presentation were indistinguishable from those of typical cases of sudden infant death syndrome (cot death), about 1 in 20 cases of which, in California, appear to result from fulminant infant botulism (Arnon, 2004). In 'classical' cases of infant botulism the first sign of the illness is usually constipation, followed by listlessness, lethargy and poor feeding. Increasing weakness occurs over 1–4 days. The infant typically shows an expressionless face, feeble cry, ptosis, poor head control, generalized weakness and hypotonia. Primary treatment is supportive care, with mechanically assisted ventilation when necessary. In the US, Botulism Immune Globulin Intravenous (human) is available and prompt treatment with this preparation is advised for infant botulism type A and type B. The use of this product has reduced the mean hospital stay per case from approximately 5.7 weeks to approximately 2.6 weeks and greatly reduced the cost per case (Arnon, 2004; Arnon *et al.*, 2006). Recovery takes several weeks, but in the absence of complications is expected to be complete.

2.4.4 *Source and transmission*

In assessing the incidence of *C. botulinum* in the environment it is important to use strictly anaerobic conditions to allow growth from low numbers of spores, to avoid heat treatment that would inactivate spores of less heat-resistant strains, and to prevent inhibition by competing microorganisms. It is particularly necessary in the case of this anaerobic bacterium to demonstrate the sensitivity of the detection/isolation method(s) used (Lund & Peck, 2000).

Clostridium botulinum has been isolated from soils in many parts of the world, but spores of other clostridia such as *C. perfringens* appear to be much more numerous. Spores of *C. botulinum* have been found in higher numbers in wet sediments and shoreline samples than in soil. The presence of the spores in soil is likely to result in contamination of plants and dissemination of the bacterium by wind, rain and building and agricultural operations.

The possibility exists of contamination of most types of food, usually with low numbers of *C. botulinum*, and the bacterium has been isolated from a wide range of foods including fish, meat, vegetables, fruits, honey, mushrooms, cheese, nuts and bamboo shoots. If foods are stored in conditions that allow germination of the spores and growth and toxin production by the vegetative bacteria, consumption may result in botulism.

In the majority of cases of infant botulism, the source of the spores is not known. Dust has been suspected in some cases and was implicated clearly in one case (Nevas *et al.*, 2005b) and a possible link to infant formula milk powder was reported for one case (Brett *et al.*, 2005). A history of consumption of honey has been reported in more than one half of the cases of infant botulism in Europe (Aureli *et al.*, 2002), the honey may have been used on pacifiers. In most of these cases samples of honey were not available for microbiological analysis. According to Arnon (2004), 35 instances were known, worldwide, in which *C. botulinum* spores were found in honey fed to infants before the onset of infant botulism, and in which the toxin type (A or B) of spores in the honey matched the toxin type (A or B) of the *C. botulinum* that caused illness of the infant. In addition to causing infant botulism, contaminated honey may also lead to infectious botulism in adults made susceptible by abdominal surgery, gastrointestinal tract abnormalities, gastric achlorhydria or recent treatment with antibiotics.

Spores of *C. botulinum* have been found in honeys from the US, Argentina, Australia, Canada, China (and Taiwan), Denmark, Finland, Italy, Norway, Spain, Japan and Central America but not in honey from the UK (Arnon, 2004; Nevas *et al.*, 2005a). Honey may be contaminated with spores from dust in the environment and there is evidence that *C. botulinum* spores can germinate and the bacteria multiply in dead bees and bee pupae.

Table 2.4.3 Foodborne botulism in the US, 1990–2000.

Source of food	Cases	% of total cases
Non-commercial, canned food	70	27
Non-commercial food, not canned	27	10
Restaurant-prepared food	25	9.5
Commercially processed food	10	4
Other or unknown processing	27	10
Non-commercial foods prepared in Alaska	103	39
Total	262	

Data from Sobel *et al.*, 2004.

2.4.5 *Outbreaks of foodborne botulism and cases of infant botulism*

Types of food and processing associated recently with foodborne botulism in the US are shown in Table 2.4.3; meat, fish and vegetables are each involved in many outbreaks and cases. Despite widespread advice to the public, home-canned food that has not been heated sufficiently to destroy spores of *C. botulinum* has been a major cause of botulism, whereas processes for the commercial canning of foods have, since the 1920s, been designed to be highly lethal to spores of *C. botulinum*. Other non-commercial foods and restaurant-made foods caused many cases, while commercial products other than canned foods caused 4% of cases. The highest proportion of cases occurred in Alaska, where all the associated foods that were identified were homemade, Alaskan native foods of aquatic origin. Many of these cases were caused by fermented products such as salmon heads, whitefish, whale skin, beaver tail and seal flipper.

The examples of outbreaks in Table 2.4.4 illustrate the varied types of food that can be involved and the deficiencies in production and preparation that can lead to outbreaks and cases. Many of these outbreaks involved foods prepared on a small scale, domestically or in restaurants. In foods in which spores of *C. botulinum* have not been inactivated, and where oxygen is depleted, e.g. by canning, bottling, or immersion in oil, any surviving spores may be able to germinate, resulting in growth of vegetative bacteria and toxin formation unless inhibition is ensured by factors such as acidification and refrigerated storage.

Infant botulism has been reported in every continent except Africa; between 60 and 100 cases/year were reported in the US between 1981 and 2003, most cases being caused by *C. botulinum* toxin type A or B (CDC, 2005). A relatively high incidence has been reported in Argentina with 150 cases, associated with *C. botulinum* type A, reported between 1984 and 1999 (Centorbi *et al.*, 1999). Cases reported in Europe are summarized in Table 2.4.5.

2.4.6 *Main methods of prevention and control*

2.4.6.1 *Foodborne botulism*

In many foods the incidence and numbers of *C. botulinum* present may be very low, nevertheless absence of the bacterium cannot be assumed. In foods where vegetative bacteria have

Table 2.4.4 Examples of outbreaks of foodborne botulism.

Country, year	Toxin type	Cases	Deaths	Food implicated	Evidence[a]	Place of supply	Factors leading to outbreak	Reference
US, 1977	B[b]	59	0	Hot sauce made with restaurant-canned peppers	M	Mexican restaurant	Peppers insufficiently heated during canning; some jars exploded after several days. Canned peppers used to make hot sauce, which was not heated sufficiently to destroy toxin	Terranova *et al.* (1978)
US, 1983	A	28	1	Sautéed onions	M	Restaurant	After cooking onions in margarine, the mixture was poured into a pan in which onions were covered by a thick layer of melted margarine, and set on a grill at ~41°C for use during the day. Spores that survived cooking probably germinated and resulted in growth of *C. botulinum* in anaerobic conditions under margarine	MacDonald *et al.* (1985)
Canada, 1985	Bp[c]	36	0	Commercial chopped garlic in soybean oil	M	Restaurant	Garlic bottled in soybean oil, with no heat treatment. The product was marked with instructions to keep refrigerated, but had been stored at room temperature	St Louis *et al.* (1988)
Canada, 1987	A	11	0	In-house bottled mushrooms	M	Hotel restaurant	Mushrooms insufficiently heated during bottling; pH high enough to allow growth of *C. botulinum*	CDC (1987); McLean *et al.* (1987)
US and Israel, 1987	E	8	1	Commercially produced, uneviscerated, salted, air-dried fish ('ribbetz' or 'kapchunka')	M	Community, ethnic food	Salt concentration probably insufficient to prevent growth of *C. botulinum* in viscera. Packaging and prolonged lack of refrigeration, would allow growth of the bacterium	Slater *et al.* (1989)
UK, 1989	Bp	27	1	Commercially produced hazelnut yoghurt	M	Community	Inadequate heat treatment of hazelnut preserve	O'Mahony *et al.* (1990)

Egypt, 1991	E	91	18	Shop-prepared, traditional, uneviscerated salted fish	M	Community	Possible modification of traditional fermentation. Uneviscerated fish allowed to putrefy before salting	Weber *et al.* (1993)
US, 1994	A	30	0	Restaurant-prepared potato dip (Skordalia)	M	Restaurant	Foil-wrapped, baked potatoes left at ambient temperature for several days before use in dip. The baking would not inactivate spores, and storage in foil wrap for several days allowed growth of *C. botulinum*	Angulo *et al.* (1998)
Italy, 1996	A	8	1	Dessert made with commercially prepared mascarpone cheese	M	Community	Production process did not involve a heat treatment that would inactivate *C. botulinum* spores in milk. Conditions in cheese allowed growth of Group I *C. botulinum* if the product was stored at a temperature higher than 12°C	Aureli *et al.* (2000)
Iran, 1997	A	27	1	Traditionally prepared cheese preserved in oil	M	Community	Spores of *C. botulinum* would survive in cheese. Immersion in oil provided anaerobic conditions allowing growth of the bacteria at the pH of the cheese (4.5–5.5)	Pourshafie *et al.* (1998)
Argentina, 1998	A	9	0	Meat roll prepared at home by a small-scale commercial producer	S	Supplied from a home	Heating of meat roll insufficient to destroy spores of *C. botulinum*. Meat roll then vacuum-packed in plastic and stored with inadequate refrigeration	Villar *et al.* (1999)
Thailand, 1998	A	13	2	Home-canned bamboo shoots	M	Sold in village	Inadequate cooking (boiling for 1 h), anaerobic conditions in can, pH (5.3–5.7) allowed growth of *C. botulinum*, during storage at ambient temperature	CDC (1999)

(*cont.*)

Table 2.4.4 *(continued)*

Country, year	Toxin type	Cases	Deaths	Food implicated	Evidence[a]	Place of supply	Factors leading to outbreak	Reference
UK, 1998	B	2	1	Home-preserved mushrooms bottled in oil in Italy	M	Family meal	Mushrooms are liable to be contaminated with *C. botulinum*; bottling in oil would not destroy spores but provided anaerobic conditions allowing germination, growth and toxin production	Brusin and Salmaso (1998)
Morocco, 1999	B	78	20	Mortadella sausage	D	Community	nr	Ouagari *et al.* (2002)
France, 2000	B	9	0	Home-canned asparagus	D	Family meal	Probable inadequate heat treatment during home-canning	Abgueguen *et al.* (2003)
US, 2001	A	16	0	Chili dish	M	Church supper	Frozen chili obtained from a salvage store, where gross mishandling occurred. The product may have undergone many cycles of thawing and freezing, during which growth and toxin formation occurred	Kalluri *et al.* (2003)
Italy, 2004	B	28	0	Home-preserved green olives in salt water	S	Restaurant	Olives received no treatment that would inactivate *C. botulinum* spores; conditions in the product would not prevent germination and growth of vegetative bacteria	Cawthorne *et al.* (2005)
Thailand, 2006	A	163	0	Home-canned bamboo shoots (in 20-L cans)	M	Religious festival	Inadequate heat treatment during canning	CDC (2006a)
US, Canada, 2006	A	6	0	Commercial carrot juice	M	Community	Juice not heated sufficiently to destroy Group I *C. botulinum*, and presumably left unrefrigerated for some time	CDC (2006b)

[a] *Evidence:* M (microbiological): identification of an organism of the same type from cases and in the suspect vehicle or vehicle ingredient(s), or detection of toxin in faeces or food; D (descriptive): other evidence, usually descriptive, reported by local investigators as indicating the suspect vehicle or food; S (statistical): a significant statistical association between consumption of the suspect vehicle and being a case.

[b] Toxin type B, not known whether produced by proteolytic or non-proteolytic *C. botulinum*.

[c] Toxin type Bp, produced by proteolytic *C. botulinum*.

nr, not reported.

Table 2.4.5 Reported cases of infant botulism in Europe.

Country	Number of reported cases	Strain and toxin type	Number of cases with a history of honey consumption
Czech Republic	1	*C. botulinum* type B	0
Denmark	2	*C. botulinum* type A	2
Finland	1	*C. botulinum* type B	0
France	1	*C. botulinum* type B	0
Germany	4	*C. botulinum* type A and untyped[a]	2
Hungary	2	*C. baratii* type F; *C. botulinum* untyped	1
Italy	20	*C. botulinum* types A and B;*C. butyricum* type E	14
Netherlands	1	*C. botulinum* untyped	1
Norway	4	*C. botulinum* type A	4
Spain	8	*C. botulinum* types A, B and untyped	7
Sweden	1	*C. botulinum* type A	1
Switzerland	1	*C. botulinum* type A	0
UK	6	*C. botulinum* types A, B, and Bf	1

Information from Aureli *et al.*, 2002; Fenicia *et al.*, 2004; Nevas *et al.*, 2005a.
[a]Untyped by monovalent antitoxins, but confirmed by polyvalent antitoxin.

been inactivated by heat or other treatment, a very low number of spores of *C. botulinum* can result in growth of the organism and toxin formation. It is essential, therefore, that foods liable to allow growth of *C. botulinum* either undergo a processing that will inactivate the spores, or that the composition and storage conditions are controlled so as to prevent growth. In foods of many types several factors are used, very often in combination, to minimize the survival of *C. botulinum* and the risk of germination and growth of the bacterium (Table 2.4.6). Foods in category 9 of this table are the subject of current concern (Peck, 2006). These are ready meals that undergo a heat treatment that does not inactivate spores of Group I, proteolytic *C. botulinum* and damages, but does not necessarily inactivate, spores of Group II strains. Conditions in packs of these foods are liable to become anaerobic and allow growth from any surviving spores. Growth from surviving spores of proteolytic strains can be prevented by storage of these foods below 12°C, which should be ensured by effective refrigeration, but surviving spores of non-proteolytic strains can give growth at 3°C. Guidelines for ensuring the safety of these foods have been issued by the UK Advisory Committee on the Microbiological Safety of Food (ACMSF, 1992, 1995) and the European Chilled Food Federation (ECFF, 1996; Peck, 2006). More recently the ACMSF concluded that for such foods where temperature is the only known controlling factor, storage at a temperature of 8°C or lower for a maximum period of 10 days should be specified (ACMSF, 2006).

A high proportion of the cases of foodborne botulism worldwide results from production of food domestically and in small catering units. In the UK, the home canning or bottling of vegetables is discouraged because of the risk of underprocessing and survival of, and growth from, spores.

Table 2.4.6 Examples of main factors used to control *C. botulinum* in foods.

Type of food	Main factors controlling *C. botulinum*	*C. botulinum* spores killed	*C. botulinum* spores inhibited	Examples of foods
Shelf-stable foods				
1	Heat process $>F_o3^a$	+	–	Low acid, canned foods, pH > 4.5 (UK) (Department of Health, 1994) or pH > 4.6 (US)
2	Heat process $<F_o3$, NaCl, nitrite	$(+)^b$	+	Shelf-stable, canned cured meats
3	Heat process $<F_o3$, high acid	(+)	+	Canned acid foods pH < 4.5 (UK) or pH < 4.6 (US) e.g. many types of canned fruit and vegetables
4	Low water activity[c] (NaCl + drying) ± nitrite. Refrigeration below 5°C during salting	–	+	Raw salted and salt-cured meats, e.g. salt pork, salt bacon, salt hams, dry-cured hams, and bacon
5	Low water activity (NaCl + drying), nitrite, pH	–	+	Fermented sausages (e.g. summer sausage, pepperoni, acidulated sausages)
Perishable foods				
6	Refrigeration alone	–	+	Raw meat, fish and vegetables
7	Refrigeration, NaCl, nitrite in many products, in some cases sorbate, benzoate, nitrate or smoke	–	+	Salted raw fish, lightly preserved fish, semi-preserved fish, hot-smoked fish, perishable cooked, cured meats
8	Low water activity (NaCl and other components), pH, refrigeration of some products, mild heat treatment	–	+	Process cheese and process cheese spreads
9	Mild heat treatment, refrigeration, limited shelf-life	(+)	+	Mildly heated, vacuum and modified atmosphere packed, chilled foods; 'sous vide' foods

Data from Hauschild, 1989; Lund & Peck, 2000.

[a] F_o3 is the equivalent of heating at a temperature of 121°C, throughout the food, for 3 min.

[b] Spores of Group II strains may be damaged or inactivated, spores of Group I strains are not inactivated.

[c] Water activity is a measure of the water available to allow growth of microorganisms.

2.4.6.2 *Infant botulism*

Spores of *C. botulinum* have been found in honey in many countries and in up to 10% of samples in the US, Italy, Japan and Argentina, with numbers in some samples greater than 10^4/kg (Lund & Peck, 2000). Global trade results in the dissemination of honeys produced in many parts of the world. Although infant botulism appears to be uncommon, in view of the possibly serious symptoms and the prolonged period that may be needed for recovery, it is important to prevent exposure of infants to a well-known source of the bacterium, such as honey. In the US, the FDA advises against the use of honey for feeding infants. In the UK since 1996, the British Honey Importers and Packers Association has advised its members to label honey 'should not be given to infants under 12 months of age', and the Food Standards Agency (FSA) recommends that other packers and suppliers should adopt this policy (FSA, 2003). Similar labels are used also in Norway, and their use has been advocated in Europe (Aureli *et al*., 2002).

Summary of measures advised in healthcare settings

1. Canned foods and other preserved foods should be obtained from approved suppliers.
2. The safety of prepared ready meals, of products such as vacuum-packed vegetables, pasteurized, chilled foods and vegetables/spices/herbs preserved in oil, should be controlled according to the guidance issued by the ACMSF in the UK and the ECFF.
3. Honey should not be given to infants under 12 months old, or to adults with recent abdominal surgery or gastrointestinal abnormalities.

References

Abgueguen, P., Delbos, V., Chennebault, J.M. *et al.* (2003) Nine cases of foodborne botulism type B in France and a literature review. *European Journal of Clinical Microbiology and Infectious Diseases* **22**, 749–752.

Advisory Committee on the Microbiological Safety of Foods (ACMSF) (1992) *Report on Vacuum Packaging and Associated Processes.* Her Majesty's Stationery Office, London.

Advisory Committee on the Microbiological Safety of Food (ACMSF) (1995) *Annual Report 1995.* Her Majesty's Stationery Office, London.

Advisory Committee on the Microbiological Safety of Food (ACMSF) (2006) *Review of Scientific Evidence to Support FSA Consultation on Vacuum-Packaged and Modified Atmosphere Packaged Foods.* ACM/777. Available from http://www.food.gov.uk/multimedia/pdfs/acm777.pdf. Accessed 5 March 2007.

Akbulut, D., Dennis, J., Gent, M. *et al.* (2005) Wound botulism in injectors of drugs: upsurge in cases in England during 2004. *Euro Surveillance* **10**(9), 172–174. Available from http://www.eurosurveillance.org/em/v10n09/1009-223.asp. Accessed 5 March 2007.

Angulo, F.J., Getz, J., Taylor, J.P. *et al.* (1998) A large outbreak of botulism: the hazardous baked potato. *Journal of Infectious Diseases* **178**, 172–177.

Arnon, S.S. (2004) Infant botulism. In: Feigin, R.D., Cherry, J.D., Demmler, G.J. & Kaplan, S. (eds). *Textbook of Pediatric Infectious Diseases*, 5th edn. W.B. Saunders Company, Philadelphia, pp. 1758–1766.

Arnon, S.S., Schechter, R., Maslanka, S.E. *et al.* (2006) Human botulism immune globulin for treatment of infant botulism. *New England Journal of Medicine* **354**, 462–471.

Aureli, P., Di Cunto, M., Maffei, A. *et al.* (2000) An outbreak in Italy of botulism associated with a dessert made with mascarpone cream cheese. *European Journal of Epidemiology* **16**, 913–918.

Aureli, P., Franciosa, G. & Fenicia, L. (2002) Infant botulism and honey in Europe: a commentary. *Pediatric Infectious Disease Journal* **21**, 866–868.

Brett, M.M., McLauchlin, J., Harris, A. *et al.* (2005) A case of infant botulism with a possible link to infant formula milk powder: evidence for the presence of more than one strain of *Clostridium botulinum* in clinical specimens and food. *Journal of Medical Microbiology* 54, 769–776.

Brusin, S. & Salmaso, S. (1998) Botulism associated with home-preserved mushrooms. *Eurosurveillance Weekly* **2**(18), 980430. Available from http://www.eurosurveillance.org/ew/1998/980430.asp#1. Accessed 5 March 2007.

Cawthorne, A., Celentano, L.P., D'Ancona, F. *et al.* (2005) Botulism and preserved green olives. *Emerging Infectious Diseases* **11**, 781–782.

Centers for Disease Control and Prevention (CDC) (1987) Restaurant-associated botulism from mushrooms bottled in-house – Vancouver, British Columbia, Canada. *Morbidity and Mortality Weekly Report* **36**(7), 103.

Centers for Disease Control and Prevention (CDC) (1999) Foodborne botulism associated with home-canned bamboo shoots – Thailand, 1998. *Morbidity and Mortality Weekly Report* **48**(21), 437–439.

Centers for Disease Control and Prevention (CDC) (2005) Summary of notifiable diseases – United States, 2003. *Morbidity and Mortality Weekly Report* **52**(54), 1–88.

Centers for Disease Control and Prevention (CDC) (2006a) Botulism from home-canned bamboo shoots – Nan province, Thailand, March 2006. *Morbidity and Mortality Weekly Report* **55**(14), 389–392.

Centers for Disease Control and Prevention (CDC) (2006b) Botulism associated with commercial carrot juice – Georgia and Florida, September 2006. *Morbidity and Mortality Weekly Report* **55**(40), 1098–1099.

Centorbi, H.J., Aliendro, O.E., Demo, N.O. *et al.* (1999) First case of infant botulism associated with honey feeding in Argentina. *Anaerobe* **5**, 181–183.

Department of Health (1994) *Guidelines for the Safe Production of Heat Preserved Foods*. HMSO, London.

European Chilled Food Federation (ECFF) (1996) *Guidelines for the Hygienic Manufacture of Chilled Foods.* The European Chilled Food Federation, Helsinki.

Fenicia, L., Anniballi, F., Pulitanò, S. *et al.* (2004) A severe case of infant botulism caused by *Clostridium botulinum* with concomitant intestinal viral infections. *European Journal of Pediatrics* **163**, 501–502.

Fenicia, L., Franciosa, G., Pourshaban, M. & Aureli, P. (1999) Intestinal toxaemia botulism in two young people, caused by *Clostridium butyricum* type E. *Clinical Infectious Diseases* **29**, 1381–1387.

Food Standards Agency (FSA) (2003) *The Honey Regulations* 2003. Guidance Notes. Available from http://www.food.gov.uk/multimedia/pdfs/honeyguidance.pdf. Accessed 5 March 2007.

Hauschild, A.H.W. (1989) *Clostridium botulinum*. In: Doyle, M.P. (ed). *Foodborne Bacterial Pathogens*. Marcel Dekker, New York, pp. 112–189.

Kalluri, P., Crowe, C., Reller, M. *et al.* (2003) An outbreak of foodborne botulism associated with food sold at a salvage store in Texas. *Clinical Infectious Disease* **37**, 1490–1495.

Lund, B.M. & Peck, M.W. (2000) *Clostridium botulinum*. In: Lund, B.M., Baird-Parker, T.C. & Gould, G.W. (eds). *The Microbiological Safety and Quality of Food.* Aspen Publishers, Gaithersburg, MD, pp. 1057–1109.

MacDonald, K.L., Spengler, R.F., Hatheway, C.L. *et al.* (1985) Type A botulism from sauteed onions. *Journal of the American Medical Association* **253**, 1275–1278.

McLean, H.E., Peck, S., Blatherwick, F.J. *et al.* (1987) Restaurant-associated botulism from in-house bottled mushrooms – British Columbia. *Canada Diseases Weekly Report* **13**, 35–36.

Nevas, M., Lindström, M., Hautamäki, K. *et al.* (2005a) Prevalence and diversity of *Clostridium botulinum* types A, B, E and F in honey produced in the Nordic countries. *International Journal of Food Microbiology* **105**, 145–151.

Nevas, M., Lindström, M., Virtanen, A. *et al.* (2005b) Infant botulism acquired from household dust presenting as sudden infant death syndrome. *Journal of Clinical Microbiology* **43**, 511–513.

O'Mahony, M.O., Mitchell, E., Gilbert, R.J. *et al.* (1990) An outbreak of foodborne botulism associated with contaminated hazelnut yogurt. *Epidemiology and Infection* **104**, 389–395.

Ouagari, Z., Chakib, A., Sodqi, M. *et al.* (2002) Le botulism à casablanca. (A propos de 11 cas). *Bulletin de la Societe de Pathologie Exotique* **95**, 272–275.

Peck, M.W. (2006) *Clostridium botulinum* and the safety of minimally heated, chilled foods: an emerging issue? *Journal of Applied Microbiology* **101**, 556–570.

Peck, M.W. & Stringer, S.C. (2005) The safety of pasteurised in-pack chilled meat products with respect to the foodborne botulism hazard. *Meat Science* **70**, 461–475.

Pourshafie, M.R., Saifie, M., Shafiee, A. *et al.* (1998) An outbreak of food-borne botulism associated with contaminated locally made cheese in Iran. *Scandinavian Journal of Infectious Disease* **30**, 92–94.

Shapiro, R.L., Hatheway, C. & Swerdlow, D.L. (1998) Botulism in the United States: a clinical and epidemiologic review. *Annals of Internal Medicine* **129**, 221–227.

Slater, P.E., Addiss, D.G., Cohen, A. *et al.* (1989) Foodborne botulism: an international outbreak. *International Journal of Epidemiology* **18**, 693–696.

Sobel, J., Tucker, N., Sulka, A. *et al.* (2004) Foodborne botulism in the United States, 1990–2000. *Emerging Infectious Diseases* **10**, 1606–1611.

St Louis, M.E., Shaun, H.S., Peck, M.B. *et al.* (1988) Botulism from chopped garlic: delayed recognition of a major outbreak. *Annals of Internal Medicine* **108**, 363–368.

Terranova, W., Breman, J.G., Locey, R.P. & Speck, S. (1978) Botulism type B: epidemiologic aspects of an extensive outbreak. *American Journal of Epidemiology* **108**, 150–156.

Villar, R.G., Shapiro, R.L., Busto, S. *et al.* (1999) Outbreak of type A botulism and development of a botulism surveillance and antitoxin release system in Argentina. *Journal of the American Medical Association* **281**, 1334–1340.

Weber, J.T., Hibbs, R.G., Jr., Darwish, A. *et al.* (1993) A massive outbreak of type E botulism associated with traditional salted fish in Cairo. *Journal of Infectious Diseases* **167**, 451–454.

World Health Organization (WHO) (2003) *WHO Surveillance Programme for Control of Foodborne Infections and Intoxications in Europe*, 8th report, 1999–2000. Available from http://www.bfr.bund.de/internet/8threport/8threp_fr.htm. Accessed 5 March 2007.

2.5 *Clostridium perfringens*

2.5.1 Importance as a cause of foodborne disease

Clostridium perfringens (previously named as *C. welchii*) causes wound infections and came into prominence as the cause of gas gangrene during the 1914–1918 war. In 1943, reports began to appear indicating that the organism caused outbreaks of foodborne disease, and proof of this came in work by Hobbs *et al.* published in 1953 (Labbé, 2000). Strains of *C. perfringens* are classified into five types (A–E) according to the extracellular toxins formed. Type A strains cause gas gangrene and are responsible for almost all the cases of *C. perfringens* foodborne illness in humans; this section will deal almost exclusively with type A strains.

In healthy adults the symptoms of food poisoning by *C. perfringens* are relatively mild, for this reason many cases and outbreaks, particularly those that result from domestic preparation of food, probably are not reported. In the elderly and in vulnerable groups the symptoms are more serious. In England and Wales in 2000, *C. perfringens* was estimated as second only to *Campylobacter* as the cause of foodborne disease, causing an estimated 84,081 cases (13% of total) and second to non-typhoidal salmonellas in lethal effect, causing 22% of deaths (Adak *et al.*, 2002). The organism is regarded in the UK as one of the five most important foodborne pathogens (FSA, 2002); in 1992–1994, it was the cause of 23 reported foodborne outbreaks in residential institutions in England and Wales (Ryan *et al.*, 1997). In the US, estimates of the annual number of cases range from 10,000 to 250,000 but there is a great deal of underreporting (Mead *et al.*, 1999), attributed in part to the anaerobic conditions required for cultivation of the organism and the perceived mild nature of the illness (Heredia & Labbé, 2001). In many European countries cases of food poisoning caused by *C. perfringens* are either not reported or greatly underreported (WHO, 2003).

2.5.2 Characteristics of the organism

Clostridium perfringens is a Gram-positive, spore-forming, non-motile, rod-shaped bacterium. It is anaerobic but will tolerate some exposure to air. Growth occurs between temperatures of 12° and 52°C (ICMSF, 1996; Peck *et al.*, 2004), and at optimum temperatures of 43–47°C the generation time in meat can be less than 10 minutes (Labbé, 2000). The pH range for growth is pH 5.0–pH 9.0, with optimum growth between pH 6.0 and 7.0. Spores are not produced readily in culture, but are produced when vegetative bacteria are consumed and reach the intestine. Types A–E are differentiated according to the exotoxins formed, and some strains of all types form a polypeptide enterotoxin (CPE) (Table 2.5.1). While type A strains are responsible for almost all *C. perfringens* illness in humans, types B–E cause disease in animals. In type A strains, formation of CPE accompanies spore formation and is the main cause of food poisoning. Fewer than 5% of a wide range of strains of *C. perfringens* studied were reported to carry the enterotoxin gene (*cpe*), which may be present either on the chromosome or on a plasmid (Sarker *et al.*, 2000). To date, the majority of food-poisoning isolates tested have carried the *cpe* gene on the chromosome, whereas many strains not associated with food poisoning carried the *cpe* gene on a plasmid.

Table 2.5.1 Properties of *Clostridium perfringens*.

Type	Major lethal exotoxins formed				Enterotoxin (CPE) formed	Pathogenicity
	Alpha	Beta	Epsilon	Iota		
A	+	−	−	−	+	*Humans* (gas gangrene; food poisoning; necrotizing enteritis of infants; septicaemia). *Animals* (necrotic enteritis of poultry; enterotoxaemia in cattle and lamb; possibly colitis)
B	+	+	+	−	+	*Animals* (lamb dysentery; enterotoxaemia of sheep, foals, goats; hemorrhagic enteritis in neonatal calves and foals)
C	+	+	−	−	+	*Humans* (enteritis necroticans (pig bel)). *Animals* (enterotoxaemia of sheep; necrotic enteritis in animals; acute toxaemia in adult sheep)
D	+	−	+	−	+	*Animals* (enterotoxaemia of sheep, lamb and calves)
E	+	−	−	+	+	*Animals* (enterotoxaemia of rabbits; canine and porcine enteritis)

Information from Labbé (2000) and Heredia & Labbé (2001).

Serotyping has been used to characterize isolates but is being superseded by molecular methods (Schalch *et al.*, 2003).

Spores of many strains are highly heat-resistant. Spores of 12 strains, obtained from foods implicated in outbreaks or from patients, when heated in phosphate buffer at 95°C for 30 minutes, showed less than a 1.5 log reduction in numbers (Peck *et al.*, 2004). Spores of food-poisoning isolates carrying a chromosomal *cpe* gene showed an average $D_{100^\circ C}$ value of 60 minutes when heated in a culture medium, whereas spores of isolates carrying a plasmid *cpe* gene showed an average $D_{100^\circ C}$ value of 1 minute (Sarker *et al.*, 2000). Thus, spores of both types are liable to survive cooking of foods, but spores of strains that carry the *cpe* gene on the chromosome and possess the higher heat-resistance, will show a higher rate of survival.

2.5.3 *Disease in humans*

Illness occurs, usually, after consumption of food containing $>10^6$ CFU of the bacteria/g of food. The incubation period before symptoms is ~8–24 hours. The enterotoxin is not usually preformed in foods in amounts that are high enough to cause clinical illness. After ingestion of the bacteria a proportion survive the acid conditions in the stomach, and in the intestine sporulation occurs with the production of enterotoxin. This increases the permeability of epithelial cells resulting in loss of fluids and electrolytes (Labbé, 2000). The illness takes the form of diarrhoea with severe abdominal pain and nausea. It is usually self-limiting and treatment with antibiotics is not recommended. Occasionally death is caused, particularly in elderly and debilitated people. Very rarely in the developed world, necrotizing enterocolitis caused by *C. perfringens* type A has been reported in previously healthy adults (Sobel *et al.*, 2005). In a foodborne outbreak of CPE-positive *C. perfringens* type A infection in the US in 2001, three patients developed severe bowel necrosis that resulted in two deaths (Bos *et al.*, 2005). The symptoms were attributed in part to drug-induced constipation and faecal impaction, resulting in prolonged exposure of colonic tissue to *C. perfringens* type A toxins.

Rarely, a severe type of necrotic enteritis in humans is caused by *C. perfringens* type C. This was described in Germany in the 1940s in chronically starved people who had eaten a large meal, possibly including poorly cooked meat (Sobel *et al.*, 2005). Cases of this illness in Germany ceased when standards of living and nutrition increased after World War II. A similar illness was recognized in Papua New Guinea in the 1960s, where it was associated with feasting on pork meat by protein-deficient persons. Intestinal necrosis was attributed to the β-toxin produced by the bacterium. Occasionally, necrotic enteritis caused by *C. perfringens* type C has occurred in adults with pre-existing chronic illnesses, and was reported in a diabetic boy following consumption of pork chitterlings (Petrillo *et al.*, 2000).

2.5.4 *Source and transmission*

Clostridium perfringens type A strains occur in soil, where they can be present at concentrations of 10^3–10^4 CFU/g; they are considered part of the normal intestinal flora of humans and many animals. Types B–E do not appear to survive for long periods in soil, but survive in the intestines of animals. Spores of type A strains are present at about 10^3–10^4/g in the normal faecal flora of most humans, but patients in chronic care institutions may carry higher numbers of spores in the gastrointestinal tract. Diagnosis of *C. perfringens* food poisoning is based on the presence of unusually high numbers of the organism and/or of enterotoxin in the stools and high numbers of the organism in the food implicated.

Only a low proportion of strains from humans and animals have been reported to carry the *cpe* gene, and in the intestines of animals a small number of *cpe*-positive strains probably co-exist with a large number of *cpe*-negative strains. The proportion of *cpe*-positive strains to total strains of *C. perfringens* in the intestinal contents of cattle, pigs and chickens varied between 1 in 10 and 1 in 10^5 (Miwa *et al.*, 1997).

The presence of *C. perfringens* in the intestines of food animals means that carcasses are liable to be contaminated with vegetative bacteria and spores during slaughter. Raw meat and poultry frequently carry between 10^1 and 10^3 *C. perfringens*/g. The numbers present depend on the conditions during slaughter and processing, and improvements in these conditions over the past 30 years in the US have led to lower levels of *C. perfringens* in meat (Kalinowski *et al.*, 2003). In other conditions, numbers up to 10^4 or 10^5 CFU/g of meat products have been reported (Rodriguez & Vargas, 2002; Stagnitta *et al.*, 2002).

In addition to their presence in meat and poultry products, in view of the presence of *C. perfringens* spores in the environment, it is not surprising that they are found in a very wide range of foods including fish, vegetables, dairy and cereal products; a relatively high incidence has been found in spices where the number usually is less than 500/g and rarely over 1000/g (ICMSF, 1998).

In a survey of American retail foods, including meat and fish, *C. perfringens* type A was detected in an average of 31% of ~900 samples (not associated with outbreaks); strains carrying a *cpe* gene were detected in ~1.4% of samples, in each case the gene was located on the chromosome (Wen & McClane, 2004).

Strains carrying the *cpe* gene were found as a low proportion of the total *C. perfringens* type A in the faeces of 18% of healthy food handlers in Finland; in a minority of strains the *cpe* gene was located on the chromosome and on the majority it was located on a plasmid (Heikinheimo *et al.*, 2006).

Foods that have been responsible mainly for illness are cooked beef and poultry, particularly where large joints are involved that have been prepared with gravy (Kalinowski *et al.*, 2003). Spores of *C. perfringens* survive cooking, and if the food remains at favourable temperatures after cooking the spores germinate forming vegetative bacteria that multiply in the food. About 10^6 CFU *C. perfringens*/g of food are required, usually, to cause illness. After refrigerated storage, pre-cooked food should be reheated to an internal temperature of 72°C (162°F) before serving, to inactivate any vegetative *C. perfringens* that may be present (EFSA, 2005).

2.5.5 *Examples of outbreaks*

Cases and outbreaks of disease are liable to occur in restaurants and in institutions such as hospitals, nursing homes, residential homes, school cafeterias and prisons, where large quantities of food are prepared several hours before they are served. The examples of reported outbreaks shown in Table 2.5.2 include incidents associated with fish and with vegetables as well as with meat products. The table emphasizes the fact that the outbreaks were caused by failure to cool food rapidly and adequately, and the use of inadequate or no reheating before serving.

2.5.6 *Main methods of prevention and control*

Spores of *C. perfringens* are liable to be present on raw foods and can survive normal cooking processes that inactivate high numbers of vegetative bacterial pathogens, such as *Salmonella* spp. and *Listeria monocytogenes.*

The major factors in prevention of foodborne infection with *C. perfringens* are rapid cooling and refrigerated storage of cooked foods, to prevent germination of surviving spores and growth of vegetative bacteria, and adequate reheating of foods to inactivate any vegetative bacteria before consumption.

Cooked foods should either be eaten immediately, or kept for a short time at a temperature higher than 63°C (145.4°F), or they should be cooled rapidly and kept below 7–8°C (ideally below 4°C to control organisms other than clostridia) (EFSA, 2005).

In the US, the guidelines for cooling cooked meat and poultry products state that during cooling the product's maximum internal temperature should not remain between 130°F (54.5°C) and 80°F (26.8°C) for more than 1.5 hours, nor between 80°F (26.8°C) and 40°F (4.5°C) for more than 5 hours (FSIS, 1999). A major objective of this guidance is to ensure that no more than a tenfold (1 log) increase in numbers of *C. perfringens* can occur during cooling of the cooked food.

Table 2.5.2 Examples of outbreaks of foodborne disease caused by *Clostridium perfringens* type A.

Place, date	Cases	Deaths	Food implicated	Evidence[a]	Where food was supplied	Factors leading to outbreak	Reference
UK, 1975	>56	nr	Boiled salmon	M	Dinner party	Fish cooked in boiling water then left to cool in the pans of water at ambient temperature overnight. Served cold with salad	Hewitt *et al.* (1986)
US, 1984	112	nr	Roast beef	D	Roast beef luncheon	Time temperature for cooling not defined. Meat served cold 18–48 h after being cooked	Gross *et al.* (1989)
US, 1985	304	0	Gravy	S	Employee banquet at factory	After heating, gravy was held at room temperature for 5.5 h before placing in a refrigerator; reheated before serving	Petersen *et al.* (1988)
US, 1986	86	0	Turkey	S	Nursing home	Frozen whole turkeys, thawed overnight in frig. Cooked, cooled at room temperature (no time/temperature data) refrigerated overnight. Slices warmed for several hours before being served	Birkhead *et al.* (1988)
Brazil, 1987	13	nr	Sausage	M	Community	Sausages maintained at room temperature during sale, not heated before consumption	Tortora & Zebral (1988)
UK, 1989	50	2	Minced beef meal	S	Psychiatric hospital	Understaffing and inadequate supervision	Pollock & Whitty (1990)
UK, 1989	17	0	Pre-cooked, vacuum-sealed, roast pork	S	Hospital	Faults in production at company that supplied ready-cooked joints. Cooked meat had only cooled to 28°C after 50 h	Regan *et al.* (1995)
US, 1990	32	0	Minestrone soup	M	Conference	Soup cooked 2 days before serving, cooled slowly before refrigeration, reheated briefly before serving. Failure to cool the soup rapidly after preparation; failure to reheat sufficiently; holding soup at warm temperature before serving	Roach & Sienko (1992)
US, 1994	~156	0	Corned beef	M	Delicatessen	Raw corned beef boiled but cooled inadequately, refrigerated then held at 49°C before serving; sandwiches held at room temperature for several hours	CDC (1994)

US, 1994	86	0	Corned beef	M	Dinner party	Large pieces cooked, cooling not specified. Meat sliced and warmed under heat lamps for ~90 min before serving	CDC (1994)
US, 1995	53	0	Roasted turkey	S	Thanksgiving meal, Juvenile detention facility	Turkey breasts cooked, piled into stockpot and placed in walk-in refrigerator. Some turkey was warm when removed from frig. Reheated (conditions unspecified) before serving	Parikh *et al.* (1997)
Australia, 1997	25	1	21 cases associated with pureed food	S	Nursing home	Food was cooked then pureed (liquidized); it may then have remained at a temperature that allowed growth of *C. perfringens*. The food was not reheated before serving	Tallis *et al.* (1999)
Italy, 1997	25	0	'Pasta a ragu' pasta with meat and tomato sauce	S	Restaurant	Undercooked meat balls left unrefrigerated for several hours before serving	Arcieri *et al.* (1999)
Germany, 1998	21	2	Minced beef heart	M	Nursing home	nr	Schalch *et al.* (2003)
Japan, 1998	30	0	Boiled spinach with fried bean curd	M	Home for senior citizens	Boiled food left at room temperature for about 16 h after cooking	Miwa *et al.* (1999)
Japan, 2001	90	nr	Boiled beans	D	Nursing home for the aged	Beans cooked in large quantities, cooled slowly, not reheated adequately before serving	Tanaka *et al.* (2003)
US, 2001	7	2	Thanksgiving meal	D	Residential care home for mentally ill patients	nr	Bos *et al.* (2005)
Sweden, 2002	64	2	Pea soup	nr	Home for the elderly	nr	De Jong *et al.* (2004)
Australia, 2003	42	nr	Suspect gravy mixed into vitaminized meals	M	Aged care facility	nr	OzFoodNet (2003)
UK, 2004	13	0	Roast lamb	D	Residential institution	nr	HPA (2005)

[a] M (microbiological): identification of an organism of the same type from cases and in the suspect vehicle or vehicle ingredient(s), or detection of toxin in faeces or food; D (descriptive): other evidence, usually descriptive, reported by local investigators as indicating the suspect vehicle or food; S (statistical): a significant statistical association between consumption of the suspect vehicle and being a case.
nr, not reported.

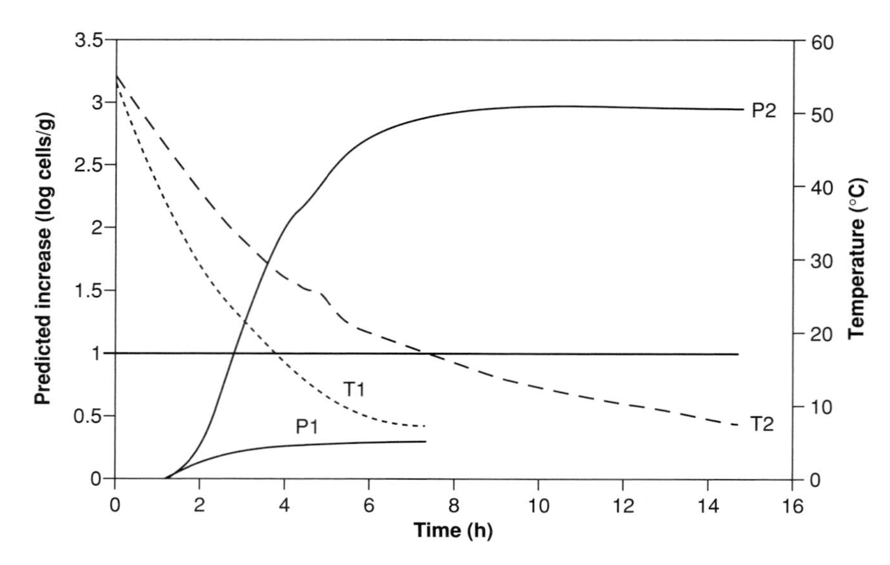

Fig. 2.5.1 The effect of rate of cooling of cooked meats on increase in number of *C. perfringens*. Calculated from Perfringens Predictor (Peck *et al.*, 2007). T1, cooling curve 1; T2, cooling curve 2; P1, predicted increase during cooling curve 1; P2, predicted increase during cooling curve 2.

The UK Food Standards Agency also advises that cooling should not allow more than a 1-log increase in numbers of *C. perfringens*. Based on published data on the effect of temperature on the rate of growth of *C. perfringens* and validation experiments with heat-resistant, CPE-producing strains from outbreaks of food poisoning (Peck *et al.*, 2004), a computer programme has been prepared that enables prediction of growth from spores of *C. perfringens* during any specified cooling curve (Peck *et al.*, 2007). Using this programme it is possible to determine whether any proposed cooling process will comply with the UK and US guidelines (Fig. 2.5.1). Following cooling, foods should be stored at a temperature of 5°C (41°F) or lower, and should be reheated thoroughly to an internal temperature of 72°C (161.5°F) before serving to inactivate bacteria of any type that may have grown in the food during storage.

Summary of measures advised in healthcare settings

Cooked foods, including meat, poultry, vegetables, casseroles, should be eaten immediately, or kept above 63°C (145.4°F) for a short time, or be cooled rapidly, maintained below 7–8°C (ideally below 4°C [39.2°F] to control organisms other than clostridia) and reheated to at least 72°C (161.5°F) before consumption.

References

Adak, G.K., Long, S.M. & O'Brien, S.J. (2002) Trends in indigenous foodborne disease and deaths, England and Wales: 1992 to 2000. *Gut* **51**, 832–841.

Arcieri, R., Dionisi, A.M., Caprioli, A. *et al.* (1999) Direct detection of *Clostridium perfringens* enterotoxin in patients' stools during an outbreak of food poisoning. *FEMS Immunology and Medical Microbiology* **23**, 45–48.

Birkhead, G., Vogt, R.L., Heun, E.M. *et al.* (1988) Characterization of an outbreak of *Clostridium perfringens* food poisoning by quantitative faecal culture and fecal enterotoxin measurement. *Journal of Clinical Microbiology* **26**, 471–474.

Bos, J., Smithee, L., McClane, B. *et al.* (2005) Fatal necrotizing colitis following a foodborne outbreak of enterotoxigenic *Clostridium perfringens* Type A infection. *Clinical Infectious Diseases* **40**, e78–e83.

Centers for Disease Control and Prevention (CDC) (1994) *Clostridium perfringens* gastroenteritis associated with corned beef served at St Patrick's Day meals – Ohio and Virginia, 1993. *Morbidity and Mortality Weekly Report* **43**(08), 137, 143–144.

De Jong, A.E.I., Rombouts, F.M. & Beumer, R.R. (2004) Effect of cooling on *Clostridium perfringens* in pea soup. *Journal of Food Protection* **57**, 352–356.

European Food Safety Authority (EFSA) (2005) Opinion of the scientific panel on biological hazards on a request from the Commission related to *Clostridium* sp. in foodstuffs. *The EFSA Journal* **199**, 1–65.

Food Safety and Inspection Services (1999) *Compliance Guidelines for Cooling Heat-Treated Meat and Poultry Products (Stabilization)*. Available from http://www.fsis.usda.gov/oa/fr/95033F-b.htm. Accessed 6 March 2007.

Food Standards Agency (FSA) (2002) *Measuring Foodborne Illness Levels*. Available from http://www.food.gov.uk/safereating/microbiology/58736. Accessed 6 March 2007.

Gross, T.P., Kamara, L.B., Hatheway, C.L. *et al.* (1989) *Clostridium perfringens* food poisoning: use of serotyping in an outbreak setting. *Journal of Clinical Microbiology* **27**, 660–663.

Health Protection Agency (HPA) (2005) General outbreaks of foodborne illness in humans, England and Wales: July to September 2004. *Communicable Disease Report CDR Weekly* **15**(10), enteric. Available from http://www.hpa.org.uk/cdr/archives/2005/cdr1005.pdf. Accessed 6 March 2007.

Heikinheimo, A., Lindstrom, M., Granum, P.E. & Korkeala, H. (2006) Humans as a reservoir for enterotoxin gene-carrying *Clostridium perfringens* type A. *Emerging Infectious Diseases* **12**, 1724–1729.

Heredia, N.L. & Labbé, R.G. (2001) *Clostridium perfringens*. In: Labbé, R.G. & Garcia, S. (eds). *Guide to Foodborne Pathogens*. John Wiley and Sons, New York, pp. 133–141.

Hewitt, J.H., Begg, N., Hewish, S. *et al.* (1986) Large outbreaks of *Clostridium perfringens* food poisoning associated with the consumption of boiled salmon. *Journal of Hygiene (Cambridge)* **97**, 71–80.

International Commission on Microbiological Specifications for Foods (ICMSF) (1996) *Clostridium perfringens*. In: *Microorganisms in Foods 5. Microbiological Specifications of Food Pathogens*. Blackie Academic and Professional, London, pp. 112–125.

International Commission on Microbiological Specifications for Foods (ICMSF) (1998) Spices, dry soups and oriental flavourings. In: *Microorganisms in Foods 6. Microbial Ecology of Food Commodities*. Blackie Academic and Professional, London, pp. 274–312.

Kalinowski, R.M., Tompkin, R.B., Bodnaruk, P.W. & Pruett, W.P., Jr. (2003) Impact of cooking, cooling, and subsequent refrigeration on the growth of survival of *Clostridium perfringens* in cooked meat and poultry products. *Journal of Food Protection* **66**, 1227–1232.

Labbé, R.G. (2000) *Clostridium perfringens*. In: Lund, B.M., Baird-Parker, T.C. & Gould, G.W. (eds). *The Microbiological Safety and Quality of Food*. Aspen Publishers, Gaithersburg, MD, pp. 1110–1135.

Mead, P.S., Slutsker, L., Dietz, V. *et al.* (1999) Foodborne illness and death in the United States. *Emerging Infectious Diseases* **5**, 607–625.

Miwa, N., Masuda, T., Terai, K. *et al.* (1999) Bacteriological investigation of an outbreak of *Clostridium perfringens* food poisoning caused by Japanese food without animal protein. *International Journal of Food Microbiology* **49**, 103–106.

Miwa, N., Nishina, T., Kubo, S. & Honda, H. (1997) Most probable numbers of enterotoxigenic *Clostridium perfringens* in intestinal contents of domestic livestock detected by nested PCR. *Journal of Veterinary Medicine Science* **59**, 557–560.

OzFoodNet Working Group (2003) OzFoodNet: enhancing foodborne disease surveillance across Australia: Quarterly report, 1 July to 30 September 2003. *Communicable Diseases Intelligence* **27**, 504–507.

Parikh, A.I., Jay, M.T., Kassam, D. *et al.* (1997) *Clostridium perfringens* outbreak at a juvenile detention facility linked to a thanksgiving holiday meal. *Western Journal of Medicine* **166**, 417–419.

Peck, M.W., Baranyi, J., Plowman, J. *et al.* (2004) *Improved Control of Clostridium Perfringens*. Final report of Food Standards Agency project B14009. Food Standards Agency, London, UK.

Peck, M.W., Plowman, J., Le Marc, Y. *et al.* (2007) *Perfringens Predictor*. Available from http://www.combase.cc. Accessed 6 March 2007.

Petersen, L.R., Mshar, G.H., Cooper, G.H., Jr. *et al.* (1988) A large *Clostridium perfringens* foodborne outbreak with an unusual attack rate pattern. *American Journal of Epidemiology* **127**, 133–144.

Petrillo, T.M., Consuelo, M., Beck-Sagué, M. *et al.* (2000) Enteritis necroticans (Pigbel) in a diabetic child. *New England Journal of Medicine* **342**, 1250–1253.

Pollock, A.M. & Whitty, P.M. (1990) Crisis in our hospital kitchens: ancillary staffing levels during an outbreak of food poisoning in a long stay hospital. *British Medical Journal* **300**, 383–385.

Regan, C.M., Syed, Q. & Tunstall, P.J. (1995) A hospital outbreak of *Clostridium perfringens* food poisoning – implications for food hygiene review in hospitals. *Journal of Hospital Infection* **29**, 69–73.

Roach, R.L. & Sienko, D.G. (1992) *Clostridium perfringens* outbreak associated with minestrone soup. *American Journal of Epidemiology* **136**, 1288–1291.

Rodriguez, E. & Vargas, M.D.G.Y.P. (2002) *Clostridium perfringens* in raw and cooked meats and its relation with the environment in Costa Rica. *Archivos Latinoamericanos de Nutricion* **52**, 155–159.

Ryan, M.J., Wall, P.G., Adak, G.K. *et al.* (1997) Outbreaks of infectious intestinal disease in residential institutions in England and Wales, 1992–4. *Journal of Infection* **34**, 49–54.

Sarker, M.R., Shivers, R.P., Sparkes, S.G. *et al.* (2000) Comparative experiments to examine the effects of heating on vegetative cells and spores of *Clostridium perfringens* isolates carrying plasmid enterotoxin genes versus chromosomal enterotoxin genes. *Applied and Environmental Microbiology* **66**, 3234–3240.

Schalch, B., Bader, L., Schau, H.-P. *et al.* (2003) Molecular typing of *Clostridium perfringens* from a disease outbreak in a nursing home: ribotyping versus pulsed-field gel electrophoresis. *Journal of Clinical Microbiology* **41**, 892–895.

Sobel, J., Mixter, C.G., Kolhe, P. *et al.* (2005) Necrotizing enterocolitis associated with *Clostridium perfringens* type A in previously healthy North American adults. *Journal of the American College of Surgeons* **201**, 48–56.

Stagnitta, P.V., Micalizzi, B. & de Guzman, A.M.S. (2002) Prevalence of enterotoxigenic *Clostridium perfringens* in meats in San Luis, Argentina. *Anaerobe* **8**, 253–258.

Tallis, G., Ng, S., Ferreira, C. *et al.* (1999) A nursing home outbreak of *Clostridium perfringens* associated with pureed food. *Australian and New Zealand Journal of Public Health* **23**, 421–423.

Tanaka, D., Isobe, J., Hosorogi, S. *et al.* (2003) An outbreak of food-borne gastroenteritis caused by *Clostridium perfringens* carrying the *cpe* gene on a plasmid. *Japanese Journal of Infectious Diseases* **56**, 137–139.

Tortora, J.C.O. & Zebral, A.A. (1988) A human food poisoning outbreak in Rio de Janeiro caused by sausage contaminated with enterotoxigenic *Clostridium perfringens* type A. *Review Microbiology São Paulo* **19**, 113–118.

Wen, Q. & McClane, B.A. (2004) Detection of enterotoxigenic *Clostridium perfringens* type A isolates in American retail foods. *Applied and Environmental Microbiology* **70**, 2685–2691.

World Health Organization (WHO) (2003) *WHO Surveillance Programme for Control of Foodborne Infections and Intoxications in Europe*, 8th report, 1999–2000. Available from http://www.bfr.bund.de/internet/8threport/8threp_fr.htm. Accessed 6 March 2007.

2.6 *Enterobacter sakazakii*

2.6.1 Importance as a cause of foodborne disease

Enterobacter sakazakii is a rare cause of invasive infection, with a high death rate, in infants up to about 2 months old; premature or low-birth-weight infants and those that are immunocompromised are at greatest risk. By 2003, approximately 50 cases had been reported worldwide in infants less than 60 days old (Iverson & Forsythe, 2003), but this is probably an underestimate as many clinical laboratories do not test for this organism and in many countries infections are not reported.

2.6.2 Characteristics of the organism

Enterobacter sakazakii is a motile, non-spore-forming, facultatively anaerobic, Gram-negative, rod-shaped bacterium. It was regarded as a yellow-pigmented form of *Enterobacter cloacae* until 1980, when it was classified as a new species (Iverson & Forsythe, 2003). The temperature range for growth is 6–47°C, with an optimum rate at 35–40°C. The heat resistance has been reported as a $D_{72°C}$ of 1.3 seconds when heated in reconstituted infant formula, other reports have given similar or lower values (EFSA, 2004). Thus, the organism is less heat-resistant than *Listeria monocytogenes*, and exposure to temperatures at or above 70°C in liquid media should provide almost instantaneous inactivation. Growth has been reported in tomato juice at pH 4.4 but not in more acidic fruit juices (Kim & Beuchat, 2005). The bacterium is relatively resistant to dry conditions, in which it can survive for long periods. A few strains produce a viscous, capsular material, which may increase resistance to dry conditions and formation of biofilms (Iverson *et al.*, 2004).

Several methods including molecular characterization, in particular by PFGE, have been used for typing isolates (Gurtler *et al.*, 2005).

2.6.3 Disease in humans

The majority of reported cases of illness caused by *E. sakazakii* are in infants less than 2 months old (Iverson & Forsythe, 2003; EFSA, 2004). Premature infants and those with underlying medical conditions are at the greatest risk of developing illness, but a healthy, full-term, newborn infant in Iceland became ill prior to discharge from hospital and suffered permanent neurological damage as a result of the infection.

The dose needed to cause illness and the incubation period before illness are unknown. Infants may be colonized and show faecal carriage for 8–18 weeks, without developing symptoms (WHO, 2004). The most common symptoms of foodborne infection in infants include bacteraemia, meningitis and necrotizing enterocolitis. Case fatality rates of >50% have been reported, but in recent years this has declined to <20% (EFSA, 2004). Most infants who survive meningitis caused by *E. sakazakii* develop long-term complications, mainly neurological (Drudy *et al.*, 2006).

A few cases of illness (usually non-gastrointestinal) in adults caused by *E. sakazakii* have been reported, in the majority of these cases the adults had underlying diseases and no evidence of foodborne transmission was reported (Lai, 2001; Iverson & Forsythe, 2003).

There is little information about the mechanism of pathogenicity.

2.6.4 *Source and transmission*

In several cases of disease the source of the bacterium has not been determined, but from 1983 onwards there has been increasing evidence implicating the consumption of powdered infant formula.

Powdered infant formula is not a sterile product. Methods of production include (a) a wet-mix process, (b) a dry-mix process, (c) a combination of (a) and (b) (WHO, 2004). In method (a), the wet mix is pasteurized or sterilized, and then dried. In method (b), individual ingredients are prepared, heat-treated as appropriate, dried and dry-blended. In method (c), part of the ingredients are processed according to (a) to produce a base powder to which the rest of the ingredients are added according to (b). After drying and blending the powder is conveyed to filling machinery and distributed into final containers. During drying and these final stages there is a risk of contamination with bacteria (EFSA, 2004). The ability of *E. sakazakii* to tolerate dry conditions allows it to survive for long periods in the final powdered product.

The bacterium has been found in commercial samples of powdered infant formula, usually at very low levels. A survey of 141 samples of powdered infant formula from 35 countries showed that 14% of samples were contaminated with *E. sakazakii*; the levels were 0.36–2.7 CFU/100 g, except for one sample which contained 66 CFU/100 g (Muytjens *et al.*, 1988). The use of newer methods may improve detection and increase the estimate of incidence (Guillaume-Gentil *et al.*, 2005; Seo & Brackett, 2005).

If infant formula is reconstituted and kept for several hours at ambient temperature low numbers of *E. sakazakii* can multiply, increasing the risk that the product cause illness; this may be the cause of reported cases of illness. The organism can survive on blenders used to prepare infant formula, if they are not cleaned well, and this can result in contamination of further batches of reconstituted formula.

Enterobacter sakazakii has been isolated also from many other sources including cheese, minced beef, sausage meat and vegetables (Leclercq *et al.*, 2002) and from several other types of food (Iverson & Forsythe, 2003). When samples were taken from the production line environment of factories or by sampling vacuum cleaner bags, the bacterium was detected in samples from eight of nine food factories (producing milk powder, cereals, chocolate, potato flour and pasta) and samples from domestic vacuum cleaner bags showed the presence of the organism in five of 16 household environments (Khandhai *et al.*, 2004).

Enterobacter sakazakii has also been isolated from plant sources including mung bean and alfalfa sprouts and from many other environmental sources (Gurtler *et al.*, 2005). There are reports that some insects, and rats can carry the bacterium (Lehner & Stephan, 2004).

The widespread occurrence of *E. sakazakii* makes it difficult to prevent contamination with very low numbers of the bacterium during production of powdered infant formula.

2.6.5 *Examples of outbreaks*

A list of reported cases of infection of infants from 1958 to 2002, including those where the source of the bacterium was not determined, was compiled by Iverson and Forsythe (2003). In recent years microbiological evidence has been obtained in some cases implicating powdered infant formula in transmission of the bacterium, reports of such cases are shown in Table 2.6.1.

Table 2.6.1 Examples of outbreaks of illness in hospital neonatal units, caused by *Enterobacter sakazakii* associated with powdered infant formula.

Country, date	Cases	Deaths	Symptoms	Food implicated	Evidence[a]	Reference
Iceland, 1986–1987 (1 hospital)	3 (full-term neonates)	1 (2 survivors had severe brain damage)	Meningitis	Powdered infant formula	M	Biering *et al.* (1989); Clark *et al.* (1990)
US, 1988 (1 hospital)	4 (pre-term neonates)	0	Sepsis, bloody diarrhoea	Powdered infant formula and blender	M	Simmons *et al.* (1989); Clark *et al.* (1990)
US, 1988 (1 hospital)	1 (6 mo old, with small bowel complications)	0	Bacteraemia (*E. sakazakii* and *Leuconostoc mesenteroides*)	Prepared infant formula and blender	M	Noriega *et al.* (1990)
Belgium, 1998 (1 hospital)	12 (pre-term neonates)	2	Necrotizing enterocolitis	Powdered infant formula	M	Van Acker *et al.* (2001)
Israel, 1999–2000 (1 hospital)	2 (1 pre-term) plus 3 colonized	0	Bacteraemia, meningitis	Prepared infant formula and blender	M	Bar-Oz *et al.* (2001); Block *et al.* (2002)
US, 2001 (1 hospital)	3 (1 pre-term neonate) plus 7 colonized	1	Meningitis	Powdered infant formula	M	CDC (2002)
France, 2004 (5 hospitals)	4, plus 5 colonized (8 pre-term or low-birth-weight)	2	2 Meningitis, 1 haemorrhagic colitis, 1 conjunctivitis	Powdered infant formula	M	Coignard & Vaillant (2006)

[a] M (microbiological): identification of an organism of the same type from cases and in the suspect vehicle or vehicle ingredient(s), or detection of toxin in faeces or food.

2.6.6 *Main methods of prevention and control*

According to the FDA, infant formulas nutritionally designed for consumption by premature or low-birth-weight infants are available only in a commercially sterile liquid form (FDA/CFSAN, 2002). 'Transition' infant formulas that are used for premature or low-birth-weight infants after hospital discharge are available commercially in both non-sterile powder form and sterile liquid form. Some other speciality formulas are only available in powder form.

The FDA recommends that powdered infant formulas should not be used in neonatal intensive care settings unless there is no alternative available (FDA/CFSAN, 2002). If the only option available to address the nutritional needs of a particular infant is a powdered formula, advice is given on procedures in the intensive care unit to reduce the risk. Other organizations also advise that, where possible, commercially available sterilized liquid products should be used instead of powdered formula, especially for high-risk infants (EFSA, 2004; WHO, 2004).

Practices used by the industry to produce safe products in powder form include:

- Selection of supplier of raw materials according to stringent criteria, e.g. for good hygienic practices, and testing of ingredients to verify the effectiveness of these criteria (WHO, 2004).
- The use of good hygienic practice and HACCP in manufacture.

Following reports of the transmission of *E. sakazakii* in powdered infant formula, the industry is adopting further measures to minimize the presence of the bacterium in these foods. Because these products are not sterile, and there is a risk of contamination during preparation, there is a need to use measures that will minimize contamination and growth of the organism in the reconstituted infant food.

To prevent such infections with *E. sakazakii* the following recommendations were made by the WHO (2004) and European Food Safety Authority (EFSA, 2004):

- Reduction of the concentration and prevalence of the organism in powdered infant formula by the use of supplier assurance schemes.
- Monitoring raw materials.
- Reduction of the level of Enterobacteriaceae in the manufacturing environment.
- Monitoring by the industry, of the incidence and concentration of Enterobacteriaceae in finished products.
- Tightening the microbiological specifications for powdered infant formula.

In accordance with these recommendations, under the present EU regulations (EC, 2005) dried infant formula and dried dietary foods for special medical purposes intended for infants below 6 months old, must show absence of Enterobacteriaceae in 10 g; if Enterobacteriaceae are detected in any sample units the batch has to be tested for *E. sakazakii*, which must be absent in 10 g samples (EC, 2005). *Salmonella* is also of great concern in these products and limits are specified (see Section 2.9).

Reconstitution of powdered infant formula in hospitals and homes was discussed by the EFSA (2004) and the following guidelines were given.

Guidelines for reconstitution, handling, storage and use at the hospital:

- Caregivers should be trained to deal with dried formula in centralized units for reconstitution and in the neonatal healthcare units.
- Good hygienic measures are essential to avoid contamination.
- Use sterile containers to reconstitute the formula under an air sterile cabinet avoiding the possibility of recontamination by the environment.
- Always reconstitute formula in hot water (>70°C), avoiding recontamination.
- If continued feeding is necessary the maximum 'hang' time (i.e. the time for which a formula is at room temperature in the feeding bag and accompanying lines during enteral feeding) should not be more than 4 hours.
- Cool the reconstituted formula rapidly to temperatures below the growth range of *E. sakazakii* (below 4–5°C).

Guidelines for reconstitution, handling, storage and use in the home:

- Good hygienic measures are essential to avoid contamination.
- Prepare powdered infant formula fresh for each meal.
- Use 'sanitized' containers to reconstitute the formula (in the home 'sanitized' means a clean container 'sterilized' by immersion in hot water or chemically).
- Always reconstitute formulae in hot water (>70°C) or water that has been boiled and cooled, avoiding recontamination.
- Cool the reconstituted formula rapidly to use temperature.
- Use the reconstituted formula immediately.
- After feeding, discard any remaining formula.

Detailed guidance on the preparation of powdered infant formula in care settings and in the home has been published by WHO/FAO (2007).

References

Bar-Oz, B., Preminger, A., Peleg, O. *et al.* (2001) *Enterobacter sakazakii* infection in the newborn. *Acta-Paediatrica* **90**, 356–358.

Biering, G., Karlsson, S., Clark, N.V.C. *et al.* (1989) Three cases of neonatal meningitis caused by *Enterobacter sakazakii* in powdered milk. *Journal of Clinical Microbiology* **27**, 2054–2056.

Block, C., Peleg, O., Minster, N. *et al.* (2002) Cluster of neonatal infections in Jerusalem due to unusual biochemical variant of *Enterobacter sakazakii. European Journal of Clinical Microbiology and Infectious Diseases* **21**, 613–616.

Centers for Disease Control and Prevention (CDC) (2002) *Enterobacter sakazakii* infections associated with the use of powdered infant formula – Tennessee, 2001. *Morbidity and Mortality Weekly Report* **51**(14), 298–300.

Clark, N.C., Hill, B.C., O'Hara, C.M. *et al.* (1990) Epidemiologic typing of *Enterobacter sakazakii* in two neonatal nosocomial outbreaks. *Diagnostic Microbiology and Infectious Diseases* **13**, 467–472.

Coignard, B. & Vaillant, V. (2006) *Infections à Enterobacter sakazakii associées à la consummation d'une preparation en poudre pour nourrissons. France, octobre à décembre 2004.* Institut de Veille Sanitaire. Available from http://www.invs.sante.fr/publications/2006/infections_e_sakazakii/infections_e_sakazakii.pdf. Accessed 6 March 2007.

Commission Regulation (EC) No 2073/2005 of 15 November 2005 on microbiological criteria for foodstuffs. *Official Journal of the European Union* **L338**, 22 December 2005. Available from http://www.food.gov.uk/multimedia/pdfs/microbiolcriteria.pdf. Accessed 6 March 2007.

Drudy, D., Mullane, N.R., Quinn, T. *et al.* (2006) *Enterobacter sakazakii*: an emerging pathogen in powdered infant formula. *Clinical Infectious Diseases* **42**, 996–1002.

European Food Safety Authority (EFSA) (2004) Microbiological risks in infant formula and follow-on formulae. *The EFSA Journal* **113**, 1–35.

FDA/CFSAN (2002) *Health Professionals Letter on Enterobacter sakazakii Infections Associated with Use of Powdered (Dry) Infant Formulas in Neonatal Intensive Care Units.* Available from http://www.cfsan.fda.gov/~dms/inf-ltr3.html. Accessed 6 March 2007.

Guillaume-Gentil, O., Sonnard, V., Kandhai, M.C. *et al.* (2005) A simple and rapid cultural method for detection of *Enterobacter sakazakii* in environmental samples. *Journal of Food Protection* **68**, 64–69.

Gurtler, J.B., Kornacki, J.L. & Beuchat, L.R. (2005) *Enterobacter sakazakii*: a coliform of increased concern to infant health. *International Journal of Food Microbiology* **104**, 1–34.

Iverson, C. & Forsythe, S. (2003) Risk profile of *Enterobacter sakazakii*, an emergent pathogen associated with infant milk formula. *Trends in Food Science and Technology* **14**, 443–454.

Iverson, C., Lane, M. & Forsythe, S.J. (2004) The growth profile, thermotolerance and biofilm formation of *Enterobacter sakazakii* grown in infant formula. *Letters in Applied Microbiology* **38**, 378–382.

Khandhai, M.C., Reij, M.W., Gorris, L.G.M. *et al.* (2004) Occurrence of *Enterobacter sakazakii* in food production environments and households. *Lancet* **363**, 39–40.

Kim, H. & Beuchat, L.R. (2005). Survival and growth of *Enterobacter sakazakii* in fresh-cut fruits and vegetables and in unpasteurized juices as affected by storage temperature. *Journal of Food Protection* **68**, 2541–2552.

Lai, K.K. (2001) *Enterobacter sakazakii* infections among neonates, infants, children and adults. *Medicine* **80**, 113–122.

Leclercq, A., Wanegue, C. & Baylac, P. (2002) Comparison of faecal coliform agar and violet red bile lactose agar for faecal coliform enumeration in foods. *Applied and Environmental Microbiology* **68**, 1631–1638.

Lehner, A. & Stephan, R. (2004) Review microbiological, epidemiological, and food safety aspects of *Enterobacter sakazakii*. *Journal of Food Protection* **67**, 2850–2857.

Muytjens, H.L., Roelofs-Willemse, H. & Jaspar, G.H.J. (1988) Quality of powdered substitutes for breast milk with regard to members of the family *Enterobacteriaceae*. *Journal of Clinical Microbiology* **26**, 743–746.

Noriega, F.R., Kotloff, K.L., Martin, M.A. & Schwalbe, R.S. (1990) Nosocomial bacteremia caused by *Enterobacter sakazakii* and *Leuconostoc mesenteroides* resulting from extrinsic contamination of infant formula. *Pediatric Infectious Disease Journal* **9**, 447–449.

Seo, K.H. & Brackett, R.E. (2005) Rapid, specific detection of *Enterobacter sakazakii* in infant formula using real time PCR assay. *Journal of Food Protection* **68**, 59–63.

Simmons, B.P., Gelfand, M.S., Haas, M. *et al.* (1989) *Enterobacter sakazakii* infections in neonates associated with intrinsic contamination of powdered infant formula. *Infection Control and Hospital Epidemiology* **10**, 398–401.

Van Acker, J., de Smet, F., Muyldermans, G. *et al.* (2001) Outbreak of necrotizing enterocolitis associated with *Enterobacter sakazakii* in powdered milk formula. *Journal of Clinical Microbiology* **39**, 293–297.

World Health Organization (WHO) (2004) *Enterobacter sakazakii and Other Microorganisms in Powdered Infant Formula: Meeting Report, MRA Series 6.* Available from http://www.who.int/foodsafety/publications/micro/enterobacter_sakazakii/en. Accessed 6 March 2007.

World Health Organization/ Food and Agriculture Organization (WHO/FAO) (2007) Safe preparation, storage and handling of powdered infant formula. Available from http://www.who.int/entity/foodsafety/publications/micro/pif_guidelines.pdf. Accessed 26 June 2007.

2.7 *Escherichia coli*

2.7.1 Introduction

Escherichia coli has been known historically as a commensal bacterium in the human gut. The pathogenicity of some strains was recognized when they were found to cause infections of the urinary tract and to be associated with wound infections. Several classes of *E. coli* cause diarrhoeal disease in humans, and some Shiga-like toxin-producing (Shiga toxin-producing; Vero cytotoxin-producing) strains pose a severe hazard for the general population.

Escherichia coli are Gram-negative, non-sporing rods that are motile by means of multiple, peritrichous flagellae. They are facultatively anaerobic and can be grown easily on laboratory media. Because of the normal presence in the gut of high numbers of non-pathogenic (commensal) strains of *E. coli*, pathogenic strains can be difficult to detect and isolate. The primary method for subdividing strains of *E. coli*, serotyping, is based on heat-stable, somatic O antigens and heat-labile, flagellar H antigens.

There are now six recognized classes of diarrhoeagenic *E. coli* (Table 2.7.1). Many clinical laboratories may not test for these bacteria or may have difficulty in identification, thus the incidence of infections is liable to be underestimated. Infections caused by EPEC, ETEC and EIEC are endemic in many developing countries, particularly affecting young children, and are associated with contaminated water supplies, inadequate sanitary facilities and poor hygiene. EAEC are important in both developing and developed countries. DAEC have been reported by some workers to have a significant association with diarrhoea, but further information about this association is needed. The importance of STEC (VTEC) was recognized first in developed countries, where outbreaks have been associated with high morbidity and mortality. There is little information on the extent of disease caused by STEC in developing countries, but it may be considerable. A brief comment is included on four of these classes, further information is given by Willshaw *et al.* (2000); the remainder of this section will focus on STEC.

Enteropathogenic E. coli (EPEC): Epidemics, mainly in infants, caused by these bacteria were frequent in Europe and North America in the 1950s and continued in Britain until the early 1970s (Willshaw *et al.*, 2000). Since then, outbreaks in infants have become rare in Britain and the US possibly as a result of improvements in hygiene, but several outbreaks have been reported in day-care centres and, occasionally, in paediatric wards (Nataro & Kaper, 1998), and sporadic cases continue to be reported. These bacteria are still a major cause of illness in infants in developing countries.

Enterotoxigenic E. coli (ETEC): Foodborne outbreaks have been reported in Japan, England and the US (Willshaw *et al.*, 2000). Foods implicated include turkey mayonnaise, imported French cheese and salad vegetables. In some outbreaks food handlers with diarrhoea have been implicated. Between 1996 and 2003, 16 outbreaks of infection by ETEC were reported in the US, affecting 2865 people (Beatty *et al.*, 2004). Three of these outbreaks occurred in cruise ships that had docked in foreign ports and one in a cruise ship that docked only in US ports. Twelve of the outbreaks were associated with restaurants or catered parties, one outbreak was attributed to fresh parsley served raw, and salads made with raw vegetables were implicated in four outbreaks. Trace-back of the parsley implicated a farm

Table 2.7.1 Classes of diarrhoeagenic *E. coli*.

E. coli class	Occurrence of disease	Main features of disease	Main method of transmission
Enteropathogenic (EPEC)	Prior to 1970s, outbreaks frequent in infants in Europe and North America. Now rare in developed countries but a major cause of potentially fatal diarrhoea in infants in developing countries	Watery diarrhoea, vomiting, fever. Often self-limiting but can be chronic	In developing countries, probable contamination of weaning foods by water supply
Enterotoxigenic (ETEC)	Endemic in developing countries, particularly affecting children and travellers. Occasional outbreaks in developed countries	Watery diarrhoea with abdominal cramps, fever, malaise, vomiting. Severe infection can cause 'rice water stools' similar to those due to *Vibrio cholerae*	Contaminated water, weaning foods and food
Enteroinvasive (EIEC)	Endemic in developing countries, particularly affecting children and travellers. Sporadic outbreaks in developed world	Diarrhoea, initially acute and watery, fever, abdominal cramps, progressing to colonic phase with bloody, mucoid stools	Contaminated food and water with subsequent person-to-person transmission
Enteroaggregative (EAEC) (EAggEC)	Affect children and adults in both developed and developing countries. Several outbreaks recognized world wide	Persistent watery diarrhoea, vomiting, dehydration, abdominal pain, sometimes bloody diarrhoea	Contaminated food and water
Diffusely adherent (DAEC)	Some reports indicate particularly important in developing countries	Persistent diarrhoea, mucus-containing watery stools, some fever and vomiting	No information
Shiga toxin-producing (STEC) or Vero cytotoxin-producing (VTEC), including enterohaemorrhagic (EHEC)	Cause of outbreaks in developed and developing countries	Mild diarrhoea, bloody diarrhoea, which may progress to haemolytic uraemic syndrome and result in renal failure. Thrombotic microangiopathy resulting in neurological symptoms may occur, usually in adults	Contaminated food and water with subsequent person-to-person transmission

Information from Nataro & Kaper, 1998; Willshaw *et al.*, 2000; Kaper *et al.*, 2004.

in Baja California, Mexico, where municipal water used in hydrocoolers was inadequately chlorinated and liable to contamination (Naimi *et al.*, 2003).

Enteroinvasive E. coli (EIEC): Outbreaks, many of which were probably foodborne, have been reported in numerous countries (Willshaw *et al.*, 2000). In an outbreak in the US in 1973 that affected at least 387 people, imported cheese was implicated (Marier *et al.*, 1973). *E. coli* O124 was the only known enteric pathogen isolated from patients and was present at levels up to 10^5–10^7 bacteria/g in a high proportion of cheese samples. Equipment used for filtering river water, used for cleaning in the factory in which the cheese was made, had been malfunctioning when the contaminated cheese was produced.

Enteroaggregative E. coli (EAEC or EAggEC): These bacteria are also an important cause of diarrhoea in developing countries (Nataro *et al.*, 1998). Information is limited, because few research laboratories are able to detect and isolate EAEC (Huang *et al.*, 2004). Four outbreaks (133 cases) probably associated with EAEC occurred in England and Wales in 1994 (Smith *et al.*, 1997) and were linked to meals in restaurants, a conference centre and a hotel. In a study of infectious intestinal disease in England and Wales, EAEC were the most commonly detected enterovirulent *E. coli* in patients who consulted their general practitioner (Tompkins *et al.*, 1999). The largest foodborne outbreak of diarrhoea caused by EAEC reported to date occurred in Japan in 1993, and involved 16 schools and a total of 2697 school children, who became ill after consuming school lunches that were all prepared in a central kitchen (Huang *et al.*, 2004).

2.7.2 *Importance of Shiga-like toxin-producing* E. coli *(STEC) (Vero cytotoxin-producing* E. coli *VTEC) as a cause of foodborne disease*

The incidence of reported disease caused by STEC is much lower than that caused by *Campylobacter* and *Salmonella*, but the severe complications and mortality resulting from infections by STEC make them of major concern (O'Brien *et al.*, 2001a). Infections are liable to be greatly underreported, because the ability to detect the organisms is limited, and surveillance systems for the bacteria are limited in many countries. Large outbreaks of disease have been reported in North America, Europe, Australia, Japan and Swaziland/South Africa; STEC infections are a major cause of childhood illness in Argentina, Brazil and Chile.

In North America, the UK and Japan, strains of the serotype O157:H7 are the most frequently reported pathogenic STEC strains (Kaper *et al.*, 2004) but in several countries other serogroups also appear to be prevalent, particularly O26, O103, O111 and O145 (WHO, 1998; Beutin *et al.*, 2004; Blanco *et al.*, 2004; Vaz *et al.*, 2004).

In the US, the reported incidence of infection with *E. coli* O157 in 1996–98 was 2.3 cases per 100,000 and the preliminary figure for ten sites monitored in 2004 was 0.9 cases per 100,000 (CDC, 2005), the number of cases caused by non-O157 strains was not reported.

In England and Wales, the reported cases of infection with *E. coli* O157 peaked at 1087 cases in 1997, with a rate of infection in England of 2.1 cases per 100,000 persons (Bolton *et al.*, 2000); in 2005, there were about 950 reported cases (HPA, 2006a). The number of cases caused by non-O157 strains was not stated, although such cases have been described (Evans *et al.*, 2002).

2.7.3 Characteristics of STEC

Typical STEC strains of *E. coli* O157:H7 differ from other *E. coli* strains in being unable to ferment sorbitol readily and failing to produce β-glucuronidase. These differences have led to the development of a culture medium on which typical O157:H7 strains are relatively easy to detect from clinical samples and food products. Some strains of Shiga toxin-producing *E. coli* O157 ferment sorbitol and may have been underrecognized. The first reported outbreak due to such strains occurred in Germany in 1988 and cases of infection have been reported in continental Europe, Australia, Scotland and England. Revised guidance on testing clinical samples for these strains has been issued in England and Wales (HPA, 2006b). Other STEC serogroups also ferment sorbitol readily and are more difficult to detect and identify than typical O157:H7 strains, a fact that has contributed to underreporting of these groups.

To characterize strains of O157 further, bacteriophage typing has been used, and molecular methods, particularly pulsed-field gel electrophoresis (PFGE) of genomic DNA.

The temperature range for growth of STEC is between 7–8°C and ~44°C with an optimum at 35–40°C (ICMSF, 1996). The bacteria are less heat-resistant than *Salmonella* spp.; heating meat to an internal temperature of 72°C for 2 minutes, and pasteurization of milk at 72°C for 15 seconds inactivate high numbers of STEC. The minimum pH for growth is about pH 4.5 in medium acidified with HCl but higher in media acidified with organic acids. The bacteria are unlikely to multiply in fruit juices, because of their acidity, but a proportion of cells may survive for weeks.

Strains of serogroups O157, O26, O103, O111 and O145 are often associated with the more severe forms of disease and are referred to frequently as enterohaemorrhagic *E. coli* (EHEC).

The pathogenicity of STEC is associated with the production of phage-encoded Shiga-like toxins (Vero cytotoxins) Stx1 and Stx2 (VT1 and VT2), protein intimin (encoded by the *eae* gene located on a 'locus of enterocyte effacement' [LEE] region of a pathogenicity island on the chromosome) which contributes to binding of the bacteria to cells of the microvilli in the intestine, fimbriae (also associated with binding to host cells), and enterohaemolysin (E-Hly) encoded by a large plasmid. A strong association has been found between the *eae* gene and the capacity of STEC isolates to cause severe disease, particularly haemolytic uraemic syndrome (HUS), in humans, but there are several reports of *eae*-negative strains associated with HUS (Boerlin *et al.*, 1999; Blanco *et al.*, 2004), suggesting that an alternative mechanism for adhesion to host cells may be important.

The Shiga-like toxins or Shiga toxins, Stx1 and Stx2, have many properties in common with Shiga toxin produced by *Shigella dysenteriae* 1. Stx1 toxins are neutralized by antibodies to Shiga toxin but Stx2 toxins are not. Stx1, Stx2 and Shiga toxin bind to the glycolipid globotriaosyl ceramide, high levels of which are present in glomerular endothelial cells of the kidney. Variants of Stx1 and particularly of Stx2 have been described (Willshaw *et al.*, 2000).

2.7.4 Disease caused by STEC in humans

Some people are infected asymptomatically. Ingestion of a low number of bacteria (~100) may cause illness, this results in a high risk of person-to-person transmission. Symptoms develop in 1–8 days, usually in 3–4 days, and include mild diarrhoea, haemorrhagic colitis (HC), abdominal pain, watery diarrhoea and bloody diarrhoea, usually without pain. If HUS does not develop, illness usually lasts about 12 days. Symptoms may progress to HUS, haemolytic anaemia and thrombocytopenia (Nataro & Kaper, 1998; Bolton *et al.*, 2000; Tarr *et al.*, 2005). Long-term complications include chronic kidney failure and neurological complications including seizure, stroke and thrombotic microangiopathy. The highest incidence of HUS occurs in infants, young children and the elderly, with deaths occurring more often in the elderly (O'Brien *et al.*, 2001a). About 8–10% of cases of illness caused by STEC lead to HUS (Willshaw *et al.*, 2000;). Treatment of HUS can require kidney dialysis and transfusions, the reported case fatality rate is from about 3–17% (Salmon *et al.*, 2000) or greater. STEC are the major cause of kidney failure in children in the UK, and in the US.

Haemolytic uraemic syndrome was described first as a distinct clinical entity by Gasser *et al.* (1955) (cited in Karmali, 1989). Tests on stored sera from patients with HUS in the Netherlands have shown that *E. coli* O157 was associated with HUS as far back as 1974 (Chart *et al.*, 1991), no stored sera from patients before this date were available. The first recorded outbreak of STEC infection in the world occurred in February and March 1982 in Oregon, US. A further outbreak occurred in Michigan in May–June 1982, when 18 people who had eaten hamburgers at a restaurant were infected. Major outbreaks followed; one in the US in 1992–93 was associated with hamburgers sold in restaurants of a fast food chain, and resulted in widespread publicity. The highest reported incidence of HUS in young children occurs in Argentina (Lopez *et al.*, 1997, 2000; Rivas *et al.*, 2006), with a reported incidence in 2002 of 12.2 cases per 100,000 children <5 years of age. STEC are reported as the major cause, with both O157 and non-O157 serogroups involved.

Bloody diarrhoea is a common feature of STEC infections but may be caused by other bacteria including *Shigella*, *Salmonella* and *Campylobacter* (Chalmers & Salmon, 2000; Talan *et al.*, 2001).

2.7.5 *Source and transmission of STEC*

Infection of humans by STEC can result from foodborne transmission, from direct contact with farm animals and their excreta, and from person-to-person transmission (Task Force, 2001). In the US, of outbreaks reported to CDC between 1982 and 2002, 52% were foodborne, 21% unknown transmission, 14% were associated with person-to-person transmission, 6% with recreational water, 3% with animal contact, 3% with drinking water and 0.3% with laboratory transmission (Rangel *et al.*, 2005). The main animal reservoirs of STEC are probably ruminants, particularly cattle, sheep, goats and also pigs; deer and rabbits can also carry the organism. Some strains may cause disease in cattle, but healthy animals frequently carry the bacteria, including O157 strains (Willshaw *et al.*, 2000).

Excretion of the bacteria in animal faeces can result in contamination of grassland, watercourses and crops that are in the vicinity of ruminants. Spreading of liquid livestock wastes on land used subsequently for production of crops or livestock grazing can disseminate STEC and other human pathogenic microorganisms.

Carcasses of animals leaving abattoirs are liable to be contaminated, and considerable efforts are being made in several countries to reduce this contamination. Carcass

Table 2.7.2 Examples of outbreaks of foodborne disease caused by STEC O157 associated with meat products.

Place, date	Cases	Number hospitalized	Number with HUS/TPP	Deaths	Food implicated	Evidence[a]	Where food was provided	Factors leading to outbreak	Reference
Canada, 1985	73 (54 primary, 19 secondary)[b]	15	12 HUS	17	Probably lunch of ham, turkey and cheese sandwiches	S	Nursing home (55 residents, 18 staff affected)	Probable contamination by food handler	Carter *et al.* (1987)
UK, 1987	26	6	1	0	Turkey roll sandwiches	S	Christening party at public house	Possible cross-contamination from raw meat; sandwiches left in warm room for some hours	Salmon *et al.* (1989)
Scotland, 1990	11 (1 primary, 10 secondary)	8	2 HUS	4	Possibly meat products brought into hospital by relative/friend	D	Psychogeriatric hospital wards (8 patients, 3 staff affected)	Not known	Kohli *et al.* (1993)
Swaziland/S. Africa, 1992[c]	Estimated 'thousands'	nr	nr	nr	Beef and water supplies	S	Community	Drought, carriage of *E. coli* O157 by cattle, contamination of surface water	Effler *et al.* (2001)
US, 1992–93	501	151	45 HUS	3	Hamburgers	M	Restaurant of fast food chain	Contamination of meat at slaughter and under-cooking	Bell *et al.* (1994); Tuttle *et al.* (1999); Willshaw *et al.* (2000)
US, 1994	23	6	2 HUS	0	Dry cured salami	M	Community	Not known	CDC (1995a)
UK, 1995	26 (24 primary, 2 secondary)	6	2 HUS	0	Cold cooked meat in sandwiches	D	Sandwiches supplied to hospital and to community	Wholesaler supplying outlets kept no record of sources of cooked meat, traceback to producer impossible	McDonnell *et al.* (1997)

US, 1995	11	0	0	0	Home-made venison jerky	M	Community	Deer carried the bacterium, meat contaminated, bacterium survived production of Jerky	Keene *et al.* (1997b)
Germany (Bavaria), 1995/96	Estimated 300–600[d]	nr	28 HUS	3	Teewurst (raw, fermented sausage, usually containing beef), Mortadella (a heat-treated sausage)	S	Community	Not known	Ammon *et al.* (1999)
Central African Republic, 1996	>108	nr	'several' HUS	4	Suspected smoked zebu meat	D	Community	Not known	Germani *et al.* (1997)
Scotland, 1996/97	512	151	27 HUS	17	Cooked meat products, gravy	D, S, M	Church hall lunch, pub birthday party, Nursing home	Cross-contamination from raw to cooked meats in butcher's shop	Pennington (1998); Willshaw *et al.* (2000); Cowden *et al.* (2001)
US, 1997	15	5	0	0	Frozen, ground beef patties	M	Community	Contamination and undercooking	CDC (1997b)
Canada, 1998	39	14	2 HUS	nr	Dry fermented Genoa salami	S and M	Community	Faults in manufacturing process	Williams *et al.* (2000)
Canada 1999	143	42	6 HUS	0	Dry-fermented salami	S and M	Community	Fermentation and drying shown previously not to inactivate *E. coli* O157 completely in salami. USDA requires additional measures to ensure safety	MacDonald *et al.* (2004)
US, 1999	13	5	3 HUS	0	Beef taco, containing ground beef	S	Fast-food restaurant chain	nr	Jay *et al.* (2004)
Spain, 2000	181	nr	6 HUS	0	Sausages	D	Meals supplied to five schools by a single caterer, and household contacts	Poor hygiene standards during production of sausages, inadequate heat treatment	Bosch (2000); Martinez *et al.* (2001)

(*cont.*)

Table 2.7.2 *(continued)*

Place, date	Cases	Number of hospitalized	Number with HUS/TPP	Deaths	Food implicated	Evidence[a]	Where food was provided	Factors leading to outbreak	Reference
US, 2000	83	6	1 HUS	0	Ground beef	M	Community	nr	Proctor *et al.* (2002)
UK, 2001	30 confirmed clinical cases	22	2 HUS	0	Cooked meats	M	Community	Cross-contamination from raw to cooked meat in butcher's shop	Rajpura *et al.* (2003)
Japan, 2002	28	nr	nr	nr	Grilled beef	M	Restaurants supplied by central restaurant	Meat contaminated and undercooked	Tsuji *et al.* (2002)
US, 2002	26	7	5 HUS	nr	Ground beef	S and M	Community	Meat contaminated and not cooked thoroughly	CDC (2002)
England, 2005	>157 (65% school children)	nr	nr	1	Sliced cooked meats	M	Meals at >40 schools supplied by a single main supplier	Practices identified that could result in contamination of cooked meat at the supplier's premises	Salmon & Outbreak Control Team (2005)
France, 2005	26[e] (24 cases children 15 mo – 9 yr old)	20	13 HUS	0	Mince beefburgers	M	Community	Meat contaminated and not cooked thoroughly	Vaillant (2005)
Netherlands, 2005	'21 confirmed corresponding to several thousand in the community'	7	0	nr	Steak tartare (raw beef product)	S	Community	Meat contaminated and eaten raw	Doorduyn *et al.* (2006)

[a]M (microbiological): identification of an organism of the same type from cases and in the suspect vehicle or vehicle ingredient(s), or detection of toxin in faeces or food; D (descriptive): other evidence, usually descriptive, reported by local investigators as indicating the suspect vehicle or food; S (statistical): a significant statistical association between consumption of the suspect vehicle and being a case.
[b]Secondary cases spread person-to-person.
[c]*Vibrio cholera* was also present.
[d]Only cases of HUS were reported, only 5–10% of persons with symptoms of VTEC infection develop HUS.
[e]Surveillance based on HUS.
nr, not reported.

contamination results in transfer of the bacteria to retail samples of raw meat, particularly of minced (ground) meat. Raw milk is also liable to be contaminated.

The high incidence of HUS in children in Argentina is associated with the report that, at least until recently, about 20% of Argentine children started to eat meat at 5 months old, and 80% had meat in their diet at least three times a week; 80% of the meat consumed was reported to be undercooked (Lopez *et al.*, 1997). Some of the first outbreaks to be recognized were attributed to the consumption of beef that had not been cooked sufficiently to kill the organism. Further outbreaks have been associated with unpasteurized or inadequately pasteurized milk and milk products and re-contaminated pasteurized milk, cooked meats, salad vegetables, unpasteurized fruit juices, vegetable sprouts and drinking water.

2.7.6 *Examples of outbreaks of foodborne infection caused by STEC*

Examples of reported outbreaks caused by STEC O157 are shown in Tables 2.7.2–2.7.4. In addition to many outbreaks in developed countries, it is clear that infection is also of major importance in less-developed countries. In 1992, STEC were associated with more than 20,000 cases of bloody diarrhoea in Malawi (Paquet *et al.*, 1993) and a large outbreak of bloody diarrhoea in Swaziland and South Africa was attributed mainly to STEC O157:NM (Effler *et al.*, 2001). In 1996, STEC caused a severe outbreak of haemorrhagic colitis in the Central African Republic; meat from zebu cattle was suspected strongly as the source of the infection (Germani *et al.*, 1997).

Several outbreaks associated with drinking water have been reported, the largest of which was in Canada in 2000, and affected more than 2300 people (Anon, 2000).

Reported foodborne outbreaks due to non-O157 STEC are shown in Table 2.7.5.

2.7.7 *Main methods of control and prevention of STEC infections*

A range of serotypes of STEC are carried by farm animals and wild animals. Measures have been suggested to reduce the incidence in farm animals, but the widespread occurrence of these bacteria makes it unlikely that reliance can be placed on elimination of the organisms from food animals.

In England and Wales, the need for the following control measures has been emphasized (ACMSF, 1995):

- to minimize contamination of carcasses at slaughter;
- to adopt a Hazard Analysis Critical Control Point (HACCP) approach in the food processing and service sectors, to prevent survival or contamination by STEC;
- to prevent cross-contamination from raw to cooked meat;
- to re-consider the sale of raw cows' milk;
- to cook beefburgers (minced or ground meat in general) thoroughly. It was advised that beefburgers should be cooked until the juices ran clear, and there are no pink bits inside, and that minced beef and beef products including beefburgers should be cooked at an internal temperature of 70°C or higher for 2 minutes, or equivalent, to destroy non-sporing microorganisms (ACMSF, 2007).

Table 2.7.3 Examples of outbreaks of foodborne disease caused by STEC O157 associated with milk products.

Place, date	Cases	Number hospitalized	Number with HUS/TPP	Deaths	Food implicated	Evidence[a]	Where food was provided	Factors leading to outbreak	Reference
England, 1991	16	13	5	0	'Live' yoghurt produced by local dairy and sold for infants and young children	S	Community	STEC of other phage types isolated from cattle on the farm. Opportunities for contamination of milk after pasteurization, no records of time and temperature of pasteurization	Morgan *et al.* (1993)
US, 1992–94	14	2	0	0	Raw milk	M	Community	Carriage of *E. coli* O157 by cows; use of unpasteurized milk	Keene *et al.* (1997a)
Scotland, 1994	22	nr	nr	1	Home-made cheese	nr	School	Cheese made from unpasteurized milk	Task Force (2001)
Scotland, 1994	>100	~ 30	9 HUS, 1 TTP	1	Contaminated, pasteurized milk	M	Community	Contaminated pipe from pasteurizer to bottling machine	Upton and Coia (1994)
Scotland, 1997	20	12 were already hospital patients	0	0	Home-baked, cream-filled cakes brought to party on wards	S	Party on geriatric continuing care wards	Possible contamination of fresh cream (made from pasteurized milk) used in cream cakes	O'Brien *et al.* (2001b)
US, 1998	55	25	nr	nr	Fresh cheese curds	S and M	Dairy plant and retail sale	Unpasteurized milk used to make cheese curds	CDC (2000)

England, 1999	114 (111 primary, 3 secondary)[b]	28	3 HUS	0	Milk pasteurized on farm	S and M	Community	Faults in pasteurization unit	Goh *et al.* (2002)
Scotland, 1999	25	nr	1 HUS	nr	Cheese made with unpasteurized goat's milk	M	Primary school	Home-made cheese, goat that supplied milk positive for outbreak organism	Curnow (1999)
Canada, 2001	5	2	2 HUS	nr	Unpasteurized goats' milk	M	Co-operative farm	Goats' milk contaminated with outbreak strain, not pasteurized	McIntyre *et al.* (2002)
Slovakia, 2001	5	5	3 HUS	0	Cream made from unpasteurized cow's milk	M	Family outbreak	Use of unpasteurized milk	Liptakova *et al.* (2004)
Denmark, 2003–04	25	2	0	0	Pasteurized, organic Jersey milk	S	Community and two daycare institutions	Post-pasteurization contamination or intermittent inadequate heat treatment	Jensen *et al.* (2006)
France, 2004	Total cases not reported[c]	2	2 HUS	0	Fresh goats' cheese made with unpasteurized milk	M	Sold from farm	Outbreak strain isolated from goat and cows (but not from cheese)	Espie *et al.* (2006b)

[a]M (microbiological): identification of an organism of the same type from cases and in the suspect vehicle or vehicle ingredient(s), or detection of toxin in faeces or food; D (descriptive): other evidence, usually descriptive, reported by local investigators as indicating the suspect vehicle or food; S (statistical): a significant statistical association between consumption of the suspect vehicle and being a case.
[b]Secondary cases spread person-to-person.
[c]Only cases of HUS were reported, only 5–10% of persons with symptoms of STEC infection develop HUS.
nr, not reported.

Table 2.7.4 Examples of outbreaks of foodborne disease caused by STEC O157 associated with vegetables and fruits.

Outbreak	Cases	Number hospitalized	Number with HUS/TPP	Deaths	Food implicated	Evidence[a]	Where food was provided	Factors leading to outbreak	Reference
US, 1995	>40	13	1 HUS	0	Lettuce	S	Community	Probable contamination of lettuce in the field	Ackers *et al.* (1998)
Canada, 1995	21	0	0	0	Imported iceberg lettuce	S	Acute-care hospital	Lettuce was heavily soiled when received in the hospital kitchen	Preston *et al.* (1997)
Japan, 1996	>8000	>390	>100 HUS	3	White radish sprouts	S	Schools, nursing home, business office, community	nr	Michino *et al.* (1999)
US and Canada, 1996	70	25	14 HUS (13 cases < 3 yr old)	1	Unpasteurized apple juice	M	Community	Orchard frequented by deer that carried *E. coli* O157:H7. Some apples may have been picked from ground	Cody *et al.* (1999)
US, 1996	>61	21	3 HUS	0	Mesclun lettuce (mixture of small red and green leaf lettuces)	S	Community, restaurant meals	Opportunities for contamination in the field and during harvesting	Hilborn *et al.* (1999)
US, 1996	14	10	5 HUS	0	Unpasteurized apple cider	S and M	Community	Some 'Drop apples' were used	CDC (1996); Hilborn *et al.* (2000)
US, Michigan and Virginia, 1997	82	36	4 HUS	0	Alfalfa sprouts	S	Community	Contamination of alfalfa in field, by cattle manure, water or deer faeces	CDC (1997a); Breuer *et al.* (2001)
Sweden, 1999	>11	nr	nr	nr	Lettuce suspected	D	Party for staff at a children's hospital	Lettuce grown with continuous fertilizing until harvest, and also irrigated	Welinder-Olsson *et al.* (2003)

US, 1999	40	7	2 HUS	0	Romaine lettuce	S	Retirement community	Not known	Rangel *et al.* (2000)
Canada, 2002	109	Cases were hospital patients and staff	0	2	Probably salads and sandwiches	S	Psychiatric hospital and other healthcare institutions	Preparation in hospital kitchen by symptomatic food handler	Bolduc *et al.* (2004)
US, 2003	20	3	0	0	Alfalfa sprouts	S	Community	Alfalfa seeds contaminated and decontamination failed	Ferguson *et al.* (2005)
US, 2003	~ 32 residents, 14 staff	9	HUS/TPP 3/32	2/32	Raw spinach	S	Nursing home	nr	Reiss *et al.* (2006)
Sweden, 2005	135	nr	11 HUS	0	Iceberg lettuce	S	Community	Outbreak strain isolated from stream used to irrigate lettuce and from cattle upstream of irrigation	Söderström *et al.* (2005, 2006)
US, 2006	199	102	31 HUS	3	Pre-packaged raw spinach	M	Community	Cattle and other animals near to fields growing spinach, possible contamination of surface or irrigation water	FDA (2007b) CDC (2006a)
US, 2006	81	26	2 HUS	0	Shredded lettuce	S	Restaurants	Outbreak strain isolated from environmental samples from dairy farms near lettuce-growing area	FDA, (2007a)

[a]M (microbiological): identification of an organism of the same type from cases and in the suspect vehicle or vehicle ingredient(s), or detection of toxin in faeces or food; D (descriptive): other evidence, usually descriptive, reported by local investigators as indicating the suspect vehicle or food; S (statistical): a significant statistical association between consumption of the suspect vehicle and being a case.
nr, not reported.

Table 2.7.5 Examples of outbreaks of foodborne disease caused by STEC serogroups other than O157.

Place, date; serotype	Cases	Number hospitalized	Number with HUS/TPP	Deaths	Food implicated	Evidence[a]	Where food was provided	Factors leading to outbreak	Reference
Italy, 1992; O111	nr[b]	9	9 HUS	1	Ground beef suspected	D	Community	Probable contamination and undercooking	Caprioli *et al.* (1994)
France, 1992–93; nr	nr[b]	4	4 HUS	1	Fresh cheese made with unpasteurized cows' and goats' milk	M	Sold from farm	Cows and goats infected with *E. coli* that formed verotoxins	Deschênes *et al.* (1996)
US, 1994; O104:H21	18	4	0	0	Pasteurized milk	S	Community	Post-pasteurization equipment in dairy contaminated with faecal coliform bacteria	CDC (1995b)
Australia, 1994–95; O111:H-	~ 200	nr	21 HUS	1	Uncooked, semi-dry, fermented sausage ('mettwurst')	M	Community	Starter cultures not used, inadequate monitoring and control of pH and water activity	Cameron *et al.* (1995); CDC (1995c); Paton *et al.* (1996); Willshaw *et al.* (2000)
Japan, 1997; O26:H11	32	1	0	0	Mixed vegetable with bean sprouts, spinach, sliced watermelon	M	Nursery	nr	Hiruta *et al.* (2000)
US, 1999; O111:H8	55	2	2 HUS	0	Salad, ice, corn cob, dinner roll	S	Youth camp	Possibly due initially to consumption of food from salad bar and subsequent spread through consumption of ice from open barrels	Brooks *et al.* (2004)
Germany, 2000; O26:H11	11	1	0	0	Beef Seemerrole suspected	S	Mother–child clinic, day-care centres, hospital	nr	Werber *et al.* (2002)
France, 2002; O148	10	2	2 HUS	0	Spitroasted mutton	M	Wedding	Mutton probably undercooked	Espie *et al.* (2006a)
Norway, 2006; O103	16	nr	10 HUS	0	Cured meat sausage consumed uncooked.	S and M	Community	nr	Schimmer *et al.* (2006)
Denmark, 2007; O26:H11	>20	nr	nr	nr	Organic, fermented, cured beef sausage	S and M	Community	nr	Ethelberg *et al.* (2007)

[a] M (microbiological): identification of an organism of the same type from cases and in the suspect vehicle or vehicle ingredient(s), or detection of toxin in faeces or food; D (descriptive): other evidence, usually descriptive, reported by local investigators as indicating the suspect vehicle or food; S (statistical): a significant statistical association between consumption of the suspect vehicle and being a case.

[b] Only cases of HUS were reported, only 5–10% of persons with symptoms of STEC infection develop HUS.

nr, not reported.

Following the outbreak in Scotland in 1996–97, further recommendations were made that included: animals should be clean when they are presented for slaughter, good practice should be used in slaughterhouses and carcasses should be treated, e.g. by steam pasteurization, to reduce the level of microorganisms present (Pennington, 1998). Other recommendations were the use of HACCP in food businesses, licensing of butchers' shops, physical separation of raw meat and unwrapped, cooked meat/meat products and other ready-to-eat foods, education of food handlers particularly those working with vulnerable groups and/or in sensitive areas such as nursing homes and day-care centres, and enforcement of these measures. The outbreak in the UK in 2001 in which a licensed butcher's counter was involved (Rajpura *et al.*, 2003) indicates that in addition to licensing of butchers' shops, clear enforcement of separation of raw and cooked meats is essential (Cree *et al.*, 2004).

Table 2.7.3 includes several outbreaks associated with unpasteurized milk or milk products, with the failure of milk pasteurization processes or with re-contamination after pasteurization. In an outbreak of infection with *E. coli* O157 PT21/28 in England affecting four people, following the consumption of unpasteurized milk, 64 of 127 cattle on the farm that supplied the milk carried *E. coli* O157 PT21/28 Stx2 with the same antibiotic-resistance type as the outbreak strain (CDSC, 2000). Properly controlled pasteurization will destroy high numbers of this bacterium and, together with prevention of post-pasteurization contamination, is the major means of preventing transmission of STEC and of other foodborne pathogens in milk. It has been advised that all raw milk and cream, including raw sheep and goats' milk for sale in England and Wales should be heat-treated, as is the legal requirement in Scotland (Task Force, 2001).

Guidance to minimize the risk of transmission of foodborne pathogens on fresh fruits and vegetables was published by the US FDA in 1998 (FDA, 1998), stimulated partly by outbreaks such as that of *E. coli* O157 in mesclun mix lettuce. The guidance emphasized that prevention of contamination should be relied upon, rather than corrective action after contamination had occurred. Contamination in the field can result from contaminated surface water or irrigation water, the presence of cattle and other animals including deer, and inadequately composted animal manure. A further action plan relating to the safety of produce has been proposed, which includes measures to prevent contamination of fresh produce at all stages of production (FDA/CFSAN, 2004a). In 2006, the FDA has been working to finalize a 'Guide to Minimize Microbial Food Safety Hazards of Fresh-cut Fruits and Vegetables'. Following the outbreak associated with spinach in 2006, further initiatives are planned to minimize the risk of another outbreak associated with leafy vegetables.

From a survey in 2003, it was concluded that in the UK salads are irrigated with water from vulnerable sources, giving contact with the edible part of the plant and limited time for die-off of pathogen in the field (Tyrrel *et al.*, 2006).

The European Chilled Food Federation (ECFF) has emphasized the risk of transmission of STEC on fresh salad vegetables and fruit, in particular the risk of transmission via agricultural practices and the use of animal or human waste as fertilizer (Goodburn *et al.*, 1998). The UK Chilled Food Association (CFA) has also published guidance for suppliers of ready-to-eat produce (fruits and vegetables) to minimize transmission of pathogenic microorganisms in the field and during and after harvest (CFA, 2002). This document sets out suitable treatments for farmyard manure and slurry, application conditions, application of sewage sludge, exclusion of animals and use of water for irrigation.

Animal manure is by far the largest proportion of organic waste used in agriculture, and needs to be composted effectively to destroy vegetative pathogens including STEC (Duffy, 2003). At present, information is lacking on the composting conditions that are needed to destroy pathogenic microorganisms, and on the practical implementation of these conditions. Draft UK guidance states that fresh manure should not be applied to land intended for growth of ready-to-eat crops unless there will be at least a 6-month period between manure application and harvest (FSA, 2002). Conditions for storage and treatment are stated and it is stipulated that batch-stored or treated manures should not be applied to land intended for growth of ready-to-eat crops, unless there will be at least a 2-month period between manure application and harvest. It has been concluded that batches of liquid livestock waste, if contaminated with bacterial pathogens, should be stored for at least 6 months to reduce levels of bacteria, but further measures would be needed to eliminate *Cryptosporidium parvum* (Hutchison *et al.*, 2005).

In addition to guidance there is a need for enforcement of control measures. In the outbreak in the US in 1996 associated with unpasteurized apple juice, the company had a written policy of accepting only hand-picked fruit, but realistically this could not be enforced, and it was not possible to know whether some fruit had been harvested from the ground (Cody *et al.*, 1999). More than 98% of juices sold in the US are reported to be pasteurized, the 2% that are not pasteurized account for all the cases of juice-based illness. The FDA has issued a Juice HACCP Regulation specifying that a 5-log reduction of 'the most pertinent microorganism', *E. coli* O157:H7, must be applied in production of juices (FDA/CFSAN, 2004b), a reduction that would result from heat pasteurization.

Following outbreaks of infection with STEC and with salmonellae associated with vegetable sprouts, the FDA issued guidance to industry and consumers to reduce the risk of disease (FDA, 1999, 2002). This included the recommendation that, just prior to sprouting, seeds should be subjected to one or more treatments that can reduce effectively, or eliminate, pathogenic bacteria, e.g. treatment with 20,000 ppm calcium hypochlorite, and that seeds should undergo a combination of treatments to achieve a 5-log reduction in levels of *Salmonella* and EHEC. FDA also recommended that spent irrigation water from sprout production should be tested for *Salmonella* and *E. coli* O157, and specified corrective actions if either of these bacteria were found (Smith, 2000). Because testing cannot ensure the absence of these bacteria, and it is difficult to rely on effective hypochlorite treatment of vegetable sprouts, the FDA advises seniors (people aged over 65 yr) and persons in high-risk groups not to eat raw alfalfa sprouts (FDA/CFSAN, 1999).

Summary of measures advised in healthcare settings

1. Foods and food components should be obtained from reputable suppliers.
2. Meat should be cooked thoroughly.
3. Cross-contamination from raw to cooked meats should be prevented. Ready-to-eat cooked meats should not be served to immunosuppressed persons.
4. Unpasteurized or inadequately pasteurized milk or milk products should not be served.
5. Raw seed sprouts should not be served to immunosuppressed persons.
6. Raw salad vegetables should be obtained from reputable suppliers and should be well washed, they should not be served to immunosuppressed persons.

7 Unpasteurized fruit juices should not be served.
8 Catering personnel should not handle food while suffering from symptoms of infection (Department of Health, 1995, 1996; FDA/CFSAN, 2005), or handle cooked food or ready-to-eat food with bare hands. Good handwashing is essential.

References

Ackers, M.-L., Mahon, B.E., Leahy, E. *et al.* (1998) An outbreak of *Escherichia coli* O157:H7 associated with leaf lettuce consumption. *Journal of Infectious Diseases* **177**, 1588–1593.

Advisory Committee on the Microbiological Safety of Food (ACMSF) (1995) *Report on Verocytotoxin-Producing Escherichia coli*. HMSO, London, pp. 148.

Advisory Committee on the Microbiological Safety of Food (ACMSF) (2007) *Ad hoc* Group on the Safe Cooking of Burgers. *Report on the safe Cooking of Burgers*. Food Standards Agency, London, 40 pp.

Ammon, A., Petersen, L.R. & Karch, H. (1999) A large outbreak of haemolytic uremic syndrome caused by an unusual sorbitol-fermenting strain of *Escherichia coli* O157:H-. *Journal of Infectious Diseases* **179**, 1274–1277.

Anon. (2000) Waterborne outbreak of gastroenteritis associated with a contaminated municipal water supply, Walkerton, Ontario, May–June 2000. *Canada Communicable Disease Report* **26**(20), 170–173.

Beatty, M.E., Bopp, J.G., Wells, J.G. *et al.* (2004) Enterotoxin-producing *Escherichia coli* O169:H41, United States. *Emerging Infectious Diseases* **10**, 518–521.

Bell, B.P., Goldoft, M., Griffin, P.M. *et al.* (1994). A multistate outbreak of *Escherichia coli* O157:H7-associated bloody diarrhea and hemolytic syndrome from hamburgers: the Washington experience. *Journal of the American Medical Association* **272**, 1349–1353.

Beutin, L., Krause, G., Zimmermann, S. *et al.* (2004) Characterization of Shiga toxin-producing *Escherichia coli* strains isolated from human patients in Germany over a 3-year period. *Journal of Clinical Microbiology* **42**, 1099–1108.

Blanco, J.E., Blanco, M., Alonso, M.P. *et al.* (2004) Serotypes, virulence genes, and intimin types of Shiga toxin (Verotoxin)-producing *Escherichia coli* isolates from human patients: prevalence in Lugo, Spain, from 1992 through 1999. *Journal of Clinical Microbiology* **42**, 311–319.

Boerlin, P., McEwen, S.A., Boerlin-Petzold, F. *et al.* (1999) Association between virulence factors of Shiga toxin-producing *Escherichia coli* and disease in humans. *Journal of Clinical Microbiology* **37**, 497–503.

Bolduc, D., Srour, L.F., Sweet, L. *et al.* (2004) Severe outbreak of *Escherichia coli* O157:H7 in health care institutions in Charlottetown, Prince Edward Island, Fall, 2002. *Canada Communicable Disease Report* **30**, 81–88.

Bosch, X. (2000) Spain's *E. coli* outbreak highlights mistakes. *Lancet* **356**, 1665.

Breuer, T., Benkel, D.H., Shapiro, R.L. *et al.* (2001) A multitstate outbreak of *Escherichia coli* O157:H7 infections linked to alfalfa sprouts grown from contaminated seeds. *Emerging Infectious Diseases* **7**, 977–982.

Brooks, J.T., Bergmire-Sweat, D., Kennedy, M. *et al.* (2004) Outbreak of Shiga toxin-producing *Escherichia coli* O111:H8 infections among attendees of a high school cheerleading camp. *Clinical Infectious Diseases* **38**, 190–198.

Cameron, S., Walker, C., Beers, M. *et al.* (1995) Enterohaemorrhagic *Escherichia coli* outbreak in South Australia associated with the consumption of mettwurst. *Communicable Diseases Intelligence* **19**, 70–71.

Caprioli, A., Luzzi, I., Rosmini, F. *et al.* (1994) Communitywide outbreak of haemolytic-uremic syndrome associated with non-O157 verotoxin-producing *Escherichia coli*. *Journal of Infectious Diseases* **169**, 208–211.

Carter, A.O., Borczyk, A.A., Carlson, J.A.K. *et al.* (1987) A severe outbreak of *Escherichia coli* O157:H7 – associated hemorrhagic colitis in a nursing home. *New England Journal of Medicine* **317**, 1496–1500.

CDSC (2000) Outbreaks of VTEC O157 infection linked to consumption of unpasteurised milk. *Communicable Disease Report CDR Weekly* **10**(23), 203, 206. Available from http://www.hpa.org.uk/cdr/archives/2000/cdr2300.pdf. Accessed 6 March 2007.

Centers for Disease Control and Prevention (CDC) (1995a) *Escherichia coli* O157:H7 outbreak linked to commercially distributed dry-cured salami–Washington and California, 1994. *Morbidity and Mortality Weekly Report* **44**(09), 157–160.

Centers for Disease Control and Prevention (CDC) (1995b) Outbreak of gastroenteritis attributable to *Escherichia coli* serotype O104:H21 – Helena, Montana, 1994. *Morbidity and Mortality Weekly Report* **44**(27), 501–503.

Centers for Disease Control and Prevention (CDC) (1995c) Community outbreak of haemolytic uremic syndrome attributable to *Escherichia coli* O111:NM – South Australia, 1995. *Morbidity and Mortality Weekly Report* **44**(29), 550–551, 557–558.

Centers for Disease Control and Prevention (CDC) (1996) Outbreak of *Escherichia coli* O157:H7 infections associated with drinking unpasteurized, commercial apple juice – British Columbia, California, Colorado and Washington, October 1996. *Morbidity and Mortality Weekly Report* **45**(44), 975.

Centers for Disease Control and Prevention (CDC) (1997a) Outbreaks of *Escherichia coli* O157:H7 infections associated with eating alfalfa sprouts – Michigan and Virginia, June–July 1997. *Morbidity and Mortality Weekly Report* **46**(32), 741–744.

Centers for Disease Control and Prevention (CDC) (1997b) *Escherichia coli* O157:H7 infections associated with eating a nationally distributed commercial brand of frozen ground beef patties and burgers – Colorado, 1997. *Morbidity and Mortality Weekly Report* **46**(33), 777–778.

Centers for Disease Control and Prevention (CDC) (2000) Outbreak of *Escherichia coli* O157:H7 infection associated with eating fresh cheese curds – Wisconsin, June 1998. *Morbidity and Mortality Weekly Report* **49**(40), 911–913.

Centers for Disease Control and Prevention (CDC) (2002) Multistate outbreak of *Escherichia coli* O157:H7 infections associated with eating ground beef – United States, June–July 2002. *Morbidity and Mortality Weekly Report* **51**(29), 637–639.

Centers for Disease Control and Prevention (CDC) (2005) Preliminary FoodNet data on the incidence of infection with pathogens transmitted commonly through food – 10 sites, United States, 2004. *Morbidity and Mortality Weekly Report* **54**(14), 352–356.

Centers for Disease Control and Prevention (CDC) (2006a) *Update on Multistate Outbreak of E. coli O157:H7 Infection from Fresh Spinach*, 6 October 2006. Available from http://www.cdc.gov/foodborne/ecolispinach/100606.htm. Accessed 6 March 2007.

Chalmers, R.M. & Salmon, R.L. (2000) Primary care surveillance for acute bloody diarrhea, Wales. *Emerging Infectious Diseases* **6**, 412–414.

Chart, H., Rowe, B., Kar, N.V.D. & Monnens, L.A.H. (1991) Serological identification of *Escherichia coli* O157 as cause of haemolytic uraemic syndrome in the Netherlands. *Lancet* **337**, 437.

Chilled Food Association (CFA) (2002) *Microbiological Guidance for Produce Suppliers to Chilled Food Producers*, 1st edn. Chilled Food Association Ltd, Kettering, UK.

Cody, S.H., Glynn, M.K., Farrar, J.A. *et al.* (1999) An outbreak of *Escherichia coli* O157:H7 infection from unpasteurized commercial apple juice. *Annals of Internal Medicine* **130**, 202–209.

Cowden, J.M., Ahmed, S., Donaghy, M. & Riley, A. (2001) Epidemiological investigation of the Central Scotland outbreak of *Escherichia coli* O157 infection, November to December, 1996. *Epidemiology and Infection* **126**, 335–341.

Cree, L., House, R. & Cowden, J.M. (2003) Has licensing improved hygiene in butchers' shops? *Communicable Disease and Public Health* **6**, 275–276.

Curnow, J. (1999) *E. coli* outbreak in Grampian. *SCIEH Weekly Report* **33**, 156.

Department of Health, UK (1995) *Food Handlers. Fitness to Work. Guidelines for Food Businesses, Enforcement Officers and Health Professionals*. Department of Health, London. Available from the Food Standards Agency, UK.

Department of Health, UK (1996) *Food Handlers. Fitness to Work. Guidelines for Food Business Managers.* Department of Health, London. Available from the Food Standards Agency, UK.

Deschênes, G., Casenave, C., Grimont, F. *et al.* (1996) Cluster of cases of haemolytic uraemic syndrome due to unpasteurized cheese. *Pediatric Nephrology* **10**, 203–205.

Doorduyn, Y., de Jager, C.M., van der Zwaluw, W.K. *et al.* (2006) Shiga toxin-producing *Escherichia coli* (STEC) O157 outbreak, the Netherlands, September–October 2005. *Euro Surveillance* **11**(7–8), 182–185.

Duffy, G. (2003) Verocytotoxigenic *Escherichia coli* in animal faeces, manures and slurries. *Journal of Applied Microbiology* **94**, 94S–103S.

Effler, P., Isaacson, M., Arntzen, L. *et al.* (2001) Factors contributing to the emergence of *Escherichia coli O157* in Africa. *Emerging Infectious Diseases* **7**, 812–819.

Espie, E., Grimont, F., Vaillant, V. *et al.* (2006a) O148 Shiga toxin-producing *Escherichia coli* outbreak: microbiological investigation as a useful complement to epidemiological investigation. *Clinical Microbiology and Infection* **12**, 992–998.

Espie, E., Vaillant, V., Mariani-Kurkjian, P. *et al.* (2006b) *Escherichia coli* O157 outbreak associated with fresh unpasteurized goats' cheese. *Epidemiology and Infection* **134**, 143–146.

Ethelberg, S., Smith, B., Torpdahl, M. *et al.* (2007) An outbreak of verocytotoxin-producing *Escherichia coli* O26:H11 caused by beef sausage, Denmark, 2007. *Euro Surveillance* **12**(5), EO70531.4. Available from http://www.eurosurveillance.org/ew/2007/070531.asp#4. Accessed 1 June, 2007.

Evans, J., Wilson, A., Willshaw, G.A. *et al.* (2002) Vero cytotoxin-producing *Escherichia coli* in a study of infectious intestinal disease in England. *Clinical Microbiology and Infection* **8**, 183–186.

Ferguson, D.D., Scheftel, J., Cronquist, A. *et al.* (2005) Temporally distinct *Escherichia coli* O157 outbreaks associated with alfalfa sprouts linked to a common seed source – Colorado and Minnesota, 2003. *Epidemiology and Infection* **133**, 439–447.

Food and Drug Administration (FDA) (1998) *Guide to Minimize Microbial Food Safety Hazards for Fresh Fruits and Vegetables.* FDA, USDA, CDC. Available from http://vm.cfsan.fda.gov/~dms/prodguid.html. Accessed 6 March 2007.

Food and Drug Administration (FDA) (1999) *Microbiological Safety Evaluations and Recommendations on Sprouted Seeds.* Available from http://www.cfsan.fda.gov/~mow/sprouts2.html. Accessed 6 March 2007.

Food and Drug Administration (FDA) (2002) Consumers advised of risks associated with eating raw and lightly cooked sprouts. Available from http://www.cfsan.fda.gov/~lrd/tpsprout.html. Accessed 26 June 2007.

Food and Drug Administration (FDA) (2007a) FDA and states closer to identifying source of *E. coli* contamination associated with illnesses at Taco John's restaurants. *FDA News*, January 12. Available from http://www.fda.gov/bbs/topics/NEWS/2007/NEW01546.html. Accessed 6 March 2007.

Food and Drug Administration (FDA) (2007b) FDA finalizes report on 2006 spinach outbreak. *FDA News* March 23. Available from http://www.fda.gov/bbs/topics/NEWS/2007/NEW01593.html. Accessed 27 June 2007.

Food and Drug Administration/Center for Food Safety and Applied Nutrition (FDA/CFSAN) (1999) *Seniors and Food Safety. Preventing Foodborne Illness. What's a Senior to Eat?* Available from http://www.cfsan.fda.gov/~dms/seniorsc.html. Accessed 6 March 2007.

Food and Drug Administration/Center for Food Safety and Applied Nutrition (FDA/CFSAN) (2004a) *Produce Safety from Production to Consumption: 2004 Action Plan to Minimize Foodborne Illness Associated with Fresh Produce Consumption.* Available from http://www.foodsafety.gov/~dms/prodpla2.html. Accessed 6 March 2007.

Food and Drug Administration/Center for Food Safety and Applied Nutrition (FDA/CFSAN) (2004b) *Juice HACCP Hazards and Controls Guidance*, 1st edn. *Final Guidance*. Available from http://www.cfsan.fda.gov/~dms/juicgu10.html. Accessed 6 March 2007.

Food and Drug Administration/Center for Food Safety and Applied Nutrition (FDA/CFSAN) (2005) *Food Code* 2005. Available from http://www.cfsan.fda.gov. Accessed 6 March 2007.

Food Standards Agency (FSA) (2002) *Managing Farm Manures for Food Safety. Guidelines for Growers to Minimise the Risks of Microbiological Contamination of Ready-to-Eat Crops*. Available from http://www.food.gov.uk/multimedia/pdfs/managingfarmmanures.pdf. Accessed 6 March 2007.

Germani, Y., Soro, B., Vohito, M. *et al.* (1997) Enterohaemorhagic *Escherichia coli* in Central African Republic. *Lancet* **349**, 1670.

Goh, S., Newman, C., Knowles, M. *et al.* (2002) *E. coli* O157 phage type 21/28 outbreak in North Cumbria associated with pasteurized milk. *Epidemiology and Infection* **129**, 451–457.

Goodburn, K. and the ECFF VTEC Working Group (1998) *VTEC and Agriculture*. Available from http://www.pinebridge.co.uk/Scireviews.htm. Accessed 2 July 2007.

Health Protection Agency (HPA) (2006a) *Vero Cytotoxin-Producing E. coli O157 Strains Examined by LEP Isolations from Humans England and Wales*, 1982–2006. Available from http://www.hpa.org.uk/infections/topics_az/ecoli/O157/data_ew.htm. Accessed 6 March 2007.

Health Protection Agency (HPA) (2006b) Sorbitol-fermenting Vero cytotoxin-producing *E. coli* O157 (VTEC O157). *Communicable Disease Report CDR Weekly* **16**(21): News. Available from http://www.hpa.org.uk/cdr/archives/2006/cdr2106.pdf. Accessed 6 March 2007.

Hilborn, E.D., Mermin, J.H., Mshar, P.A. *et al.* (1999) A multistate outbreak of *Escherichia coli* O157:H7 infections associated with consumption of mesclun lettuce. *Archives of Internal Medicine* **159**, 1758–1764.

Hilborn, E.D., Mshar, P.A., Fiorentino, T.R. *et al.* (2000) An outbreak of *Escherichia coli* O157:H7 infections and haemolytic uraemic syndrome associated with consumption of unpasteurized apple cider. *Epidemiology and Infection* **124**, 31–36.

Hiruta, N., Murase, T. & Okamura, N. (2000) An outbreak of diarrhoea due to multiple antimicrobial-resistant Shiga toxin-producing *Escherichia coli* O26:H11 in a nursery. *Epidemiology and Infection* **127**, 221–227.

Huang, D.B., Okhuysen, P.C., Jiang, Z.-D. & Dupont, H.L. (2004) Enteroaggregative *Escherichia coli*: an emerging enteric pathogen. *American Journal of Gastroenterology* **99**, 383–389.

Hutchison, M.L., Walters, L.D., Moore, A. & Avery, S.M. (2005) Declines of zoonotic agents in liquid livestock wastes stored in batches on-farm. *Journal of Applied Microbiology* **99**, 58–65.

International Commission on Microbiological Specifications for Foods (ICMSF) (1996) Intestinally pathogenic *Escherichia coli*. In: *Microbiological Specifications of Food Pathogens*. Blackie Academic and Professional, London, 126–140.

Jay, M.T., Garrett, V., Mohle-Boetani, J.C. *et al.* (2004) A multistate outbreak of *Escherichia coli* O157:H7 infection linked to consumption of beef tacos at a fast-food chain. *Clinical Infectious Diseases* **39**, 1–7.

Jensen, C., Ethelberg, S., Gervelmeyer, A. *et al.* (2006) First general outbreak of Verocytotoxin-producing *Escherichia coli* O157 in Denmark. *Euro Surveillance* **11**(2), 55–58.

Kaper, J.B., Nataro, J.P. & Mobley, H.L.T. (2004) Pathogenic *Escherichia coli*. *Nature Reviews Microbiology* **2**, 123–140.

Karmali, M.A. (1989) Infection by verotoxin-producing *Escherichia coli*. *Clinical Microbiology Reviews* **2**, 15–38.

Keene, W.E., Hedberg, K., Herriott, D.E. *et al.* (1997a) A prolonged outbreak of *Escherichia coli* O157:H7 infections caused by commercially distributed raw milk. *Journal of Infectious Diseases* **176**, 815–818.

Keene, W.E., Sazia, E., Kok, J. *et al.* (1997b) An outbreak of *Escherichia coli* O157:H7 infections traced to jerky made from deer meat. *Journal of the American Medical Association* **277**, 1229–1231.

Kohli, H.S., Chaudhuri, A.K.R., Todd, W.T.A. *et al.* (1993) The Hartwoodhill hospital *E. coli* O157 outbreak. *Communicable Diseases and Environmental Health in Scotland* **27**, 8–11.

Liptakova, A., Siegfried, L., Rosocha, J. *et al.* (2004) A family outbreak of haemolytic uremic syndrome and haemorrhagic colitis caused by verocytotoxigenic *Escherichia coli* O157 from unpasteurized cow's milk in Slovakia. *Clinical Microbiology and Infection* **10**, 576–578.

Lopez, E.L., Contrini, M.M., Sanz, M. *et al.* (1997) Perspectives on Shiga-like toxin infections in Argentina. *Journal of Food Protection* **60**, 1458–1462.

López, E.L., Prado-Jiménez, V. O'Ryan-Gallardo, M. & Contrini, M.M. (2000) *Shigella* and Shiga toxin-producing *Escherichia coli* causing bloody diarrhoea in Latin America. *Infectious Disease Clinics of North America* **14**, 41–65.

MacDonald, D.M., Fyfe, M., Paccagnella, A. *et al.* (2004) *Escherichia coli* O157:H7 outbreak linked to salami, British Columbia, Canada, 1999. *Epidemiology and Infection* **132**, 283–289.

Marier, R., Wells, J.G., Swanson, R.C. *et al.* (1973) An outbreak of enteropathogenic *Escherichia coli* foodborne disease traced to imported French cheese. *Lancet* **ii**, 1376–1378.

Martinez, A., Oliva, J.M., Panella, E. *et al.* (2001) *Outbreak of E. coli O157:H7 Infection in Spain*. *Eurosurveillance Weekly* **5**(1), 010104. Available from http://www.eurosurveillance.org/ew/2001/010104.asp#1. Accessed 6 March 2007.

McDonnell, R.J., Rampling, A., Crook, S. *et al.* (1997) An outbreak of Vero cytotoxin producing *Escherichia coli* O157 infection associated with takeaway sandwiches. *Communicable Disease Report* **7**(Review no. 13), R201–R205.

McIntyre, L., Fung, J., Paccagnella, A. *et al.* (2002) *Escherichia coli* O157 outbreak associated with the ingestion of unpasteurized goat's milk in British Columbia, 2001. *Canada Communicable Disease Report* **28-01**, 1 January 2002.

Michino, H., Araki, K., Minami, S. *et al.* (1999) Massive outbreak of *Escherichia coli* O157:H7 infection in schoolchildren in Sakai City, Japan, associated with consumption of white radish sprouts. *American Journal of Epidemiology* **150**, 787–796.

Morgan, D., Newman, C.P., Hutchinson, D.N. *et al.* (1993) Verotoxin producing *Escherichia coli* O157 infections associated with the consumption of yoghurt. *Epidemiology and Infection* **111**, 181–187.

Naimi, T.S., Wicklund, J.H., Olsen, S.J. *et al.* (2003) Concurrent outbreaks of *Shigella sonnei* and enterotoxigenic *Escherichia coli* infections associated with parsley: implications for surveillance and control of foodborne illness. *Journal of Food Protection* **66**, 535–541.

Nataro, J.P. & Kaper, J.B. (1998) Diarrheagenic *Escherichia coli*. *Clinical Microbiology Reviews* **11**, 142–201.

Nataro, J.P., Steiner, T. & Guerrant, R.L. (1998) Enteroaggregative *Escherichia coli*. *Emerging Infectious Diseases* **4**, 251–261.

O'Brien, S.J., Adak, G.K. & Reilly, W.J. (2001a) The task force on *E. coli* O157 final report: the view from here. *Communicable Disease and Public Health* **4**, 154–156.

O'Brien, S.J., Murdoch, P.S., Riley, A.H. *et al.* (2001b) A foodborne outbreak of vero cytotoxin-producing *Escherichia coli* O157:H – phage type 8 in hospital. *Journal of Hospital Infection* **49**, 167–172.

Paquet, C., Perea, W., Grimont, F. *et al.* (1993) Aetiology of haemorrhagic colitis epidemic in Africa. *Lancet* **342**, 175.

Paton, A.W., Ratcliff, R.M., Doyle, R.M. *et al.* (1996) Molecular microbiological investigation of an outbreak of haemolytic–uremic syndrome caused by dry fermented sausage contaminated with Shiga-like toxin-producing *Escherichia coli*. *Journal of Clinical Microbiology* **34**, 1622–1627.

Pennington, H. (1998) Factors involved in recent outbreaks of *Escherichia coli* O157:H7 in Scotland and recommendations for its control. *Journal of Food Safety* **18**, 383–391.

Preston, M., Borczyk, A., Davidson, R. *et al.* (1997) Hospital outbreak of *Escherichia coli* O157:H7 associated with a rare phage type – Ontario. *Canada Communicable Disease Report* **23**, 33–36.

Proctor, M.E., Kurzynski, T., Koschmann, C. *et al.* (2002) Four strains of *Escherichia coli* O157:H7 isolated from patients during an outbreak of disease associated with ground beef: importance of evaluating multiple colonies from an outbreak-associated product. *Journal of Clinical Microbiology* **40**, 1530–1533.

Rajpura, A., Lamden, K., Forster, S. *et al.* (2003) Large outbreak of infection with *Escherichia coli* O157 PT21/28 in Eccleston, Lancashire, due to cross contamination at a butcher's counter. *Communicable Disease and Public Health* **6**, 279–284.

Rangel, J.M., Rossiter, S., Stadler, D. *et al.* (2000) Outbreak of *Escherichia coli* O157:H7 infections linked to Romaine lettuce in a Pennsylvania retirement community [Abstract 515]. *Clinical Infectious Diseases* **31**, 301.

Rangel, J.M., Sparling, P.H., Crowe, C. *et al.* (2005) Epidemiology of *Escherichia coli* O157:H7 outbreaks, United States, 1982–2002. *Emerging Infectious Diseases* **11**, 603–609.

Reiss, G., Kunz, P., Koin, D. & Keefe, E.B. (2006) *Escherichia coli* O157:H7 infection in nursing homes: review of literature and report of recent outbreak. *Journal of the American Geriatrics Society* **54**, 680–684.

Rivas, M., Miliwebsky, E., Chinen, I. *et al.* (2006) Characterization and epidemiologic subtyping of Shiga toxin-producing *Escherichia coli* strains isolated from hemolytic uremic syndrome and diarrhea cases in Argentina. *Foodborne Pathogens and Disease* **3**, 88–96.

Salmon, R. & Outbreak Control Team (2005) Outbreak of verotoxin producing *E. coli* O157 infections involving over forty schools in south Wales, September 2005. *Euro Surveillance* **10**(10), E05 1006. Available from http://www.eurosurveillance.org/ew/2005/051006.asp#1. Accessed 6 March 2007.

Salmon, R. & Subcommittee (2000) Guidelines for the control of infection with Vero cytotoxin producing *Escherichia coli* (VTEC). *Communicable Disease and Public Health* **3**, 14–23.

Salmon, R.L., Farrell, I.D., Hutchinson, J.G.P. *et al.* (1989) A christening party outbreak of haemorrhagic colitis and haemolytic uraemic syndrome associated with *Escherichia coli* O157:H7. *Epidemiology and Infection* **103**, 249–254.

Schimmer, B. & Outbreak Investigation Team (2006) Outbreak of haemolytic uraemic syndrome in Norway: update. *Euro Surveillance* **11**(4), E060406.2. Available from http://www.eurosurveillance.org/ew/2006/060406.asp#2. Accessed 6 March 2007.

Smith, H.R., Cheasty, T. & Rowe, B. (1997) Enteroaggregative *Escherichia coli* and outbreaks of gastroenteritis in UK. *Lancet* **350**, 814–815.

Smith, M. (2000) *Microbial Testing of Spent Irrigation Water During Sprout Production*. Available from http://www.cfsan.fda.gov/~acrobat/sprouts3.pdf. Accessed 6 March 2007.

Söderström, A., Lindberg, A. & Andersson, Y. (2005) EHEC 0157 outbreak in Sweden from locally produced lettuce, August–September 2005. *Euro Surveillance* **10**(9), E050922.1. Available from http://www.eurosurveillance.org/ew/2005/050922.asp#1. Accessed 6 March 2007.

Söderström, A., Lindberg, A. & Andersson, Y. (2006) *E. coli* O157, lettuce – Sweden (West Coast) 2005. *Promed*, 16 September 2006.

Talan, D.A., Moran, G.J., Newdow, M. *et al.* (2001) Etiology of bloody diarrhea among patients presenting to United States emergency departments: prevalence of *Escherichia coli* O157:H7 and other enteropathogens. *Clinical Infectious Diseases* **32**, 573–580.

Task Force (2001) *Task Force on E. coli O157*. Final report. Available from http://www.food.gov.uk/multimedia/pdfs/ecolitaskfinreport.pdf. Accessed 6 March 2007.

Tarr, P.I., Gordon, C.A. & Chandler, W.L. (2005) Shiga-toxin-producing *Escherichia coli* and haemolytic uraemic syndrome. *Lancet* **365**, 1073–1086.

Tompkins, D.S., Hudson, M.J., Smith, H.R. *et al.* (1999) A study of infectious intestinal disease in England: microbiological findings in cases and controls. *Communicable Disease and Public Health* **2**, 108–113.

Tsuji, H., Oshibe, T., Hamada, K. *et al.* (2002) An outbreak of enterohaemorrhagic *Escherichia coli* caused by ingestion of contaminated beef at grilled-meat restaurant chain stores in the Kinki district in Japan: epidemiological analysis by pulsed field gel electrophoresis. *Japanese Journal of Infectious Diseases* **55**, 91–92.

Tuttle, J., Gomez, T., Doyle, M.P. *et al.* (1999) Lessons from a large outbreak of *Escherichia coli* O157:H7 infections: insights into the infectious dose and method of widespread contamination of hamburger patties. *Epidemiology and Infection* **122**, 185–192.

Tyrrel, S.F., Knox, J.W. & Weatherhead, E.K. (2006) Microbiological water quality requirements for salad irrigation in the United Kingdom. *Journal of Food Protection* **69**, 2029–2035.

Upton, P. & Coia, J.E. (1994) Outbreak of *Escherichia coli* O157 infection associated with pasteurised milk supply. *Lancet* **344**, 1015.

Vaillant, V. for French multi-agency outbreak investigation team (2005) Outbreak of *E. coli* O157:H7 infection associated with a brand of beefburgers in France. *Euro Surveillance* **10**(11), E051103.1.3. Available from http://www.eurosurveillance.org/ew/2005/051103.asp#1. Accessed 6 March 2007.

Vaz, T.M.I., Irino, K., Kato, M.A.M.F. *et al.* (2004) Virulence properties and characteristics of Shiga toxin-producing *Escherichia coli* in São Paulo, Brazil, from 1976 through 1999. *Journal of Clinical Microbiology* **42**, 903–905.

Welinder-Olsson, C., Stenqvist, K., Badenfors, M. *et al.* (2003) EHEC outbreak among staff at a children's hospital – use of PCR for verocytotoxin detection and PFGE for epidemiological investigation. *Epidemiology and Infection* **132**, 43–49.

Werber, D., Fruth, A., Liesegang, A. *et al.* (2002) A multistate outbreak of Shiga toxin-producing *Escherichia coli* O26:H11 infections in Germany, detected by molecular subtyping surveillance. *Journal of Infectious Diseases* **186**, 419–422.

Williams, R.C., Isaccs, S., Decou, M.L. *et al.* (2000) Illness outbreak associated with *Escherichia coli* O157:H7 in Genoa salami. *Canadian Medical Association Journal* **162**, 1409–1413.

Willshaw, G.A., Cheasty, T. & Smith, H.R. (2000) *Escherichia coli*. In: Lund, B.M., Baird-Parker, T.C. & Gould, G.W. (eds). *The Microbiological Safety and Quality of Food.* Aspen Publishers, Gaithersburg, MD, pp. 1136–1177.

World Health Organization (WHO) (1998) *Zoonotic non-O157 Shiga toxin-producing Escherichia coli (STEC).* Report of a WHO Scientific working group. Berlin, Germany, WHO/CSR/APH/98.8.

2.8 *Listeria monocytogenes*

2.8.1 *Importance as a cause of foodborne disease*

From the late 1940s to 1980, a series of outbreaks of listeriosis in humans were reported from many countries, but the means of transmission was unproven. The first outbreak that was shown to be foodborne occurred in 1981 and was attributed to consumption of coleslaw mix, prepared from cabbage grown in fields fertilized with both raw and composted manure from a flock of sheep in which listeriosis had occurred recently (Schlech *et al.*, 1983).

Listeria monocytogenes is of concern because of the severity of the disease caused in susceptible persons; in terms of disease burden it is regarded in the UK as one of the five most important foodborne pathogens (Adak *et al.*, 2002; FSA, 2002). Listeriosis caused by foodborne transmission is statutorily notifiable as food poisoning in the UK; reporting in European countries has been variable. In the Netherlands and Portugal listeriosis was not notifiable in 2000 (WHO, 2003); in Germany only congenital cases of listeriosis were notified prior to 2001, when all listeriosis became notifiable (Koch & Stark, 2006). In European countries that reported incidence in 1999–2001, the reported rates were between 0.73 and 0.03 per 100,000 persons (De Valk *et al.*, 2005). In the US in 2003, the incidence was estimated as 0.27 cases per 100,000 (CDC, 2005). An increase in reported listeriosis since 2000 has occurred in the UK (HPA, 2006), in Germany (Koch & Stark, 2006) and, following implementation of more active surveillance in 2005, in the Netherlands (Doorduyn *et al.*, 2006).

2.8.2 *Characteristics of the organism*

Listeria monocytogenes is a non-sporing, facultatively anaerobic, Gram-positive, rod-shaped bacterium. It has a few peritrichous flagellae and is motile at temperatures of 20–25°C but not at 37°C. *Listeria* spp. can grow readily on/in laboratory media and can multiply at temperatures between about 0 and +45°C with an optimum between 30 and 37°C.

The bacterium is relatively heat-resistant; heating meat to a temperature, throughout, of 70°C for 2 minutes is necessary to inactivate *L. monocytogenes* that may be present, giving an estimated 7 log inactivation (Farber & Peterkin, 2000). Properly controlled pasteurization of milk is considered to reduce the number of *L. monocytogenes* occurring in raw milk to levels that do not pose an appreciable risk to human health (Lou & Yousef, 1999) giving at least a 3-log reduction (Farber & Peterkin, 2000). The minimum HTST milk pasteurization involves heating at 71.7°C for 15 seconds, but many dairies use temperatures above the minimum legal limit.

Listeria monocytogenes can survive for long periods in up to 10% (w/v) NaCl, and can survive well in the environment. The ability of the bacterium to multiply in certain foods kept under refrigeration, in particular in mould-ripened soft cheeses and on prepared cooked meats where the organism can reach high numbers in the food before the end of shelf-life, is of particular concern. In otherwise favourable conditions, *L. monocytogenes* can multiply at pH values as low as 4.3, but in many mildly acidic foods such as types of yoghurt and cottage cheese, the presence of organic acids including lactic and acetic acid tends to inhibit the bacteria at pH values lower than 5.0 (Hicks & Lund, 1991).

Strains of *L. monocytogenes* have been differentiated into 13 serotypes, of which serotypes 1/2a, 1/2b and 4b have been associated most frequently with listeriosis in humans. Strains can be characterized further by phage typing and by molecular methods, particularly pulsed field gel electrophoresis (PFGE) of DNA. The great majority of typical strains of *L. monocytogenes* are considered to be pathogenic in humans, although the virulence of strains differs. The food industry and government must continue to treat all *L. monocytogenes* strains as potentially pathogenic, particularly because several strains can sometimes be isolated from the same food, depending on method of isolation, and because the presence of any strain of *L. monocytogenes* indicates that conditions for production of the food can allow the presence of virulent strains.

2.8.3 *Disease in humans*

The incubation period for the development of listeriosis ranges from 24 hours to 91 days (Farber & Peterkin, 2000). In healthy adults, infection with *L. monocytogenes* may result only in mild gastrointestinal symptoms. Unless cases with such mild symptoms are reported as part of an outbreak they are unlikely to be noticed. The bacterium causes invasive disease particularly in vulnerable groups of the population, including infants and neonates, people who are immunosuppressed, and the elderly (over 60–65), with a fatality rate of 20–30% (WHO, 2004). Gastrointestinal symptoms such as nausea, vomiting and diarrhoea may precede the more serious symptoms. In pregnant women, infection usually causes a mild, flu-like illness but the organism can cross the placenta and gain access to the fetus, leading to abortion, stillbirth or delivery of an acutely ill baby.

Worldwide figures for reported human listeriosis for the years 1989 and 1990 showed that 43% and 34.3%, respectively, of the cases were maternal/neonatal (Farber & Peterkin, 2000). In England and Wales, non-pregnancy-associated cases of infection with *L. monocytogenes* have greatly outnumbered pregnancy-associated cases since 1991 (HPA, 2006) (Fig. 2.8.1). Between January 1995 and 31 December 1999, of 543 reported cases of listeriosis in England and Wales, 452 were non-pregnancy-associated, of which 326 (72%) had major medical conditions, 140 (31%) had received immunosuppressive therapy and 97 (21%) were aged 60 years or over (Smerdon *et al.*, 2001). The relationship between age, malignancy and listeriosis in the UK is shown in Figure 2.8.2.

In non-perinatal cases, the main symptoms are septicaemia and/or meningitis. In the US, *L. monocytogenes* is reported to be the fifth most common cause of bacterial meningitis and the most common cause of meningitis in patients with underlying malignancies. Serious symptoms have been reported also in apparently healthy adults. For example, in the outbreak in Switzerland in 1984–87, attributed to contaminated Vacherin Mont d'Or cheese, of 57 non-pregnant adults affected there were 22 cases of meningoencephalitis; the proportion of these with an underlying medical condition (including age >65 yr as an underlying medical condition) was 45%, the remaining 55% were aged between 18 and 65 and were apparently free of pre-existing medical condition (Bula *et al.*, 1995).

In invasive, foodborne infections *L. monocytogenes* appears to enter the host via intestinal epithelial cells or the M cells of Peyer's patches (Farber & Peterkin, 2000). The bacteria

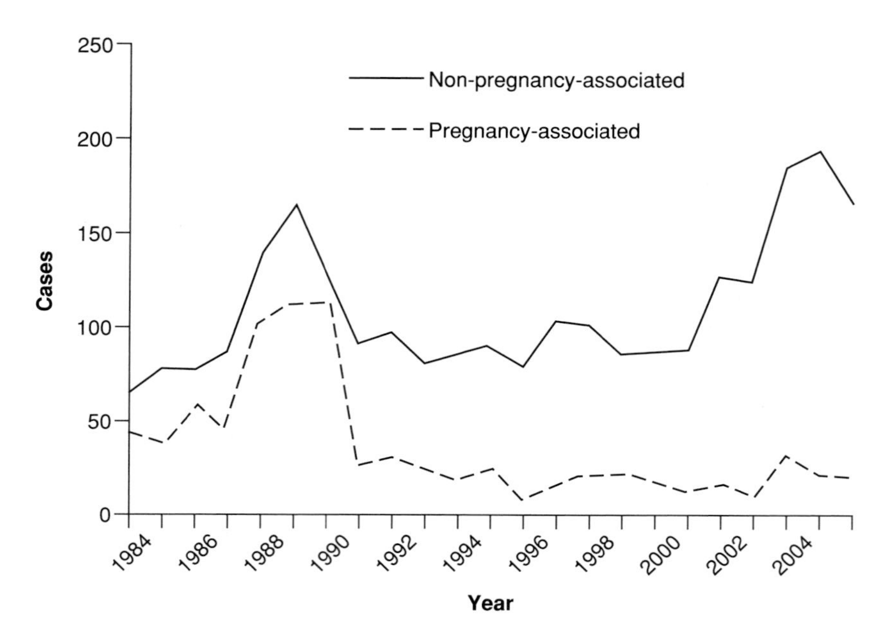

Fig. 2.8.1 *Listeria monocytogenes* pregnancy and non-pregnancy human cases in residents of England and Wales, 1983–2005 (HPA, 2006).

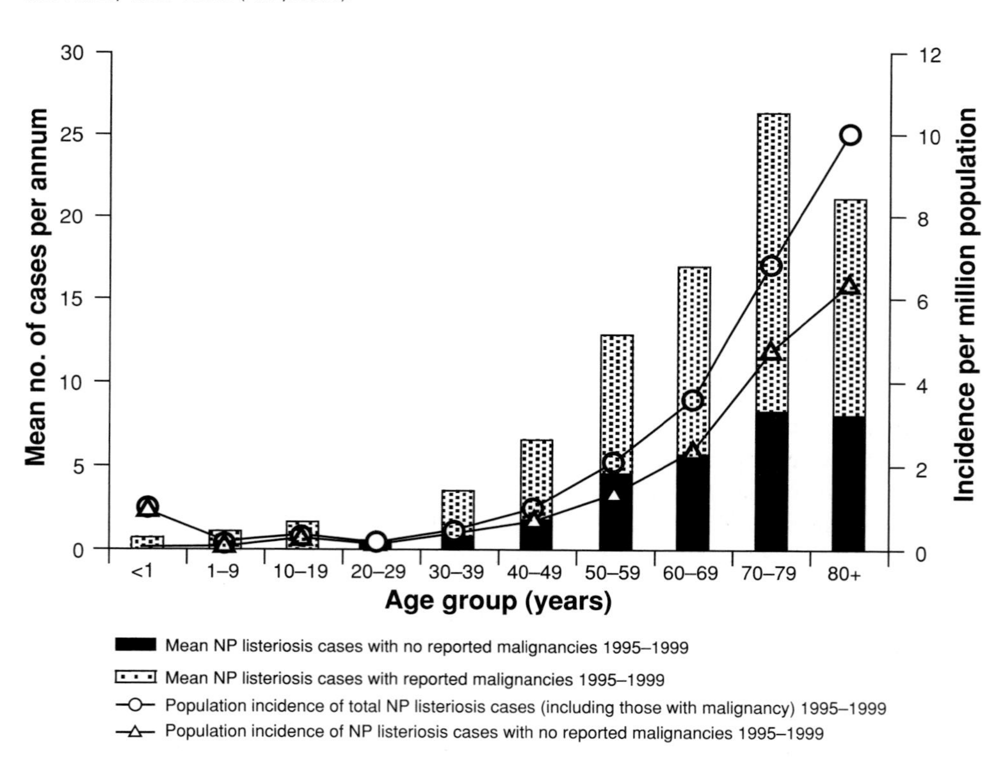

Fig. 2.8.2 Age distribution and incidence* of reported non-pregnancy (NP) associated listeriosis, with reported malignant neoplasms, 1995–1999, England and Wales. *Incidence rates calculated as the mean annual age-specific number of NP listeriosis cases reported in England and Wales/age-specific population estimates in 1998 for England and Wales. Office of National Statistics (ONS). Expressed as cases per million population. (Reprinted from McLauchlin *et al.*, 2004. Copyright (2004) with permission from Elsevier).

Table 2.8.1 Relative susceptibilities to *L. monocytogenes* for different sub-populations, based on French epidemiological data and American data.

Condition	Relative susceptibility
Based on French data	
Transplant	2584
Cancer – blood	1364
AIDS	865
Dialysis	476
Cancer – pulmonary	229
Cancer – gastrointestinal and liver	211
Non-cancer liver disease	143
Cancer – bladder and prostate	112
Cancer – gynaecological	66
Diabetes, insulin-dependent	30
Diabetes, non-insulin dependent	25
Alcoholism	18
Perinatals	14[a]
Over 65 yr old	7.5
Less than 65 yr, no other condition (reference population)	1

Modified from WHO, 2004.
[a] Based on American data.

are capable of intracellular growth and cell-to-cell movement, becoming bloodborne and causing septicaemia.

The effect of underlying conditions on risk of listeriosis in Denmark in 1989–90 was reported by Jensen *et al.* (1994). A report in 2004 concluded that, based on susceptibility information from the US, the elderly (60 yr and older) were 2.6 times more susceptible than the general healthy population, while perinatals were 14 times more susceptible (WHO, 2004). Estimates of the effect of conditions that compromise the immune system on susceptibility are shown in Table 2.8.1. The risk to people with pre-existing medical conditions is illustrated by a report on 84 sporadic cases in Australia between 1983 and 1992 (Paul *et al.*, 1994). Seventy-two of the cases were patients with an underlying medical condition, 13 of these were perinatal and 59 were non-perinatal, and of the latter 39% were over the age of 60, 46% were taking at least 10 mg of prednisolone daily, 29% were taking concurrent azathioprine or cyclosporine, and 20% had undergone chemotherapy for malignancy. Cases of serious infection with *L. monocytogenes* have been associated with treatment with tumour necrosis factor alpha (TNFalpha) – neutralizing agents infliximab, and etanercept for Crohn's disease or for rheumatoid arthritis (Kamath *et al.*, 2002; Slifman *et al.*, 2003).

The minimum infectious dose of *L. monocytogenes* is uncertain and will depend on the virulence of the strain and the immune status of the infected person. It is considered that the vast majority of cases have resulted from consumption of high numbers of the bacterium, and foods where the level of the pathogen was higher than 100 CFU/g (WHO, 2004).

The greatest probability of illness resulting from a low dose of *L. monocytogenes* was that calculated from an outbreak in Finland, associated with the consumption of butter. This outbreak, in 1998–99, resulted in 25 cases most of whom developed severe symptoms and six of whom died (Lyytikäinen *et al.*, 2000; Maijala *et al.*, 2001). The majority of the cases (15/25) occurred in a tertiary care hospital among immunocompromised people, mostly haematological or organ transplant patients, while ten occurred in hospitals in other parts of Finland. The outbreak was attributed to butter (prepared from pasteurized cream) that was contaminated with a type 3a strain of *L. monocytogenes*, a type that had not been implicated in outbreaks previously and had been regarded as of very low pathogenicity. Many samples of butter from the hospital kitchen and from the dairy and a wholesale store were analysed for *L. monocytogenes*. The outbreak strain was detected in all samples from the hospital kitchen and in many samples from a wholesale store, other outlets and the dairy. Most of the butter consumed was in 7 g packs in which the level of listeria detected in samples from the hospital kitchen and in most other samples tested was <100 CFU/g (range 5–79 CFU/g); the levels in two samples (of different batches from the hospital samples) from another outlet and from a wholesale store were 325 CFU/g and 11,000 CFU/g, respectively. Using the contamination level found in hospital samples, the estimated number of *L. monocytogenes* consumed by patients was between 14 and 2200 CFU/day, whereas using the maximum level of contamination found in a wholesale sample, the maximum number of *L. monocytogenes* consumed would be between 1.21×10^4 and 6.6×10^5 CFU/day (Maijala *et al.*, 2001).

This outbreak indicates the possibility that repeated consumption of a food contaminated with a low level of *L. monocytogenes* may have serious consequences for severely immunocompromised people.

2.8.4 *Source and transmission*

Listeria monocytogenes occurs in the faeces of healthy humans, where it may be transient; estimates of its incidence differ widely, the average frequency of carriage is reported as 2–6% (Farber & Peterkin, 2000). The bacterium causes disease in most domestic animals, in particular in sheep, goats and cattle; it is often shed in faeces and milk by animals, including those that fail to show symptoms of infection (Wesley, 1999). The organism has been reported, sometimes in relatively high numbers, in sewage and sewage plant effluent, river water, trade effluent, canals and lakes, and occurs, probably in lower numbers, on vegetation. Multiplication of the bacterium in poorly prepared silage can cause listeriosis in cattle and sheep that subsequently feed on the silage. The bacterium can enter the water supply as a result of faecal contamination from animals, release of sewage effluent, and runoff from agricultural land; freshwater and seawater in coastal areas can be contaminated (Farber & Peterkin, 2000).

Listeria monocytogenes occurs in/on a wide range of foods, the numbers present may vary greatly. In 21 reported studies of the incidence of the bacterium in raw milk in 14 countries, between 0 and 81% of samples were contaminated (Farber & Peterkin, 2000). Differences in virulence of strains, and in numbers of the bacteria present, may contribute to the observation that despite an incidence of *L. monocytogenes* in foods of 1–10%, the reported incidence of listeriosis remains relatively low. The majority of cases of listeriosis result from consumption of contaminated food, the risk is highest when strains become established in an environment in which ready-to-eat (RTE) foods can become contaminated (e.g. after cooking and before packaging, or during fermentation and ripening of certain cheeses) and growth of the bacterium occurs before the food is eaten by a susceptible population (Tompkin, 2002).

Infection can also result from contact with animals and, rarely, from aerosol transmission. Neonatal listeriosis can result from cross-infection in hospitals. Most cases of listeriosis in humans are reported as sporadic cases, but a proportion of these cases may, in fact, be associated with outbreaks that are unrecognized because they are diffuse in time or geographically (FDA/CFSAN, 2003a).

2.8.5 *Examples of outbreaks of foodborne listeriosis*

Examples of outbreaks are shown in Tables 2.8.2 and 2.8.3. Many of these outbreaks were caused by serotype 4b strains, but a high proportion of apparently sporadic cases of listeriosis are caused by serotype 1/2a strains. In several countries, serotype 1/2a may be replacing serotype 4b as the predominant serotype in human listeriosis (Lukinmaa *et al.*, 2003). In four clusters of cases in the UK between 1999 and 2004, the food vehicles were sandwiches sold in hospitals (Gillespie *et al.*, 2005). Previous studies have shown that sandwiches were contaminated frequently with *L. monocytogenes* and may be 'high-risk' foods for vulnerable groups. Following the outbreak in Australia in 2005, associated with RTE cold meats supplied to hospitals, the health department re-issued a warning to those at high risk of listeriosis, including pregnant women and those with impaired immune systems, not to consume RTE cold meats, cold cooked chicken, pâté or raw and chilled RTE seafoods.

2.8.6 *Main methods of prevention and control*

Foods, including meat and fish should be cooked thoroughly (Chapter 8) to inactivate *L. monocytogenes*. For example, cooking that achieves an internal temperature of at least 70°C for 2 minutes is sufficient to inactivate a high number of *L. monocytogenes* (10^6 reduction).

Properly controlled pasteurization of milk (heating every particle of milk at 71.7°C for 15 s) reduces *L. monocytogenes* that may be present in raw milk to safe levels (Lou & Yousef, 1999). Where outbreaks of infection have been associated with pasteurized milk, evidence has pointed to failure to apply the required heat treatment or to post-pasteurization contamination.

Listeria monocytogenes is of particular concern in RTE foods in which the bacterium can multiply. Such foods, or their ingredients, may have received a listericidal treatment (e.g. pasteurization of milk for cheese production) or cooking (for frankfurters and RTE, sliced meats), but subsequent processing (e.g. fermentation and ripening of cheeses, slicing and packaging of cooked meats) may allow a risk of contamination. In the US, on the basis of risk per serving, the following categories of food were assessed as high risk: delicatessen meats, frankfurters not reheated, pâté and meat spreads, unpasteurized fluid milk, smoked seafood and cooked RTE crustaceans (FDA/CFSAN, 2003a). Deli meats had a moderate contamination frequency with high contamination levels, they support the relatively rapid

Table 2.8.2 Examples of outbreaks of febrile gastroenteritis caused by *Listeria monocytogenes*.

Place, date	Serotype	Cases	Food implicated	Evidence[a]	Where food was provided	Factors leading to outbreak	Reference
Italy, 1993	1/2a	18[b]	Rice salad	S	Private supper	Contamination in kitchen possible	Salamina *et al.* (1996)
US, 1994	1/2b	45[c]	Chocolate milk	M	Picnic	Post-pasteurization contamination of milk, which was then left for periods unrefrigerated	Dalton *et al.* (1997)
Finland, ~1998	1/2a	5	Vacuum-packed, cold-smoked rainbow trout	M	Community	The product had been stored at ~10°C in retail store for several days and contained 1.9×10^5 CFU *L. monocytogenes*/g	Miettinen *et al.* (1999)
Italy, 1997	4b	1566[d]	Salad containing corn	M	2 primary schools and university canteen served by the same caterer	Corn from cans left at room temperature for several hours, during which listeria could multiply to high numbers	Aureli *et al.* (2000)
New Zealand, 2000	1/2	32	Ready-to-eat meats (corned beef, luncheon meat, corned silverside)	M	Community	Possibly inadequate heat treatment; products opened and re-packed after cooking, allowing possible contamination; long shelf life	Sim *et al.* (2002)
Japan, 2001	1/2b	38	Washed-type cheese	M	Community	*L. monocytogenes* isolated from environment and cheese samples	Makino *et al.* (2005)
Sweden, 2001	1/2a	48[e]	Raw milk cheese	M	Local summer farm	The cheese, made on the farm, contained high numbers of the outbreak bacterium	Carrique-Mas *et al.* (2003)
US, 2001	1/2a	16	Precooked, sliced turkey breast in sandwiches	M	Birthday party catered by delicatessen	Package of turkey breast stored for days in a delicatessen walk-in refrigerator with a temperature of 11–12°C	Frye *et al.* (2002)

[a]M (microbiological): identification of an organism of the same type from cases and in the suspect food or food ingredient, or detection of toxin in faeces or food; S (statistical): a significant statistical association between consumption of the suspect food and being a case.
[b]Four cases were hospitalized.
[c]'The non-invasive illness was not benign, many persons were bedridden and 4 were hospitalized.' Three additional cases were of invasive illness.
[d]Two hundred ninety-two children hospitalized.
[e]One person was hospitalized, a 64-year-old woman with chronic arthritis on immunosuppressive therapy with corticosteroids, who developed diarrhoea and a septic *L. monocytogenes* joint abscess.

growth of *L. monocytogenes* under refrigerated storage, and often are stored for extended periods. The risk of contamination occurs after cooking and before final packaging. Since this assessment, a rule has been introduced in the US that establishments producing RTE meat and poultry products must adopt methods to control *L. monocytogenes* that involve (1) both a post-cooking treatment (in-pack pasteurization) and a growth inhibitor (e.g. potassium lactate, sodium diacetate) or (2) either a post-cooking treatment or a growth inhibitor or (3) sanitation measures only. The frequency of verification by the US Food Safety and Inspection Service (FSIS) will depend on which of these three methods is adopted by an establishment (FSIS, 2006). In the UK, factors controlling the safety of delicatessen meats (RTE, refrigerated, cooked meats) include control of cooking temperature and time, segregation of cooked meat during production, hygiene control, temperature control during transport and sale, and a relatively short shelf-life (K. Goodburn pers. com; Peck *et al.*, 2006). Nevertheless, because *L. monocytogenes* can grow slowly at temperatures of 0–5°C it is important that RTE foods should be stored at as low a temperature as possible below 5°C, to minimize growth of this bacterium.

Listeria monocytogenes can survive for long periods in the environment; it can become established in processing equipment in factories and be very difficult to eliminate (ICMSF, 2002; Tompkin, 2002). This is a particular problem in the production of RTE foods such as packaged, sliced meats and soft cheeses, where contamination from processing equipment has occurred at a late stage in production, and when no subsequent listericidal treatment is applied. It has been suggested that in commercial food processes that include a step such as cooking, it should be possible to keep the incidence of product contamination to $<0.5\%$ (Tompkin, 2002). End product testing is of little value for detecting lots with this low level of contamination, since even if 60 samples per lot (batch) were tested, assuming random distribution of the organism a lot with 0.5% of defective samples (e.g. containing *L. monocytogenes*) would be accepted on about 19 occasions in 20 (ICMSF, 2002). Control depends mainly, therefore, on preventing the establishment and multiplication of *L. monocytogenes* in the environment of food production. Detailed accounts of the procedures and systems necessary to control *L. monocytogenes* during commercial food production have been published (ICMSF, 2002; Tompkin, 2002). In the US, the Listeria Action Plan (FDA/CFSAN, 2003b) has identified six areas for action that focus on:

- guidance for processors preparing RTE foods and for retail, food service and institutional establishments;
- training and technical assistance for industry and food safety regulators;
- enhanced information for healthcare providers and consumers;
- review of enforcement and regulation;
- enhanced disease surveillance;
- further refinement of risk assessment.

This action plan emphasized that consumer practices such as keeping refrigerators at 40°F (4.5°C) or lower, and using perishable pre-cooked or RTE food as soon as possible, have a significant role in reducing illness.

An FAO/WHO risk assessment consultation panel proposed that the number of *L. monocytogenes* in RTE foods should not exceed 100 CFU/g *at the point of consumption* (Szabo *et al.*, 2003). This criterion is used in an EC regulation (EC, 2005). For RTE foods able to

Table 2.8.3 Examples of outbreaks of invasive, foodborne listeriosis.[a]

Place, date; serotype	Cases	Perinatal/ non-perinatal	Deaths	Stillbirths/ miscarriages/ infant deaths	Food implicated	Evidence[b]	Where food was provided	Factors leading to outbreak	Reference
US, 1979; 4b	20 (15 nosocomial)	0/20	3	0	Raw celery. tomatoes, lettuce	S	8 hospitals	Not known	Ho *et al.* (1986)
New Zealand, 1980; 1/2a	22	22/0	0	6	Raw seafood (finfish and molluscs)	D	Community	Possible contamination of seafood growing areas	Lennon *et al.* (1984)
Canada, 1981; 4b	41	34/7	2	15	Coleslaw mix (cabbage and carrot)	M	Community	Cabbage grown in fields fertilized with composted and raw manure from a flock of sheep in which listeriosis had occurred	Schlech *et al.* (1983)
US, 1983; 4b	49	7/42	12	2	Pasteurized milk[c]	S	Community	Listeriosis had occurred in cattle supplying the milk. Several serotypes were isolated from the raw milk. Probable post-pasteurization contamination	Fleming *et al.* (1985); Ryser (1999)
Switzerland, 1983–1987; 4b	122	65/57	18	13	Vacherin Mont'd Or cheese	M	Community	Cheese made with unpasteurized milk	Bille (1990); Bula *et al.* (1995)
US, 1985; 4b	142	93/49	18	30	Mexican-style cheese	M	Community	Cheese made with unpasteurized milk	Linnan *et al.* (1988); Ryser (1999)
US, 1986–1987; 4b,1/2b,1/2a	36	4/32	14	2	Ice cream, salami, Brie cheese	S	Community	Not reported	Schwartz *et al.* (1989)
England and Wales, 1987–1989; 4b, 4bX	355[e]	185/129	94		Meat pate	S	Community	Probable post-process contamination after cooking	McLauchlin *et al.* (1991); McLauchlin *et al.* (2004)
Denmark, 1989–1990; 4b	26	3/23	6	0	Blue mould cheese or hard cheese	D	Community	nr	Jensen *et al.* (1994); Ryser (1999)

Australia, 1990; 1/2a	11	11/0	0	6	Processed meats or pate	D & M	Community	nr	Watson & Ott (1990); Ryser (1999)
France, 1992–93; 4b	279	92/187	56	29	Pork tongue in jelly	M	Community	nr	Jacquet *et al.* (1995); Ryser (1999)
France, 1993; 4b	38	31/7	1	10	Rillettes (ham meat cooked with fat) and other meat products	M	Community	Epidemic strain established in factory, no lethal step in production, growth of listeria during storage at 8°C	Goulet *et al.* (1998); Ryser (1999)
Sweden, 1994–1995; 4b	8 or 9	3/6	1	1	'Gravad' or cold-smoked rainbow trout	M	Community	Epidemic strain established in factory of producer for some time	Ericsson *et al.* (1997); Tham *et al.* (2000)
France, 1995; 4b	36	9/27	4	7	Soft cheese (Brie de Meaux)	M	Community	Cheese made with unpasteurized milk	Goulet *et al.* (1995); Ryser (1999); De Buyser *et al.* (2001)
France, 1997; 4b	14	nr	0	0	Pont l'Eveque cheese	M	Community	Cheese made with unpasteurized milk	Ryser (1999)
Finland, 1998–1999; 3a	25	0/25	6	0	Soured cream butter	M	Tertiary care hospital	Epidemic strain in dairy environment and product	Lyytikäinen *et al.* (2000)
US, 1998–1999; 4b	108	13/95	14	4	Heated frankfurters	S and M	Community	Outbreak strain probably colonized equipment in plant producing frankfurters and deli meats. Alteration to equipment in an area critical for frankfurter production led to increased listeria contamination of food surfaces	CDC (1998, 1999); Mead *et al.* (2006)

(*cont.*)

Table 2.8.3 *(continued)*

Place, date; serotype	Cases	Perinatal/ non-perinatal	Deaths	Stillbirths/ miscarriages/ infant deaths	Food implicated	Evidence[b]	Where food was provided	Factors leading to outbreak	Reference
UK, 1999; 4b	2, one outpatient, one in-patient	0/2	0	0	Cheese and cheese and salad sandwiches (2 cases)	M	2 hospital shops supplied by same caterer	Hygiene procedures in catering premises needed improvement. Patient strain of *L. monocytogenes* isolated from sandwiches and sandwich manufacturing environment	CDSC (1999); Graham *et al.* (2002)
France, 1999–2000; 4b	10	3/7	2	1	Rillettes, a pate-like meat product	M	Community	Tests by manufacturer showed *L. monocytogenes* at <10 CFU/g in finished product towards the end of shelf life on several occasions from Sept. to Dec. 1999. This level of contamination was allowed	De Valk *et al.* (2001)
France, 1999–2000; 4b	32	9/23	5	5	Jellied pork tongue and other deli meats	S	Community	Possible cross-contamination from pork tongue to other products	De Valk *et al.* (2001)
US, 1999; 1/2a	11	2/9	nr	nr	Pate	nr	Community	nr	FDA/CFSAN (2003a)
US, 2000; 1/2a	30	8/22	4	3	Pre-cooked, sliced turkey Deli meat,	M	Community	Probable persistence of the listeria strain in the processing plant	CDC (2000); Olsen *et al.* (2005)
US, 2000–2001; nr	12	10/2	0	5	Homemade Mexican-style cheese	M	Community	Cheese made with raw milk	CDC (2001)
Canada, 2002; nr	17	3/14	nr	nr	Cheese	M	Community	Cheese made with unpasteurized milk, contamination via milk or environment	Gaulin *et al.* (2003)
France, 2002; nr	8	0/8	1	0	Spreadable sausage made with raw pork and fat	M	nr	nr	Goulet *et al.* (2002)

US, 2002; nr	53	3/50	8	3	Pre-cooked, sliced turkey Deli meat	M[d]	Community	Possible establishment of the bacterium in the factory environment	CDC (2002a, b)
UK, 2003; 4b	17	11/6	2	3	Dairy butter in sandwiches	M	Community	Intermittent low level contamination of butter supplied to sandwich outlets	ACMSF (2003); Gillespie *et al.* (2005)
UK, 2003; 1/2a	2 cancer outpatients	0/2	0	0	Pre-packed sandwiches	M	Hospital canteen	Patient strains isolated from sandwiches and environmental sites in sandwich manufacturing factory	ACMSF (2003); Gillespie *et al.* (2005)
UK, 2003; 1/2a	5	5/0	0	0	Sandwiches	M	Hospital retailer	Patient strain recovered from factory where sandwiches were prepared	HPA (2003); Dawson *et al.* (2006)
Switzerland, 2005; 1/2a	10 (2 additional non-invasive)	2/8	3	2	Neuenburg Tomme soft cheese	M	Community	Up to 32,000 CFU of *L. monocytogenes*/g found in cheese	Bille *et al.* (2006)
Australia, 2005; nr	4	0/4	2	0	Ready-to-eat meat products supplied to hospitals	nr	2 hospitals	*Listeria* found on cutting equipment at the supplier's factory and in the hospital kitchens	ProMED-mail (2005)

[a] Information also obtained from Ryser (1999) and McLauchlin *et al.* (2004).
[b] M (microbiological): identification of an organism of the same type from cases and in the suspect food or food ingredient, or detection of toxin in faeces or food; D (descriptive): other evidence reported by local investigators as indicating the suspect food; S (statistical): a significant statistical association between consumption of the suspect food and being a case.
[c] Flaws in evidence implicating pasteurized milk were described by Ryser (1999).
[d] Outbreak strain isolated from environment of the producing factory.
[e] Information was not available to classify 41 patients.
nr, not reported.

support the growth of *L. monocytogenes*, other than those intended for infants and for special medical purposes, the number must be lower than 100 CFU/g throughout the shelf-life. The organism must be absent in 25 g before the product leaves the immediate control of the producing food business operator, when he is unable to demonstrate that the product will not exceed the limit of 100 CFU/g throughout the shelf-life (EC, 2005). For fresh, precut, packaged lettuce with a storage time of not more than 7 days at a maximum temperature not greater than 8°C, the number of *L. monocytogenes* at the end of production must not exceed 13 CFU/10 g in order to comply with a limit of not more than 100 CFU/g at the point of consumption (Szabo *et al.*, 2003). For RTE foods intended for infants, and RTE foods for special medical purposes, European regulations specify absence of *L. monocytogenes* in 25 g (EC, 2005).

The outbreak in Finland in 1998–99 (Table 2.8.3), which affected immunosuppressed hospital patients, may have resulted from the prolonged consumption of contaminated butter resulting in an intake of between 14 and 2200 CFU of *L. monocytogenes*/day (based on samples from the hospital kitchen) or between 2.2×10^4 and 3.1×10^5 CFU/day (based on the highest level detected in a wholesale sample). This indicates the possibility that even low levels of *L. monocytogenes* in a food, a relatively small amount of which was consumed regularly, could be a hazard to vulnerable persons.

To avoid the risk of listeriosis, in 1989, the Chief Medical Officer advised all doctors in England that pregnant women should avoid eating soft, ripened cheeses such as brie, camembert and blue vein types, and should reheat cook-chilled meals and RTE poultry 'until they are piping hot'. He recommended that immunocompromised patients, transplant recipients, patients with leukaemias and lymphoid malignancies, and those on immunosuppressive drugs including oral and systemic steroids, should also follow this advice. Later in 1989, these vulnerable groups were advised to avoid eating any type of paté. In 1991, a booklet on Food Safety published by the then Ministry of Agriculture, Fisheries and Food repeated the above advice to 'pregnant women and anyone with a low resistance to infection'.

Jensen *et al.* (1994) concluded that 'the immense importance of predisposing immunodeficiencies' in listeriosis was confirmed by their work, and that 'persons with a high risk of listeriosis should be individually informed about food hygiene and risk foods'. Doorduyn *et al.* (2006) recommended that advice in the Netherlands that pregnant women should avoid high-risk foods, should be extended to immunocompromised patients, and in the UK, Gillespie *et al.* (2006) stated that '... dietary advice on the avoidance of high-risk foods should be provided routinely to the elderly and immunocompromised and not just to pregnant women'.

Because *L. monocytogenes* can grow at refrigerator temperatures and is found in RTE foods, the US FDA has issued specific advice to at-risk consumers, including pregnant women, older adults (aged over 65) and people with weakened immune systems. People in these groups are advised (FDA/CFSAN, 2003c):

- *Do not eat* hot dogs and luncheon meats, unless they are reheated until steaming hot.
- *Do not eat* soft cheese such as Feta, Brie and Camembert cheeses, blue-veined cheeses, queso blanco, queso fresco and Panela unless it is labelled as made with pasteurized milk.

- *Do not eat* refrigerated patés or meat spreads. Canned or shelf-stable patés and meat spreads *may be eaten*.
- *Do not eat* refrigerated smoked seafood, unless it is contained in a cooked dish such as a casserole. Refrigerated smoked seafood, such as salmon, trout, whitefish, cod, tuna or mackerel, is most often labelled as 'nova-style', 'lox', 'kippered', 'smoked' or 'jerky'. These products are found in the refrigerator section or sold at deli counters of grocery stores and delicatessens. Canned or shelf-stable smoked seafood *may be eaten*.
- *Do not drink* raw (unpasteurized) milk, *do not eat* foods that contain unpasteurized milk.

Summary of measures advised in healthcare settings

1. Obtain foods from a reputable supplier.
2. Meats and fish should be cooked thoroughly, meats should be heated at an internal temperature of at least 72°C for at least 2 minutes.
3. Potentially hazardous foods should be stored at <5°C.
4. Refrigerated cooked meats, refrigerated pâtés and meat spreads, smoked seafood should not be served to vulnerable groups unless in a cooked meal.
5. Soft cheeses made with unpasteurized milk should not be served.
6. Unpasteurized milk should not be served.

References

Adak, G.K., Long, S.M. & O'Brien, S.J. (2002) Trends in indigenous foodborne disease and deaths, England and Wales: 1992 to 2000. *Gut* **51**, 832–841.

Advisory Committee on the Microbiological Safety of Food (ACMSF) (2003) *Recent Trends in Listeriosis in the UK.* ACM/667. Available from http://www.food.gov.uk/multimedia/pdfs/acm667.pdf. Accessed 6 March 2007.

Aureli, P., Fiorucci, G.C., Caroli, D. *et al.* (2000) An outbreak of febrile gastroenteritis associated with corn contaminated by *Listeria monocytogenes*. *New England Journal of Medicine* **342**, 1236–1241.

Bille, J. (1990) Epidemiology of human listeriosis in Europe, with special reference to the Swiss outbreak. In: Miller, A.J., Smith, J.L. & Somkuti, G.A. (eds). *Foodborne Listeriosis.* Elsevier, Amsterdam, pp. 71–74.

Bille, J., Blanc, D.S., Schmid, H. *et al.* (2006) Outbreak of human listeriosis associated with tome cheese in Northwest Switzerland, 2005. *Euro Surveillance* **11**(6), 91–93.

Bula, C., Bille, J. & Glauser, M.P. (1995) An epidemic of food-borne listeriosis in Western Switzerland: description of 57 cases involving adults. *Clinical Infectious Diseases* **20**, 66–72.

Carrique-Mas, J.J., Hökeberg, I., Andersson, Y. *et al.* (2003) Febrile gastroenteritis after eating on-farm manufactured fresh cheese – an outbreak of listeriosis? *Epidemiology and Infection* **130**, 79–86.

Centers for Disease Control and Prevention (CDC) (1998) Multistate outbreak of listeriosis – United States, 1998. *Morbidity and Mortality Weekly Report* **47**(50), 1085–1086.

Centers for Disease Control and Prevention (CDC) (1999) Update: Multistate outbreak of listeriosis – United States, 1998–1999. *Morbidity and Mortality Weekly Report* **47**(51and 52), 1117–1118.

Centers for Disease Control and Prevention (CDC) (2000) Multistate outbreak of listeriosis – United States, 2000. *Morbidity and Mortality Weekly Report* **49**(50), 1129–1130.

Centers for Disease Control and Prevention (CDC) (2001) Outbreak of listeriosis associated with homemade Mexican-style cheese – North Carolina, October 2000–January 2001. *Morbidity and Mortality Weekly Report* **50**(26), 560–562.

Centers for Disease Control and Prevention (CDC) (2002a) Public Health Dispatch: Outbreak of listeriosis – Northeastern US, 2002. *Morbidity and Mortality Weekly Report* **51**(42), 950–951.

Centers for Disease Control and Prevention (CDC) (2002b) Press release. *Update: Listeriosis outbreak investigation.* 21 November 2002.

Centers for Disease Control and Prevention (CDC) (2005) Preliminary FoodNet data on the incidence of infection with pathogens transmitted commonly through food – 10 sites, United States, 2004. *Morbidity and Mortality Weekly Report* **54**(14), 352–356.

Communicable Disease Surveillance Centre (CDSC) (1999) Listeriosis linked to retail outlets in hospitals – caution needed. *Communicable Disease Report (CDR) Weekly* **9**(25). Available from http://www.hpa.org.uk/cdr/archives/1999/cdr2599.pdf. Accessed 6 March 2007.

Dalton, C.B., Austin, C.C., Sobel, J. *et al.* (1997) An outbreak of gastroenteritis and fever due to *Listeria monocytogenes* in milk. *New England Journal of Medicine* **336**, 100–105.

Dawson, S.J., Evans, M.R.W., Willby, D. *et al.* (2006) Listeria outbreak associated with sandwich consumption from a hospital retail shop, United Kingdom. *Euro Surveillance* **11**(6), 89–91.

De Buyser, M.-L., Dufour, B., Maire, M. & Lafarge, V. (2001) Implication of milk and milk products in food-borne diseases in France and different industrialised countries. *International Journal of Food Microbiology* **67**, 1–17.

De Valk, H., Jacquet, C., Goulet, V. *et al.* (2005) Surveillance of listeria infections in Europe. *Euro Surveillance* **10**(10), 251–255.

De Valk, H., Vaillant, V., Jacquet, C. *et al.* (2001) Two consecutive nationwide outbreaks of listeriosis in France, October 1999–February 2000. *American Journal of Epidemiology* **154**, 944–950.

Doorduyn, Y., de Jager, C.M., van der Zwaluw, W.K. *et al.* (2006) First results of the active surveillance of *Listeria monocytogenes* infections in the Netherlands reveal higher than expected incidence. *Euro Surveillance* **11**(4) E060420.4. Available from http://www.eurosurveillance.org/ew/2006/060420.asp#4. Accessed 6 March 2007.

Ericsson, H., Eklöw, A., Daniellson-Tham, M.-L. *et al.* (1997) An outbreak of listeriosis suspected to have been caused by rainbow trout. *Journal of Clinical Microbiology* **35**, 2904–2907.

European Commission (EC) (2005) Commission Regulation (EC) No. 2073/2005 on microbiological criteria for foodstuffs. *Official Journal of the European Union* **L338/1-26**. Available from http://www.food.gov.uk/multimedia/pdfs/microbiolcriteria.pdf. Accessed 6 March 2007.

Farber, J.M. & Peterkin, P.I. (2000) *Listeria monocytogenes*. In: Lund, B.M., Baird-Parker, T.C. & Gould, G.W. (eds). *The Microbiological Safety and Quality of Food*, Vol. 2. Aspen Publishers, Gaithersburg, MD, pp. 1178–1232.

FDA/CFSAN (2003a) *Quantitative Assessment of the Relative Risk to Public Health from Foodborne Listeria monocytogenes Among Selected Categories of Ready-to-Eat Foods*. Available from http://www.cfsan.fda.gov/~dms/lmr2-toc.html. Accessed 6 March 2007.

FDA/CFSAN (2003b) *Reducing the Risk of Listeria monocytogenes FDA/CDC 2003 Update of the Listeria Action Plan*. Available from http://www.cfsan.fda.gov/~dms/lmr2plan.html. Accessed 6 March 2007.

FDA/CFSAN (2003c) *Consumer Advisory. How to Safely Handle Refrigerated Ready-to-Eat Foods and Avoid Listeriosis*. Updated October 2003. Available from http://www.cfsan.fda.gov/~dms/adlister.html. Accessed 6 March 2007.

Fleming, D.W., Cochi, S.L., MacDonald, K.L. *et al.* (1985) Pasteurized milk as a vehicle of infection in an outbreak of listeriosis. *New England Journal of Medicine* **312**, 404–407.

Food Safety and Inspection Service (FSIS) (2006) *Compliance Guidelines to Control Listeria monocytogenes in Post-Lethality Exposed Ready-to-Eat Meat and Poultry Products*. Available from http://www.fsis.usda.gov/oppde/rdad/FRPubs/97013F/LM_Rule_Compliance_Guidelines_May_2006.pdf. Accessed 6 March 2007.

Food Standards Agency (FSA) (2002) *Measuring Foodborne Illness Levels*. Available from http://www.food.gov.uk/safereating/microbiology/58736. Accessed 6 March 2007.

Frye, D.M., Zweig, R., Sturgeon, J. *et al.* (2002) An outbreak of febrile gastroenteritis associated with delicatessen meat contaminated with *Listeria monocytogenes. Clinical infectious Diseases* **35**, 943–949.

Gaulin, C., Ramsey, D., Ringuette, L. & Ismail, J. (2003) First documented outbreak of *Listeria monocytogenes* in Quebec, 2002. *Canada Communicable Disease Report* **29**, 181–186.

Gillespie, I., McLauchlin, J., Adak, B. *et al.* (2005) *Changing Pattern of Human Listeriosis in England and Wales 1993–2004.* ACM/753. Report for the Advisory Committee on the Microbiological Safety of Food (ACMSF) Available from http://www.food.gov.uk/multimedia/pdfs/acm753.pdf. Accessed 6 March 2007.

Gillespie, I., McLauchlin, J., Grant, K.A. *et al.* (2006) Changing pattern of listeriosis in England and Wales, 2001–2004. *Emerging Infectious Diseases* **12**, 1361–1366.

Goulet, V., Jacquet, C., Vaillant, V. *et al.* (1995) Listeriosis from consumption of raw-milk cheese. *Lancet* **345**, 1581–1582.

Goulet, V., Martin, P. & Jacquet, C. (2002) Cluster of listeriosis cases in France. *Eurosurveillance Weekly* **6**(27), 020704. Available from http://www.eurosurveillance.org/ew/2002/020704.asp#6. Accessed 6 March 2007.

Goulet, V., Rocourt, J., Rebiere, I. *et al.* (1998) Listeriosis outbreak associated with the consumption of rillettes in France in 1993. *Journal of Infectious Diseases* **177**, 155–160.

Graham, J.C., Lanser, S., Bignardi, G. *et al.* (2002) Hospital acquired listeriosis. *Journal of Hospital Infection* **51**, 136–139.

Health Protection Agency (HPA) (2003) Cluster of pregnancy associated listeria cases in the Swindon area. *Communicable Disease Report CDR Weekly* **13**(50). Available from http://www.hpa.org.uk/cdr/archives/2003/cdr5003.pdf. Accessed 6 March 2007.

Health Protection Agency (HPA) (2006) *Listeria – Epidemiological Data.* Available from http://www.hpa.org.uk/infections/topics_az/listeria/data_ew_gr.htm. Accessed 6 March 2007.

Hicks, S.J. & Lund, B.M. (1991) The survival of *Listeria monocytogenes* in cottage cheese. *Journal of Applied Bacteriology* **70**, 308–374.

Ho, J.L., Shanda, K.N., Friedland, G. *et al.* (1986) An outbreak of type 4b *Listeria monocytogenes* infection involving patients from eight Boston hospitals. *Archives of Internal Medicine* **146**, 520–524.

ICMSF (International Commission on Microbiological Specifications for Foods) (2002) *Microorganisms in Foods 7. Microbiological Testing in Food Safety Management.* Kluwer Academic/Plenum Publishers, New York, pp. 161.

Jacquet, C., Catimel, B., Brosch, R. *et al.* (1995) Investigations related to the epidemic strain involved in the French listeriosis outbreak in 1992. *Applied and Environmental Microbiology* **61**, 2242–2246.

Jensen, A., Frederiksen, W. & Gerner-Smidt, P. (1994) Risk factors for listeriosis in Denmark. *Scandinavian Journal of Infectious Diseases* **26**, 171–178.

Kamath, B.M., Mamula, P., Balassano, R.N. & Markowitz, J.E. (2002) Listeria meningitis after treatment with infliximab. *Journal of Pediatric Gastroenterology and Nutrition* **34**, 410–412.

Koch, J. & Stark, K. (2006) Significant increase of listeriosis in Germany – epidemiological patterns 2001–2005. *Euro Surveillance* **11** (6), 85–88.

Lennon, D., Lewis, B., Mantell, C. *et al.* (1984) Epidemic perinatal listeriosis. *Pediatric Infectious Diseases* **3**, 30–34.

Linnan, M.J., Mascola, L., Lou, X.D. *et al.* (1988) Epidemic listeriosis associated with Mexican-style cheese. *New England Journal of Medicine* **319**, 823–828.

Lou, Y. & Yousef, A.E. (1999) Characteristics of *Listeria monocytogenes* important to food processors. In: Ryser, E.T. & Marth, E.H. (eds). *Listeria, Listeriosis and Food Safety*, 2nd edn. Marcel Dekker, New York, pp. 131–224.

Lukinmaa, S., Mitetinen, M., Nakari, U.-M. *et al.* (2003). *Listeria monocytogenes* isolates from invasive infections: variation of sero- and geno-types during an 11-year period in Finland. *Journal of Clinical Microbiology* **41**, 1694–1700.

Lyytikäinen, O., Autlo, T., Maijala, R. *et al.* (2000) An outbreak of *Listeria monocytogenes* serotype 3a infections from butter in Finland. *Journal of Infectious Diseases* **181**, 1838–1841.
Maijala, R., Lyytikainen, O., Johansson, T. *et al.* (2001) Exposure to *Listeria monocytogenes* within an epidemic caused by butter in Finland. *International Journal of Food Microbiology* **70**, 97–109.
Makino, S.-I., Kawamoto, K., Takeshi, K. *et al.* (2005) An outbreak of food-borne listeriosis due to cheese in Japan, during 2001. *International Journal of Food Microbiology* **104**, 189–196.
McLauchlin, J., Hall, S.M., Velani, S.K. & Gilbert, R.J. (1991) Human listeriosis and paté: a possible association. *British Medical Journal* **303**, 773–775.
McLauchlin, J., Mitchell, R.T., Smerdon, W.J. & Jewell, K. (2004) *Listeria monocytogenes* and listeriosis: a review of hazard characterization for use in microbiological risk assessment of foods. *International Journal of Food Microbiology* **92**, 15–33.
Mead, P.S., Dunne, E.F., Graves, L. *et al.* (2006) Nationwide outbreak of listeriosis due to contaminated meat. *Epidemiology and Infection* **134**, 744–751.
Miettinen, M.K., Siitonen, A., Heiskanen, P. *et al.* (1999) Molecular epidemiology of an outbreak of febrile gastroenteritis caused by *Listeria monocytogenes* in cold-smoked rainbow trout. *Journal of Clinical Microbiology* **37**, 2358–2360.
Olsen, S.J., Patrick, M., Hunter, S.B. *et al.* (2005) Multistate outbreak of *Listeria monocytogenes* infection linked to delicatessen turkey meat. *Clinical Infectious Diseases* **40**, 962–967.
Paul, M.L., Dwyer, D.E., Chow, C. *et al.* (1994) Listeriosis – a review of eighty-four cases. *Medical Journal of Australia* **160**, 489–493.
Peck, M.W., Goodburn, K.E., Betts, R.P. & Stringer, S.C. (2006) *Microbial Risk Management – Clostridium botulinum in Vacuum and Modified Atmosphere Packed (MAP) Chilled Foods* (B13006). Report to the Food Standards Agency.
ProMED-mail (2005) 13 December. Listeriosis, nosocomial – Australia (South Australia).
Ryser, E.T. (1999) Foodborne listeriosis. In: Ryser, E.T. & Marth, E.H. (eds). *Listeria, Listeriosis and Food Safety*, 2nd edn. Marcel Dekker, New York, pp. 299–358.
Salamina, G., Dalle Donne, E., Niccolini, A. *et al.* (1996) A foodborne outbreak of gastroenteritis involving *Listeria monocytogenes*. *Epidemiology and Infection* **117**, 429–436.
Schlech, W.F., Lavigne, P.M., Bortolussi, R.A. *et al.* (1983) Epidemic listeriosis – evidence for transmission by food. *New England Journal of Medicine* **308**, 203–206.
Schwartz, B., Hexter, D., Broome, C.V. *et al.* (1989) Investigation of an outbreak of listeriosis: new hypotheses for the etiology of epidemic *Listeria monocytogenes* infections. *Journal of Infectious Diseases* **159**, 680–685.
Sim, J., Hood, D., Finnie, L. *et al.* (2002) Series of incidents of *Listeria monocytogenes* non-invasive febrile gastroenteritis involving ready-to-eat meats. *Letters in Applied Microbiology* **35**, 409–413.
Slifman, N.R., Gershon, S.K., Lee, J.-H. *et al.* (2003) *Listeria monocytogenes* infection as a complication of treatment with tumor necrosis factor [alpha]-neutralizing agents. *Arthritis and Rheumatism* **48**, 319–324.
Smerdon, W.J., Jones, R., McLauchlin, J. & Reacher, M. (2001) Surveillance of listeriosis in England and Wales 1995–1999. *Communicable Disease and Public Health* **4**, 188–193.
Szabo, E.A., Simons, L., Coventry, M.J. & Cole, M.B. (2003) Assessment of control measures to achieve a food safety objective of less than 100 CFU of *Listeria monocytogenes* per gram at the point of consumption for fresh, precut Iceberg lettuce. *Journal of Food Protection* **66**, 256–264.
Tham, W., Ericsson, H., Loncarevic, S. *et al.* (2000) Lessons from an outbreak of listeriosis related to vacuum-packed gravid and cold-smoked fish. *International Journal of Food Microbiology* **62**, 173–175.
Tompkin, R.B. (2002) Control of *Listeria monocytogenes* in the food-processing environment. *Journal of Food Protection* **65**, 709–725.
Watson, C. & Ott, K. (1990) *Listeria* outbreak in Western Australia. *Communicable Diseases Intelligence* **24**, 9–12.

Wesley, I.V. (1999) Listeriosis in animals. In: Ryser, E.T. & Marth, E.H. (eds). *Listeria, Listeriosis and Food Safety*, 2nd edn. Marcel Dekker, New York, pp. 39–73.

World Health Organization (WHO) (2003) *WHO Surveillance Program for Control of Foodborne Infections and Intoxications in Europe*. 8th report 1999–2000. Available from http://www.bfr.bund.de/internet/8threport/8threp_fr.htm. Accessed 6 March 2007.

World Health Organization (WHO) (2004) *Risk Assessment of Listeria monocytogenes in Ready-to-Eat Foods*, MRA Series 4. Interpretative Summary. Available from http://www.who.int/foodsafety/publications/micro/mra_listeria/en/. Accessed 6 March 2007.

2.9 *Salmonella*

2.9.1 *Importance as a cause of foodborne disease*

Worldwide, typhoid fever caused by *Salmonella enterica* subspecies *enterica* serotype Typhi (*S.* Typhi) transmitted in contaminated water or sometimes in food, was estimated to have resulted in >21 million illnesses and 216,510 deaths during 2000, while paratyphoid fever caused by *S.* Paratyphi resulted in >5 million illnesses (Crump *et al.*, 2004). Regions with a high incidence of typhoid fever (>100 cases per 100,000 persons/yr) include south-central Asia and south-east Asia, those of medium incidence (10–100/100,000 cases/yr) include the rest of Asia, Africa, Latin America, the Caribbean, Oceania except Australia and New Zealand, those of low incidence (<10/100,000 cases/yr) include Europe, North America and the rest of the developed world.

Non-typhoid salmonellas continue to be a major cause of foodborne disease worldwide. In European countries that reported incidence rates in 2000, these were higher than 100 per 100,000 persons in some countries and between 20 and 30 per 100,000 in England and Wales, France and the Netherlands (WHO, 2003). In the US in 2003, the incidence was estimated as 14.5 cases per 100,000 (CDC, 2004).

2.9.2 *Characteristics of the organisms*

The genus *Salmonella* is divided taxonomically into only two species, *S. enterica* and *S. bongori*. *Salmonella enterica* is divided into six subspecies, *enterica*, *salamae*, *arizonae*, *diarizonae*, *houtenae* and *indica*. At least 2523 serotypes are recognized within the genus *Salmonella* (Popoff *et al.*, 2003) (Table 2.9.1). Most of the salmonellas that cause gastroenteritis in humans belong to the group *S. enterica* subsp. *enterica*; serotypes (based on O, somatic, and H, flagellar, antigens) are named, usually, according to the geographical place where the serotype was first isolated. Serotypes belonging to other subspecies are designated by their antigenic formulae following the subspecies name (Brenner *et al.*, 2000). More than 1500 serotypes of *S. enterica* subsp. *enterica* have been described, almost all are

Table 2.9.1 Number of serotypes included in each species and subspecies of *Salmonella*.

Species	Subspecies	Number of serotypes
S. enterica	*enterica*	1492
	salamae	500
	arizonae	95
	diarizonae	331
	houtenae	71
	indica	13
S. bongori		21
Total		2523

Information from Popoff *et al.*, 2003.
All name-bearing serotypes (e.g. *S.* Enteritidis, *S.* Typhimurium, *S.* Typhi) belong to the subspecies *enterica*. Serotypes in the remaining subspecies of *S. enterica* and in *S. bongeri* are identified by antigenic formula.

considered potentially pathogenic for humans. Between October and December 2003, 119 different serotypes from humans were reported to the *Salmonella* Reference Laboratory for England and Wales (HPA, 2004). The Vi capsular antigen is associated with virulence in strains of *S. enterica* serotype Typhi and is present in some strains of *S. enterica* serotype Dublin (Parry *et al.*, 2002).

For epidemiological purposes, phage typing is used to subdivide several serotypes, including *S. enterica* subsp. *enterica* serotypes Enteritidis (*S.* Enteritidis) and Typhimurium (*S.* Typhimurium), while subdivision may be done also on the basis of sensitivity to bacteriocins, resistance to antibiotics, the presence of plasmid DNA, and molecular typing, particularly by pulsed field gel electrophoresis (PFGE).

Salmonellas are facultatively anaerobic, Gram-negative, non-spore-forming, rod-shaped bacteria. The majority of strains are motile by peritrichous flagellae. The bacteria grow readily on laboratory media with optimal growth at 35–37°C and a temperature range for growth from approximately 7 to 49°C. The great majority of salmonellae are inactivated readily by heat. The pasteurization treatments for whole liquid egg used in several countries (e.g. heating at 64°C for 2.5 min in England) give between 1000- and 10,000-fold inactivation of most salmonella (ICMSF, 1998), while pasteurization of milk at 72°C for 15 seconds (England) gives about 10^5-fold inactivation (D'Aoust, 2000). During freezing and drying of foods and in low moisture foods, such as peanut butter and chocolate, where the bacteria are protected by food components, they can survive for long periods. In otherwise optimal conditions, the pH range for growth is approximately 3.6–9.5 (D'Aoust, 2000). Salmonellae can also survive in relatively acidic foods such as fruit juices for an extended time.

2.9.3 *Disease in humans*

2.9.3.1 *Typhoid and paratyphoid fevers*

Symptoms can be mild or severe and include fever with a temperature as high as 39–40°C, headache, malaise, anorexia, constipation or diarrhoea, rose-coloured spots on the chest area and enlarged spleen and liver (Parry *et al.*, 2002). The incubation period is usually 1–2 weeks. Complications occur in 10–15% of patients. Paratyphoid fever gives similar symptoms but the disease is generally milder. The incidence of bacteraemia is much higher than in disease caused by non-typhoid salmonellas (Mandal & Brennand, 1988).

2.9.3.2 *Non-typhoid salmonellas*

The infective dose in man has often been considered to be of the order of 10^6 bacteria, but in some cases very low numbers have caused illness, for example 100 *S.* Typhimurium in kosher egg-based ice cream and 100 *S.* Eastbourne in chocolate (ICMSF, 1996). The infectious dose will depend on factors such as whether infection is associated with a meal or between meals, and whether the food vehicle is protective against the action of stomach acids, for example in a high fat product such as chocolate.

Acute symptoms of disease include nausea, vomiting, abdominal pain, diarrhoea, mild fever and chills. Symptoms usually start within 6–72 hours of infection, depending on the infecting dose. In cases of uncomplicated disease symptoms usually subside within 5 days, but the patient may continue to excrete the bacterium for several months. In

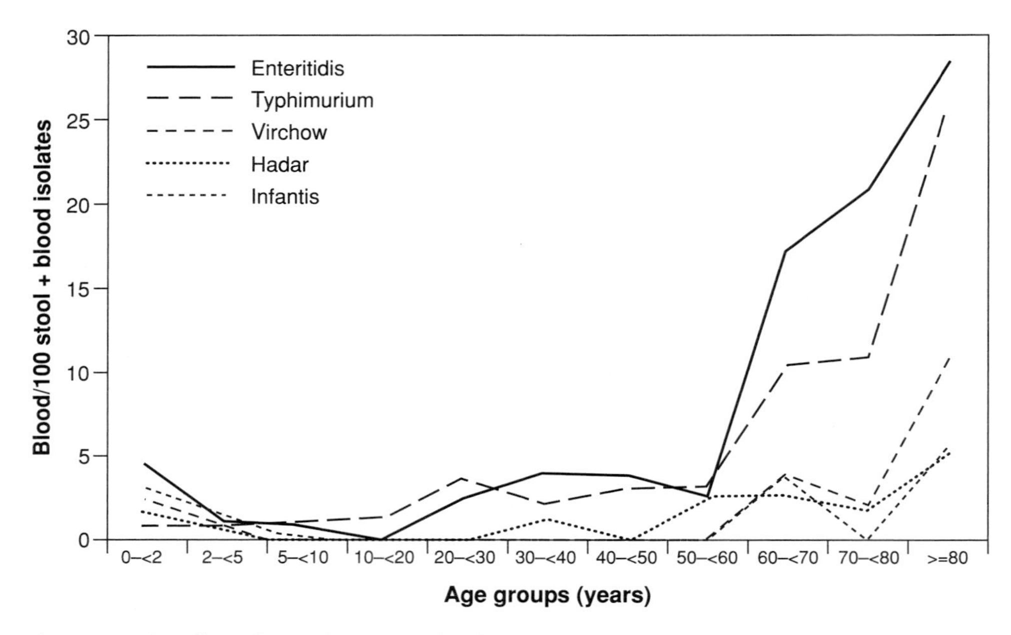

Fig. 2.9.1 The effect of age of patient and *Salmonella enterica* serotype on blood invasiveness ratio, estimated as number of blood isolates per 100 stool + blood isolates. (Reprinted with permission from Weinberger *et al.*, 2004. Copyright Cambridge University Press)

general, the illness is self-limiting. In some circumstances the infection becomes systemic; serotypes that mainly cause enteritis in previously healthy adults may result in septicaemia in young or elderly patients or those with underlying disease (D'Aoust, 2000). In a UK hospital unit, *S.* Dublin, *S.* Infantis, *S.* Virchow, *S.* Panama and *S.* Newport were the most invasive serotypes in humans (Mandal & Brennand, 1988). In elderly people who have developed septicaemia due to *S.* Dublin there can be a mortality rate of 15%, and in infections with *S.* Enteritidis in hospital/nursing home outbreaks a mortality of 3.6% has been reported, with the elderly particularly affected (FDA/CFSAN, 2003). Invasiveness of non-typhoid salmonellas is reported to increase with the age of the patient, with *S.* Enteritidis and *S.* Typhimurium showing the highest invasiveness in persons aged over 60 years (Fig. 2.9.1).

Chronic sequelae including reactive arthritis, Reiter's syndrome and ankylosing spondylitis can follow systemic infections with non-typhoid salmonellas (D'Aoust, 2000). These symptoms may occur 3–4 weeks after the start of acute symptoms. Of persons who developed gastroenteritis due to *S.* Enteritidis after attending a Thanksgiving banquet in the US, 29% developed reactive arthritis and 3% developed Reiter's syndrome; persons with more severe illness and longer diarrhoea were at increased risk of reactive arthritis (Dworkin *et al.*, 2001). Following an outbreak in Denmark involving mainly medical doctors and their spouses, 19% of patients developed reactive arthritis (Locht *et al.*, 2002).

In the UK, the US and many European countries the main serotypes associated with foodborne gastroenteritis are *S.* Enteritidis and *S.* Typhimurium (WHO, 2003). The most

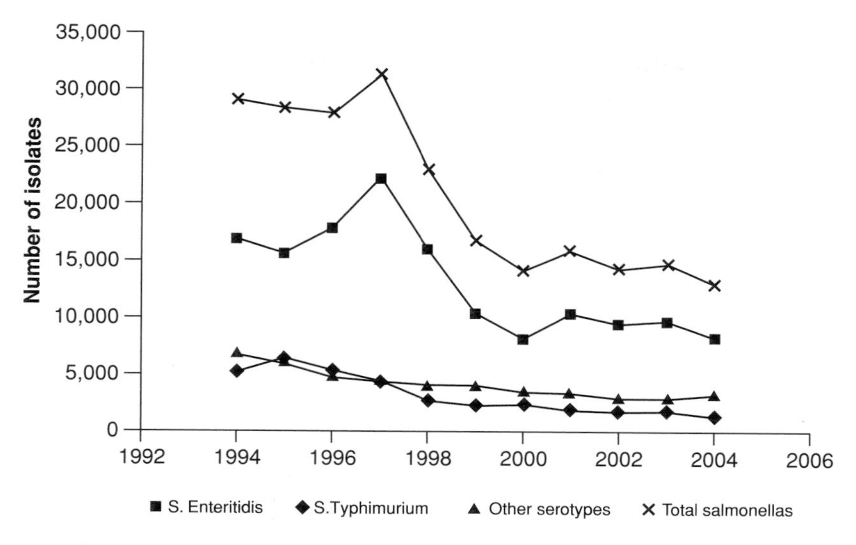

Fig. 2.9.2 *Salmonella* in humans (excluding *S.* Typhi and *S.* Paratyphi). Faecal and lower gastrointestinal tract isolates, England and Wales, 1981–2004. (Data from UK Health Protection Agency, 2007)

common salmonellas isolated from human sources in England and Wales and in the US are shown in Figures 2.9.2 and 2.9.3.

The occurrence of strains of salmonella resistant to antimicrobial drugs has become of increasing importance (Old & Threlfall, 1998). Strains of particular concern are *S.* Typhimurium definitive phage type 104 (DT 104) a high proportion of strains of which

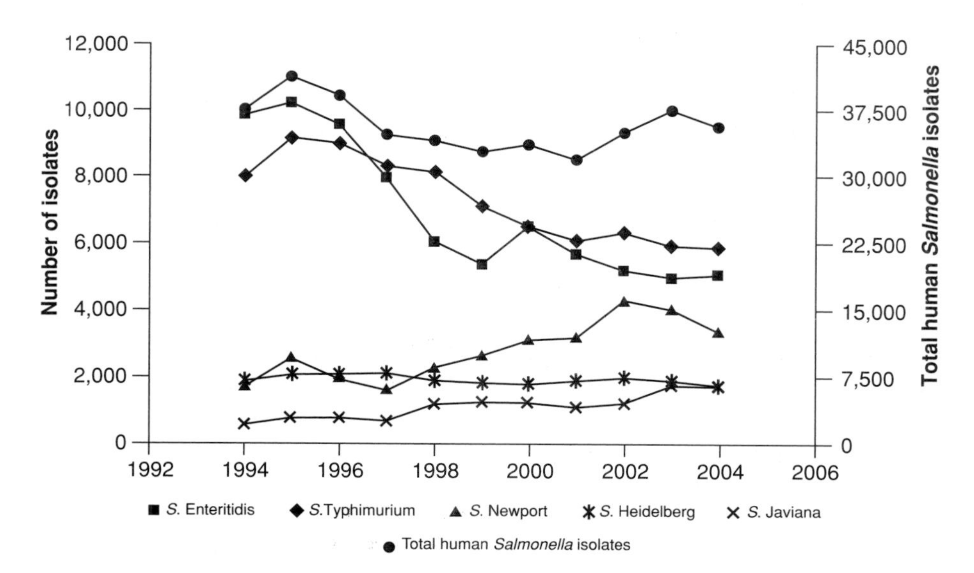

Fig. 2.9.3 Most common *Salmonella* isolates from human sources. Left axis, *S.* Enteritidis; *S.* Typhimurium; *S.* Newport; *S.* Heidelberg; *S.* Javiana; right axis, total human *Salmonella* isolates. (Data from US Centers for Disease Control. *Salmonella* Annual Summary 2004)

are resistant to ampicillin, chloramphenical, streptomycin, suphonamides and tetracycline (Helms *et al.*, 2005). Some strains show increasing resistance to quinolones, this is of concern because fluoroquinolones are important in the treatment of severe salmonella infections in humans.

2.9.4 *Source and transmission*

Salmonella Typhi is adapted specifically to humans and transmitted in the faeces and urine of infected people; some strains of *S.* Paratyphi B are adapted to humans and rarely isolated from animals, these strains cause enteric fever in humans. Widespread infection by these bacteria occurs when supplies of clean water are unavailable or are disrupted, and facilities for sanitation and hygiene are lacking. After recovery from typhoid or paratyphoid fever a small proportion of people become carriers who, if they handle food, are liable to transmit the bacterium. Other strains of *S.* Paratyphi B (biotype Java strains) cause gastroenteritis in humans and are isolated from animals and food (Threlfall *et al.*, 1999). In endemic areas risk factors for infection with *S.* Typhi and *S.* Paratyphi B include eating food prepared outside the home, e.g. ice cream or iced drinks from street vendors, drinking contaminated water, poor housing and inadequate hygiene (Parry *et al.*, 2002).

Non-typhoid salmonellas can be present in the intestinal tract of farm and wild animals, including mammals, birds (particularly poultry), reptiles, amphibians and arthropods. In the production of animals reared for human food, much of the animal feed used is manufactured commercially. Animal feed is liable to be contaminated with *S. enterica* as well as with other pathogens, and can be a major source of entry of these pathogens into the food supply (Crump *et al.*, 2002). Contamination of feed leads to the establishment of the microorganisms in food animals and subsequently to cases of foodborne disease in humans. The presence of *S. enterica* in the animal host commonly causes little or no disease, but the bacteria are excreted in large numbers in faeces. This results in cross-infection of other animals, and may cause contamination of the environment including grassland and fruit, vegetable and cereal crops. At slaughter of animals and poultry, cross-contamination of carcasses can occur.

Estuaries and coastal waters can be contaminated with salmonellas from human and agricultural wastes. Fish and shellfish from such areas can be contaminated, as well as fish and shellfish produced by aquaculture, a high proportion of which originate in Asia.

During the 1980s and 1990s, *S.* Enteritidis replaced *S.* Typhimurium as the main reported cause of foodborne salmonellosis in the UK, the US and much of Europe. Transmission of these two serotypes has been associated particularly with eggs and egg products and with red and white meats. Since the early 1900s, *S.* Typhimurium and other salmonellas have been known to occur on the outside of shell eggs, and contamination of egg products from the outside of the shell has resulted in human salmonellosis from consumption of raw or undercooked egg products. *Salmonella* Enteritidis is able to infect the ovaries of healthy hens and contaminate the eggs before the shell has been formed, thus even stringent cleaning of the outside of eggs fails to reduce the internal contamination. The reason for the increase in the incidence of S. Enteritidis within eggs in the late 1980s is not known; eradication of the antigenically similar *S.* Gallinarum and *S.* Pullorum (pathogens of poultry but rarely of humans) from poultry may have allowed the wider spread of *S.* Enteritidis, possibly associated with transmission from rodents (ACMSF, 2001; Cogan & Humphrey, 2003). In surveys of *Salmonella* in eggs between 1995 and 1997, the main phage type isolated from UK shell eggs was *S*. Enteritidis PT4, and from imported eggs it was PT21. In the US, the most common phage types that emerged initially were PTs 8 and 13a, but more recently the number of cases due to PT4 has increased.

Following the widespread adoption in Britain of stringent controls to eliminate *S*. Enteritidis from poultry, the incidence has fallen markedly since 1998, and other serotypes predominate that do not appear to invade the reproductive system of poultry. An average of 4.0% of fresh chicken and 10.4% of frozen chicken on retail sale in the UK in 2001 was contaminated with salmonellas (FSA, 2003). The frequency of contamination in frozen UK chicken was lower than that in imported frozen samples (8.3% in UK chicken, 13.6% in non-UK chicken).

On the basis of reports of incidents of *Salmonella* in livestock, and of isolations from humans in the UK in 2005, the most common serotypes were: Cattle, *S*. Dublin; Sheep, *S. enterica* subsp. *diarizonae*; Pigs, *S*. Typhimurium; Poultry, *S*. Indiana; Humans, *S*. Enteritidis (VLA, 2006).

While infection by salmonellas has been associated primarily with the consumption of animal-derived foods, the widespread distribution of these bacteria in the environment has resulted in outbreaks associated with a broad range of foods including raw meats, poultry, eggs, milk and dairy products, fish, shrimp, frogs legs, yeast, coconut, sauces and salad dressing, cake mixes, cream-filled desserts and toppings, dried gelatin, peanut butter, cocoa, chocolate (FDA/CFSAN, 2003), nuts, fruits including cantaloupes and tomatoes, unpasteurized fruit juices, raw vegetable sprouts and lettuce (Horby *et al.*, 2003). Rarely, infection of infants has been associated with the use of powdered infant formula, and *Salmonella* together with *Enterobacter sakazaii* are the microorganisms of greatest concern in infant formula (EFSA, 2004).

The main ways in which non-typhoid salmonellas contaminate foods and cause illness are:

1 via animal products including meat, poultry, eggs and milk
2 by contamination of fruit and vegetables during growth and harvest
3 by cross-contamination in the food service environment

2.9.5 *Examples of outbreaks of foodborne salmonellosis in healthcare settings*

In developing countries, typhoid and paratyphoid fever occur mainly as a result of unsafe drinking water, inadequate sewage disposal and flooding. Cases of infection with *S*. Paratyphi B in the UK are rare with two outbreaks associated with biotype Java reported between 1988 and 1994, one cluster of 50 cases associated with a reception and 12 cases in 1990 probably associated with consumption of cold meats (CDSC, 1994). Outbreaks of infection with *S*. Paratyphi B biotype Java occurred in France in 1990 and 1993 associated with the consumption of cheese made from unpasteurized goats' milk, the outbreak in 1993 affected more than 273 people (Desenclos *et al.*, 1996; Threlfall *et al.*, 1999).

Numerous outbreaks of salmonellosis have been reported in healthcare settings, examples are listed in Tables 2.9.2–2.9.4. The association of outbreaks caused by *S*. Enteritidis with

Table 2.9.2 Examples of outbreaks of foodborne disease caused by *Salmonella enterica* serotype Enteritidis in healthcare settings and institutions.

Place, date	Phage type	Total cases	Number hospitalized	Deaths	Food implicated	Evidence[a]	Where food was provided	Factors leading to outbreak	Reference
US, 1987	8	404	Nosocomial	9	Mayonnaise made with raw eggs	S and M	Hospital	Ovarian infection with *S. enteritidis* in hens in farm corporation that supplied the eggs	Telzak *et al.* (1990)
US, 1989	28 and *S.* Schwarz-engrund	50	11	4	nr	D	Nursing home	Meals prepared in single kitchen; inadequate cooling of foods, inadequate cleaning of food equipment, raw meats thawed over ready-to-eat produce	Taylor *et al.* (1993)
UK, 1990	4	109 (101 patients, 8 staff) (plus 44 asymptomatic)	Nosocomial	1	Beef rissole made with raw liquid egg	S	Hospital for mentally handicapped people	Inadequate heating during deep fat frying of rissole. Salmonella isolated from pooled egg sample	Evans *et al.* (1996)
UK, date not reported	7A	29	2	0	Banana custard meringue	S	2 adjoining nursing homes for the elderly	Meringue prepared with whites of 8 raw shell eggs and soft-cooked	Holtby & Stenson (1992)
UK, 1993	4	29 (21 patients, 8 staff)	Nosocomial	0	nr	D	2 hospitals with shared catering	Infected, asymptomatic food handlers	Dryden *et al.* (1994)
Mexico, 1994	nr	97 (hospital employees)	nr	0	Egg-covered meat plate	S	Hospital cafeteria	Inadequate heating of egg	Molina-Gamboa *et al.* (1997)
Denmark, 1995	6	35	Nosocomial	nr	nr	nr	Hospital	nr (contaminated eggs implied)	Skibsted *et al.* (1998)

Germany, 1995	14b/n.c.	99 (58 symptomatic. Patients, nursery school children and staff)	Nosocomial	nr	Vanilla pudding	D	University hospital	Probable cross-contamination from turkey	Kistemann *et al.* (2000)
UK, 1995	5a	27	2	2	Several foods	S	Residential home for elderly people	Use of raw shell eggs to prepare mayonnaise used in prawn vol-au-vents	Hansell *et al.* (1998)
Ireland, 1996	nr	65	Nosocomial	0	Chocolate mousse cakes made with unpasteurized eggs	S, M	Psychiatric hospital	Use of raw eggs in a product that was not subsequently cooked. Inadequate cold storage space, desserts stored for long periods at room temperature. Food and healthcare staff working while having diarrhoea, because of staff shortages	Grein *et al.* (1997)
Belgium, 1997	nr	88 (63 patients, 6 staff, 12 visitors)	Nosocomial	nr	Raw ground beef and raw eggs	M	University hospital	Serving of raw beef and raw eggs	Sion *et al.* (2000)
Germany, 1998	nr	17 (15 symptomatic). Staff of surgical unit and anaesthetic department	0	0	Suspected Chinese meal	D	Staff dinner in hospital	Meal prepared at home by nurse on previous day and brought to ward for staff dinner	Metz *et al.* (2001)
Spain, 1998	nr	22	Nosocomial	7	Egg-containing food consumed by index patient	D, M	Tertiary care teaching hospital	Major reconstruction was occurring in the hospital kitchen, and adherence to correct hygiene measures could not be guaranteed. Evidence that eggs were a probable source of the salmonella	Guallar *et al.* (2004)

(*cont.*)

Table 2.9.2 *(continued)*

Place, date	Phage type	Total cases	Number hospitalized	Deaths	Food implicated	Evidence[a]	Where food was provided	Factors leading to outbreak	Reference
Brazil, 1999–2000	nr	8	Nosocomial	3	Commercial enteral feed	M	Hospital	Feed contained lyophilized egg albumen, which may have been the source of the salmonella	Matsuoka *et al.* (2004)
Netherlands, 2001	6	82	Nosocomial	5	Bavaroise	S	Hospital and nursing home served by same hospital kitchen	Raw eggs used to prepare bavaroise, probably underheated, no temperatures checked during preparation, temperature of refrigerator apparently too high	Bruins *et al.* (2003)
England, 2002	4	18 (12 residents, 6 staff)	nr	1	Egg dish	M	Nursing home	Eggs (non-Lion quality) infected with outbreak strain and undercooked	PHLS (2002)
England, 2002	6a	29	Nosocomial	nr	Imported eggs	M	Hospital A	Imported eggs infected with outbreak strain	PHLS (2003)
England, 2002	1	88	Nosocomial	1	nd	D	Hospital A	Suspected initial foodborne infection with subsequent cross-infection	PHLS (2003)
Slovakia, 2003 (serotype not specified)	nr	>279	53	nr	Cream cakes made with fresh eggs	D	Nursery schools	Food for all schools prepared in central kitchen. Raw eggs probably contaminated	ProMED-mail (2003)
Greece, 2005	nr	133	86 inpatients, 31 visitors, 16 employees	2	Probably roast chicken	M	University hospital	Cross-contamination via kitchen equipment, faults in food-handling practices.	Gikas *et al.* (2007)

[a]M (microbiological): identification of an organism of the same type from cases and in the suspect food or food ingredient, or detection of toxin in faeces or food; D (descriptive): other evidence reported by local investigators as indicating the suspect food; S (statistical): a significant statistical association between consumption of the suspect food and being a case.

nr, not reported.

Table 2.9.3 Examples of outbreaks of foodborne disease caused by *Salmonella enterica* serotype Typhimurium in healthcare settings and institutions.

Place, date	Phage type	Cases	Number hospitalized	Deaths	Food implicated	Evidence[a]	Where food was provided	Factors leading to outbreak	Reference
US, 1981	nr	22	3	0	Home-made ice cream	D	School party	Ice cream made with raw eggs, some from local farms	CDC (1981)
UK, 1984	49	>400	Nosocomial	19	Cold roast beef	M	Psychiatric hospital	Bad working practices in hospital kitchen. Cross-contamination of cooked beef from chicken, inadequate refrigerated storage	DHSS (1986); Pennington (2003)
UK, 1995	104	26	nr	nr	Unpasteurized milk	M	Nursery school	Outbreak strain present in bulk milk tank and in farm dairy herd	Djuretic *et al.* (1997)
Australia, 1996	Non-serotypable (typed by PFGE)	52 (17 patients, 5 spouses, 3 babies, 27 staff)	Nosocomial	1	Sandwiches	S	Teaching hospital	Possible cross-contamination and inadequate storage. Outbreak strain isolated from assorted sandwiches	McCall *et al.* (1999)
France, 1999–2000	104	35	29 of patients were already in hospitals or care homes	3	Hamburgers from a single producer	S and M	Hospitals, retirement homes, school canteens	Hamburgers contaminated with the outbreak strain and probably undercooked	Haeghebaert *et al.* (2003)
Australia, 2001	135	18 (16 residents, 2 staff)	3	0	Rice pudding, meat-based potato pie	S and M	Aged care facility	Raw shell eggs whisked into rice pudding immediately before serving; and incorporated into potato topping of pie before lightly browning	Tribe *et al.* (2002)
Australia, 2002	135a	16	0	0	Egg sandwiches suspected	D	Child care centre	Lack of quality assurance for eggs	McCall *et al.* (2003)
Finland, 2005	104B	>56	nr	nr	Iceberg lettuce imported from Spain	M[b]	Cafeteria in Nursing school, restaurant	nr	Takkinen *et al.* (2005)

[a]M (microbiological): identification of an organism of the same type from cases and in the suspect food or food ingredient, or detection of toxin in faeces or food; D (descriptive): other evidence reported by local investigators as indicating the suspect food; S (statistical): a significant statistical association between consumption of the suspect food and being a case.
[b]The kitchen of the cafeteria kept frozen food samples from dishes served each day, thus the lettuce could be tested for salmonella.
nr, not reported.

Table 2.9.4 Examples of outbreaks of foodborne disease caused by other *Salmonella enterica* serotypes in healthcare settings and institutions.

Place, date	*S. enterica* serotype	Cases	Number hospitalized	Deaths	Food implicated	Evidence[a]	Where food was provided	Factors leading to outbreak	Reference
UK, 1992	Livingstone	21	Nosocomial	0	Cheese sandwiches	D	Psychogeriatric hospital	Cheese sandwiches stored on unwashed metal trays at 7°C, then left on trolleys at >30°C for 6 h before distribution	Djuretic *et al.* (1997)
US, 1993	Heidelberg (and *C. jejuni*)	>93 (one or both bacteria)	31	1	Pureed diet including chopped liver salad	S	Nursing home in a medical centre complex	Lapses in food handling. Ground, cooked livers were placed in a bowl containing raw chicken liver juices. Liver salad stored with inadequate refrigeration	Layton *et al.* (1997)
US, 1993–94	Senftenberg	>22	Nosocomial, 18 patients, 4 employees	0	Deli-turkey, lettuce, cauliflower, cottage cheese	S	Hospital	Possible cross-contamination from turkey; deficiencies in kitchen hygiene	L'Ecuyer *et al.* (1996)
UK, 1994	Virchow	10 (1 patient, 9 staff)	Nosocomial	0	Turkey sandwiches	S	Private hospital	Probable contamination by infected food handler. Deficiencies found in kitchen practices; temperature in sandwich refrigerator 15.3°C	Maguire *et al.* (2000)
France, 1996	Heidelberg and Manhatten	72	Nosocomial, patients and staff affected	0	Bacon	S and M	Hospital	Attributed to insufficient preservation of bacon	Simon *et al.* (1998)
US, 1998–99	Baildon	86	16	3	Raw tomatoes	S	Restaurants, outlets of fast-food restaurant, two nursing homes	Probable contamination on farm or during packing. After harvest, fruit is submerged in water; this may allow bacteria to be drawn into the fruit. Chlorine added to water but concn. Not checked	Cummings *et al.* (2001)

France, 1999	Paratyphi B	8	Nosocomial	0	Hamburgers supplied by outside producer	M	Hospital	Outbreak strain identified in samples of frozen-pre-cooked hamburgers	Haeghebaert *et al.* (2001)
Wales, 2000	Indiana	17 (staff, relatives, patients)	Nosocomial	0	Egg mayonnaise sandwiches	S	Hospital	Probable pasteurization failure of egg roll used to make sandwiches	Mason *et al.* (2001)
Japan, 2001	Brandenberg and Corvallis	7	Nosocomial	0	4 lunch items	M	Hospital	nr	Hamada & Tsuji (2001)
US, 2001	Saintpaul	11	Nosocomial	0	Enteral feeding formula	S	University-affiliated children's hospital	Possible introduction of salmonella during preparation in hospital formula preparation room	Bornemann *et al.* (2002)
US, 2002	Javiana	>141	3	0	Diced tomatoes	S	Theme park used for US Transplant Games	Many cases were transplant recipients and were receiving immunosuppressive therapy. Tomatoes were diced and washed in water with variable chorine level	CDC (2002b); Srikantiah *et al.* (2005)
France, 2005	Worthington	45	Nosocomial	0	Powdered milk, added to food for elderly patients	M	3 hospitals	Reason for contamination not reported	InVS Outbreak investigation Group (2005)

[a] M (microbiological): identification of an organism of the same type from cases and in the suspect food or food ingredient, or detection of toxin in faeces or food; D (descriptive): other evidence reported by local investigators as indicating the suspect food; S (statistical): a significant statistical association between consumption of the suspect food and being a case.
nr, not reported

eggs is illustrated in Table 2.9.2. Pooling raw shell eggs increases the risk of contamination with salmonellas. Probable cross-contamination from turkey was implicated in the outbreak caused by *S*. Enteritidis in Germany in 1995.

Eggs and poultry were also implicated in several of the outbreaks caused by *S*. Typhimurium shown in Table 2.9.3 including the large outbreak in the UK in 1984, linked to probable cross-contamination. Use of unpasteurized milk from an infected dairy herd caused illness in children at a nursery school in the UK in 1995.

Transmission on raw vegetables is illustrated by the outbreak of infection with *S*. Typhimurium in Finland associated with imported lettuce (Table 2.9.3) and outbreaks in the US caused by *S*. Baildon and *S*. Javiana associated with raw tomatoes (Table 2.9.4).

Food handlers have been implicated in some outbreaks, for example an outbreak in the UK in 1993 (Table 2.9.2) that affected 22 people in two hospitals was probably caused by an asymptomatic food handler. Some individuals may excrete salmonellas for weeks, months or longer, with little or no evidence of illness. Unless good hygienic practices are observed these individuals may contaminate food during preparation.

Salmonella Dublin, which is particularly invasive in humans, was the predominant serotype isolated from cattle in the UK in 2005 and the previous 6 years (VLA, 2006). This serotype is seldom associated with human infection in the UK, but in 1989, a community outbreak of infection with *S*. Dublin affected 42 people in south east England, in which the organism was transmitted in an imported, Irish soft cheese made with unpasteurized cow's milk; three cases were admitted to hospital (Maguire *et al.*, 1992). In November–December 1995, an outbreak of *S*. Dublin infection in France affected 25 people of whom 12 were admitted to hospital and five, who had coexisting chronic illness, died (Vaillant *et al.*, 1996). The organism was transmitted in a raw milk cheese, produced in the Doubs region of France; the same type of cheese had been implicated in an outbreak in Switzerland in November 1995.

Raw vegetable sprouts have become increasingly popular and regarded as a healthy food, but community outbreaks of salmonellosis in many countries have been associated with consumption of raw vegetable sprouts (Taormina *et al.*, 1999; CDC, 2002a; Duynhoven *et al.*, 2002). An outbreak caused by *S*. Havana in the US in 1998, that affected 18 people, resulted in a high incidence of extraintestinal infections, hospitalization of four patients and one death (Backer *et al.*, 2000). The problem of contamination of the seeds of vegetable sprouts with salmonellas and other pathogens has been recognized widely.

Outbreaks of salmonellosis in infants, associated with contaminated, powdered infant formula are shown in Table 2.9.5.

2.9.6 *Main methods of prevention and control of foodborne infection by salmonellas*

Prevention of typhoid and paratyphoid fevers in developing countries depends on provision of safe water supplies and proper sanitation systems; in the interim a wider use of vaccination programmes may be necessary (Parry *et al.*, 2002). Health education regarding hand-washing after toilet use and before food preparation, and exclusion of disease carriers from food handling are also essential.

Table 2.9.5 Examples of outbreaks of *Salmonella* spp. associated with powdered infant formula.

Place, date	*S. enterica* serotype	Cases	Number hospitalized	Deaths	Food implicated	Evidence[a]	Where food was provided	Factors leading to outbreak	Reference
UK, 1985-6	Ealing	46 infants and 16 non-infants with symptoms	7 infants	1	Infant dried milk	M	Community	Holes and cracks in spray-drier. *S.* Ealing isolated from insulation material and powder under a hole	Rowe *et al.* (1987)
Canada, US, 1993	Tennessee	3	nr	nr	Powdered infant formula	M	Community	*S.* Tennessee isolated from production equipment at factory	CDC (1993)
Spain, 1994	Virchow	>48	nr	nr	Powdered infant formula	M	Hospital and community	nr	Usera *et al.* (1996, 1998)
UK, France 1996–97	Anatum	17	4	0	Powdered infant formula	M	Community	nr	Anon. (1997); Threlfall *et al.* (1998)
France, 2005	Agona	21	nr	nr	Powdered infant formula	D	Community	nr	Espie *et al.* (2005)

[a]M (microbiological): identification of an organism of the same type from cases and in the suspect food or food ingredient, or detection of toxin in faeces or food; D (descriptive): other evidence reported by local investigators as indicating the suspect food; S (statistical): a significant statistical association between consumption of the suspect food and being a case.
nr, not reported.

To prevent foodborne infection with non-typhoid salmonellas, in many countries action is being taken at critical points in the food chain, from the farm to the consumer. Following an outbreak of salmonellosis in Germany attributed to contaminated paprika powder added to potato chips, the investigators concluded that snacks contaminated with between 0.04 and 0.45 salmonellae/g caused a nationwide outbreak with an estimated 1000 cases (Lehmacher *et al.*, 1995). This outbreak emphasizes the fact that bacteriological inspection of end-product samples is likely not to be sufficient to ensure bacteriological safety of food, and that a HACCP system must consider the whole of production including the production of raw materials in their country of origin.

Controls are necessary on the farm to minimize the carriage of salmonellas by food animals. To reduce the risk of transmission in animal feed, tests and controls for salmonella are used at several points in the production process in the UK (VLA, 2006). Statutory monitoring takes place for the presence of salmonella in processed animal protein destined for livestock feed and Codes of Practice are available for the control of salmonella in production of final feed for livestock and during storage, handling and transport (VLA, 2006). The manufacturing process for feed is expected to reduce salmonellas but recontamination may occur after processing; a combination of heat treatment and addition of organic acid is used by primary poultry breeders to ensure as far as possible a salmonella-free feed. Any standards for salmonella-free animal feed need to be implemented and enforced stringently (Crump *et al.*, 2002).

The minimization of *Salmonella* in food animals requires the use of salmonella-free stock and animal feed, biosecurity to protect feed and water from contamination by wild animals and birds, hygienic disposal of waste, and design of intensive rearing areas to enable effective cleaning between uses and to exclude rodents. Outside areas should be protected from contamination, for example fresh slurry should not be applied to land where it could contaminate drinking water supplied to animals or grazing land, and the siting of refuse tips close to farms should be avoided.

Controls during transportation of animals, and hygienic practices in abattoirs and during further processing, are designed to minimize contamination with salmonellas and other zoonotic organisms.

In spite of measures to reduce the carriage of salmonellas in live animals, and to reduce contamination in the abattoir and during further processing, it is inevitable that some raw meats will be contaminated with salmonellas. It is essential, therefore, that persons who prepare food should follow good hygienic practices to prevent cross-contamination, should store food at low temperature to prevent growth of any viable salmonellas and should cook food thoroughly.

In the UK, statutory regulations require compulsory testing of breeder, layer, and broiler flocks that supply chicks to the poultry industry (Cogan & Humphrey, 2003). Vaccination of poultry against *S.* Enteritidis was started in the UK in 1996 and is now used widely. For years previously the poultry industry has used vaccines to control major infectious diseases of poultry in both breeding birds and commercial layers. The British Egg Industry Council has produced a Lion Code of Practice that sets standards for biosecurity and bacteriological testing, and states that birds destined for Lion Quality egg-producing flocks must be vaccinated against *S.* Enteritidis (British Egg Industry Council, 2005). More than 80% of eggs on sale (those with the Lion mark) in the UK were reported to originate from such flocks (Cogan & Humphrey, 2003). A survey of UK-produced eggs on retail sale in 2003 showed that only 0.34% were contaminated with *Salmonella*; 78% (3) of isolates were *S.* Enteritidis and none were known vaccine strains (FSA, 2004).

In countries of the EU and Norway, the incidence of salmonellas in flocks of laying hens ranged from 0 to 79% (Editorial team, 2006). From August 2006, EU legislation sets targets for the reduction of *S.* Enteritidis and *S.* Typhimurium in laying hens, and from January 2008 vaccination of laying hens will be required in countries where more than 10% of commercial egg-laying hens carry these salmonellas (Editorial team, 2006). The vaccination must be distinguishable from infection. Europe-wide salmonella reduction targets for other animals are envisaged.

In the US, stringent measures have been taken to eliminate *S.* Enteritidis from flocks of laying poultry (Braden, 2006). The USDA tests breeder flocks that produce egg-laying poultry to check that they are free from *S.* Enteritidis. If commercial flocks are infected with the bacterium the eggs are diverted for processing as pasteurized liquid egg or powdered egg, to kill the salmonellas (CDC, 2005). The FDA proposes action to reduce further the contamination of eggs with *S.* Enteritidis (FDA, 2004). The action applies to producers with 3000 or more laying hens that produce eggs for retail sale and do not process their eggs with a treatment such as pasteurization. The measures include: provisions for procurement of chicks and pullets, a biosecurity programme, a pest and rodent control programme, cleaning and disinfection of poultry houses that have had an environmental sample or egg test positive for *S.* Enteritidis, refrigerated storage of eggs on the farm, producer testing of the environment for *S.* Enteritidis in poultry houses, and the identification of a person responsible for prevention of contamination at each farm. Some US producers use in-shell pasteurization of table eggs, a process in which the eggs are heated for a prolonged period at a temperature that reduces microbial contamination but preserves the liquid nature of the egg contents (Braden, 2006).

Despite the controls by the egg industry to prevent contamination with salmonellas, contamination with salmonellas continues to be a problem, particularly with eggs that are not Lion Quality and with some imported eggs (Mitchell *et al.*, 2002; O'Brien *et al.*, 2004).

To prevent contamination of vegetables, particularly salad vegetables, and fruits, avoidance of contamination in the field is of major importance; guidelines to minimize contamination have been produced (FDA/CFSAN, 2004). In the case of vegetable sprouts, a Code of Practice produced in the UK (Brown and Oscroft, 1989) and guidance in the US (FDA, 1999a, b;) emphasize the need to disinfect seed by carefully controlled treatment with hypochlorite. Other methods of disinfection are being investigated.

It is essential that caterers, and consumers must contribute to the prevention of salmonellosis in humans. In 1998, the UK Department of Health repeated the Chief Medical Officer's advice of 1988 to the public to avoid eating raw eggs or uncooked foods made from them, such as home-made mayonnaise, mousse or ice cream (DH, 1998). The advice included the statement that: 'Although the risk of harm to any healthy individual is small, it is advisable for vulnerable people such as the elderly, the sick, babies and pregnant women to consume only eggs which have been cooked until the white and the yolk are solid'.

In 1988, the Chief Medical Officer in the UK advised that in institutions with high-risk groups raw shell eggs should be replaced with pasteurized egg in recipes for products that would not be cooked or would only be lightly cooked. Hospitals were reminded of this advice by O'Brien and Ward (2002).

In the US, the FDA advice about eggs includes the following instructions (FDA/CFSAN, 2002):

- 'To prevent illness from bacteria: keep eggs refrigerated, cook eggs until yolks are firm, and cook foods containing eggs thoroughly'.
- *Establishments that serve highly susceptible populations (young children, elderly persons and individuals with weakened immune systems), such as day-care centres, nursing homes and hospitals, should take the following additional precautions:*
  - *In NO case should soft-cooked eggs, soufflés or meringues or other foods that contain raw or undercooked eggs be served in these facilities.*
  - As a general rule, treated eggs or pasteurized egg products should be used in any recipe that calls for combining more than one egg ('pooling') and for any recipe, preparation or serving procedure that involves holding eggs or egg-containing foods before or after cooking.
  - To add an extra margin of safety, in addition to all of the above precautions, buyers can specify that suppliers provide eggs produced only from flocks managed under a *Salmonella* Enteritidis programme that is recognized by a state regulatory agency or a state poultry association.

Powdered infant formula is not a sterile product. Measures to control the safety of this product during production, reconstitution and feeding are outlined in Section 2.6.6. Under the present EU regulations (EC, 2005) powdered infant formula must show absence of *Enterobacteriaceae* in 10 g; if *Enterobacteriaceae* are detected in any sample units, the batch has to be tested for *E. sakazakii* (see Section 2.6.6) and for *Salmonella*, which must be absent in 30 samples, each of 25 g, from a batch.

Summary of measures advised in healthcare settings

1. Meat, poultry, eggs, seafood, salad vegetables and berry fruits should be obtained from approved sources.
2. Eggs should be cooked until the white and the yolk are solid; in recipes using raw shell eggs these should be replaced with pasteurized egg for any products that will not be cooked, or will only be lightly cooked.
3. Unpasteurized or inadequately pasteurized milk or milk products should not be served.
4. Raw seed sprouts should not be served to immunosuppressed persons.
5. Raw salad vegetables should be obtained from approved sources and should be well washed; they should not be served to highly immunosuppressed persons.
6. Unpasteurized fruit juices should not be served.
7. Catering personnel should not handle food while suffering from symptoms of infection, or handle cooked food or ready-to-eat food with bare hands (DH, 1995, 1996). Good handwashing is essential.
8. Guidelines for the use of powdered infant formulas should be followed (EFSA, 2004).

References

Advisory Committee on the Microbiological Safety of Food (ACMSF) (2001) *Second Report on Salmonella in Eggs.* The Stationery Office, London.

Anon. (1997) Preliminary report of an international outbreak of *Salmonella anatum* infection linked to infant formula dried milk. *Eurosurveillance Monthly* **2**(3), 22–24.

Backer, H.D., Mohle-Boetani, J.C., Werner, S.B. *et al.* (2000) High incidence of extra-intestinal infections in a *Salmonella* Havana outbreak associated with alfalfa sprouts. *Public Health Reports* **115**, 339–345.

Bornemann, R., Zerr, D.M., Heath, J. *et al.* (2002) An outbreak of *Salmonella* serotype Saintpaul in a children's hospital. *Infection Control and Hospital Epidemiology* **23**, 671–676.

Braden, C.R. (2006) *Salmonella enterica* serotype Enteritidis and eggs: a national epidemic in the United States. *Clinical Infectious Diseases* **43**, 512–517.

Brenner, F.W., Villar, R.G., Angulo, F.J. *et al.* (2000) *Salmonella* nomenclature. *Journal of Clinical Microbiology* **38**, 2465–2467.

British Egg Industry Council (2005) *Summary of Lion Quality Code of Practice.* Available from http://www.britegg.co.uk/lionquality05/summarylionqual.pdf. Accessed 1 March 2007.

Brown, K.L. & Oscroft, C.A. (eds) (1989) *Guidelines for the Hygienic Manufacture, Distribution and Retail Sale of Sprouted Seeds, with Particular Reference to Mung Beans.* Technical Manual No 25. Campden Food and Drink Research Association.

Bruins, M.J., Fernandes, T.M.A., Ruijs, G.J.H.M. *et al.* (2003) Detection of a nosocomial outbreak of salmonellosis may be delayed by application of a protocol for rejection of stool cultures. *Journal of Hospital Infection* **54**, 93–98.

Centers for Disease Control and Prevention (CDC) (1981) Salmonellosis from homemade ice cream – Georgia. *Morbidity and Mortality Weekly Report* **30**, 467–468.

Centers for Disease Control and Prevention (CDC) (1993) *Salmonella* serotype Tennessee in powdered milk products and infant formula – Canada and United States, 1993. *Morbidity and Mortality Weekly Report* **42**(26), 516–517.

Centers for Disease Control and Prevention (CDC) (2002a) Outbreak of *Salmonella* serotype Kottbus infections associated with eating alfalfa sprouts – Arizona, California, Colorado, and New Mexico, February–April, 2001. *Morbidity and Mortality Weekly Report* **51**(01), 7–9.

Centers for Disease Control and Prevention (CDC) (2002b) Outbreak of *Salmonella* serotype Javiana infections – Orlando, Florida, June 2002. *Morbidity and Mortality Weekly Report* **51**(31), 683–684.

Centers for Disease Control and Prevention (CDC) (2004) Preliminary FoodNet data on the incidence of infection with pathogens transmitted commonly through food – selected sites, United States, 2003. *Morbidity and Mortality Weekly Report* **53**(16), 338–343.

Centers for Disease Control and Prevention (CDC) (2005) *Disease Information. Salmonella enteritidis.* Available from http://www.cdc.gov/ncidod/dbmd/diseaseinfo/salment_g.htm. Accessed 7 March 2007.

Cogan, T.A. & Humphrey, T.J. (2003) The rise and fall of *Salmonella* Enteritidis in the UK. *Journal of Applied Microbiology* **94**, 114S–119S.

Communicable Disease Surveillance Centre (CDSC) (1994) An outbreak of *Salmonella paratyphi* B infection in France. *Communicable Disease Report CDR Weekly* **4**(35). Available from http://www.hpa.org.uk/cdr/archives/1994/cdr3594.pdf. Accessed 1 March 2007.

Crump, J.A., Griffin, P.M. & Angulo, F.J. (2002) Bacterial contamination of animal feed and its relationship to human foodborne illness. *Clinical Infectious Diseases* **35**, 859–865.

Crump, J.A., Luby, S.P. & Mintz, E.D. (2004) The global burden of typhoid fever. *Bulletin of the World Health Organization* **82**, 346–353.

Cummings, K., Barrett, E., Mohle-Boetani, J.C. *et al.* (2001) A multistate outbreak of *Salmonella enterica* serotype Baildon associated with domestic raw tomatoes. *Emerging Infectious Diseases* **7**, 1046–1048.

D'Aoust, J.-Y. (2000) *Salmonella.* In: Lund, B.M., Baird-Parker, T.C. & Gould, G.W. (eds). *The Microbiological Safety and Quality of Food*, Vol. 2. Aspen Publishers, Gaithersburg, MD, pp. 1233–1299.

Department of Health (DH) (UK) (1995) *Food Handlers. Fitness to Work. Guidelines for Food Businesses, Enforcement Officers and Health Professionals*. Department of Health, London. Available from the Food Standards Agency, UK.

Department of Health (DH) (UK) (1996) *Food Handlers. Fitness to Work. Guidelines for Food Business Managers.* Department of Health, London. Available from the Food Standards Agency, UK.

Department of Health (DH) (UK) (1998) Expert advice repeated on *Salmonella* and raw egg. *News Release* 98/138.

Department of Health and Social Security (DHSS) (1986) *Report of the Committee of Inquiry into an Outbreak of Food Poisoning at Stanley Royd Hospital.* HMSO, London.

Desenclos, J.-C., Bouvet, P., Benz-Lemoine, E. *et al.* (1996) Large outbreak of *Salmonella enterica* serotype paratyphi B infection caused by a goats' milk cheese, France, 1993: a case finding and epidemiological study. *British Medical Journal* **312**, 91–94.

Djuretic, T., Wall, P.G. & Nichols, G. (1997) General outbreaks of infectious intestinal disease associated with milk and dairy products in England and Wales: 1992 to 1996. *Communicable Disease Report* **Review 7**(Review number 3), R41–R45.

Dryden, M.S., Keyworth, N., Gabb, R. & Stein, K. (1994) Asymptomatic food handlers as the source of nosocomial salmonellosis. *Journal of Hospital Infection* **28**, 195–208.

Duynhoven, T.H.P. van, Widdowson, M.-A., Jager, C.M. de *et al.* (2002) *Salmonella* enterica serotype Enteritidis phage type 4b outbreak associated with bean sprouts. *Emerging Infectious Diseases* **8**, 440–443.

Dworkin, M.S., Shoemaker, P.C., Goldoft, M.J. & Kobayashi, J.M. (2001) Reactive arthritis and Reiter's syndrome following an outbreak of gastroenteritis caused by *Salmonella enteritidis*. *Clinical Infectious Diseases* **33**, 1010–1014.

EC (2005) Commission Regulation (EC) No. 2073/2005 of 15 November 2005 on microbiological criteria for foodstuffs. *Official Journal of the European Union* **L338**, 22 December 2005. Available from http://www.food.gov.uk/multimedia/pdfs/microbiolcriteria.pdf. Accessed 7 March 2007.

Editorial team (2006) Reducing salmonella in European egg-laying hens: EU targets now set. *Euro Surveillance* **11**(8), E060810.3. Available from http://www.eurosurveillance.org/ew/2006/060810.asp#3. Accessed 7 March 2007.

Espié, E., Weill, F.-X., Brouard, C. *et al.* (2005) Nationwide outbreak of *Salmonella enterica* serotype Agona infections in infants in France, linked to infant milk formula, investigations ongoing. *Euro Surveillance* **10**(3), E050310.1. Available from http://www.eurosurveillance.org/ew/2005/050310.asp#1. Accessed 7 March 2007.

European Food Safety Authority (EFSA) (2004) Microbiological risks in infant formula and follow-on formula. *The EFSA Journal* **113**, 1–35.

Evans, M.R., Hutchings, P.G., Ribeiro, C.D. & Westmorland, D. (1996) A hospital outbreak of salmonella food poisoning due to inadequate deep-fat frying. *Epidemiology and Infection* **116**, 155–160.

Food and Drug Administration (FDA) (1999a) *Recommendations on Sprouted Seeds. Reducing Microbial Food Safety Hazards for Sprouted Seeds and Guidance for Industry*. P 57893-57902. Office of the Federal Register, U.S. Government Printing Office, Washington, DC.

Food and Drug Administration (FDA) (1999b) (updated 2005) Guidance for industry. Reducing microbial food safety hazards for sprouted seeds. Available from http://www.cfsan.fda.gov/~dms/sprougd1.html. Accessed 27 June 2007.

Food and Drug Administration (FDA) (2004) FDA proposes further action to improve farm-to-table shell egg safety. *FDA News* 20 September 2004.

Food and Drug Administration/Center for Food Safety and Applied Nutrition (FDA/CFSAN) (2002) Assuring the safety of eggs and menu and deli items made from raw, shell eggs. Available from http://www.cfsan.fda.gov/~dms/fs-eggs2.html. Accessed 7 March 2007.

Food and Drug Administration/Center for Food Safety and Applied Nutrition (FDA/CFSAN) (2003) *Salmonella* spp. In: *Foodborne Pathogenic Microorganisms and Natural Toxins Hand Book*. Available from http://www.cfsan.fda.gov/~mow/chap1.html. Accessed 7 March 2007.

Food and Drug Administration/Center for Food Safety and Applied Nutrition (FDA/CFSAN) (2004) *Produce Safety from Production to Consumption: 2004 Action Plan to Minimize Foodborne Illness Associated with Fresh Produce Consumption*. Available from http://www.foodsafety.gov/~dms/prodpla2.html. Accessed 7 March 2007.

Food Standards Agency (FSA) (2003) *UK-Wide Survey of Salmonella and Campylobacter Contamination of Fresh and Frozen Chicken on Retail Sale*. Available from http://www.food.gov.uk/multimedia/webpage/111802. Accessed 7 March 2007.

Food Standards Agency (FSA) (2004) *Report of the Survey of Salmonella Contamination of UK Produced shell Eggs on Retail Sale*. Available from http://www.food.gov.uk/multimedia/pdfs/fsis5004report.pdf. Accessed 7 March 2007.

Gikas, A., Kritsotakis, E.I., Maraki, S. *et al.* (2007) A nosocomial outbreak of *Salmonella Enterica* serovar Enteritidis in a University hospital in Greece: the importance of establishing HACCP systems in hospital catering. *Journal of Hospital Infection* **62**, 194–196.

Grein, T., O'Flanagan, D., McCarthy, T. & Prendergast, T. (1997) An outbreak of *Salmonella enteritidis* food poisoning in a psychiatric hospital in Dublin, Ireland. *Eurosurveillance Monthly* **2**(11), 84–86.

Guallar, C., Ariza, J., Dominguez, M.A. *et al.* (2004) An insidious nosocomial outbreak due to *Salmonella enteritidis*. *Infection Control and Hospital Epidemiology* **25**, 10–15.

Haeghebaert, S., Duche, L. & Desenclos, J.C. (2003) The use of case-crossover design in a continuous common source food-borne outbreak. *Epidemiology and Infection* **131**, 809–813.

Haeghebaert, S., Duché, L., Gilles, C. *et al.* (2001) Minced beef and human salmonellosis: review of the investigation of three outbreaks in France. *Eurosurveillance Monthly* **6**(2), 21–26.

Hamada, K. & Tsuji, T. (2001) *Salmonella* Brandenberg and *S*. Corvallis involved in a food poisoning outbreak in a hospital in Hygo prefecture. *Japanese Journal of Infectious Diseases* **54**, 195–196.

Hansell, A.L., Sen, S., Sufi, F. & McCallum, A. (1998) An outbreak of *Salmonella enteritidis* phage type 5a in a residential home for elderly people. *Communicable Disease and Public Health* **1**, 172–175.

Health Protection Agency (HPA) (2004) Salmonella serotypes recorded in the Health Protection Agency salmonella data set October to December 2003. *Communicable Disease Report CDR Weekly* **14**(11), enteric. Available from http://www.hpa.org.uk/cdr/archives/2004/cdr1104.pdf. Accessed 7 March 2007.

Health Protection Agency (HPA) (2007) *Infections. Topics. Salmonella. Epidemiological Data*. Available from http://www.hpa.org.uk/infections/topics_az/salmonella/data_human_gr.htm. Accessed 7 March 2007.

Helms, M., Ethelberg, S., Mølbak, K. & the DT104 Study Group (2005) International *Salmonella* Typhimurium DT104 infections, 1992–2001. *Emerging Infectious Diseases* **11**, 859–867.

Holtby, I. & Stenson, P. (1992) Food poisoning in two homes for the elderly. *Communicable Disease Report* **2**(Review no. 11), R125.

Horby, P.W., O'Brien, S.J., Adak, G.K. *et al.* (2003) A national outbreak of multi-resistant *Salmonella enterica* serovar Typhimurium definitive phage type (DT) 104 associated with consumption of lettuce. *Epidemiology and Infection* **130**, 169–178.

International Commission on Microbiological Specifications for Foods (ICMSF) (1996) *Microbiological Specifications of Food Pathogens*. Blackie Academic and Professional, London, pp. 218.

International Commission on Microbiological Specifications for Foods (ICMSF) (1998) *Microbial Ecology of Food Commodities*. Blackie Academic and Professional, London, pp. 497.

InVS, Outbreak Investigation Group (2005) Outbreak of *Salmonella* Worthington infection in elderly people due to contaminated milk powder, France, January–July 2005. *Euro Surveillance* **10**(7), E050721.3. Available from http://www.eurosurveillance.org/ew/2005/050721.asp#3. Accessed 7 March 2007.

Kistemann, T., Dangendorf, F., Krizek, L. *et al.* (2000) GIS-supported investigation of a nosocomial *Salmonella* outbreak. *International Journal of Hygiene and Environmental Health* **203**, 117–126.

Layton, M.C., Calliste, S.G., Gomez, T.M. *et al.* (1997) A mixed foodborne outbreak with *Salmonella heidelberg* and *Campylobacter jejuni* in a nursing home. *Infection Control and Hospital Epid*emiology **18**, 115–121.

L'Ecuyer, P.B., Diego, J., Murphy, D. *et al.* (1996) Nosocomial outbreak of gastroenteritis due to *Salmonella senftenberg*. *Clinical Infectious Diseases* **23**, 734–742.

Lehmacher, A., Bockemuhl, J. & Aleksic, S. (1995) Nationwide outbreak of human salmonellosis in Germany due to contaminated paprika and paprika-powdered potato chips. *Epidemiology and Infection* **115**, 501–511.

Locht, H., Molbak, K. & Krogfelt, K.A. (2002) High frequency of reactive joint symptoms after an outbreak of *Salmonella enteritidis*. *Journal of Rheumatology* **29**, 767–771.

Maguire, H., Cowden, J., Jacob, M. *et al.* (1992) An outbreak of *Salmonella* Dublin infection in England and Wales associated with soft unpasteurised cows' milk cheese. *Epidemiology and Infection* **109**, 389–396.

Maguire, H., Pharoah, P., Walsh, B. *et al.* (2000) Hospital outbreak of *Salmonella virchow* possibly associated with a food handler. *Journal of Hospital Infection* **44**, 261–266.

Mandal, B.K. & Brennand, J. (1988) Bacteraemia in salmonellosis: a 15-year retrospective study from a regional infectious diseases unit. *British Medical Journal* **297**, 1242–1243.

Mason, B.W., Williams, N., Salmon, R.L. *et al.* (2001) Outbreak of *Salmonella indiana* associated with egg mayonnaise sandwiches at an acute NHS hospital. *Communicable Disease and Public Health* **4**, 300–304.

Matsuoka, D.M., Costa, S.F., Mangini, C. *et al.* (2004) A nosocomial outbreak of *Salmonella enteritidis* associated with lyophilized enteral nutrition. *Journal of Hospital Infection* **58**, 122–127.

McCall, B.J., Bell, R.J., Neill, A.S. *et al.* (2003) An outbreak of *Salmonella* Typhimurium phage type 135a in a child care centre. *Communicable Diseases Intelligence* **27**, 257–259.

McCall, B.J., McCormack, J.G., Stafford, R. & Towner, C. (1999) An outbreak of *Salmonella typhimurium* at a teaching hospital. *Infection Control and Hospital Epidemiology* **20**, 55–56.

Metz, R., Jahn, B., Kohnen, W. *et al.* (2001) Outbreak of *Salmonella* Enteritidis gastrointestinal infections among medical staff due to contaminated food prepared outside the hospital. *Journal of Hospital Infection* **48**, 324–325.

Mitchell, R., Little, C., Ward, L. & Surman, S. (2002) Public health investigation of *Salmonella* Enteritidis in raw shell eggs in England and Wales. *Eurosurveillance Weekly* **6**(50), 021212. Available from http://www.eurosurveillance.org/ew/2002/021212.asp#3. Accessed 7 March 2007.

Molina-Gamboa, J.D., Ponce-de-Leon-Rosales, S., Guerrero-Almeida, M.L. *et al.* (1997) *Salmonella* gastroenteritis outbreak among workers from a tertiary care hospital in Mexico City. *Revista de Investigacion Clinica* **49**, 349–353.

O'Brien, S., Gillespie, I., Charlett, A. *et al.* (2004) National case-control study of *Salmonella* enteritidis phage type 14b infection in England and Wales implicates eggs used in the catering trade. *Eurosurveillance Weekly* **8**(8), 040219. Available from http://www.eurosurveillance.org/ew/2004/040219.asp#1. Accessed 7 March 2007.

O'Brien, S. & Ward, L. (2002) Nosocomial outbreak of *Salmonella* Enteritidis PT6a (Nx, CpL) in the United Kingdom. *Eurosurveillance Weekly* **6**(43), 021024. Available from http://www.eurosurveillance.org/ew/2002/021024.asp#20. Accessed 7 March 2007.

Old, D.C. & Threlfall, E.J. (1998) *Salmonella*. In: Collier, L., Balows, A. & Sussman, M. (eds). *Topley and Wilson's Microbiology and Microbial Infections*, 9th edn, Vol. 2: Systematic Bacteriology (volume editors Balows, A. & Duerden, B.I.), Arnold, London. pp. 969–997.

Parry, C.M., Hien, T.T., Dougan, G. *et al.* (2002) Typhoid fever. *New England Journal of Medicine* **347**, 1770–1782.

Pennington, H. (2003) *When Food Kills*. Oxford University Press, Oxford, UK, pp. 45–55.

ProMED-mail (2003) *Salmonellosis, Foodborne – Serbia (Belgrade).* Archive number 20031216.3071. Available from http://www.promedmail.org. Accessed 7 March 2007.

Popoff, M.Y., Bockemuhl, J. & Gheesling, L.L. (2003) Supplement 2001 (no 45) to the Kauffmann-White scheme. *Research in Microbiology* **154**, 173–174.

Public Health Laboratory Service (PHLS) (2002) Outbreak of *Salmonella* Enteritidis in a nursing home: risk from eggs must not be forgotten. *Communicable Disease Report CDR Weekly* **12**(14). Available from http://www.hpa.org.uk/cdr/archives/2002/cdr1402.pdf. Accessed 7 March 2007.

Public Health Laboratory Service (PHLS) (2003) *Salmonella* Enteritidis outbreak in a London hospital – update. *Communicable Disease Report CDR Weekly* **13**(8). Available from http://www.hpa.org.uk/cdr/archives/2003/cdr0803.pdf. Accessed 7 March 2007.

Rowe, B., Begg, N.T., Hutchinson, D.N. *et al.* (1987) *Salmonella* Ealing infections associated with consumption of infant dried milk. *Lancet* **II**, 900–903.

Simon, L., Rabaud, C., Abballe, X. & Hartmann, P. (1998) A hospital intestinal salmonellosis outbreak due to *Salmonella enterica* ser. Heidelberg and ser. Manhattan, at the Nancy University Hospital. *Medecine et Maladies Infectieuses* **28**, 221–223.

Sion, C., Garrino, M.-G., Glupczynski, Y. *et al.* (2000) Nosocomial outbreak of *Salmonella enteritidis* in a University Hospital. *Infection Control and Hospital Epidemiology* **21**, 182–183.

Skibsted, U., Baggesen, D.L., Dessau, R. & Lisby, G. (1998) Random amplification of polymorphic DNA (RAPD), pulsed-field gel electrophoresis (PGFE) and phage typing in the analysis of a hospital outbreak of *Salmonella* Enteritidis. *Journal of Hospital Infection* **38**, 207–216.

Srikantiah, P., Bodager, D., Toth, B. *et al.* (2005) Web-based investigation of multistate salmonellosis outbreak. *Emerging Infectious Diseases* **11**, 610–612.

Takkinen, J., Nakari, U.M., Johannson, T. *et al.* (2005) A nationwide outbreak of multiresistant *Salmonella* Typhimurium var. Copenhagen DT104B infection in Finland due to contaminated lettuce from Spain, May 2005. *Euro Surveillance* **10**(6), EO50630.1. Available from http://www.eurosurveillance.org/ew/2005/050630.asp#1. Accessed 7 March 2007.

Taormina, P.J., Beuchat, L.R. & Slutsker, L. (1999) Infections associated with eating seed sprouts: an international concern. *Emerging Infectious Diseases* **5**, 626–634.

Taylor, J.L., Dwyer, D.M., Groves, C. *et al.* (1993) Simultaneous outbreak of *Salmonella enteritidis* and *Salmonella schwarzengrund* in a nursing home: association of *S. enteritidis* with bacteremia and hospitalization. *Journal of Infectious Diseases* **167**, 781–782.

Telzak, E.E., Budnick, L.D., Greenberg, M.S.Z. *et al.* (1990) A nosocomial outbreak of *Salmonella enteritidis* infection due to the consumption of raw eggs. *New England Journal of Medicine* **323**, 394–397.

Threlfall, E.J., Ward, L.R., Hampton, M.D. *et al.* (1998) Molecular fingerprinting defines a strain of *Salmonella enterica* serotype Anatum responsible for an international outbreak associated with formula-dried milk. *Epidemiology and Infection* **121**, 289–293.

Threlfall, J., Ward, L. & Old, D. (1999) Changing the nomenclature of salmonella. *Communicable Disease and Public Health* **2**, 156–157.

Tribe, I.G., Cowell, D., Cameron, P. & Cameron, S. (2002) An outbreak of *Salmonella* Typhimurium phage type 135 infection linked to the consumption of raw shell eggs in an aged care facility. *Communicable Disease Intelligence* **26**, 38–39.

Usera, M.A., Echeita, A., Aladueña, A. *et al.* (1996) Interregional foodborne salmonellosis outbreak due to powdered infant formula contaminated with lactose-fermenting *Salmonella virchow*. *European Journal of Epidemiology* **12**, 377–381.

Usera, M.A., Rodriguez, A., Echeita, A. & Cano, R. (1998) Multiple analysis of a foodborne outbreak caused by infant formula contaminated by an atypical *Salmonella virchow* strain. *European Journal of Clinical Microbiology and Infectious Diseases* **17**, 551–555.

Vaillant, V., Haeghebaert, S., Desenclos, J.C. *et al.* (1996) Outbreak of *Salmonella* Dublin infection in France, November–December 1995. *Euro Surveillance* **1**(2), 9–10.

Veterinary Laboratories Agency (VLA) (2006) *Salmonella in Livestock Production in GB*. 2005 Report. Available from http://www.defra.gov.uk. Accessed 7 March 2007.

Weinberger, M., Andorn, N., Agmon, V. *et al.* (2004) Blood invasiveness of *Salmonella* enterica as a function of age and serotype. *Epidemiology and Infection* **132**, 1023–1028.

World Health Organization (WHO) (2003) *WHO Surveillance Programme for Control of Foodborne Infections and Intoxications in Europe*, 8th report, 1999–2000. Available from http://www.bfr.bund.de/internet/8threport/8threp_fr.htm. Accessed 7 March 2007.

2.10 *Shigella* spp.

2.10.1 *Importance as a cause of foodborne disease*

Shigella infections have been estimated to affect 164.7 million people annually worldwide, and to cause at least 80 million cases of bloody diarrhoea and 700,000 deaths each year (Kotloff *et al.*, 1999; WHO, 2005b). Ninety-nine per cent of infections occur in developing countries, 69% of episodes and 61% of all deaths attributable to shigellosis were in children less than 5 years old. In Asia, shigellas have been estimated to cause 91 million cases of disease and 414,000 deaths annually (WHO, 2005a). Many outbreaks are associated with contaminated water supplies and lack of adequate sanitary facilities.

In industrialized countries such as the UK and the US, the majority of infections occur in children under 5 years old (Newman, 1993; Gupta *et al.*, 2004; von Seidlein *et al.*, 2006). Transmission is mainly person-to-person facilitated by the low infectious dose of the bacteria and difficulty of hygiene control in young children. The bacterium is often spread in day-care centres and subsequently in communities.

Foodborne infection is probably greatly under-detected and under-reported. This is due partly to the low infectious dose and to the difficulty in isolation of low numbers of the bacterium from food. In the US, shigellosis has been estimated to affect about 300,000 people annually, with about 20% of cases attributable, probably, to foodborne transmission (Mead *et al.*, 1999; FDA/CFSAN, 2003). In England, the reported mean incidence of shigella infections between 1993 and 1996 was 27 per 100,000 persons/year (Wheeler *et al.*, 1999) with an unknown proportion of foodborne, whereas in parts of Asia the overall incidence was estimated as 210 cases per 100,000 persons/year (von Seidlein *et al.*, 2006).

An increase in plasmid-encoded resistance of shigellas to antimicrobial drugs, has been reported over many years and is of particular concern in developing countries, where antibiotics may be sold without prescription and where there are few options for treatment of multiresistant shigella infections (Kotloff *et al.*, 1999; von Seidlein *et al.*, 2006).

2.10.2 *Characteristics of the organisms*

Shigellas are Gram-negative, non-motile, non-sporing, rod-shaped bacteria that have been described as metabolically inactive biotypes of *Escherichia coli* (Smith, 1987). There are four species, differentiated serologically, *Shigella dysenteriae*, *S. flexneri*, *S. boydii* and *S. sonnei*, which all cause disease in humans, but differ in the nature of the disease and in the geographical prevalence (Table 2.10.1). The species are divided into serotypes and subtypes based on lipopolysaccharide O-antigens. Shigellas have no natural hosts other than humans and some other primates including monkeys and chimpanzees; long-term human carriers, usually after symptomatic infection, are considered to be rare but seem to be one of the reservoirs for maintenance and spread of the bacterium.

Although shigellas are sometimes considered to be fragile, they can survive in foods and on surfaces for several weeks. Limited information is available on limits for growth, strains of *S. sonnei* and *S. flexneri* were reported to grow at temperatures from 6 to 47°C and 8 to 45°C, respectively (ICMSF, 1996). These bacteria are killed readily by heating

Table 2.10.1 Prevalence and characteristics of *Shigella* spp.

Species	Serotypes	Geographic distribution	Clinical effect	Outcome
S. dysenteriae	15	Mainly south Asia and sub-Saharan Africa, often in areas of political upheaval and natural disaster	Cause severe cases of dysentery, high mortality if untreated. Serotype 1 forms Shiga toxin	Requires antibiotic therapy
S. flexneri	6, 15 subtypes	Frequent in developing countries, also found in developed countries	Less severe symptoms of dysentery	Requires antibiotic therapy
S. boydii	18	Rare in developed countries, found in Indian subcontinent	Biochemically identical to *S. flexneri*, distinguished by serology. Causes some dysentery	Requires antibiotic therapy
S. sonnei	1	Main cause of shigellosis in developed countries, also common in developing countries	Produces mildest form of shigellosis, watery diarrhoea	Usually self-limiting

Data from Kotloff *et al.*, 1999; Lampel *et al.*, 2000; WHO, 2005b.

at about 63°C (Smith, 1987; ICMSF, 1996). The pH range for growth was reported as 4.9–9.34 for *S. sonnei* and 5.0–9.19 for *S. flexneri* and the maximum concentrations of NaCl for growth were 5.18% and 3.78% for *S. sonnei* and *S. flexneri*, respectively. Shigellas have been reported to survive well in distilled water at 4°C, and on refrigerated produce, to survive or multiply on several prepared foods, and to multiply after inoculation onto fresh cut papaya, jicama and watermelon (Warren *et al.*, 2006).

Very low numbers of shigellas in foods can pose a public health risk. Detection of such low numbers and isolation from foods by cultural methods takes several days and available methods are relatively insensitive, as shigellas are liable to be overgrown by other bacteria. Several PCR methods for detection of shigellas in food have been developed (Warren *et al.*, 2006). Methods of typing include biotyping, phage typing, antimicrobial susceptibility testing, plasmid profiling and molecular methods, particularly PGFE (von Seidlein *et al.*, 2006).

2.10.3 *Disease in humans*

The minimum infective dose has been estimated as 10 CFU for *S. dysenteriae* and 500 CFU for *S. sonnei* (Warren *et al.*, 2006). Following infection, symptoms develop usually within 12–24 hours and start with fever, aches and fatigue. Diarrhoea follows, initially as watery stools, this may progress to dysentery and bloody diarrhoea in the case of infections with *S. flexneri* and *S. dysenteriae*, but not usually with *S. sonnei*. Dysentery is often accompanied by severe cramps. Severe dehydration is not usually a problem, but anorexia often occurs and is particularly important in developing countries where the patient may be malnourished before infection. In epidemics caused by *S. dysenteriae* type 1 in Africa and Central America about 5–15% of cases have been fatal (CDC, 2005). Less severe shigellosis, if untreated, is usually self-limiting and clinical illness lasts for 1–2 weeks, although it may last up to 1 month, after which the patient recovers. In the case of infection with *S. sonnei* the bacterium is excreted for between 1 and 4 (or more) weeks in untreated cases, even if the person is asymptomatic (Jewell *et al.*, 1993).

Mild disease resulting in diarrhoea involves the proximal small intestine whereas more severe dysentery involves the colon (Smith, 1987). All four species of *Shigella* penetrate epithelial cells of the intestine by attaching to the surface, after which they are engulfed by invagination of the epithelial cell membrane. The shigellae multiply within the epithelial cells and spread to adjacent cells causing tissue damage. Dysentery results when shigellae invade the epithelial cells of the colonic mucosa and spread from cell to cell resulting in necrosis and release of dead epithelial cells and blood into the colon (Lampel *et al.*, 2000). *Shigella dysenteriae* type 1 forms Shiga toxin, which is encoded chromosomally and is similar to the toxins formed by Shiga toxin-producing *E. coli*. Infection with *S. dysenteriae* type 1 can result in haemolytic uraemic syndrome (HUS) (Section 2.7.4), which results in death in about 10% of cases. Immunity to shigellas is serotype-specific (Kotloff *et al.*, 1999).

Further complications that can result from shigellosis, in addition to HUS, include septicaemia, Reiter's syndrome and reactive arthritis.

In six Asian countries shigellosis was associated most frequently with *S. flexneri* or *S. sonnei* (von Seidlein *et al.*, 2006). The majority of cases of shigellosis in the UK and the US are caused by *S. sonnei* and the fewest by *S. dysenteriae*. Shigellosis often occurs in men who have sex with men and in persons affected by AIDS (FDA/CFSAN, 2003).

2.10.4 *Source and transmission*

Humans are a major reservoir of shigellas, which may be harboured by people recovering from illness and by asymptomatic carriers, including children under the age of 5 years in developing areas (Lampel & Maurelli, 2001). In conditions where sanitation is lacking or inadequate and when safe drinking water is not available, transmission of shigellas is common. In the tropics, flies may be vectors of the bacterium (Lampel *et al.*, 2000).

The low infective dose of shigellas facilitates transmission from person to person. In developed countries, outbreaks of shigellosis are frequent in infants attending day-care centres, and the organisms can be endemic in institutions where conditions are crowded and hygiene is inadequate.

In foodborne outbreaks of shigellosis in the US between 1961 and 1982, poor personal hygiene of food handlers was the main contributing factor (Smith, 1987). Outbreaks that result from foodborne transmission are liable to result in extensive, secondary person-to-person transmission, which may obscure the initial cause of the outbreak.

2.10.5 *Examples of outbreaks*

The examples of foodborne outbreaks outlined in Tables 2.10.2 and 2.10.3 show that two major causative factors were involved (1) contamination of food at source, for example salad vegetables in the field or shellfish during growth and (2) contamination by food handlers.

Table 2.10.2 Examples of outbreaks of foodborne disease caused by *Shigella sonnei*.

Place, date	Cases	Number hospitalized	Deaths	Food implicated	Evidence[a]	Where food was provided	Factors leading to outbreak	Reference
US, 1983	>100	~ 25	0	Lettuce	S	2 University cafeterias supplied by the same company	Contamination on the farm or in the warehouse where the lettuce was stored	Martin *et al.* (1986)
US, 1986	347	66	0	Shredded lettuce, commercially prepared and distributed	S	Restaurants	Evidence that one of the three workers responsible for slicing the lettuce had gastroenteritis	Davis *et al.* (1988)
US, 1986	24	6	0	Raw oysters	S	Restaurants	A crew member of harvesting boat was an asymptomatic carrier of *S. sonnei*; contents of pails used as toilets on boat were often dumped overboard	Reeve *et al.* (1989)
US, 1988	>3000	nr	nr	Uncooked tofu salad	S	Outdoor music festival	Volunteer food handlers suffered shigellosis shortly before the festival. Limited access to soap and water for handwashing	Lee *et al.* (1991)
US, 1988	~ 240	nr	nr	Sandwiches	S	Commercial airline	Sandwiches prepared in flight kitchen; food handlers worked while affected by diarrhoeal illness, handwashing was inadequate, there was bare hand contact with fruits and vegetables	Hedberg *et al.* (1992)
UK, 1992	100	nr	nr	Prawn dishes	S	School party	Possible contamination by a food handler in the school, or by caterer who supplied the prawns	Jewell *et al.* (1993)
Norway, 1994 (also Sweden)	110	nr	nr	Iceberg lettuce imported from Spain	S	Community	Probable contamination in the field by incompletely treated sewage effluent or polluted water	Kapperud *et al.* (1995)

UK, 1994	>40	nr	nr	Iceberg lettuce imported from Spain	S	Community	Probably faecally contaminated water used for irrigation or for cooling lettuce	Frost *et al.* (1995)
Spain, 1995–96	>200	nr	nr	Fresh cheese made from pasteurized milk	S	Community	A food handler at the cheese factory had diarrhoea on the day when the cheese implicated was produced. The outbreak strain was isolated from the food handler. Food handlers used bare hands in production of cheese after pasteurization of the milk	Garcia-Fulgueiras *et al.* (2001)
Denmark, 1998	>29	nr	nr	'Baby maize' imported from Thailand	S	Community (retail)	nr	Mølbak and Niemann (1998)
US, Canada, 1998	486	nr	0	Fresh parsley imported from Mexico	S	Restaurants, food fair	At farm in Mexico, water used for cooling parsley was unchlorinated and susceptible to contamination. Sanitary facilities on farm limited	CDC (1999) Naimi *et al.* (2003)
US, 1998	>300	nr	nr	Imported cilantro	nr	nr	Originated from the above farm in Mexico	Naimi *et al.* (2003)
US, 2000	406	14	0	Refrigerated dip: beans, salsa, guacamole, nacho cheese, sour cream	M	Community	Employee with gastroenteritis, inadequate hygiene control, inadequate cleaning and sanitation of equipment, inadequate refrigeration of product	Kimura *et al.* (2004)
Japan, 2001	28	nr	nr	Oysters imported from Korea	M	Community	nr	Terajima *et al.* (2004)
Lithuania, 2004	36	23	0	Unpasteurized milk curds	S	Community	*S. sonnei* isolated from cases; outbreak strain isolated from the dairy owner and a worker who handled unpasteurized cheese curds at a market where cheese was sold	Zagrebneviene *et al.* (2005)

[a] M (microbiological): identification of an organism of the same type from cases and in the suspect vehicle or vehicle ingredient(s) or detection of toxin in faeces or food; S (statistical): a significant statistical association between consumption of the suspect vehicle and being a case.
nr, not reported.

Table 2.10.3 Examples of outbreaks of foodborne disease caused by *Shigella flexneri* and *S. dysenteriae*.

Place, date	Cases	Number hospitalized	Deaths	Food implicated	Evidence[a]	Where food was provided	Factors leading to outbreak	Reference
S. flexneri								
The Netherlands, 1984	> 59	nr	14 (from home for elderly)	Shrimp cocktail made from peeled and frozen shrimps imported from Far East	S	28 cases were in a home for elderly people	Attributed to recontamination in a severely contaminated, local environment, after cooking	Kayser & Mossel (1984)
Caribbean, 1989	84	13	0	Potato salad	S	Cruise ship	Probable contamination by food handler. Toilet facilities for galley crew were limited	Lew *et al.* (1991)
US, 1992	44	21	0	Tossed salads	S	Chain of restaurants, salads prepared at central kitchen	Food handlers who prepared salad worked while affected by diarrhoeal illness. Salad components added by hand	Dunn *et al.* (1995)
UK, 1998	36	5	0	Fruit salad	S	Supermarket pick and mix counter	nr	O'Brien (1998)
US, 1998	25	3	0	nr	M	Restaurant	Employees who prepared food were positive for outbreak strain; inspection showed inadequate handwashing	Trevejo *et al.* (1999)
US, 2001	>306	22	0	Sliced tomatoes	S	Several restaurants	Hand-sorted, over-ripe, bruised, 'special grade' tomatoes used, which had not been washed and were stored in a warm place. Contamination by food handlers suspected	Reller *et al.* (2006)
S. dysenteriae								
Thailand, 1992	50	nr	0	Coconut milk dessert	S	Pre-school/elementary school	Preparation of dessert involved kneading grated coconut by hand, no plastic gloves, no handwashing policy. Food handler who prepared dessert had diarrhoea just prior to outbreak	Hoge *et al.* (1995)

[a] M (microbiological): identification of an organism of the same type from cases and in the suspect vehicle or vehicle ingredient(s), or detection of toxin in faeces or food; S (statistical): a significant statistical association between consumption of the suspect vehicle and being a case.
nr, not reported.

2.10.6 *Main methods of control and prevention*

In developing countries and in disaster areas improvement in sanitary conditions and provision of safe water are required to prevent shigellosis. Efforts are being made to develop effective vaccines (Kotloff *et al.*, 1999; WHO 2005b). Good personal hygiene is essential. As flies can transfer shigellas from human faecal material to food the use of suitable flytraps may reduce the incidence of shigellosis.

Guidelines for the prevention of person-to-person transmission of shigellas in nurseries and infant schools emphasize standards of hygiene including the provision of adequate toilet facilities, thorough hand washing by children after use of toilets, and the exclusion of children with diarrhoea until symptom-free (PHLS Working Group, 1993). Measures to prevent transmission during preparation and consumption of food have been published (WHO, 2005b).

To prevent contamination of salad vegetables in the field and during harvest, similar measures are needed to those that apply to other enteric pathogens including Shiga toxin-producing *E. coli* (Section 2.7.7).

Shellfish from waters that are contaminated with human waste are liable to contamination with *Shigella* spp.; the use of shellfish from approved areas should avoid such contamination.

In the England and Wales, guidelines state that food handlers who suffer gastroenteritis should be excluded from food handling and not return to this work until 48 hours after cessation of symptoms (DH, 1995, 1996). In the US, food workers who have symptoms of *Shigella* infection are required to be excluded from a food establishment; a food worker who is diagnosed with an infection from *Shigella* spp. and is asymptomatic is required to be excluded from a food establishment serving a highly susceptible population but restricted if they are not serving a highly susceptible population (FDA, 2005).

In view of the possible long excretion of shigellas and the low infective dose, it is essential that food handlers practise good hygiene, including thorough hand washing after use of the toilet, and avoid bare-hand contact with ready-to-eat foods.

Summary of measures advised in healthcare settings

1 Foods, particularly salad vegetables and seafood, should be obtained from reputable suppliers.
2 Raw salad vegetables and fruit should be well washed; only undamaged fruit and vegetables that can be peeled and eaten immediately should be served to immunosuppressed persons.
3 Seafood should be cooked thoroughly.
4 Catering personnel should not handle food while suffering from symptoms of infection, or for a subsequent period of at least 48 hours, and should not handle cooked food or ready-to-eat food with bare hands. Good handwashing is essential.

References

Centers for Disease Control and Prevention (CDC) (1999) Outbreaks of *Shigella sonnei* infection associated with eating fresh parsley – United States and Canada, July–August 1998. *Morbidity and Mortality Weekly Report* **48**(14), 285–289.

Centers for Disease Control and Prevention (CDC) (2005) *Shigellosis*. Available from http://www.cdc.gov/ncidod/dbmd/diseaseinfo/shigellosis_t.htm. Accessed 7 March 2007.

Davis, H., Taylor, J.P., Perdue, J.N. *et al.* (1988) A shigellosis outbreak traced to commercially distributed shredded lettuce. *American Journal of Epidemiology* **128**, 1312–1321.

Department of Health (DH) (UK) (1995) *Food Handlers. Fitness to Work. Guidelines for Food Businesses, Enforcement Officers and Health Professionals*. Department of Health, London. Available from the Food Standards Agency, UK.

Department of Health (DH) (UK) (1996) *Food Handlers. Fitness to Work. Guidelines for Food Business Managers*. Department of Health, London. Available from the Food Standards Agency, UK.

Dunn, R.A., Hall, W.N., Altamirano, J.V. *et al.* (1995) Outbreak of *Shigella flexneri* linked to salad prepared at a central commissary in Michigan. *Public Health Reports* **110**, 580–586.

FDA (2005) *Food Code* 2005. Available from http://www.cfsan.fda.gov/~dms/foodcode.html. Accessed 7 March 2007.

FDA/CFSAN (2003) *Shigella* spp. In: *Foodborne Pathogenic Microorganisms and Natural Toxins Handbook: The "Bad Bug Book"*. Available from http://www.cfsan.fda.gov/~mow/chap19.html. Accessed 7 March 2007.

Frost, J.A., McEvoy, M.B., Bentley, C.A. *et al.* (1995) An outbreak of *Shigella sonnei* infection associated with consumption of iceberg lettuce. *Emerging Infectious Diseases* **1**, 26–28.

Garcia-Fulgueiras, A., Sanchez, S., Guillen, J.J. *et al.* (2001) A large outbreak of *Shigella sonnei* gastroenteritis associated with consumption of fresh pasteurized milk cheese. *European Journal of Epidemiology* **17**, 533–538.

Gupta, A., Polyak, C.S., Bishop, R.D. *et al.* (2004) Laboratory-confirmed shigellosis in the United States, 1989–2002: epidemiologic trends and patterns. *Clinical Infectious Diseases* **38**, 1372–1377.

Hedberg, C.W., Levine, W.C., White, K.E. *et al.* (1992) An international foodborne outbreak of shigellosis associated with a commercial airline. *Journal of the American Medical Association* **268**, 3208–3212.

Hoge, C.W., Bodhidatta, L., Tungtaem, C. & Echeverria, P. (1995) Emergence of nalidixic acid resistant *Shigella dysenteriae* type 1 in Thailand: an outbreak associated with consumption of a coconut milk dessert. *International Journal of Epidemiology* **24**, 1228–1232.

International Commission on Microbiological Specifications for Foods (ICMSF) (1996) *Shigella*. In: *Microorganisms in Foods 5. Microbiological Specifications of Food Pathogens*. Blackie Academic and Professional, London, pp. 280–298.

Jewell, J.A., Warren, R.E. & Buttery, R.B. (1993) Foodborne shigellosis. *Communicable Disease Report* **3**(Review no. 3), R42–R44.

Kapperud, G., RórVik, L.M., Hasseltvedt, V. *et al.* (1995) Outbreak of *Shigella sonnei* infection traced to imported Iceberg lettuce. *Journal of Clinical Microbiology* **33**, 609–614.

Kayser, A. & Mossel, D.A.A. (1984) Intervention sensu Wilson: the only valid approach to microbiological safety of food. *International Journal of Food Microbiology* **1**, 1–4.

Kimura, A.C., Johnson, K., Palumbo, M.S. *et al.* (2004) Multistate shigellosis outbreak and commercially prepared food, United States. *Emerging infectious Diseases* **10**, 1147–1149.

Kotloff, K.L., Winickojj, J.P., Clemens, J.D. *et al.* (1999) Global burden of *Shigella* infections: implications for vaccine development and implementation of control strategies. *Bulletin of the World Health Organization* **77**, 651–666.

Lampel, K.A., Madden, J.M. & Wachsmuth, I.K. (2000) *Shigella* species. In: Lund, B.M., Baird-Parker, T.C. & Gould, G.W. (eds). *The Microbiological Safety and Quality of Food*. Aspen Publishers, Gaithersburg, MD, pp. 1300–1316.

Lampel, K.A. & Maurelli, A.T. (2001) *Shigella* species. In: Doyle, M.P., Beuchat, L.R. & Montville, T.J. (eds). *Food Microbiology. Fundamentals and Frontiers*, 2nd edn. ASM Press, Washington, DC, pp. 247–261.

Lee, L.A., Ostroff, S.M., McGee, H.B. *et al.* (1991) An outbreak of shigellosis at an outdoor music festival. *American Journal of Epidemiology* **133**, 608–615.

Lew, J.F., Swerdlow, D.L., Dance, M.E. *et al.* (1991) An outbreak of shigellosis aboard a cruise ship caused by a multiple-antibiotic-resistant strain of *Shigella flexneri*. *American Journal of Epidemiology* **134**, 413–420.

Martin, D.L., Gustafson, T.L., Pelosi, J.W. *et al.* (1986) Contaminated produce – a common source for two outbreaks of *Shigella* gastroenteritis. *American Journal of Epidemiology* **124**, 299–305.

Mead, P.S., Slutsker, L., Dietz, V. *et al.* (1999) Food-related illness and death in the United States. *Emerging Infectious Diseases* **5**, 607–622.

Mølbak, K. & Niemann, J. (1998) *Outbreak in Denmark of Shigella sonnei Infection Related to Uncooked 'Baby Maize' Imported from Thailand. Eurosurveillance Weekly* **2**(33), 980813. Available from http://www.eurosurveillance.org/ew/1998/980813.asp#1. Accessed 7 March 2007.

Naimi, T.S., Wicklund, J.H., Olsen, S.J. *et al.* (2003) Concurrent outbreaks of *Shigella sonnei* and enterotoxigenic *Escherichia coli* infections associated with parsley: implications for surveillance and control of foodborne illness. *Journal of Food Protection* **66**, 535–541.

Newman, C.P.S. (1993) Surveillance and control of *Shigella sonnei* infection. *Communicable Disease Report, CDR Review* **3**(5), R63–R68.

O'Brien, S. (1998) *Shigella flexneri Outbreak in South East England. Eurosurveillance Weekly* **2**(34), 980820. Available from http://www.eurosurveillance.org/ew/1998/980820.asp#3. Accessed 7 March 2007.

PHLS Working Group (1993) Revised guidelines for the control of *Shigella sonnei* infection and other infective diarrhoeas. *Communicable Disease Report* **3** (review no. 5), R69–R70.

Reeve, G., Martin, D.L. & Pappas, J. (1989) An outbreak of shigellosis associated with the consumption of raw oysters. *New England Journal of Medicine* **321**, 224–227.

Reller, M.E., Nelson, J.M., Mølbak, K. *et al.* (2006) A large, multiple-restaurant outbreak of infection with *Shigella flexneri* serotype 2a traced to tomatoes. *Clinical Infectious Diseases* **42**, 163–169.

Smith, J.L. (1987) *Shigella* as a foodborne pathogen. *Journal of Food Protection* **50**, 788–801.

Terajima, J., Tamura, K., Hirose, K. *et al.* (2004) A multi-prefectural outbreak of *Shigella sonnei* infections associated with eating oysters in Japan. *Microbiology and Immunology* **48**, 49–52.

Trevejo, R.T., Abbott, S.L., Wolfe, M.J. *et al.* (1999) An untypeable *Shigella flexneri* strain associated with an outbreak in California. *Journal of Clinical Microbiology* **37**, 2352–2353.

von Seidlein, L., Kim, D.R., Ali, M. *et al.* (2006) A multicentre study of *Shigella* diarrhoea in six Asian countries: disease burden, clinical manifestations, and microbiology. *PLoS Medicine* **3**, 1556–1569.

Warren, B.R., Parish, M.E. & Schneider, K.R. (2006) *Shigella* as a foodborne pathogen and current methods for detection in food. *Critical Reviews in Food Sciences and Nutrition* **46**, 551–567.

Wheeler, J.G., Sethi, D. & Cowden, J.M. (1999) Study of infectious intestinal disease in England: rates in the community, presenting to general practice, and reported to national surveillance. *British Medical Journal* **318**, 1046–1050.

World Health Organization (WHO) (2005a) Shigellosis: disease burden, epidemiology and case management. *Weekly Epidemiological Record* **80**, 94–99.

World Health Organization (WHO) (2005b) *Guidelines for the Control of Shigellosis, Including Epidemics due to Shigella dysenteriae Type 1*. WHO, Geneva. Available from http://whqlibdoc.who.int/publications/2005/9241592330.pdf. Accessed 7 March 2007.

Zagrebneviene, G., Jasulaitiene, V., Morkunas, B. *et al.* (2005) *Shigella sonnei* outbreak due to consumption of unpasteurized milk curds in Vilnius, Lithuania, 2004. *Euro Surveillance* **10**(12), E051201.3. Available from http://www.eurosurveillance.org/ew/2005/051201.asp#3. Accessed 7 March 2007.

2.11 *Staphylococcus aureus*

2.11.1 Importance as a cause of foodborne disease

Staphylococcal food poisoning is caused by the consumption of food that contains enterotoxin produced by strains of *Staphylococcus aureus,* and occasionally other *Staphylococcus* spp. The illness usually lasts for a short time; sporadic cases are unlikely to be reported and do not feature generally in national surveillance. In some European countries however, including France, the Netherlands and Norway, *S. aureus* was reported as an important cause of outbreaks of foodborne disease in 1999 and 2000 (WHO, 2003). In England and Wales, four outbreaks were reported in 1999 affecting 81 people and in 2000 no outbreaks were reported; in 2001, six outbreaks, and from 2002 to 2006, 0–2 outbreaks were reported each year (HPA, 2007). In the US, *Staphylococcus* was estimated to cause about 1.3% (185,000 cases) of illness caused by known foodborne pathogens annually, but most of these cases were not reported (Mead *et al.*, 1999).

2.11.2 Characteristics of the organism

Staphylococcus aureus was shown to cause foodborne illness in 1914, and in 1930 the illness was attributed to bacterial toxin (Baird-Parker, 2000). The organism causes a range of symptoms in man including infections of skin and wounds, urinary tract infections, pneumonia and bacteraemia, and intoxications such as toxic shock syndrome and food poisoning. It also causes infections in animals, including bovine mastitis (Smyth *et al.*, 2005). The bacterium is a non-motile, Gram-positive coccus that divides in more than one plane to form clusters of cells; it is facultatively anaerobic, but grows best in aerobic conditions. Most strains will grow at temperatures between 7 and 48°C, with optimum growth at 35–40°C. The organism is not very heat-resistant, and in neutral phosphate buffer strains were inactivated at 54–60°C with D values (time required to reduce the number of survivors by tenfold) at 56°C between 0.60 and 2.55 minutes (Bergdoll, 1989); heat resistance is liable to be greater in foods. The bacteria survive well during freezing and drying. In optimum conditions, *S. aureus* can grow in the presence of >20% NaCl; in foods stored at <20°C, growth is prevented by 15% NaCl (Baird-Parker, 2000). Growth can occur in the pH range 4–10, but is extremely slow at the extreme ends of the range.

Strains of *S. aureus* from different animal sources can be grouped into four biotypes according to their biochemical characteristics (Baird-Parker, 2000). Biotype A strains are associated with humans, biotype B with pigs and poultry, biotype C with cattle and sheep, biotype D with hares. Phage typing has been used widely for epidemiological purposes, different sets of phages are needed to type human and animal strains. Genetic techniques, particularly PFGE, are now used for strain discrimination.

Food poisoning is caused when the bacteria multiply in food and form enterotoxins (SEs), which are single-chain proteins with a molecular weight of 26–30 kDa. On the basis of antigenic differences, about 18 SEs have been recognized and designated SEA, SEB, SEC1, SEC2, SEC3, SED, SEE and more recently SEG to SER and SEU (Jørgensen *et al.*, 2005). SEA is the enterotoxin associated most frequently with foodborne disease, but SEB, SEC and SED may also cause food poisoning; further information is needed on the involvement of other SEs. Commercial kits are available for detection of SEA to SEE, and immunological methods are used to detect other toxins. The presence of genes for SEs can be detected by PCR. Genes encoding some of the SEs are located on mobile genetic elements including phages, plasmids and pathogenicity islands. The SEs are superantigens, able to stimulate non-specific T-cell proliferation (Alouf & Müller-Alouf, 2003).

About 50% of strains of *S. aureus* from healthy people, and a higher proportion from people with staphylococcal infections, have been reported as enterotoxin producers (Bergdoll, 1989). Strains from humans mainly produce SEA, SEB and SEC. Very few strains of *S. aureus* from animals have the genes *sea*, *seb*, *see* or *seh* for toxin production or produce the SEA, SEB, SEE or SEH toxins, but there are some reports that the *sea*, *seb* or *seh* genes were present frequently in isolates from bulk-tank samples of milk from sheep and goats, and in bovine strains (Smyth *et al.*, 2005). The *sec* gene has been reported frequently in strains from cattle, sheep and goats (Smyth *et al.*, 2005); the reported frequency of the *sed* gene in bovine isolates varies, and it occurs in a low proportion of isolates from sheep, goats and poultry.

The SEs are resistant to proteolytic enzymes such as pepsin, trypsin, chymotrypsin, rennin and papain, they are also heat-resistant (Bergdoll, 1989) and are not inactivated by pasteurization of milk (72°C for 15 s). SEA is inactivated gradually by heating at 100°C, and heat processing used generally in production of commercially canned food will destroy the amount of enterotoxin usually present in foods involved in food poisoning outbreaks (0.5–10 μg/100 g food), but in some cases enterotoxin has survived commercial sterilization of canned food (Baird-Parker, 2000). Consumption of about 1 μg of SEA can cause illness FDA/CFSAN, 2005); the presence of this amount requires *S. aureus* to multiply to more than 10^5 bacteria/g of food.

Methicillin-resistant *S. aureus* (MRSA) is a major cause of hospital infection via wounds and mucous membranes. Most strains of MRSA appear to be hospital-acquired (HA-MRSA), but community-acquired MRSA (CA-MRSA) strains, which differ phenotypically and genotypically from HA-MRSA, have been reported in many countries and are becoming important in healthcare settings in some countries (Deurenberg *et al.*, 2007; Maree *et al.*, 2007).

2.11.3 *Disease in humans*

The symptoms of staphylococcal food poisoning include nausea, vomiting, abdominal cramps and diarrhoea, and most often occur between 1 and 8 hours (usually 2–4 h) after the consumption of contaminated food. Generally recovery occurs within 2 days. In some cases vomiting may be severe and frequent and accompanied by diarrhoea, with blood present in the vomitus and stool (Bergdoll, 1989). In the acute phase the temperature may be reduced, the pulse rate increased and cold sweats and prostration may occur, together with mild headache, muscular cramping (usually of the leg muscles) and dehydration. The illness is seldom fatal, but deaths have been reported, including those of two children, aged 3 and 4 years, after each had consumed 125 mL of milk, from a goat with staphylococcal mastitis, that contained a relatively high concentration of enterotoxin (Bergdoll, 1989). In outbreaks in France associated with milk and milk products, 16% of cases associated with *S. aureus* were hospitalized, compared with 6.7% of cases associated with *Salmonella* (De Buyser *et al.*, 2001).

2.11.4 *Source and transmission*

Staphylococcus aureus occurs widely on the skin and mucous membranes of warm-blooded animals. In humans it is carried in the nose and throat, on the hands and under the fingernails. On healthy skin the bacterium is transient, but if skin is damaged, for example by frequent immersion of hands in hot water, *S. aureus* can become established and multiply to high numbers (Baird-Parker, 2000). The organism can be found in the nostrils and perineum of about 50% of the population. In patients and staff in hospitals, nasal carriage may be as high as 60–80% (Hobbs & Roberts, 1993). The organism has been isolated at least one of four times during a year from 37% of food handlers in Germany and 74% in Japan (Bergdoll, 1989). Nasal carriage of *S. aureus* was found in 30% of 47 food handlers in Brazil; the majority of the handlers harboured multiple strains (Acco *et al.*, 2003). Infected wounds, skin lesions and boils may be a source of the organism, and coughing and sneezing may spread the bacterium. There is, therefore, a high risk of contamination of food by food handlers.

Raw meats and dairy products are liable to be contaminated with both human and animal strains of *S. aureus*. In products that are heat-treated, these bacteria will usually be inactivated. The organism is an important cause of mastitis in cattle, sheep and goats (Bergdoll, 1989). In bovine mastitis the condition may be chronic and is commonly subclinical, with the bacterium surviving in the host for long periods of time (Smyth *et al.*, 2005). There is a risk of enterotoxin production in milk, particularly that from mastitic animals, and resulting illness if the milk is consumed. In Korea, 1.3% (12 of 894) of milk samples from dairy cattle contained methicillin-resistant strains of *S. aureus,* nine of the samples came from cows that showed signs of mastitis, which was often treated in Korea with antibiotics including ampicillin and penicillin but rarely methicillin (Lee, 2003).

Foods that require handling are liable to be contaminated with *S. aureus*; toxin formation can occur if the food is maintained at temperatures that allow growth, and the organism is not inhibited by other bacteria. This is particularly likely in the case of foods that are re-contaminated after cooking. In the UK between 1969 and 1990 the majority of foods implicated in reported incidents were meat products (53%), followed by poultry products (22%), dairy products (8%), fish and shellfish (7%), boiled eggs and egg dishes (3.5%) (Wieneke *et al.*, 1993). In unpublished outbreaks, excluding listeriosis, associated with milk and milk products (particularly cheese) in France between 1992 and 1997, and in which the aetiologic agent was confirmed, *S. aureus* was by far the most frequent pathogen involved (De Buyser *et al.*, 2001).

The bacterium survives well in the environment, it can become established on processing equipment in food factories and act as a source of contamination or recontamination; this has been a problem particularly in poultry processing plants.

2.11.5 *Examples of outbreaks*

Food handlers are considered as the main source of contamination leading to foodborne outbreaks, the majority of which have been associated with the presence of SEA (Bergdoll, 1989; Kérouanton *et al.*, 2007)

Examples of outbreaks, foods involved and causative factors are shown in Table 2.11.1.

The risk of transfer of MRSA by food in hospitals was illustrated by an outbreak in the Netherlands (Kluytmans *et al.*, 1995; Chapter 4). There was evidence that food prepared by a dietary worker for a neutropenic patient (who was severely immunocompromised and had taken antacids and oral ciprofloxacin) caused the initial case of MRSA septicaemia. Infection spread to other patients in the haematology unit and in a surgical unit, resulting in five deaths. A community-acquired outbreak of illness caused by MRSA in the US, which affected three family members, was attributed to contamination of coleslaw prepared in a delicatessen by a food handler who was a nasal carrier of the outbreak strain, which she may have acquired while visiting an infected relative in a nursing home (Jones *et al.*, 2002).

2.11.6 *Main methods of prevention and control*

Control of raw materials is important; raw materials that may be contaminated with high numbers of *S. aureus* should not be used for further processing. Although the bacteria may be killed by processing, any SEs present may not be removed or destroyed. Introduction of the contaminated material into a food-processing environment may allow *S. aureus* to become established in that environment. If raw materials are used in a process such as production of raw-milk cheese or fermented sausage, further multiplication may occur.

Consumption of unpasteurized milk or milk products should be avoided, particularly if there is a risk that they are obtained from animals affected by mastitis.

Food processing equipment should be cleaned thoroughly to prevent the establishment of *S. aureus*.

Susceptible foods should be refrigerated promptly (to <7°C or lower) after preparation, in order to prevent growth of any staphylococci present; this is important whether or not the food will be cooked, because some enterotoxin produced during growth will probably survive cooking.

Food handlers with lesions on exposed skin (hands, face, neck or scalp) that are actively weeping or discharging should be excluded from work until the lesions have healed (DH, 1995, 1996). Nasal carriers of *S. aureus* need to be excluded from work as food handlers if they are implicated as a source of an outbreak; they can be treated with antistaphylococcal cream in order to eliminate the bacterium (Hobbs & Roberts, 1993). Food handlers should wash their hands thoroughly before preparation of food, practise good hygiene, and minimize bare hand contact with cooked foods and uncooked foods, such as cream, that will support growth of the bacteria.

Summary of measures advised in healthcare settings

1. Obtain food from reputable sources.
2. Keep food-processing equipment clean to prevent establishment of *S. aureus*.
3. Refrigerate susceptible foods promptly to <7°C to prevent growth of *S. aureus*, and preferably to <5°C to prevent growth of other pathogens.

Table 2.11.1 Examples of outbreaks of foodborne disease caused by *Staphylococcus aureus*.

Place, date	Enterotoxin	Cases	Deaths	Food implicated	Evidence[a]	Where food was provided	Factors leading to outbreak	Reference
Flight Rio de Janeiro to New York City, 1976	D	~ 76	nr	Chocolate eclairs	M	Airplane flight	Eclairs left unrefrigerated for 12 h before flight. *S. aureus* 10^9/g isolated from éclairs and enterotoxin D detected	Bergdoll (1989)
England, 1983	A	36 (5 hospitalized, 3 of whom needed IV fluid replacement)	0	Vanilla slices	M	Community supplied by bakery	Eczematous lesion on the hand of a member of the bakery staff yielded heavy, pure growth of *S. aureus*, but was not the phage type isolated from the vanilla slices and did not yield enterotoxin. Slices kept at ambient temperature (17.3–28.9°C) for >19 h. This method had been used by bakery for 25 years	Fenton *et al.* (1984)
France, 1983	A and D	20	nr	Ewe's milk cheese made with raw milk	M	nr	Contamination from producer; outbreak strain isolated from nose and hands	De Buyser *et al.* (2001)
UK (cases also reported in Italy and Luxemberg), (1984)	A	47 (5 hospitalized)	0	Lasagne made in Italy	M	School, restaurant, pub, community	Inadequately pasteurized egg mixed into raw dough pasta which was dried slowly at a temperature allowing growth of *S. aureus*	Woolaway *et al.* (1986)
Scotland, 1984	A	27	nr	Ewe's milk cheese made with raw milk	D	Hotel dinner	Staphylococcal mastitis present in flock of ewes. Milk warmed to 32°C and held at this temperature before addition of starter culture. *S. aureus* present intermittently in milk, of 3 strains tested 2 produced SEA and 1 SEC	Bone *et al.* (1989)
US, 1985	A	>850	nr	2% chocolate milk (pasteurized milk)	M	Supplied to school children	Chocolate milk held for several hours in a tank that was not adequately cooled, before pasteurization, which would not destroy the toxin	Bergdoll (1989); Evenson *et al.* (1988)

US, 1989	A	99 (18 hospitalized)	0	Canned mushrooms from China	M	University cafeteria; hospital cafeteria; restaurant; pizzeria	Mushrooms hand-picked at small farms in China and manually graded. Stored without refrigeration in plastic bags or in brine, which would allow growth of *S. aureus*. Toxin may have survived the canning process, post-processing contamination may have occurred through microleaks in cans during cooling	Levine *et al.* (1996)
US, 1990	A. *S. aureus* and *Salmonella infantis* involved	215	0	Turkey	M	Prison	After cooking, the turkey was held at room temperature during deboning. Contamination may have been caused by the handler responsible for deboning, who had skin lesions on arms that yielded *S. aureus*. The turkey was held for hours at room temperature and in a food warmer at unchecked temperature	Meehan *et al.* (1992)
US, 1990	A	>65 (12 hospitalized)	0	Ham	M	Two schools supplied by another school kitchen	Cooked hams contaminated by a food handler who peeled casings from the hams. Contaminated ham held at 10–49°C for at least 15 h because of improper procedures, and warmed to about 60°C before serving	Richards *et al.* (1993)
Australia, 1997	+, type not reported	42	nr	Curry egg and pasta	D	Old people's home	nr	Cowell *et al.* (2002)
US, 1997	A	18	0	Pre-cooked ham	D	Retirement party	Pre-cooked ham handled and sliced while hot, slicer not first cleaned adequately, large quantity of ham may not have cooled quickly and was served cold next day	CDC (1997)
Brazil, 1998	A	~ 4000 (396 hospitalized)	16	Meal of chicken, roast beef, rice and beans	M	Ordination ceremony for Catholic priest	Eight food handlers were all positive for *S. aureus* on hands. Foods left at room temperature for 24 h	do Carmo *et al.* (2004)
Brazil, 1998	A	180 (~40 hospitalized)	0	Vegetable salad with mayonnaise, broiled chicken, pasta in tomato sauce	M	Luncheon at school	Outbreak strain isolated from 4 asymptomatic food handlers, who probably contaminated the food	Colombari *et al.* (2007)

(cont.)

Table 2.11.1 *(continued)*

Place, date	Enterotoxin	Cases	Deaths	Food implicated	Evidence[a]	Where food was provided	Factors leading to outbreak	Reference
Brazil, 1999	A, B, C	50	nr	Homemade Minas white cheese made with unpasteurized milk	D	nr	Staphylococci isolated from milk and cheese samples. Food handlers may have contributed to contamination of cheese	do Carmo *et al.* (2002)
Brazil, 1999[b]	C, D	328	nr	Raw milk	D	Community	Suspected that staphylococci were associated with mastitis in cows	do Carmo *et al.* (2002)
Australia, 2000	+, type not reported	~ 50 (2 hospitalized)	0	Pre-prepared meal of chicken meat and salad	M	Club meal attended by elderly people	Food handlers at manufacturing premises handling both raw and cooked meat; after cooking, chicken maintained at temperatures allowing growth of *S. aureus*; further handling provided risk of contamination	Cowell *et al.* (2002)
Japan, 2000	A, H	13,420	nr	Powdered skim milk used in low-fat milk and drink-type yoghurts	D	Community	During production of powdered skim milk operations at the factory were stopped for 3 h or longer because of a power cut, and processing was delayed for over 9 h	Asao *et al.* (2003); Ikeda *et al.* (2005)
Taiwan, 2000	A, B	10	0	nr	M	High school	Outbreak strains isolated from hand lesion of a food handler	Wei and Chiou (2002)
US, date not given	C[c]	3	0	Shredded pork barbeque and coleslaw from convenience store	M	Family	Outbreak strain isolated from food handler involved in preparation of coleslaw. The food handler had several times visited an elderly relative who had a staphylococcal infection and died	Jones *et al.* (2002)
Norway, 2003	H	8 (5 children, 3 adults)	0	Mashed potato prepared with raw milk	M	Kindergarten	Mashed potato prepared on the previous day, storage not specified, reheated before serving. Contained $\sim 8 \times 10^8$ CFU *S. aureus*/g and enterotoxin H. Raw milk contained outbreak strain	Jørgensen *et al.* (2005)

[a] M (microbiological): identification of an organism of the same type from cases and in the suspect vehicle or vehicle ingredient(s), or detection of toxin in faeces or food; D (descriptive): other evidence, usually descriptive, reported by local investigators as indicating the suspect vehicle or food.
[b] *Staphylococcus* non-aureus.
[c] Methicillin-resistant (MRSA) strain.
nr, not reported.

4 Unpasteurized milk or milk products should not be served.
5 Food handlers with weeping or discharging lesions on exposed skin should be excluded from this work. Catering personnel should not handle cooked food or ready-to-eat food with bare hands. Good handwashing is essential.

References

Acco, M., Ferreira, F.S., Henriques, J.A.P. & Tondo, E.C. (2003) Identification of multiple strains of *Staphylococcus aureus* colonizing nasal mucosa of food handlers. *Food Microbiology* **20**, 489–493.

Alouf, J.E. & Müller-Alouf, H. (2003) Staphylococcal and streptococcal superantigens: molecular, biological and clinical aspects. *International Journal of Medical Microbiology* **292**, 429–440.

Asao, T., Kumeda, Y., Kawai, T. *et al.* (2003) An extensive outbreak of staphylococcal food poisoning due to low-fat milk in Japan, estimation of enterotoxin A in the incriminated milk and powdered skim milk. *Epidemiology and Infection* **130**, 33–40.

Baird-Parker, T.C. (2000) *Staphylococcus aureus*. In: Lund, B.M., Baird-Parker, T.C. & Gould, G.W. (eds). *The Microbiological Safety and Quality of Food*. Aspen Publishers, Gaithersburg, MD, pp. 1317–1335.

Bergdoll, M.S. (1989) *Staphylococcus aureus*. In: Doyle, M.P. (ed). *Foodborne Bacterial Pathogens*. Marcel Dekker, New York and Basel, pp. 463–523.

Bone, F.J., Bogie, D. & Morgan-Jones, S.C. (1989) Staphylococcal food poisoning from sheep milk cheese. *Epidemiology and Infection* **103**, 449–458.

Centers for Disease Control and Prevention (CDC) (1997) Outbreak of staphylococcal food poisoning associated with precooked ham – Florida, 1997. *Morbidity and Mortality Weekly Report* **46**(50), 1189–1191.

Colombari, V., Mayer, M.D.B., Laicini, Z.M. *et al.* (2007) Foodborne outbreak caused by *Staphylococcus aureus*: phenotypic and genotypic characterization of strains of food and human sources. *Journal of Food Protection* **70**, 489–493.

Cowell, N.A., Hansen, M.T., Langley, A.J. *et al.* (2002) Outbreak of staphylococcal enterotoxin food poisoning. *Communicable Diseases Intelligence* **26**, 574–575.

De Buyser, M.-L., Dufour, B., Maire, M. & Lafarge, V. (2001) Implication of milk and milk products in food-borne diseases in France and in different industrialised countries. *International Journal of Food Microbiology* **67**, 1–17.

Department of Health, UK (DH) (1995) *Food Handlers. Fitness to Work. Guidelines for Food Businesses, Enforcement Officers and Health Professionals*. Department of Health, London.

Department of Health (UK) (DH) (1996) *Food Handlers Fitness to Work. Guidelines for Food Business Managers*. Department of Health, London.

Deurenberg, R.H., Vink, C., Kalenic, S. *et al.* (2007) The molecular evolution of methicillin-resistant *Staphylococcus aureus*. *Clinical Microbiology and Infection* **13**, 222–235.

do Carmo, L.S., Cummings, C., Linardi, V.R. *et al.* (2004) A case study of a massive staphylococcal food poisoning incident. *Foodborne Pathogens and Disease* **1**, 241–246.

do Carmo, L.S., Dias, R.S., Linardi, V.R. *et al.* (2002) Food poisoning due to enterotoxigenic strains of *Staphylococcus* present in Minas cheese and raw milk in Brazil. *Food Microbiology* **19**, 9–14.

Evenson, M.L., Hinds, M.W., Bernstein, R.S. & Bergdoll, M.S. (1988) Estimation of human dose of staphylococcal enterotoxin A from a large outbreak of staphylococcal food poisoning involving chocolate milk. *International Journal of Food Microbiology* **7**, 311–316.

FDA/CFSAN (2005) *Staphylococcus aureus*. In: *Foodborne Pathogenic Microorganisms and Natural Toxins Handbook*. Available from http://www.cfsan.fda.gov/~mow/chap3.html. Accessed 7 March 2007.

Fenton, P.A., Dobson, K.W., Eyre, A. *et al.* (1984) Unusually severe food poisoning from vanilla slices. *Journal of Hygiene* **93**, 377–380.

Health Protection Agency (HPA) (2007) *Staphylococcus aureus Food Poisoning England and Wales, 1992–2006.* Available from http://www.hpa.org.uk/infections/topics_az/staphylo/food/data_ew.htm. Accessed 7 March 2007.

Hobbs, B.C. & Roberts, D. (1993) *Food Poisoning and Food Hygiene*, 6th edn. Edward Arnold, London, pp. 60, 169.

Ikeda, T., Tamate, N., Yamaguchi, K. & Makino, S. (2005) Mass outbreak of food poisoning disease caused by small amounts of staphylococcal enterotoxins A and H. *Applied and Environmental Microbiology* **71**, 2793–2795.

Jones, T.F., Kellum, M.E., Porter, S.S. *et al.* (2002) An outbreak of community-acquired foodborne illness caused by methicillin-resistant *Staphylococcus aureus. Emerging Infectious Diseases* **8**, 82–84.

Jørgensen, H.J., Mathisen, T., Løvseth, A. *et al.* (2005) An outbreak of staphylococcal food poisoning caused by enterotoxin H in mashed potato made with raw milk. *FEMS Microbiology Letters* **252**, 267–272.

Kérouanton, A., Hennekinne, J.A., Letertre, C. *et al.* (2007) Characterization of *Staphylococcus aureus* strains associated with food poisoning outbreaks in France. *International Journal of Food Microbiology* **115**, 369–375.

Kluytmans, J., van Leeuwen, W. & Goessens, W. *et al.* (1995) Food-initiated outbreak of methicillin-resistant *Staphylococcus aureus* analyzed by pheno- and genotyping. *Journal of Clinical Microbiology* **33**, 1121–1128.

Lee, J.H. (2003) Methicillin (Oxacillin)-resistant *Staphylococcus aureus* strains isolated from major food animals and their potential transmission to humans. *Applied and Environmental Microbiology* **69**, 6489–6494.

Levine, W.C., Bennet, R.W., Choi, Y. *et al.* (1996) Staphylococcal food poisoning caused by imported canned mushrooms. *Journal of Infectious Diseases* **173**, 1263–1267.

Maree, C.L., Daum, R.S., Boyle-Vavra, S. *et al.* (2007) Community-associated methicillin-resistant *Staphylococcus aureus* isolates causing healthcare-associated infections. *Emerging Infectious Diseases* **13**, 236–242.

Mead, P.S., Slutsker, L., Dietz, V. *et al.* (1999) Food-related illness and death in the United States. *Emerging Infectious Diseases* **5**, 607–625.

Meehan, P.J., Atkeson, T., Kepner, D.E. & Melton, M. (1992) A foodborne outbreak of gastroenteritis involving two different pathogens. *American Journal of Epidemiology* **136**, 611–616.

Richards, M.S., Rittman, M., Gilbert, T.T. *et al.* (1993) Investigation of a staphylococcal food poisoning outbreak in a centralized school lunch program. *Public Health Reports* **108**, 765–771.

Smyth, D.S., Hartigan, P.J., Meaney, W.J. *et al.* (2005) Superantigen genes encoded by the *egc* cluster and SaPlbov are predominant among *Staphylococcus aureus* isolates from cows, goats, sheep, rabbits and poultry. *Journal of Medical Microbiology* **54**, 401–411.

Wei, H.L. & Chiou, C.S. (2002) Molecular subtyping of *Staphylococcus aureus* from an outbreak associated with a food handler. *Epidemiology and Infection* **128**, 15–20.

Wieneke, A.A., Roberts, D. & Gilbert, R.J. (1993) Staphylococcal food poisoning in the United Kingdom, 1969–90. *Epidemiology and Infection* **110**, 519–531.

Woolaway, M.C., Bartlett, C.L.R., Wieneke, A. *et al.* (1986) International outbreak of staphylococcal food poisoning caused by contaminated lasagne. *Journal of Hygiene* **96**, 67–73.

World Health Organization (WHO) (2003) *WHO Surveillance Programme for Control of Foodborne Infections and Intoxications in Europe*, 8th report, 1999–2000. Available from http://www.bfr.bund.de/internet/8threport/8threp_fr.htm. Accessed 7 March 2007.

2.12 *Vibrio* spp.

2.12.1 *Importance as a cause of foodborne disease*

Twelve *Vibrio* species are human pathogens, of which eight are known to be food-associated (Oliver & Kaper, 2001; Morris, 2003). The major foodborne and waterborne pathogenic vibrios are *Vibrio cholerae*, *V. parahaemolyticus* and *V. vulnificus*.

Worldwide, between 1817 and the present time seven or possibly eight pandemics caused by toxigenic *V. cholerae* have been recorded (Table 2.12.1). Outbreaks take place predominantly in conditions of inadequate sanitation, shortage of clean water and crowded living conditions. In Europe and North America, infection with *V. cholerae* is found mainly in travellers returning from abroad. Nontoxigenic *V. cholerae* primarily causes gastroenteritis but has not been associated with epidemics.

Vibrio parahaemolyticus was described first as the cause of an outbreak of foodborne illness in Japan in 1950 and is a major cause of outbreaks in Japan and Asia. It is an important cause of seafood-associated gastroenteritis in the US and in New Zealand; outbreaks have been reported in South America and Europe. Occasionally, *V. parahaemolyticus* can cause a skin infection if an open wound is exposed to warm seawater.

Vibrio vulnificus was described first in 1976 and is one of the most invasive and rapidly lethal of human pathogenic bacteria. Infections have been reported in the US, Australia, Northern Europe (Strom & Paranjpye, 2000), Israel (Bisharat *et al.*, 1999), Japan (Fukushima & Seki, 2004) and Taiwan (Hsueh *et al.*, 2004). Ingestion of *V. vulnificus* can cause gastroenteritis in healthy individuals, but in persons with an underlying chronic disease, particularly liver disease, it causes primary septicaemia (i.e. septicaemia without an obvious focus of infection) (Morris, 2003) with a high mortality, it also causes infection through contamination of wounds.

Table 2.12.1 Cholera pandemics.

Pandemic	Dates	Origin	Spread	Biotype
I	1817–1823	Indian subcontinent		
II	1829–1851	Indian subcontinent		
III	1852–1859	Indian subcontinent		
IV	1863–1879	Indian subcontinent		
V	1881–1896	Indian subcontinent		Classical
VI	1899–1923/5	Indian subcontinent		O1 Classical
VII	1961–Present	Indonesia	SE Asia, Asia, Middle East, India, Africa, S America, Central America, Zaire	O1 El Tor
VIII	1992–Present	Indian subcontinent	India, Bangladesh, Nepal, Burma, Pakistan, Thailand, China, Malaysia, Saudi Arabia	O139 serotype

Modified from Faruque *et al.*, 1998; Kaysner, 2000; Sack *et al.*, 2004.

Table 2.12.2 Limits for growth of *Vibrio* spp.

	V. cholerae	*V. parahaemolyticus*	*V. vulnificus*
Temperature optimum (°C)	37	37	37
Temperature range for growth (°C)	10–43	5–43	8–43
NaCl optimum (%)	0.5	3	2.5
NaCl range for growth (%)	0.1–4.0	0.5–10.0	0.5–~6.0
pH range for growth	5.0–9.6	4.8–11	5–10

Data from ICMSF, 1996; Oliver and Kaper, 2001.

2.12.2 *Characteristics of the organisms*

Vibrios are facultatively anaerobic, Gram-negative, non-sporing, straight or curved, rod-shaped, bacteria. In liquid media they are motile by a single, polar, sheathed flagellum.

Vibrios occur in aquatic environments and growth is stimulated by sodium ions. *Vibrio cholerae* does not require addition of NaCl for growth in nutrient broth, as its obligate requirement for sodium ions can be met by trace amounts in most media constituents (FDA/CFSAN, 2004). *V. parahaemolyticus* requires 0.5–1% (w/v) NaCl and 80% of strains will grow in the presence of 8% NaCl, *V. vulnificus* requires about 1% NaCl and 65% of strains will grow in the presence of 6% NaCl (Oliver & Kaper, 2001). The temperature range for growth of these vibrios is from between 5 and 10°C to 43°C with an optimum at 37°C (ICMSF, 1996) (Table 2.12.2.). In optimum conditions, the doubling times of *V. cholerae* and of *V. parahaemolyticus* may be as low as 18 and 9 minutes, respectively. Vibrios are very sensitive to heat, irradiation, drying, and acid pH, and tend to die rapidly in these conditions (ICMSF, 1996). In cool, moist conditions the organisms may survive for many days. Vibrios can survive in frozen shellfish, the extent of survival depending on the conditions used for freezing and the storage temperature.

Vibrio cholerae can be divided into more than 150 serogroups on the basis of lipopolysaccharide, O, antigens, of which O1 and O139 are associated with cholera. Serogroup O1 can be subdivided into the Ogawa, the Inaba, and the very rare Hikojima serotypes (Kaysner, 2000). Serogroup O1 strains can also be divided into two biotypes, classical and El Tor. The former are non-haemolytic whereas the latter produce a beta-haemolysin detectable on sheep blood agar plates. O139 strains also form a beta-haemolysin. The most virulent strains possess a polysaccharide capsular, Vi, antigen. Pathogenicity is associated with several virulence factors, in particular production of cholera toxin (Sack *et al.*, 2004).

Vibrio parahaemolyticus can be serotyped on the basis of lipopolysaccharide, O, and acidic polysaccharide, K, antigens. Twelve O groups and 65 K groups are recognized; usually strains can be grouped according to O antigens but some are untypeable by the K antigen. Strains are designated Kanagawa-positive or -negative according to whether they produce a thermostable, direct hemolysin (TDH) active on fresh human red blood cells; usually this is correlated with pathogenicity in humans. A thermostable related haemolysin (TRH) has also been found in strains causing gastroenteritis and many clinical strains produce both TDH and TRH (FDA/CFSAN, 2004); other possible virulence factors have been described. There is no clear correlation between serology and pathogenicity.

Vibrio vulnificus has been differentiated into three biotypes. Biotype 1 is the reported cause of most infections in humans, whereas biotype 2 strains cause disease in eels and rarely cause infection in humans. An outbreak of infections among Israeli fish workers and consumers was attributed to a new strain, biotype 3 (Bisharat *et al.*, 1999). Factors that may be associated with virulence include the presence of a polysaccharide capsule, production of haemolysin, and elastase, collagenase, lipase and protease enzymes (Kaysner, 2000).

2.12.3 *Disease in humans*

Ingestion of *V. cholerae* O1 or O139 can result in mild or no symptoms. The incubation period before symptoms occur is usually between several hours and 5 days and the organisms can cause severe illness with vomiting and profuse, watery diarrhoea ('rice-water stools'), followed by suppressed renal function, thirst, and leg and abdominal cramping (Kaysner, 2000). Rapid loss of body fluids can result in dehydration and shock, which can be fatal within hours unless treatment is given. In previously healthy persons who receive replacement fluids and salts, recovery can occur in 1–6 days. Information from natural infections indicates that illness was caused by ingestion of 10^2–10^3 bacteria (Kaper *et al.*, 1995). The bacteria adhere to the surface of the small intestine where they multiply and form cholera enterotoxin (CT), which causes massive loss of fluid and salts from cells; invasion of the mucosa can occur resulting in systemic infection. Complications occur in 10–15% of patients.

Most *V. cholerae* non-O1/O139 strains lack the gene for cholera toxin, but can cause gastroenteritis and a mild form of cholera in healthy individuals and septicaemia in persons with underlying diseases, such as cirrhosis, diabetes or hemachromatosis, or who are subject to immunosuppressive therapy (Madden, 2004). These bacteria can also cause wound infections.

Many strains of *V. parahaemolyticus* are not pathogenic; on the West Coast and Gulf Coast of the US 3% and 0.2–0.3%, respectively, of strains were reported pathogenic (Madden, 2004). Ingestion of pathogenic *V. parahaemolyticus* causes nausea, vomiting, abdominal cramps, fever and diarrhoea (usually watery, sometimes bloody). Symptoms usually occur within 24 hours, typically the illness is self-limited and lasts for about 3 days. More severe disease occurs in persons with compromised immune systems. The infective dose in volunteers was estimated as between 10^5 and 10^6 bacteria, but outbreaks in the US in 1997–1998 indicated that it may be much lower (Daniels *et al.*, 2000; Madden, 2004). The bacterium can also cause skin infections if an open wound is exposed to warm seawater.

Ingestion of *V. vulnificus* can result in gastroenteritis in healthy persons and primary septicaemia in susceptible persons. Conditions that cause susceptibility include liver cirrhosis, hepatitis, metastatic cancer and liver transplantation (Strom & Paranjpye, 2000). Haematological conditions that result in elevated serum iron levels also predispose to infection as do chemotherapy, AIDS, and chronic diseases such as diabetes mellitus, renal disease, chronic intestinal disease, steroid dependency or low gastric acid. Following ingestion, susceptible persons develop primary septicaemia with sudden onset of fever and chills, often with vomiting, diarrhoea, abdominal pain and pain in the extremities (Strom & Paranjpye, 2000).

Within 24 hours of the onset of illness cutaneous lesions appear and septicaemic shock may occur. The lesions often become necrotic (necrotizing fasciitis) and surgical debridement or amputation may be necessary. Primary septicaemia is fatal in 60–75% of patients. The infective dose for gastroenteritis in healthy people is not known, but for susceptible persons fewer than 100 bacteria can probably cause septicaemia (FDA/CFSAN, 2005a). Infection through wounds results in symptoms that are similar, but differ in their timing and severity; the fatality rate for wound infections is 20–30%.

2.12.4 *Source and transmission*

Vibrios are present naturally in coastal waters in temperate and tropical regions worldwide. Important factors that affect their numbers are salinity and temperature. Pathogenic vibrios are most numerous at water temperatures between 20 and 30°C, and are isolated infrequently when water temperatures are lower than 10° or higher than 26–30°C (Motes *et al.*, 1998). Salinity levels of 2–5 parts per thousand (ppt) favour *V. cholerae*, up to 25 ppt favours *V. vulnificus*, while *V. parahaemolyticus* has been isolated from a salinity of 30 ppt (Duan & Su, 2005). Vibrios are able to digest chitin, this enables them to attach to, and colonize, zooplankton and crustaceans, whose shells contain a high proportion of chitin. They may also accumulate in the tissues of filter-feeding, bivalve shellfish such as clams, oysters and mussels and are present on the external surface and in the digestive tract of fishes.

Vibrio cholerae is present normally in estuarine waters. Non-O1/O139 strains are most numerous, but toxin-producing O1 strains also appear to survive for long periods in these environments. In regions where epidemics of cholera have occurred, discharge of human waste is liable to increase the numbers of this organism in coastal waters. *Vibrio cholerae* is also endemic in certain arid and inland areas of Africa that are distant from coastal waters (Kaper *et al.*, 1995).

Contaminated water supplies are mainly responsible for transmission of *V. cholerae* in developing countries, but in both developing and developed countries foodborne transmission can occur. This may result from the use of contaminated water in food preparation or from irrigation of vegetables with water contaminated by sewage (Shuval, 1993). Many types of seafood have been implicated also in transmission (Kaysner, 2000).

Vibrio parahaemolyticus is widespread in estuarine and coastal waters, sediment, suspended particles, plankton, and many species of fish and shellfish (Oliver & Kaper, 2001). The numbers are influenced by water temperature, salinity, and association with certain plankton, and apparently are not related to numbers of faecal coliforms or other indicators of faecal contamination. Occasionally this organism has been isolated from fresh-water or non-marine fish (Madden, 2004). There is evidence that new O3:K6 strains were present and may persist in French coastal waters (Quilici *et al.*, 2005) and may occur in the UK coastal environment (Martinez-Urtaza *et al.*, 2005).

V. vulnificus also occurs widely in estuarine environments, the numbers present rising with increase in water temperature between 15 and 22°C (Oliver & Kaper, 2001). Oysters from the US Gulf Coast harbour about 1000 *V. vulnificus*/g in the warmer months of April to October and usually fewer than 10/g during the remaining months (WHO, 2005).

In a survey of ~347 lots of oysters (harvested from coasts in the US and Canada) sampled at retail level throughout the US, *V. parahaemolyticus* was detected in 73% of samples; in

2.6% of samples the numbers present were $>10^5$/g (Cook *et al.*, 2002). *Vibrio vulnificus* was detected in 63% of these samples; in 4.1% the numbers were $>10^5$/g.

A high proportion of cases of infection with *V. parahaemolyticus* and *V. vulnificus* and some cases of infection with *V. cholerae* are associated with the consumption of seafood that is raw or has been re-contaminated after cooking.

2.12.5 *Examples of outbreaks*

Many outbreaks of cholera are the direct result of contamination of water supplies in regions that lack sewage systems, or where services have been disrupted by conflict or natural disasters. A wide range of foods have been implicated in outbreaks (Kaysner, 2000), some documented outbreaks are shown in Table 2.12.3.

Forty foodborne outbreaks of *V. parahaemolyticus* infection reported in the US between 1973 and 1998 are listed by Daniels *et al.* (2000). Examples of outbreaks in several countries, including large outbreaks in the US in 1998, are shown in Table 2.12.4. The outbreak in Alaska in 2004 was associated with rising temperatures in seawater, and extended by 1000 km the northernmost documented source of oysters that caused illness. There is evidence that local outbreaks in the US have coincided with large increases in sporadic cases nationally (CDC, 2006).

Because *V. vulnificus* mainly infects vulnerable people, most reports of infection are of individual cases rather than outbreaks. Most of the cases of foodborne infection with *V. vulnificus* in the US have resulted from consumption of raw oysters by susceptible individuals (CDC, 1996), but a few cases appear to have involved consumption of raw clams and cooked shrimp (Strom & Paranjpye, 2000). In Japan, foodborne infection occurs mainly in people who have eaten slices of raw fish (sashimi) and vinegary fish and rice (sushi) made with raw fish and shellfish caught in an inland sea or gulf during the summer (Fukushima & Seki, 2004).

2.12.6 *Main methods of control and prevention*

V. cholerae. Reduction and prevention of cholera in the developing world and in disaster areas requires the provision of safe drinking water, effective sewage disposal and hygienic preparation of food. In other areas, thorough cooking of seafood and avoidance of recontamination can reduce the risk of transmission.

V. parahaemolyticus. In the US, the National Shellfish Sanitation Program has issued guidance on a range of measures to minimize the sale of shellfish contaminated with vibrios or other microorganisms pathogenic to humans (FDA/CFSAN, 2005c). In states that are liable to be affected by *V. parahaemolyticus* these measures include monitoring shellfish growing waters for this organism. Depending on circumstances, shellfish growing areas should be closed, and harvesting prevented. Wholesalers, retailers and consumers should be advised about the potential problem, and it is recommended that shellfish should not be consumed raw during periods when they are liable to be contaminated with *V. parahaemolyticus*.

Table 2.12.3 Examples of outbreaks of foodborne disease caused by *Vibrio cholerae*.

Place, date	Serotype	Cases	Number hospitalized	Deaths	Food implicated	Evidence[a]	Where food was supplied	Factors leading to outbreak	Reference
Israel, 1970	O1 El Tor	176	nr	nr	Uncooked vegetables	D	Community	Cholera brought to city by visitors. Wastewater from sewers used to irrigate vegetables grown outside the city	Shuval *et al.* (1985)
Italy, 1973	O1 El Tor	278	nr	nr	Uncooked mussels and other seafood	S	Community	Inadequate treatment of sewage; unsanitary production, harvesting and marketing of seafood, consumption of undercooked products	Baine *et al.* (1974)
Portugal, 1974	O1 El Tor	nr	2467	48	Raw/undercooked cockles, spring water, commercially bottled water	S	Community	*V. cholerae* may have been brought to the area by visitors or people returning. Sewage from coastal towns discharged into sea resulting in contamination of shellfish. Spring may have been contaminated with river water	Blake *et al.* (1977)
US, 1977	O1 El Tor	8	nr	0	Boiled/steamed shrimp	M	Community	Some of boiled crabs had been held without refrigeration for approx. 6 h	Bradford *et al.* (1978)
Gilbert Islands, 1978	O1 El Tor	nr	572	19	Raw salt fish, sardines and clams from lagoon	D	Community	Lagoon subject to repeated faecal contamination with *V. cholerae*	McIntyre *et al.* (1979)
US, 1981	O1 El Tor	15	2	0	Cooked rice	S	Oil rig	Cooked rice rinsed with water (contaminated) and held for several hours at warm temperature	Johnston *et al.* (1983)

Singapore, 1982	O1 El Tor	37	nr	nr	Seafood	S	Canteen on construction site	Food handlers infected with *V. cholerae* O1. Lack of hygiene in kitchen, packets of food kept at room temperature for at least 4 h before sale	Goh *et al.* (1984)
Thailand, 1987	O1 El Tor	15	0	0	Uncooked pork	S	Food served at a funeral	A butcher who prepared the pork was an asymptomatic carrier. The pork was stored overnight without refrigeration	Swaddiwudhipong *et al.* (1990)
Thailand, 1988	O1 El Tor	71	5	0	Uncooked beef	S	Community	Beef may have been contaminated by an infected butcher or from flies. Some of the beef was kept for several hours without refrigeration	Swaddiwudhipong *et al.* (1992)
US, 1991	O1 El Tor	8	5	0	Crab meat	S	Private house	Contaminated, crabmeat transported illegally from Ecuador	CDC (1991); Finelli *et al.* (1992)
US, 1991	O1 El Tor	4	0	0	Frozen, fresh coconut milk imported from Thailand	M	Private house	Method of production allowed contamination; no decontamination step	Taylor *et al.* (1993)
Airline flight (1992)	O1 El Tor	75	10	1	Cold seafood salad served between Lima, Peru and Los Angeles, US	S	Airline flight	nr	Eberhart-Phillips *et al.* (1996)
Zambia, 2003–04	O1, El Tor	4343	nr	154	Raw vegetables	S	Community	nr	CDC (2004), Dubois *et al.* (2006)

[a]M (microbiological): identification of an organism of the same type from cases and in the suspect food or food ingredient, or detection of toxin in faeces or food; D (descriptive): other evidence reported by local investigators as indicating the suspect food; S (statistical): a significant statistical association between consumption of the suspect food and being a case.
nr, not reported.

Table 2.12.4 Examples of outbreaks of foodborne disease caused by *Vibrio parahaemolyticus*.

Place, date	Serotype	Cases	Number of hospitalized	Deaths	Food implicated	Evidence[a]	Place of supply	Factors leading to outbreak	Reference
US, Canada, 1997	O4:K12, O1:K6	209	nr	1	Raw or undercooked oysters	M	Restaurants and community	High temperatures in coastal sea surface and failure to cook oysters	CDC (1998)
France, 1997	nr	44	nr	0	Sauce made with imported seafood	S	Military regiment	Seafood heated inadequately	Lemoine *et al.* (1999)
Chile, 1997–98	O3:K6, O1:K56	298	nr	nr	Shellfish	M	Community	High temperature in coastal water and shellfish heated inadequately	Córdova *et al.* (2002); González-Escalona *et al.* (2005)
US, 1998	O3:K6	416	nr	nr	Oysters	nr	nr	nr	Daniels *et al.* (2000)
US, 1998	O3:K6	23	nr	0	Raw or undercooked shellfish	D	Community	High temperatures in surface water, shellfish not heated adequately	CDC (1999)
Spain, 1999	O4:K11	64	9	0	Raw oysters	D	Street market	Oysters not cooked	Lozano-León *et al.* (2003); Martinez-Urtaza *et al.* (2004)

US (Alaska), 2004	O6:K18	62	nr	0	Raw oysters	S, M	Cruise ship	Mean water temperatures were above 15°C for a longer period than in previous years. Oysters not cooked	McLaughlin *et al.* (2005)
Chile, 2004	O3:K6, O4:K12	~ 1500	nr	nr	Shellfish	nr	nr	Summer temperatures higher than normal	González-Escalona *et al.* (2005)
Mexico, 2004	O3:K6	~ 1225	nr	nr	Raw or undercooked shrimp	D	Community	High water temperature, shrimp not cooked thoroughly	Cabanillas-Beltrán *et al.* (2006)
Spain, 2004	O3:K6	80	nr	nr	Boiled crabs	D	Restaurant	Live crabs imported from UK, processed under unhealthy conditions and stored at room temperature for several hours before consumption	Martinez-Urtaza *et al.* (2005)
US, 2006	O4:K12	177	3	0	Raw or cooked shellfish, mainly oysters, clams, also lobster, scallops, crab, shrimp	D	Restaurants, seafood markets, recreational harvesting	Presumed increased levels of the bacterium in shellfish harvest areas	CDC (2006)

[a]M (microbiological): identification of an organism of the same type from cases and in the suspect food or food ingredient, or detection of toxin in faeces or food; D (descriptive): other evidence reported by local investigators as indicating the suspect food; S (statistical): a significant statistical association between consumption of the suspect food and being a case.
nr, not reported.

Reports of infection with *V. parahaemolyticus* in Europe, to date, are rare and in the EU measures to control the microbiological safety of shellfish do not include specific criteria for *V. parahaemolyticus* and *V. vulnificus* (EC, 2004, 2005). Research has started in Europe to assess the prevalence of *V. parahaemolyticus* in seafood.

V. vulnificus. The US National Shellfish Sanitation Program includes a management plan to reduce infection by *V. vulnificus* (FDA/CFSAN, 2005c). This includes education of at-risk groups of the public that consumption of raw, untreated oysters can be life-threatening (Acheson, 2005), and post-harvest measures, by industry, to reduce the levels of *V. vulnificus* in oysters including: immediate chilling of oysters after harvest, freezing, heat treatment, treatment by hydrostatic pressure, or irradiation (pending approval).

Heating oysters at 50°C for 5 minutes is stated to give a decrease of over 10,000-fold in the number of viable *V. parahaemolyticus* and *V. vulnificus* (FDA/CFSAN, 2005b). The measures designed to reduce foodborne infection with these bacteria will also reduce infection with other vibrio species.

Summary of measures advised in healthcare settings

1 Ensure that shellfish have been harvested from approved areas and are chilled promptly.
2 Avoid exposure of vegetables in the field to contaminated irrigation water.
3 Avoid cross-contamination to other foods.
4 Ensure that seafood is cooked thoroughly before serving, and that recontamination is avoided.
5 Apply employee health policy; handwashing; no bare hand contact with RTE foods.

References

Acheson, D.W.K. (2005) *Letter to Health Professionals Regarding the Risk of Vibrio vulnificus Septicemia Associated with the Consumption of Raw Oysters*. Available from http://www.cfsan.fda.gov/~dms/vvltr2.html. Accessed 9 March 2007.

Baine, W.B., Mazzotti, M., Greco, D. *et al.* (1974) Epidemiology of cholera in Italy. *Lancet* **ii**, 1370–1374.

Bisharat, N., Agmon, V., Finkelstein, R. *et al.* (1999) Clinical, epidemiological, and microbiological features of *Vibrio vulnificus* biogroup 3 causing outbreaks of wound infection and bacteraemia in Israel. Israel Vibrio Study group. *Lancet* **354**, 1421–1424.

Blake, P.A., Rosenberg, M.L., Costa, J.B. *et al.* (1977) Cholera in Portugal, 1974. 1. Modes of transmission. *American Journal of Epidemiology* **105**, 337–343.

Bradford, H.B., Jr. & Caraway, C.T. (1978) Follow-up on *Vibrio cholerae* serotype Inaba infection – Louisiana. *Morbidity and Mortality Weekly Report* **27**, 388–389.

Cabanillas-Beltrán, H., LLausás-Magaña, E., Romero, R. *et al.* (2006) Outbreak of gastroenteritis caused by the pandemic *Vibrio parahaemolyticus* O3:K6 in Mexico. *FEMS Microbiology Letters* **265**, 76–80.

Centers for Disease Control and Prevention (CDC) (1991) Epidemiologic notes and reports cholera – New Jersey and Florida. *Morbidity and Mortality Weekly Report* **40**(17), 287–290.

Centers for Disease Control and Prevention (CDC) (1996) *Vibrio vulnificus* infections associated with eating raw oysters – Los Angeles, 1996. *Morbidity and Mortality Weekly Report* **45**(29), 621–624.

Centers for Disease Control and Prevention (CDC) (1998) Outbreak of *Vibrio parahaemolyticus* infections associated with eating raw oysters – Pacific Northwest, 1997. *Morbidity and Mortality Weekly Report* **47**(22), 457–462.

Centers for Disease Control and Prevention (CDC) (1999) Outbreak of *Vibrio parahaemolyticus* infection associated with eating raw oysters and clams harvested from Long Island Sound – Connecticut, New Jersey, and New York, 1998. *Morbidity and Mortality Weekly Report* **48**(03), 48–51.

Centers for Disease Control and Prevention (CDC) (2004) Cholera epidemic associated with raw vegetables – Lusaka, Zambia, 2003–2004. *Morbidity and Mortality Weekly Report* **53**(34), 783–786.

Centers for Disease Control and Prevention (CDC) (2006) *Vibrio parahaemolyticus* infections associated with consumption of raw shellfish – Three States, 2006. *Morbidity and Mortality Weekly Report* **55**(31), 854–856.

Cook, D.W., O'Leary, P., Hunsucker, J.C. *et al.* (2002) *Vibrio vulnificus* and *Vibrio parahaemolyticus* in U.S. retail shell oyster: a national survey from June 1998 to July 1999. *Journal of Food Protection* **65**, 79–87.

Córdova, J.L., Astorga, J., Silva, W. & Riquelme, C. (2002) Characterization by PCR of *Vibrio parahaemolyticus* isolates collected during the 1997–1998 Chilean outbreak. *Biological Research* **35**, 433–440.

Daniels, N.A., MacKinnon, L., Bishop, R. *et al.* (2000) *Vibrio parahaemolyticus* infections in the United States, 1973–1998. *Journal of Infectious Diseases* **181**, 1661–1666.

Duan, J. & Yi-Cheng Su. (2005) Occurrence of *Vibrio parahaemolyticus* in two Oregon oyster-growing bays. *Journal of Food Science* **70**, M58–M63.

Dubois, A.E., Sinkala, M., Kalluri, P. *et al.* (2006) Epidemic cholera in urban Zambia: hand soap and dried fish as protective factors. *Epidemiology and Infection* **134**, 1226–1230.

Eberhart-Phillips, J., Besser, R.E., Tormey, M.P. *et al.* (1996) An outbreak of cholera from food served on an international aircraft. *Epidemiology and Infection* **116**, 9–13.

EC (2004) Regulation (EC) No. 853/2004 of the European Parliament and of the Council of 29 April 2004 laying down specific hygiene rules for food of animal origin. *Official Journal of the European Union* **L226/22**, 25 June 2004. Available from http://www.food.gov.uk/multimedia/pdfs/h2ojregulation.pdf. Accessed 9 March 2007.

EC (2005) Commission Regulation (EC) No. 2073/2005 of 15 November 2005 on microbiological criteria for foodstuffs. *Official Journal of the European Union* **L338**, 22 December 2005. Available from http://www.food.gov.uk/multimedia/pdfs/microbiolcriteria.pdf. Accessed 9 March 2007.

Faruque, S.M., Albert, M.J. & Mekalanos, J.J. (1998) Epidemiology, genetics and ecology of toxigenic *Vibrio cholerae*. *Microbiology and Molecular Biology Reviews* **62**, 1301–1314.

FDA/CFSAN (2004) *Bacteriological Analytical Manual*, Chapter 9, *Vibrio*. Available from http://www.cfsan.fda.gov/~ebam/bam-9.html. Accessed 9 March 2007.

FDA/CFSAN (2005a) *Vibrio vulnificus*. In: *Foodborne Pathogenic Microorganisms and Natural Toxins Handbook*. Available from http://www.cfsan.fda.gov/~mow/chap10.html. Accessed 9 March 2007.

FDA/CFSAN (2005b) *Quantitative Risk Assessment on the Public Health Impact of Pathogenic Vibrio parahaemolyticus in Raw Oysters*. Available from http://www.cfsan.fda.gov/~acrobat/vpra.pdf. Accessed 9 March 2007.

FDA/CFSAN (2005c) *National Shellfish Sanitation Program. Guide for the Control of Molluscan Shellfish 2005. Guidance Documents Chapter IV. Naturally Occurring Pathogens*. Available from http://www.cfsan.fda.gov/~ear/nss3-44.html. Accessed 9 March 2007.

Finelli, L., Swerdlow, D., Mertz, K. *et al.* (1992) Outbreak of cholera associated with crab bought from an area with epidemic disease. *Journal of Infectious Disease* **166**, 1433–1435.

Fukushima, H. & Seki, R. (2004) Ecology of *Vibrio vulnificus* and *Vibrio parahaemolyticus* in brackish environments of the Sada river in Shimane prefecture, Japan. *FEMS Microbiology Ecology* **48**, 221–229.

Goh, K.T., Lam, S., Kumarapathy, S. & Tan, J.L. (1984) A common source foodborne outbreak of cholera in Singapore. *International Journal of Epidemiology* **13**, 210–215.

González-Escalona, N., Cachicas, V., Acevedo, C. *et al.* (2005) *Vibrio parahaemolyticus* diarrhea, Chile, 1998 and 2004. *Emerging Infectious Diseases* **11**, 129–131.

Hsueh, P.-R., Lin, C.-Y., Tang, H.-J. *et al.* (2004) *Vibrio vulnificus* in Taiwan. *Emerging Infectious Diseases* **10**, 1363–1368.

International Commission on Microbiological Specifications for Foods (ICMSF) (1996) *Vibrio cholerae, Vibrio parahaemolyticus, Vibrio vulnificus*. In: *Microorganisms in Foods 5. Microbiological Specifications of Food Pathogens*. Chapters 22, 23, 24. Blackie Academic and Professional, London, pp. 414–439.

Johnston, J.M., Martin, D.L., Perdue, J. *et al.* (1983) Cholera on a Gulf Coast oil rig. *New England Journal of Medicine* **309**, 523–526.

Kaper, J.B., Morris, J.G., Jr. & Levine, M.M. (1995) Cholera. *Clinical Microbiology Reviews* **8**, 48–86.

Kaysner, C.A. (2000) *Vibrio* species. In: Lund, B.M., Baird-Parker, T.C. & Gould, G.W. (eds). *The Microbiological Safety and Quality of Food*, Vol. 2. Aspen Publishers, Gaithersburg, MD, pp. 1336–1362.

Lemoine, T., Germanetto, P. & Giraud, P. (1999) Toxi-infection alimentaire collective a *Vibrio parahaemolyticus*. *Bullein Epidemiologie Hebdomadaire* **10,** 37–38.

Lozano-León, A., Torres, J., Osorio, C.R. & Martínez-Urtaza, J. (2003) Identification of *tdh*-positive *Vibrio parahaemolyticus* from an outbreak associated with raw oyster consumption in Spain. *FEMS Microbiology Letters* **226**, 281–284.

Madden, J.M. (2004) *Vibrio* species. In: *Bacteria Associated with Foodborne Diseases. An IFT Scientific Status Summary*. Available from http://www.foodprocessing.com/whitepapers/2004/4.html. Accessed 9 March 2007.

Martinez-Urtaza, J., Lozano-Leon, A., DePaola, A. *et al.* (2004) Characterization of pathogenic *Vibrio parahaemolyticus* isolates from clinical sources in Spain and comparison with Asian and North American pandemic isolates. *Journal of Clinical Microbiology* **42**, 4672–4678.

Martinez-Urtaza, J., Simental, L., Velasco, D. *et al.* (2005) Pandemic *Vibrio parahaemolyticus* O3:K6, Europe. *Emerging Infectious Diseases* **11**, 1319–1320.

McIntyre, R.C., Tira, T., Flood, T. & Blake, P.A. (1979) Modes of transmission of cholera in a newly infected population on an atoll: Implications for control measures. *Lancet* **i**, 311–314.

McLaughlin, J.B., DePaola, A., Bopp, C. *et al.* (2005) Outbreak of *Vibrio parahaemolyticus* gastroenteritis associated with Alaskan oysters. *New England Journal of Medicine* **353**, 1463–1470.

Morris, J.G., Jr. (2003) Cholera and other types of vibriosis: a story of human pandemics and oysters on the half shell. *Clinical Infectious Diseases* **37**, 272–280.

Motes, M.L., DePaola, A., Cook, D.W. *et al.* (1998) Influence of temperature and salinity on *Vibrio vulnificus* in North Gulf and Atlantic Coast oysters (*Crassostrea virginica*). *Applied and Environmental Microbiology* **64**, 1459–1465.

Oliver, J.D. & Kaper, J.B. (2001) *Vibrio* species. In: Doyle, M.P., Beuchat, L.R. & Montville, T.J. (eds). *Food Microbiology. Fundamentals and Frontiers*, 2nd edn. ASM Press, Washington, DC, Chapter 13, pp. 263–300.

Quilici, M.-L., Robert-Pillot, A., Picart, J. & Fournier, J.-M (2005) Pandemic *Vibrio parahaemolyticus* O3:K6 spread, France. *Emerging Infectious Diseases* **11**, 1148–1149.

Sack, D.A., Sack, R.B., Nair, G.B. & Siddique, A.K. (2004) Cholera. *Lancet* **363**, 223–233.

Shuval, H.I. (1993) Investigation of typhoid fever and cholera transmission by raw wastewater irrigation in Santiago, Chile. *Water Science and Technology* **27**, 167–174.

Shuval, H.I., Yekutiel, P. & Fattal, B. (1985) Epidemiological evidence for helminth and cholera transmission by vegetables irrigated with wastewater: Jerusalem – a cases study. *Water Science and Technology* **17**, 433–442.

Strom, M.S. & Paranjpye, R.N. (2000) Epidemiology and pathogenesis of *Vibrio vulnificus*. *Microbes and Infection* **2**, 177–188.

Swaddiwudhipong, W., Akarasewi, P., Chayaniyayodhin, T. *et al.* (1990) A cholera outbreak associated with eating uncooked pork in Thailand. *Journal of Diarrhoeal Disease Research* **8**, 382–383.

Swaddiwudhipong, W., Jirakanuisun, R. & Rodklai, A. (1992) A common source foodborne outbreak of El Tor cholera following the consumption of uncooked beef. *Journal of the Medical Association of Thailand* **75**, 413–416.

Taylor, J.L., Tuttle, J., Pramukul, T. *et al.* (1993) An outbreak of cholera in Maryland associated with imported, commercial frozen fresh coconut milk. *Journal of Infectious Diseases* **167**, 1330–1335.

WHO (2005) *Risk Assessment of Vibrio vulnificus in Raw Oysters: Interpretative Summary and Technical Report*. Microbial Risk Assessment Series 8.

2.13 *Yersinia enterocolitica* and *Y. pseudotuberculosis*

2.13.1 *Importance as a cause of foodborne disease*

Yersinia enterocolitica. Few countries in Europe reported cases of yersiniosis in 2000 to the WHO. Of those that gave information, the reported incidence in Belgium was 6 cases per 100,000 persons and in that in Norway was 3.2 cases per 100,000 (WHO, 2003). In England and Wales in 2000, 24 cases were reported (HPA, 2007), an incidence of <0.05 cases per 100,000 persons. The reported annual incidence in Finland during 1995–1999 was 11.2–17.4 cases per 100,000 persons (Fredriksson-Ahomaa *et al.*, 2001), that in Germany in 2002 was 9.1 cases per 100,000, varying from 4 to 27 cases per 100,000 persons in different regions (Fredriksson-Ahomaa *et al.*, 2004) and that in Sweden was equivalent to 6–6.7 cases per 100,000 (Thisted Lambertz & Danielsson-Tham, 2005). In the US, the reported annual incidence was about 0.85 cases per 100,000, the estimated total number of cases was assumed to be 38 times this (Mead *et al.*, 1999). Outbreaks have been reported in Japan since 1972 (Maruyama, 1987). In addition to foodborne disease, *Y. enterocolitica* is an important cause of often fatal bacteraemia following transfusion of contaminated blood or blood products (CDC, 1997; Leclercq *et al.*, 2005).

Yersinia pseudotuberculosis. Outbreaks in humans have been reported occasionally, particularly from Finland and Japan (Tsubokura *et al.*, 1989). The means of transmission to humans has been unproven until recently, when outbreaks were reported that were associated with contaminated, fresh produce. An annual infection rate of one case per 100,000 persons was reported in Finland in 1999 (Niskanen *et al.*, 2002).

2.13.2 *Characteristics of the organisms*

Yersinia pseudotuberculosis was isolated first in 1883, whereas *Y. enterocolitica* was described first (as *Bacterium enterocoliticum*) in 1939. *Yersinia* spp. are facultatively anaerobic, Gram-negative, non-sporing rods or coccobacilli. Eleven species of *Yersinia* have been described three of which, *Y. pestis* (the cause of plague, and transmitted by flea bites or aerosols), *Y. pseudotuberculosis* and *Y. enterocolitica*, are pathogenic in humans. *Yersinia pseudotuberculosis* and *Y. enterocolitica* are non-motile when grown at 37°C but motile at 22–29°C.

Strains of *Y. enterocolitica* have been differentiated into biotypes (BT), based on ability to metabolize various substrates, and serotypes, based on cell wall antigens (Table 2.13.1). Most pathogenic strains are in biotypes 1B, 2, 3, 4, 5. Most strains of BT 1A have been considered generally as non-pathogenic, except in patients with underlying disorders. The majority of reported infections worldwide have been caused by BT 4, serotype O:3; BT 2, serotype O:9; BT 2 or 3 serotype O:5,27; or BT 1B serotype O:8 strains (Robins-Browne, 2001). Serological cross-reactions may occur between *Y. enterocolitica* and *Brucella abortus*, salmonella species and other bacteria and possibly thyroid-tissue antigens (Nesbakken, 2000). Biotyping, serotyping, antimicrobial susceptibility, phage typing and, increasingly, molecular methods, particularly PFGE, have been used to characterize isolates (Fredriksson-Ahomaa *et al.*, 2006).

Table 2.13.1 Relationship between biotype, O-serotype and pathogenicity of *Y. enterocolitica*.

Y. enterocolitica biotype	Serotype
1A	O:4; O:5; O:6.30; O:6,31; O:7,8; O:7,13; O:10; O:14; O:16; O:21; O:22; O:25; O:37; O:41,42; O:46; O:47; O:57; NT[a]
1B	**O:4,32; O:8; O:13a,13b**; O:16; **O:18; O:20; O:21 (Tacoma);** O:25; O:41,42; NT
2	**O:5,27; O:9**; O:27.
3	**O:1,2,3; O:3; O:5,27**.
4	**O:3**
5	**O:2,3**

Modified from Nesbakken, 2000; Robins-Browne, 2001.
[a] Not typeable.
Serotypes that include strains considered as primary pathogens are in bold.

The temperature range for growth of *Y. enterocolitica* is 0–44°C, with maximum growth rate at 28–30°C; the bacterium probably multiplies more rapidly at 0–5°C than any other foodborne pathogenic bacterium and can grow in a range of foods at refrigeration temperatures. For example, virulent strains of serotypes O:5,27, O:8 and O:9, inoculated into pasteurized skim milk stored at 4°C, multiplied and were not suppressed by the other bacteria present (Amin & Draughon, 1987). For the first few days, the number of yersinias increased about tenfold each day, and by a factor of 10^4–10^6 in 2 weeks, which was the shelf life of the milk.

Yersinias can survive for prolonged periods in frozen foods, even after repeated freezing and thawing, and can grow in the presence of >5% but not 7% NaCl and over a pH range of approximately 4–10. *Yersinia enterocolitica* is sensitive to heat, and high numbers are destroyed by pasteurization of milk (ICMSF, 1996).

Strains of *Y. pseudotuberculosis* have been differentiated into serotypes O:1 to O:14; serotypes O:1 to O:5 have been isolated in Europe and the Far East and almost all are pathogenic to humans; serotypes O:6 to O:14 were isolated from animals and environments in the Far East but not from clinical samples (Fukushima *et al.*, 2001). Antigenic relationships have been shown between some serotypes and serotypes of *Salmonella*, *Escherichia coli* and *Enterobacter cloacae*. Growth occurs at temperatures between 4 and 42°C; resistance to heating is similar to that of *Y. enterocolitica*. The organism can grow over a pH range from approximately 4–10.

Difficulties in isolation have limited the information available on pathogenic *Y. enterocolitica* and *Y. pseudotuberculosis*; the increasing availability of PCR methods should provide improved information on the incidence and epidemiology of yersiniosis (Fredriksson-Ahomaa *et al.*, 2006).

Pathogenicity of *Y. enterocolitica* Biotypes 1B, 2, 3, 4, 5, has been associated with several virulence factors. These include two chromosomal genes *ail* (attachment invasion locus) and *inv* (invasion) coding for adherence and invasion of epithelial cells, and a 70–75 kb plasmid pYV (plasmid for *Yersinia* virulence) (Robins-Browne, 2001; Fredriksson-Ahomaa *et al.*, 2006). Biotype 1B strains that are highly virulent carry a 'pathogenicity island'; involved in an iron acquisition system, these strains are more virulent for mice infected orally than strains that lack the pathogenicity island, and may be more virulent for humans (Robins-Browne, 2001).

Strains lacking the pYV plasmid, including BT 1A strains, have generally been considered avirulent, but there is evidence that some BT 1A strains lacking the pYV plasmid may cause disease resembling yersiniosis caused by plasmid-bearing strains (Tennant *et al.*, 2003).

Pathogenicity of European strains of *Y. pseudotuberculosis* was associated with the presence of a pYV plasmid, and in the case of some serotype O1 strains with a high-pathogenicity island, while pathogenicity of Far Eastern strains was associated with presence of a pYV plasmid and production of a superantigenic toxin, with or without the presence of a high-pathogenicity island (Fukushima *et al.*, 2001).

2.13.3 *Disease in humans*

Yersinia enterocolitica. The median infective dose is probably greater than 10^4 CFU. Symptoms develop within 4–7 days and may last for 1–3 weeks or for several months. In a study of Canadian children with enterocolitis the organism was excreted in faeces for 14–97 days (mean 42) (Cover & Aber, 1989). In children less than 5 years old the main symptoms are diarrhoea, low-grade fever, abdominal pain and enterocolitis, often accompanied by a sore throat (Robins-Browne, 2001). Occasionally, complications may occur. In older children and adolescents, symptoms often include pseudoappendicitis, due to acute inflammation of the terminal ileum or mesenteric lymph nodes, accompanied by fever and little or no diarrhoea. Pseudoappendicitis has led to appendectomies before the cause of the symptoms has been diagnosed. In adults pharyngitis may occur in association with gastrointestinal illness or independently. Focal infection may occur in numerous extraintestinal sites in the absence of detectable bacteraemia (Cover & Aber, 1989).

Complications include autoimmune reactions, particularly reactive arthritis, Reiter's syndrome and uveitis, all associated especially with the presence of the human leukocyte antigen B27 (HLA-B27), and erythema nodosum (Robins-Browne, 2001). Bacteraemia is rare, except in people who are immunocompromised or affected by iron overload states, particularly if these are managed by treatment with desferrioxamine B; other risk factors include blood dyscrasias, malnutrition, chronic liver disease, chronic liver failure or diabetes mellitus (Nesbakken, 2000; Robins-Browne, 2001). The case fatality for bacteraemia can be 30–60%.

Yersinia pseudotuberculosis. The symptoms are similar to those caused by *Y. enterocolitica* (Nesbakken, 2000) and can be followed by reactive arthritis (Hannu *et al.*, 2003).

2.13.4 *Source and transmission*

Yersinia enterocolitica occurs in the intestinal tracts of many species of mammals, birds, fish, shellfish and amphibians and is also found in the environment, particularly in soil, on vegetation, and in lakes, rivers and streams. Many of the strains found in these sources lack pathogenicity markers, their significance for human and animal health is not clear and they are considered generally to be non-pathogenic. Some farm animals, including sheep, cattle and deer, can suffer symptoms as a result of infection with *Y. enterocolitica* or *Y. pseudotuberculosis*, but evidence of transmission to humans from these animals is limited (Robins-Browne, 2001). Apparently healthy pigs often harbour strains that are pathogenic to humans, mainly bioserotype 4/0:3 and sometimes serotype O:9, and there is evidence that pigs are a major source of strains infecting humans. The organisms appear to colonize the tonsils, tongues and throats of pigs, and are found also in the faeces and intestinal mucosa (Verhaegen *et al.*, 1998). In Belgium, meat dishes including pork were liable to be undercooked, and a case–control study showed that the occurrence of yersiniosis was associated with the practice of feeding young children with raw or undercooked pork (Tauxe *et al.*, 1987). Following this work, a media campaign in Belgium to dissuade people from this practice, combined with changes in slaughtering to minimize contamination of carcasses with yersinia, has resulted in a decrease in the number of reported infections with *Y. enterocolitica* in Belgium (Verhaegen *et al.*, 1998).

Edible pig offal was reported as an important means of transmission of *Y. enterocolitica* bioserotype 4/O:3 to humans in Finland (Fredriksson-Ahomaa *et al.*, 2001) and this bioserotype was found frequently in butchers' shops in Germany, particularly in pork products (Fredriksson-Ahomaa *et al.*, 2004). Using PCR methods following enrichment, pathogenic *Y. enterocolitica* was detected in a range of pig products (Fredriksson-Ahomaa *et al.*, 2006), including 25% of 225 samples of minced pork samples from shops in Finland (Fredriksson-Ahomaa *et al.*, 1999), up to 79% of 350 samples of chitterlings (pork intestines) and up to 38% of 350 minced pork samples in the US (Boyapalle *et al.*, 2001), and in 10% of 91 raw pork samples in Sweden (Thisted Lambertz & Danielsson-Tham, 2005).

Using cultural methods, faecal carriage of *Y. enterocolitica* was found at slaughter in 6.3% of cattle, 10.7% of sheep and 26.1% of pigs in the UK (McNally *et al.*, 2004). The predominant biotype isolated from these animals was 1A, which is generally assumed to be non-pathogenic. Biotype 1A was also the major biotype identified from humans with diarrhoea in this study. Whether the human isolates were the cause of disease or secondary colonizers is not known, neither was it reported whether the patients were particularly susceptible because of underlying disease. The major, recognized pathogenic bioserotypes isolated from humans in the UK in the same period were 3/O:9 and 4/O:3; bioserotype 3/0:9 formed 11% of isolates from pigs but was not found in isolates from sheep or cattle, and bioserotype 4/0:3 formed 5% of isolates from pigs, only 1% of isolates from sheep, and was not found in isolates from cattle. The pathogenic bioserotype 3/O:5,27 formed 1% of isolates from humans and was isolated from a high proportion of pigs and sheep and a lower proportion of cattle.

Yersinia pseudotuberculosis can be transmitted by rodents and other infected animal species and by wild birds. Serotype O:3 strains have been associated with disease in animals such as sheep, goats, pigs and less frequently cattle (Seimiya *et al.*, 2005). Pathogenic bioserotype 2/O:3 strains were isolated by culture from the tonsils of 4% of fattening pigs in Finland; *Y. pseudotuberculosis* was found in between 2 and 6% of tonsils or oral cavity of pigs in Germany and Japan, and bioserotype 2/O:3 has been isolated from cattle in Brazil (Niskanen *et al.*, 2002). Pathogenic serotype O:2 was reported on 1.6% of 128 samples of iceberg lettuce on a farm in Finland, and possibly resulted from use of irrigation water contaminated with faeces of wild animals (Niskanen *et al.*, 2003). Transmission appears possible on pork products, and may occur via water or plant foods contaminated by animals or birds.

Table 2.13.2 Examples of outbreaks of foodborne disease caused by *Yersinia enterocolitica*.

Place, date	Serotype	Cases[a]	Number hospitalized	Food implicated	Evidence[b]	Where food was provided	Factors leading to outbreak	Reference
US, 1976	O:8	>222	36 (16 appendectomies)	Chocolate milk	S and M	School cafeterias	Contamination probably occurred at dairy when chocolate syrup was mixed with previously pasteurized milk by hand with a perforated metal stirring rod	Black *et al.* (1978)
Japan, 1980	O:3	1051	nr	Milk	M	Pre-primary and junior high school	nr	Maruyama (1987)
US , 1981	O:8	159	7 (5 appendectomies)	Suspension of powdered milk; turkey chow mein	S	Summer camp	Probable contamination by food handlers	Shayegani *et al.* (1983)
US, 1981–1982	O:8, O:Tacoma	50	15 (2 appendectomies; 1 partial colectomy)	Tofu packed in untreated spring water	M	Community	Tofu was steamed but untreated spring water was added before the package was sealed. Inadequate hygiene practices in production plant	Aulisio *et al.* (1983); Tacket *et al.* (1985)
US, 1982	O:8	16	nr	Bean sprouts, well water	M	Brownie camp	Immersion of bean sprouts in contaminated well water	Cover & Aber (1989)
US, 1982	O:13	>172	(17 appendectomies)	Pasteurized milk	S	Community	Possible contamination on outside of bottles. Outbreak strain found on milk crate on a pig farm where outdated milk was fed to pigs	Aulisio *et al.* (1982); Tacket *et al.* (1984)

Hungary, 1983	O:3	8	nr	Pork cheese	M	Domestic	Pork cheese was 'a meat product containing small pieces of boiled chitterlings[c] stuffed into a coat prepared from the stomach of the hog and heated in boiling water'	Marjai *et al.* (1987)
UK, 1984-5	O:10K, O:6,30	36 (mostly mild symptoms	Nosocomial	Pasteurized milk	M	Paediatric ward of hospital	Pasteurized milk supplied in bottles. Washing of bottles often unsatisfactory; contamination of bottle filler valve may have occurred	Greenwood & Hooper, (1990); Green wood *et al.* (1990)
US, 1988–1989	O:3	15	7	Pork chitterlings[c,d]	M	Community	Infants affected. Households involved in cleaning raw pork intestines[c]	Lee *et al.* (1990)
US, 1995	O:8	10	3 (1 appendectomy)	Pasteurized milk	S	Community	Post-pasteurization contamination of milk. Pigs also kept at the dairy. Milk bottles washed with untreated well water before filling	Ackers *et al.* (2000)
US, 2002	O:3	12	4	Pork chitterlings[d]	D	Community	Infants affected. Households involved in cleaning raw pork intestines[d]	Jones *et al.* (2003)
Japan, 2004	O:8	42	nr	Salad containing apples, cucumbers, ham, potatoes, carrots, mayonnaise	M	Nursery school	nr	Sakai *et al.* (2005)

[a] No deaths reported.

[b] M (microbiological): identification of an organism of the same type from cases and in the suspect food or food ingredient, or detection of toxin in faeces or food; D (descriptive): other evidence reported by local investigators as indicating the suspect food; S (statistical): a significant statistical association between consumption of the suspect food and being a case.

[c] Chitterlings: pig intestines.

[d] Outbreak was attributed to exposure during preparation of chitterlings, not to consumption.

nr, not reported.

2.13.5 *Examples of outbreaks*

Between 1971 and 1981 outbreaks of yersiniosis were reported from several countries, but in most cases the means of transmission of the bacteria was not detected (Nesbakken, 2000). In the Auvergne region of France, an increase in the incidence of *Y. enterocolitica* 0:9 infections in humans between 1990 and 1998 followed an increase in the incidence of yersiniosis in cattle (Gourdon *et al.*, 1999). Most cases of human infection may be considered sporadic (Verhaegen *et al.*, 1998).

Examples of foodborne outbreaks caused by *Y. enterocolitica* are shown in Table 2.13.2. Two of these outbreaks were associated with the preparation, not the consumption, of pork chitterlings. This preparation can involve cleaning the chitterlings in a sink for several hours and considerable risk of cross-contamination and transmission to other members of the household including infants. A third outbreak was associated with consumption of pork cheese (made with boiled chitterlings) from which *Y. enterocolitica* was isolated; probably the chitterlings were heated inadequately or were re-contaminated after heating and before inclusion in the final product. Three outbreaks were associated with use of contaminated water in food production.

Outbreaks of suspected foodborne infection with *Y. pseudotuberculosis* in Japan have been reported since 1977 (Tsubokura *et al.*, 1989). Reported foodborne outbreaks are shown in Table 2.13.3. Further outbreaks have been reported in Finland; although a food source was not detected, eating outside the home was identified as a risk factor (Jalava *et al.*, 2004).

2.13.6 *Main methods of prevention and control*

It may be possible, by management practices, to reduce the number of herds of pigs that carry *Y. enterocolitica* (Nesbakken, 2000). In Norway, changes in practice during the slaughter of pigs, in order to minimize contamination of the carcass from the intestine, rectum, tonsils and tongue, were followed by a marked reduction in reported human yersiniosis, and a similar result occurred in Sweden (Nesbakken, 2000).

Salad vegetables and fruit should be protected from contamination by animals in the field and from irrigation with polluted water supplies. Consumption of untreated water supplies that may have been contaminated by wild animals or birds should be avoided, it is also important to prevent contamination of drinking water and water used in food production.

Summary of measures in healthcare settings

1. Obtain foods from reputable sources. Vegetables and fruit to be eaten raw should be obtained from sources where contamination of soil, surface water and irrigation water from animal sources is prevented; contamination during storage should be prevented.
2. Salad vegetables and fruit should be washed thoroughly.
3. Avoid the use of meat products liable to be highly contaminated (e.g. pork chitterlings).
4. Follow good hygienic practice, clean and disinfect equipment used to handle raw meat.
5. Cook meat products, particularly pork, thoroughly.

Table 2.13.3 Examples of outbreaks of foodborne disease caused by *Yersinia pseudotuberculosis*.

Place, date	Serotype	Cases	Number hospitalized	Deaths	Food implicated	Evidence[a]	Where food was provided	Factors leading to outbreak	Reference
Finland, 1997	O:3	38	6 (4 appendectomies)	0	Food prepared in school kitchen	D	School	nr	Pebody *et al.* (1997)
Finland, 1998	O:3	47	16 (5 appendectomies)	1	Iceberg lettuce	S	Community	Wildlife, particularly roe deer, had access to water used, untreated, for spray irrigation of lettuce in field	Nuorti *et al.* (2004)
Canada, 1998	O:1b	74	nr	nr	A brand of homogenized milk suspected but not proven	S	Community	nr	Nowgesic *et al.* (1999)
Finland, 2003	O:1	111[b] con-firmed (possibly 558)	9 (One appendectomy)	0	Raw, grated carrots	S, M[c]	Day-care centre and school supplied by the same institutional kitchen	Carrots had been harvested in autumn 2002 and stored in open containers in unenclosed barn, accessible to rodents	Jalava *et al.* (2006)
Finland, 2004	O:1	125	nr	nr	Raw carrots	S	School; community	Contamination by wild animals on the farm, possibly during storage	Takkinen *et al.* (2004)

[a] M (microbiological): identification of an organism of the same type from cases and in the suspect food or food ingredient, or detection of toxin in faeces or food; D (descriptive): other evidence reported by local investigators as indicating the suspect food; S (statistical): a significant statistical association between consumption of the suspect food and being a case.

[b] Sixty-one cases developed erythema nodosum, one developed reactive arthritis.

[c] No implicated carrots were available for culture, but outbreak strain was isolated from soil samples containing carrot residue, taken from the areas of washing, peeling and production line at the production farm.

nr, not reported.

References

Ackers, M.-L., Schoenfeld, S., Markman, J. *et al.* (2000) An outbreak of *Yersinia enterocolitica* O:8 infections associated with pasteurized milk. *Journal of Infectious Diseases* **181**, 1834–1837.

Amin, M.K. & Draughon, F.A. (1987) Growth characteristics of *Yersinia enterocolitica* in pasteurized, skim milk. *Journal of Food Protection* **50**, 849–852.

Aulisio, C.G., Lanier, J.M. & Chappel, M.A. (1982) *Yersinia enterocolitica* O:13 associated with outbreaks in three Southern States. *Journal of Food Protection* **45**, 1263.

Aulisio, C.G., Stanfield, J.T., Weagant, S.D. & Hill, W.E. (1983) Yersiniosis associated with tofu consumption: serological, biochemical and pathogenicity studies of *Yersinia enterocolitica* isolates. *Journal of Food Protection* **46**, 226–239.

Black, R.E., Jackson, R.J., Tsai, T. *et al.* (1978) Epidemic *Yersinia enterocolitica* infection due to contaminated chocolate milk. *New England Journal of Medicine* **298**, 76–79.

Boyapalle, S., Wesley, I.V., Hurd, H.S. & Gopal, R.P. (2001) Comparison of culture, multiplex, and 5'nuclease polymerase chain reaction assays for the rapid detection of *Yersinia enterocolitica* in swine and pork products. *Journal of Food Protection* **64**, 1352–1361.

Centers for Disease Control and Prevention (CDC) (1997) Red blood cell transfusions contaminated with *Yersinia enterocolitica* – United States, 1991–1996, and initiation of a national study to detect bacteria-associated transfusion reactions. *Morbidity and Mortality Weekly Report* **46**(24), 553–566.

Cover, T.L. & Aber, R.C. (1989) *Yersinia enterocolitica*. *New England Journal of Medicine* **321**, 16–24.

Fredriksson-Ahomaa, M., Hallanvuo, S., Korte, T. *et al.* (2001) Correspondence of genotypes of sporadic *Yersinia enterocolitica* bioserotype 4/O:3 strains from human and porcine sources. *Epidemiology and Infection* **127**, 37–47.

Fredriksson-Ahomaa, M., Hielm, S. & Korkeala, H. (1999) High prevalence of yadA-positive *Yersinia enterocolitica* in pig tongues and minced meat at the retail level in Finland. *Journal of Food Protection* **62**, 123–127.

Fredriksson-Ahomaa, M., Koch, U., Bucher, M. & Stolle, A. (2004) Different genotypes of *Yersinia enterocolitica* 4/O:3 strains widely distributed in butcher shops in the Munich area. *International Journal of Food Microbiology* **95**, 89–94.

Fredriksson-Ahomaa, M., Stolle, A. & Korkeala, H. (2006) Molecular epidemiology of *Yersinia enterocolitica* infections. *FEMS Immunology and Medical Microbiology* **47**, 315–329.

Fukushima, H., Matsuda, Y., Seki, R. *et al.* (2001) Geographical heterogeneity between Far Eastern and Western Countries in prevalence of the virulence plasmid, the superantigen *Yersinia pseudotuberculosis*-derived mitogen, and the high-pathogenicity island among *Yersinia pseudotuberculosis* strains. *Journal of Clinical Microbiology* **39**, 3541–3547.

Gourdon, F., Beytout, J., Reynaud, A. *et al.* (1999) Human and animal epidemic of *Yersinia enterocolitica* O:9, 1989–1997, Auvergne, France. *Emerging Infectious Diseases* **5**, 719–721.

Greenwood, M.H. & Hooper, W.L. (1990) Excretion of *Yersinia* spp. associated with consumption of pasteurized milk. *Epidemiology and Infection* **104**, 345–350.

Greenwood, M.H., Hooper, W.L. & Rodhouse, J.C. (1990) The source of *Yersinia* spp. in pasteurized milk: an investigation at a dairy. *Epidemiology and Infection* **104**, 351–360.

Hannu, T., Mattile, L., Nuorti, J.P. *et al.* (2003) Reactive arthritis after an outbreak of *Yersinia pseudotuberculosis* serotype O:3 infection. *Annals of the Rheumatic Diseases* **62**, 866–869.

Health Protection Agency (HPA) (2007) *Yersinia spp. Laboratory Reports of all Isolations England and Wales, 1981–2006*. Available from http://www.hpa.org.uk/infections/topics_az/yersinia/data.htm. Accessed 12 March 2007.

International Commission on Microbiological Specifications for Foods (ICMSF) (1996) *Yersinia enterocolitica*. In: *Microorganisms in Foods 5. Microbiological Specifications of Food Pathogens*. Blackie Academic and Professional, London, pp. 458–478.

Jalava, K., Hakkinen, M., Valkonen, M. *et al.* (2006) An outbreak of gastrointestinal illness and erythema nodosum from grated carrots contaminated with *Yersinia pseudotuberculosis*. *Journal of Infectious Diseases* **194**, 1209–1216.

Jalava, K., Hallanvuo, S., Nakari, U.-M. *et al.* (2004) Multiple outbreaks of *Yersinia pseudotuberculosis* infections in Finland. *Journal of Clinical Microbiology* **42**, 2789–2791.

Jones, T.F., Buckingham, S.C., Bopp, C.A. *et al.* (2003) From pig to pacifier: chitterling-associated yersiniosis outbreak among black infants. *Emerging Infectious Diseases* **9**, 1007–1009.

Leclercq, A., Martin, L. Vergnes, M.L. *et al.* (2005) Fatal *Yersinia enterocolitica* biotype 4 serovar O:3 sepsis after red blood cell transfusion. *Transfusion* **45**, 814–818.

Lee, L.A., Gerber, A.R., Lonsway, D.R. *et al.* (1990) *Yersinia enterocolitica* O:3 infections in infants and children, associated with the household preparation of chitterlings. *New England Journal of Medicine* **322**, 984–987.

Marjai, E., Kalman, I., Kajary, A. *et al.* (1987) Isolation from food and characterization by virulence tests of *Yersinia enterocolitic*a associated with an outbreak. *Acta Microbiologica Hungarica* **34**, 97–109.

Maruyama, T. (1987) *Yersinia enterocolitica* infection in humans and isolation of the organism from pigs in Japan. *Contributions to Microbiology and Immunology* **9**, 48–55.

McNally, A., Cheasty, T., Fearnley, C. *et al.* (2004) Comparison of the biotypes of *Yersinia enterocolitica* isolated from pigs, cattle and sheep at slaughter and from humans with yersiniosis in Great Britain during 1999–2000. *Letters in Applied Microbiology* **39**, 103–108.

Mead, P.S., Slutsker, L., Dietz, V. *et al.* (1999) Food-related illness and death in the United States. *Emerging Infectious Diseases* **5**, 607–625.

Nesbakken, T. (2000) *Yersinia* species. In: Lund, B.M., Baird-Parker, T.C. & Gould, G.W. (eds). *The Microbiological Safety and Quality of Food*. Aspen Publishers, Gaithersburg, MD, pp. 1363–1393.

Niskanen, T., Fredriksson-Ahomaa, M. & Korkeala, H. (2002) *Yersinia pseudotuberculosis* with limited genetic diversity is a common finding in tonsils of fattening pigs. *Journal of Food Protection* **65**, 540–545.

Niskanen, T., Fredriksson-Ahomaa, M. & Korkeala, H. (2003) Occurrence of *Yersinia pseudotuberculosis* in iceberg lettuce and environment. *Advances in Experimental Medicine and Biology* **529**, 383–385.

Nowgesic, E., Fyfe, M., Hockin, J. *et al.* (1999) Outbreak of *Yersinia pseudotuberculosis* in British Columbia – November 1998. *Canada Communicable Disease Report* **25–11**. Available from http://www.phac-aspc.gc.ca/publicat/ccdr-rmtc/99vol25/dr2511ea.html. Accessed 12 March 2007.

Nuorti, J.P., Niskanen, T., Hallanvuo, S. *et al.* (2004) A widespread outbreak of *Yersinia pseudotuberculosis* O:3 infection from iceberg lettuce. *Journal of Infectious Diseases* **189**, 766–774.

Pebody, R., Leino, T., Holmstrom, P. *et al.* (1997) Outbreak of *Yersinia pseudotuberculosis* infection in central Finland. *Eurosurveillance Weekly* **1**(21), 970918. Available from http://www.eurosurveillance.org/ew/1997/970918.asp#3. Accessed 12 March 2007.

Robins-Browne, R.M. (2001) *Yersinia enterocolitica*. In: Doyle, M.P., Beuchat, L.R. & Montville, T.J. (eds). *Food Microbiology. Fundamentals and Frontiers*, 2nd edn. ASM Press, Washington, DC, pp. 215–245.

Sakai, T., Nakayama, A., Hashida, M. *et al.* (2005) Outbreak of food poisoning by *Yersinia enterocolitica* serotype O:8 in Nara prefecture: the first case report in Japan. *Japanese Journal of Infectious Diseases* **58**, 257–258.

Seimiya, Y.M., Sasaki, K., Satoh, C. *et al.* (2005) Caprine enteritis associated with *Yersinia pseudotuberculosis* infection. *Journal of Veterinary Medical Science* **67**, 887–890.

Shayegani, M., Morse, D., DeForge, I. *et al.* (1983) Microbiology of a major foodborne outbreak of gastroenteritis caused by *Yersinia enterocolitica* serogroup O:8. *Journal of Clinical Microbiology* **17**, 35–40.

Tacket, C.O., Ballard, J., Harris, N. *et al.* (1985) An outbreak of *Yersinia enterocolitica* infections caused by contaminated tofu (soybean curd) *American Journal of Epidemiology* **121**, 705–711.

Tacket, C.O., Narain, J.P., Sattin, R. *et al.* (1984) A multistate outbreak of infections caused by *Yersinia enterocolitica* transmitted by pasteurized milk. *Journal of the American Medical Association* **251**, 483–486.

Takkinen, J., Kangas, S., Hakkinen, M. *et al.* (2004) *Yersinia pseudotuberculosis* infections traced to raw carrots in Finland. *Eurosurveillance Weekly* **8**(41), 041007. Available from http://www.eurosurveillance.org/ew/2004/041007.asp#2. Accessed 12 March 2007.

Tauxe, R.V., Vandepitte, J., Wauters, G. *et al.* (1987) *Yersinia enterocolitica* infections and pork: the missing link. *Lancet* **i**, 1129–1132.

Tennant, S.M., Grant, T.H. & Robins-Browne, R.M. (2003) Pathogenicity of *Yersinia enterocolitica* biotype 1A. *FEMS Immunology and Medical Microbiology* **38**, 127–137.

Thisted Lambertz, S. & Danielsson-Tham, M.-L. (2005) Identification and characterization of pathogenic *Yersinia enterocolitica* isolates by PCR and pulsed-field gel electrophoresis. *Applied and Environmental Microbiology* **71**, 3674–3681.

Tsubokura, M., Otsuki, K., Sato, K. *et al.* (1989) Special features of distribution of *Yersinia pseudotuberculosis* in Japan. *Journal of Clinical Microbiology* **27**, 790–791.

Verhaegen, J., Charlier, J., Lemmens, P. *et al.* (1998) Surveillance of human *Yersinia enterocolitica* infections in Belgium: 1967–1996. *Clinical Infectious Diseases* **27**, 59–64.

World Health Organization (WHO) (2003) *WHO Surveillance Programme for Control of Foodborne Infections and Intoxications in Europe*, 8th report, 1999–2000. Available from http://www.bfr.bund.de/internet/8threport/8threp_fr.htm. Accessed 12 March 2007.

2.14 Foodborne viruses

The most important food- and water-borne viruses affecting people in the Western world appear to be norovirus and hepatitis A virus (Koopmans & Duizer, 2004). This section will concentrate on these, with a brief mention of other relevant viruses, which were reviewed by Carter (2005).

Awareness of foodborne viral infections is increasing, but laboratory reports underestimate greatly the true incidence of these infections; a shortage of expert laboratory diagnostic facilities has hampered recognition of the incidence of foodborne viral illness (ACMSF, 1998). Wider availability of improved methods of detection should clarify the role of food and water in transmission. Viruses do not replicate in food, which merely acts as a means of transmission.

2.14.1 *Importance of norovirus as a cause of foodborne disease in humans*

The name Norwalk virus was given to the first recognized norovirus, which caused an outbreak of acute gastroenteritis, termed 'winter vomiting disease' that occurred in Norwalk, Ohio, in 1968 (CDC, 2006c). The group of similar viruses have also been called 'small round structured viruses' (SRSVs) because of their appearance under the electron microscope.

Noroviruses are the most commonly identified cause of infectious intestinal disease in Western Europe and North America; they account for an estimated 6% of all infectious intestinal diseases in England and 11% in the Netherlands (Lopman *et al.*, 2003a), and are the most common cause of outbreaks of gastrointestinal infection. Even the most developed surveillance systems greatly underestimate viral gastroenteritis (Lopman *et al.*, 2003b).

In England and Wales between 1992 and 2000, 1877 outbreaks of norovirus (NoV) infection were reported, the majority occurred in hospitals and residential care facilities (Table 2.14.1), with a high proportion in elderly-care and geriatric units (Lopman *et al.*, 2003a). In 2002 in six European regions, reported outbreaks were predominantly in healthcare facilities (including hospitals, residential homes and nursing homes) (Fig. 2.14.1) (Lopman *et al.*, 2004a). The proportion of infections that are associated with foodborne transmission is not known. Some outbreaks are probably initiated by contaminated food

Table 2.14.1 Settings of 1877 reported norovirus outbreaks in England and Wales, 1992–2000.

40%	Hospitals
39%	Residential facilities
7.8%	Hotels
6%	Food outlets
4%	Schools
3.9%	Others (private homes, holiday camps, military bases)

Data from Lopman *et al.*, 2003a.

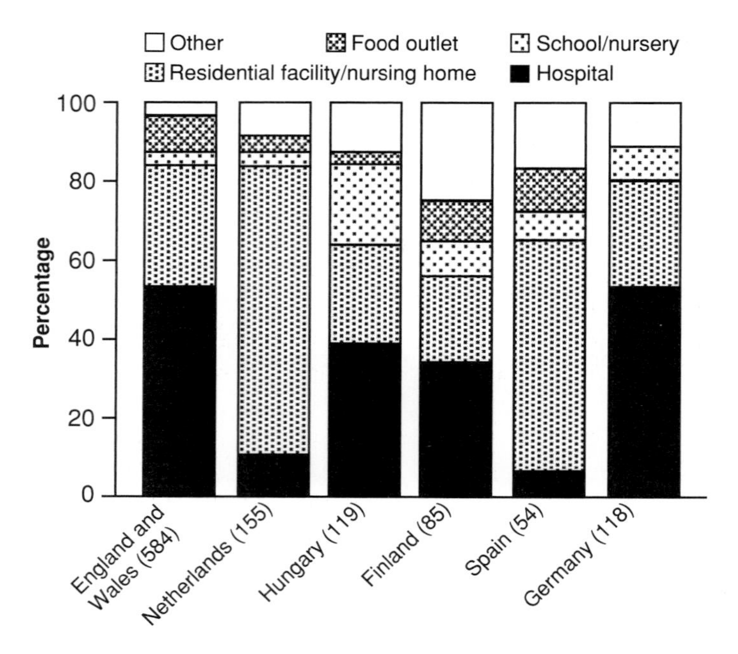

Fig. 2.14.1 Settings of norovirus outbreaks in 2002 for six European regions. (Regions were included only if settings were reported for 50 or more outbreaks.) (Reprinted from Lopman *et al.*, 2004a. Copyright (2004) with permission from Elsevier)

and spread rapidly by person-to-person transmission, which can be extensive in institutions such as hospitals and residential care homes, and which may mask the primary cause of the outbreak (Koopmans *et al.*, 2003).

In the US, it is thought that at least 50% of all foodborne outbreaks of gastroenteritis can be attributed to NoV (Widdowson *et al.*, 2005; CDC, 2006c). Among 232 outbreaks of illness due to NoV reported between July 1997 and June 2000, 36% were associated with restaurants or catered meals and 23% with nursing homes (Table 2.14.2). In the US, norovirus outbreaks not thought to be food- or water-borne were not reportable to CDC up to 2006 (Widdowson, M.-A., 2006, pers. comm.).

Table 2.14.2 Settings of 232 norovirus outbreaks reported to CDC from July 1997 to June 2000.

36%	Restaurants and catered meals
23%	Nursing homes
13%	Schools
10%	Vacation settings or cruise ships
18%	Other/not given

Data from CDC, 2006c.

2.14.2 *Characteristics of norovirus*

Noroviruses are a group of single-stranded RNA, non-enveloped viruses in the family *Caliciviridae* (CDC, 2006c). Three human norovirus genogroups are recognized at present (GI, GII and GIV). GI also contains porcine strains, GIII and GV contain bovine and murine strains (Estes *et al.*, 2006). Genogroups G1 and GII can be subdivided into 8 and 19 genotypes, respectively. Strains of NoV have been named according to the geographic setting of outbreaks with which they were associated, e.g. Norwalk, Hawaii, Snow Mountain, Southampton, Taunton, Bristol, Lordsdale, etc. Genogroup II.4 strains have predominated in outbreaks reported from many countries since 1996 (Lopman *et al.*, 2004a). As yet there is no evidence that NoV can be transmitted between animals and humans but, because of the variability of RNA viruses such as NoV, this may occur in the future (Koopmans & Duizer, 2004).

Human NoV cannot yet be propagated in cell culture, although culture of mouse NoV was reported in 2004. Electron microscopy following concentration of the virus has been used for detection of NoV in stools, but the number of virus particles present may be too low to detect (Caul, 2000). The development of molecular techniques, particularly those based on the PCR (reverse transcriptase (RT)-PCR) (ACMSF, 1998; CDC, 2001; Carter, 2005), in the 1990s has enabled improved detection in stool specimens and facilitated detection in some foods particularly shellfish (Seymour & Appleton, 2001). Improved methods have been developed for recovery of NoV from other foods, enabling RT-PCR and sequencing, and providing further evidence for foodborne transmission (Schwab *et al.*, 2000; Kobayashi *et al.*, 2004; Boxman *et al.*, 2006).

Investigations of factors affecting the probable survival of NoV have been made using as a model a related virus, feline calicivirus (FCV), which can be cultured on feline kidney cells and the number of infective units determined (Slomka & Appleton, 1998; Bidawid *et al.*, 2004).

Noroviruses are resistant to acid pH, refrigeration, and freezing, and may remain infective after heating at 60°C for 30 minutes (Caul, 2000; Seymour & Appleton, 2001; Koopmans & Duizer, 2004; Carter, 2005). They survive well on inanimate surfaces, and treatment with a free residual chlorine level of 10 mg/L is required for inactivation. Noroviruses are not inactivated by 0.5–1.0 mg free residual chlorine/L, which is the concentration present in drinking water in the UK.

2.14.3 *Disease in humans caused by norovirus*

The number of NoV required to cause infection may be as low as 10 viral particles (CDC, 2006c). Infection can be asymptomatic (Gallimore *et al.*, 2004). The incubation period is usually 24–48 hours but can be less than 12 hours. Symptoms include vomiting (which can be projectile and severe), watery non-bloody diarrhoea, abdominal cramps, nausea, malaise and headaches, and last usually for 24–60 hours (Caul, 2000; CDC, 2006c). In general, symptoms are mild and patients recover within 3 days, but in hospitalized patients, the young, the elderly, and persons with chronic diseases, the symptoms may be more severe (Lopman *et al.*, 2004b; Mattner *et al.*, 2006). In patients who are severely immunosuppressed

Table 2.14.3 Worldwide endemicity of hepatitis A infection.

Hepatitis A endemicity	Regions by epidemiological pattern	Average age of patients (yr)	Most likely mode of transmission
Very high	Africa, parts of South America, the Middle East and of south-east Asia	Under 5	Person-to-person Contaminated food and water
High	Brazil's Amazon basin, China, Latin America	5–14	Person-to-person Outbreaks/contaminated food or water
Intermediate	Southern and Eastern Europe, some regions of the Middle East	5–24	Person-to-person Outbreaks/ contaminated food or water
Low	Australia, US, Western Europe	5–40	Common source outbreaks
Very low	Northern Europe and Japan	Over 20	Exposure during travel to high endemicity areas, uncommon source

Data from WHO, 2000.

infection can lead to chronic diarrhoea and shedding of the virus for months to years (Estes *et al.*, 2006).

Shedding of the virus in faeces may occur before the development of symptoms, usually it begins with the onset of symptoms and may continue for 2 weeks after symptoms have stopped (Parashar *et al.*, 1998; CDC, 2006c). Faecal shedding has been reported in people with asymptomatic infections, but it is not clear whether the numbers of virus shed are sufficient to cause further infections.

Noroviruses replicate in the mucous epithelium of the small intestine, causing mild atrophy of the villi (Caul, 2000). The immunological response to these viruses appears to be short-term, and predominantly genotype-specific (Koopmans & Duizer, 2004).

2.14.4 *Importance of hepatitis A virus as a cause of foodborne disease in humans*

In many developing countries, including most of Africa, India, the Far East and South America, infection with hepatitis A virus (HAV) is endemic (CDC, 1999; WHO, 2000) (Table 2.14.3). In these areas transmission occurs generally by the faecal–oral route through person-to-person contact or ingestion of contaminated food or water. Most infections occur in young children, who generally remain symptomless; infection in adults usually results in symptoms, which can be severe. Infection generally confers lifelong immunity to all strains of the virus (Jacobsen & Koopman, 2004). In countries where infection is endemic most adults are immune and outbreaks are rare. In developed countries improved access to clean water and sanitation has been accompanied by a reduction in incidence of HAV infection, a rise in the average age at infection, and a decline in immunity in children and adults (Crowcroft *et al.*, 2001; Jacobsen & Koopman, 2004). Seroprevalence of antibodies to HAV is low in the US, Canada, Western Europe, Australia and New Zealand and very low in Scandinavia. In the UK, the majority of the population under 50 years of age are

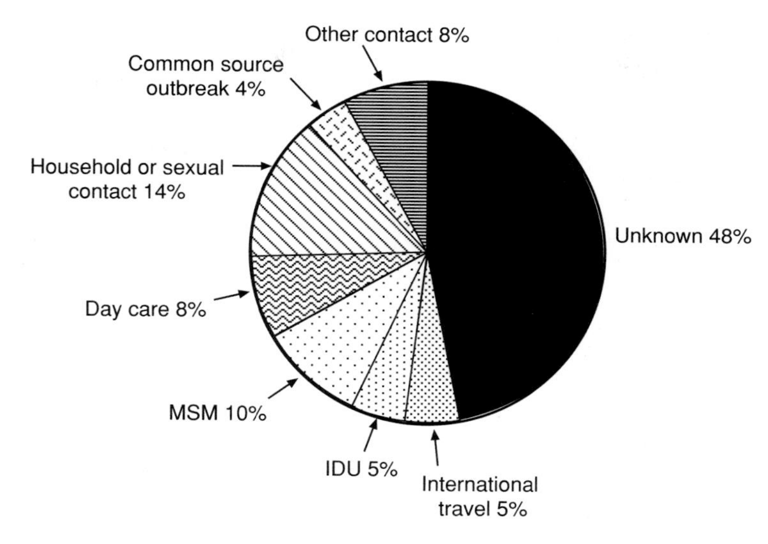

Fig. 2.14.2 Reported risk factors among persons with hepatitis A, United States, 1990–2000. Data are from the Viral Hepatitis Surveillance Program, Centers for Disease Control and Prevention. IDU, injection drug use; MSM, men who have sex with men. (Modified from Fiore, 2004)

now susceptible to HAV (Crowcroft *et al.*, 2001). While a lower proportion of adults over 60 years old are susceptible, in this age group the symptoms of disease are more severe. The decline in immunity in developed countries means that adults remain vulnerable in the event of exposure to the virus.

Disease caused by hepatitis A is notifiable statutorily in the UK. The number of reported cases of infection has decreased gradually in the past decade. In England and Wales, 1194 cases of hepatitis A infection were notified in 2003 (HPA, 2004), an incidence of about 2.02 cases per 100,000, but underreporting of cases may be high (Matin *et al.*, 2006). Since 1999, the reported incidence in the US has decreased markedly (Fiore, 2004; CDC, 2006b), the proportion of these cases that are foodborne is unknown, but could be considerable; for most cases the source of infection is not proven (Fig. 2.14.2).

2.14.5 *Characteristics of hepatitis A virus*

Hepatitis A viruses are single-stranded RNA, non-enveloped viruses in the family *Picornaviridae.* The genome organization differs from that of the *Caliciviridae*, which includes the noroviruses (Koopmans *et al.*, 2002). HAV can be grown in tissue culture; the wild-type virus is difficult to grow but attenuated strains have been adapted to cell culture and used to develop vaccines (WHO, 2000). Seven genotypes of HAV have been described, four of these occur in humans and the other three have been found in captive monkeys (Koopmans *et al.*, 2002). Human strains of HAV can induce disease in non-human primates, and monkeys have been found infected naturally with HAV strains that were closely related, antigenically, to human HAV strains (WHO, 2000). Whether humans are susceptible to HAV strains from naturally infected monkeys is not known. Only one serotype of HAV is known.

Diagnosis of HAV infection in humans has relied on serological detection of HAV-specific IgM or IgG (Caul, 2000). Methods based on RT-PCR are used now to detect the virus in clinical samples and in food.

HAV is resistant to acidity, drying and freezing and can survive for long periods in the environment; it is probably more heat-resistant than most non-sporing bacterial pathogens. Heating at 60°C for 30 minutes, cooking solid foods at 70°C for 2 minutes (or equivalent treatment) and pasteurization of liquids such as milk gave less than 100-fold inactivation but boiling at 100°C gave more than 10,000-fold inactivation (Koopmans & Duizer, 2004).

2.14.6 *Disease caused by hepatitis A virus in humans*

The minimum number of HAV required to cause disease in humans is unknown (Fiore, 2004). Infection is commonly asymptomatic in children under 6 years old. Among older children and adults, symptoms usually develop between 15 and 50 days (average 28 days) after infection and include fever, headache, fatigue, nausea and abdominal discomfort, followed by hepatitis and jaundice, which occurs in the majority of symptomatic patients, but illness is generally self-limiting. Prolonged or relapsing disease lasting up to 6 months has been reported in 10–15% of patients. The rate of overall case fatality has been reported as 0.3%, but among persons older than 50 years as 1.8% (Fiore, 2004).

Following infection, the virus reaches the small intestine, where some replication occurs before systemic spread and infection of the liver. Damage to liver epithelial cells results, probably, from an immune response. In persons who develop clinical symptoms the virus is shed in large numbers in the faeces; this begins between 1 and 3 weeks before symptoms appear and faeces remain infective for a further 1–2 weeks after jaundice develops (WHO, 2000). Faecal excretion of HAV persists longer in children and in immunocompromised persons (up to 4–5 mo after infection) than in otherwise healthy adults.

Infection results in the development of protective anti-HAV IgM, IgA and IgG antibodies. IgM antibodies decline to non-detectable levels within about 6 months of infection and IgA also decline substantially within 12 months. IgG antibodies persist for years and provide long-term immunity (WHO, 2000).

2.14.7 *Source and transmission of norovirus and hepatitis A virus*

Present evidence indicates that humans are the only reservoir of NoV and HAV and there is no evidence of a hazard associated with animals. Food- and water-borne infection results, therefore, from contamination of food or water with human faecal material or vomitus.

The low infectious dose and the spread of NoV by sudden, projectile vomiting result in person-to-person spread, which is particularly liable to occur in confined communities including nursing and residential homes and cruise ships. Such person-to-person spread commonly extends outbreaks initiated by contamination of food or of the environment. In reports up to 2000 a much lower proportion of outbreaks was attributed to foodborne transmission in England and Wales than in the US (Tables 2.14.4 and 2.14.5).

HAV is spread by the faecal–oral route either directly from person-to-person or by ingestion of contaminated food or water. Outbreaks of HAV infection among illicit drug users

Table 2.14.4 Reported primary mode of transmission of norovirus in 1877 outbreaks in England and Wales, 1992–2000.

85%	Person-to-person
5%	Foodborne
5%	Foodborne followed by person-to-person
5%	Unknown
1%	Waterborne

Data from Lopman *et al.*, 2003a.

and men who have sex with men have been recognized in the US, Canada, UK, Europe and Australia for many years (Crowcroft, 2003; CDC, 2006d).

Recognition of foodborne transmission of HAV can be difficult for several reasons; in particular, after the long incubation period (2–6 weeks) patients may have difficulty in recall of food history, the food implicated may no longer be available, and detection of the virus in foods can be difficult (Fiore, 2004).

In the US, CDC operates the Calicinet system for reporting outbreaks of foodborne illness due to enteric viruses. In Europe a surveillance network for enteric viruses, in particular NoV, is being built up by the European Consortium on Foodborne Viruses (Koopmans *et al.*, 2003).

Following excretion in faeces, NoV and HAV can survive for long periods in the environment. Coastal waters where shellfish are grown are liable to be contaminated with sewage; if treatment processes have been applied they are only partially effective in removal of viruses (ACMSF, 1998; Lees, 2000). Bivalve molluscs (including oysters, mussels, cockles, scallops and clams) are filter feeders and concentrate particles, including microorganisms, that are in their environment (Lees, 2000; Croci *et al.*, 2003). Many outbreaks caused by NoV and HAV have been associated with consumption of raw or lightly cooked shellfish (e.g. Lopalco *et al.*, 2005). Light cooking, such as steaming, does not inactivate NoV or HAV; it is difficult to inactivate these viruses by domestic cooking of shellfish without impairing the flavour, but this can be done commercially (Section 2.14.10).

Salad vegetables and fruit that are consumed raw can be contaminated with NoV and HAV in the field during growth as a result of contamination of irrigation water or surface water, or the use of untreated or inadequately treated sewage sludge (or possibly night soil) as fertilizer. Hepatitis A virus was detected on market lettuce in Costa Rica, possibly as a result

Table 2.14.5 Reported mode of transmission in 232 outbreaks of norovirus gastroenteritis reported to CDC from July 1997 to June 2000.

57%	Foodborne
23%	Not determined
16%	Person-to-person
3%	Waterborne

Data from CDC, 2006c.

of the discharge of untreated sewage into river water used to irrigate crops (Hernandez *et al.*, 1997). There is also a risk of contamination by field workers if on-site sanitary facilities are not provided. Salad vegetables and fruit, imported into countries with a low prevalence of HAV from communities where the transmission of HAV remains common, can be sources of exposure (Wheeler *et al.*, 2005). Raspberries have also been associated with outbreaks of NoV infection in several countries.

Outbreaks of infection with NoV or HAV arising from contamination of food by infected food handlers are common (Appleton, 2000; Fiore, 2004). Norovirus and hepatitis A accounted for 60% (49) of the outbreaks of foodborne disease believed to have resulted from contamination of food by food workers in the US between 1975 and 1998 (Guzewich & Ross, 1999). Food handlers may not have symptoms of illness or may have recovered from symptoms but still shed the virus; they may have contact with family members, particularly children, who show symptoms of gastroenteritis. The fact that HAV is shed in faeces 1–3 weeks before symptoms occur and for 1–2 weeks after jaundice develops poses a particular risk. In the US in 2003, approximately 8% of adults reported with hepatitis A were identified annually as food handlers, indicating that thousands of food handlers have hepatitis A each year (CDC, 2003a; Fiore, 2004), but surveillance in the US indicates that most food handlers infected with HAV do not transmit the virus to consumers.

2.14.8 *Examples of foodborne outbreaks caused by norovirus*

More recent outbreaks that have been reported in detail are shown in Tables 2.14.6–2.14.8.

The outbreak that affected hospitals in England and Wales necessitated the closure of the four hospitals to admissions for 10 days, causing major disruption to services (Lo *et al.*, 1994).

2.14.9 *Examples of foodborne outbreaks caused by hepatitis A virus*

In the UK, at the present time, reported outbreaks of foodborne HAV infection are rare. In 1978 and in 1980/81, widespread outbreaks were traced to inadequately cooked mussels and cockles and between 1987 and 1988, five further outbreaks were attributed to consumption of molluscan shellfish (ACMSF, 1998). In 1985, several outbreaks were traced to frozen raspberries, probably contaminated by workers during picking or weighing. During the 1980s, five further reported outbreaks were attributed to contamination of food by infected food handlers.

Outbreaks associated with HAV are shown in Tables 2.14.9–2.14.11.

In most outbreaks the evidence of foodborne transmission was epidemiological/statistical. An epidemic of diarrhoea in Shanghai in December 1987 and previous epidemics of HAV attributed to consumption of clams, led to testing of clams for HAV before the recognition of the outbreak in 1988 (Halliday *et al.*, 1991). In some outbreaks it was possible to obtain microbiological evidence, despite the long incubation time between infection and development of symptoms, because samples of stored, frozen foods were available (Conaty *et al.*, 2000; Sanchez *et al.*, 2002; Calder *et al.*, 2003).

Table 2.14.6 Examples of outbreaks of foodborne disease caused by norovirus associated with shellfish.

Place, date	Cases[a]	Number hospitalized	Food implicated	Evidence[b]	Where food was provided	Comments and factors leading to outbreak	Reference
US, 1993	190	nr	Raw oysters	S and M	Community	Sewage from harvesting boats dumped overboard while harvesting; >1 harvester ill	Berg *et al.* (2000)
US, 1994–95	131	5	Cooked and raw oysters	S	Community	Sewage from 16 harvesting boats dumped overboard while harvesting. Cooking oysters failed to prevent illness	McDonnell *et al.* (1997)
US, 1996	75	nr	Raw oysters	S	Social gatherings	Malfunction of sewage disposal system on oil rig; discharge of sewage by harvesters	Berg *et al.* (2000)
Australia, 1997	>97	0	Raw oysters	S	Restaurants, sports and community clubs, private function	Probable sewage contamination of growing area	Stafford *et al.* (1997)
US, 1996–97	153	nr	Raw oysters	S	Community	Discharge of sewage (>1 harvester ill) by boats harvesting oysters	Berg *et al.* (2000)
Denmark, Sweden, 1997	>650	nr	Raw, imported oysters	M	Community	Original source could not be traced. Oysters met the standards for *E. coli*. Both norovirus and enterovirus in oysters and stool samples	Christensen *et al.* (1998)
UK, 1997	>9	nr	Raw oysters	S	Hotel lunch	Oysters from grade B area, depurated	Ang (1997)
US, 1998	171	nr	Raw or undercooked oysters	S and M	nr	Substandard septic systems along the shoreline and overboard waste discharge from boats	Shieh *et al.* (2000)
France, 2000	>14	nr	Oysters	S and M	Household	Contamination in growing area	Le Guyader *et al.* (2003)
France, Italy, 2002–3	>290	nr	Oysters	S and M	Community	Contamination of oyster production area	Doyle *et al.* (2004)
US, 2006	8	nr	Raw frozen oysters on the half shell, from South Korea	M	Private event	nr	FDA (2006)

[a] No deaths reported.
[b] M (microbiological): identification of an organism of the same type from cases and in the suspect vehicle or vehicle ingredient(s), or detection of toxin in faeces or food; D (descriptive): other evidence, usually descriptive, reported by local investigators as indicating the suspect vehicle or food; S (statistical): a significant statistical association between consumption of the suspect vehicle and being a case.
nr, not reported.

Table 2.14.7 Examples of outbreaks of foodborne disease caused by norovirus associated with fruit and vegetables.

Place, date	Cases[a]	Number hospitalized	Food implicated	Evidence[b]	Where food was provided	Comments and factors leading to outbreak	Reference
Finland, 1998	108	nr	Imported, frozen raspberries	S and M	Canteens of large company	Possible contamination in the field or before freezing	Ponka *et al.* (1999)
Sweden, 2001	30	nr	Cakes containing frozen, whole raspberries	S and M	Community	nr	Le Guyader *et al.* (2004)
Arabian Gulf, 2003	37	0	Salad	S	British Royal Fleet Auxiliary ship	Five different viruses detected. Faecal contamination of fresh salad delivered to the ship suspected	Gallimore *et al.* (2005)
France, 2005	75	0	Frozen raspberries	S	School canteen	nr	Cotterelle *et al.* (2005)
Denmark, 2005	>1000 (6 outbreaks)	Many already patients, + at least 23 served by meals-on-wheels	Imported, frozen raspberry pieces (not distributed to retail markets) and whole, frozen raspberries	S	1 hospital, 2 nursing homes, meals-on-wheels service, private function, clothing company	Raspberries grown on several small farms in Poland, frozen and packed by exporting company. Evidence indicated contamination during growth and/or harvesting	Falkenhorst *et al.* (2005)
Sweden, 2006	43	0	Raspberries imported from China	D	2 private parties, 1 school, 1 meeting	nr	Hjertqvist *et al.* (2006)

[a] No deaths reported.
[b] M (microbiological): identification of an organism of the same type from cases and in the suspect vehicle or vehicle ingredient(s), or detection of toxin in faeces or food; D (descriptive): other evidence, usually descriptive, reported by local investigators as indicating the suspect vehicle or food; S (statistical): a significant statistical association between consumption of the suspect vehicle and being a case.
nr, not reported.

The outbreak in Italy in 1996–97 was attributed to initial foodborne transmission followed by person-to-person spread (Malfait *et al.*, 1999), but the outbreak in Shanghai (1988) was attributed mainly to foodborne transmission (Halliday *et al.*, 1991).

2.14.10 *Main methods of prevention and control of foodborne infection by norovirus and hepatitis A virus*

For both of these viruses the main means of transmission in foodborne outbreaks were (1) raw or under-cooked shellfish, (2) salad vegetables and fruits and (3) food handlers. In the US, >56% of foodborne outbreaks of NoV have been associated with eating salads, sandwiches, or fresh produce, foods that require handling but no subsequent heating (Widdowson *et al.*, 2005). Prevention of waterborne transmission is considered in Chapter 7.

The main methods of preventing foodborne infections by these viruses are (a) prevent sewage contamination of shellfish beds, and control the harvesting, treatment and microbiological quality of shellfish, (b) avoid consumption of uncooked or undercooked shellfish, (c) prevent contamination of fruit and salad vegetables in the field and during harvesting, and wash thoroughly and (d) prevent transmission by food handlers.

(a) *Prevent sewage contamination of shellfish beds, and control the harvesting, treatment and microbiological quality of shellfish.* In the EU and the US, controls on the production of live bivalve molluscs include classification of growing areas in terms of the numbers of faecal indicator bacteria in shellfish (EU) or in water (US) in the growing area, the use of purification treatments (relaying or depuration) to reduce low-level contamination, and prohibition of harvesting if contamination is above specified levels (Table 2.14.12). Norovirus and HAV survive better than *E. coli* during purification treatments, however, and shellfish that complied with the regulations for *E. coli* have been associated with outbreaks of foodborne viral infection (Dore *et al.*, 2003). There is a need for criteria for pathogenic viruses in live bivalve molluscs when analytical methods have been developed sufficiently, this is recognized in the EC regulations.

(b) *Avoid consumption of uncooked or undercooked shellfish.* Live bivalve molluscs from category B and C areas that have not been purified and do not meet the criteria for direct consumption may be processed by sterilization in hermetically sealed containers or subjected to a heat treatment, equivalent to heating in boiling water, or in steam under pressure, sufficiently that the internal temperature of the flesh is not less than 90°C and this temperature is maintained for 90s (EC, 2004a). This treatment inactivates non-sporing bacterial pathogens and reduces the concentration of hepatitis A virus by 10^4 infectious units (Millard *et al.*, 1987), while NoV is more heat-sensitive (Slomka & Appleton, 1998). It is difficult to apply this heat treatment to large batches without rendering some shellfish unpalatable, but commercially this can be done using continuous-flow machinery.

To prevent infection with NoV and HAV from consumption of shellfish, these should not be eaten raw but should be cooked thoroughly. Cooking in the home or restaurant may reduce contamination with HAV and NoV but is unlikely to be as effective as a well-controlled, commercial process. In the outbreak of NoV infection associated

Table 2.14.8 Examples of outbreaks of foodborne disease caused by norovirus associated with food handlers.

Place, date	Cases[a]	Number hospitalized	Food implicated	Evidence[b]	Where food was provided	Comments and factors leading to outbreak	Reference
UK, 1991	65 (37 patients, 28 staff + secondary cases)	Nosocomial	Sandwiches	S	Geriatric hospital	Probable contamination of sandwiches by food handler	Stevenson, *et al.* (1994)
Wales, 1990	67	1	Coronation chicken	S	Conference centre	Food handler suffering from mild gastroenteritis used bare hands to prepare food	Patterson *et al.* (1993)
Wales, year not stated	195	81 were already hospital patients	Salad items	S	4 hospitals	Contamination by pre-symptomatic food handler in central hospital kitchen	Lo *et al.* (1994)
UK, 1996	47	0	Potato salad	S	Hotel, wedding reception	Kitchen assistant vomited into vegetable-preparation sink, attempted disinfection but did not report incident. Next day sink was used to drain potatoes for salad	Patterson *et al.* (1997)
US, 1997	85	nr	Sandwiches	S and M	Lunch catered by private company	Contamination by food handler(s)	Parashar *et al.* (1998)
Canada, 1998	48	0	Salad	S	Restaurant	Salad handled with bare hands by pre-symptomatic food handler	Gaulin *et al.* (1999)
US, 1998	125	23	Deli ham sandwiches	M	University cafeteria deli bar	Contamination by food handler	Daniels *et al.* (2000)
Sweden, 1999	400 including secondary spread	nr	Pumpkin salad	S and M	30 day-care centres served by the same caterer	Contamination by food handler	Gotz *et al.* (2002)

Sweden, 2000	147	2	Specific food not identified	S	Hotel located at University hospital	Chef and other staff became sick while preparing meals	Johansson *et al.* (2002)
Japan, 2001	52	nr	Rolled cabbage and macaroni in lunch box	S and M	Lunch boxes delivered to employees of 11 companies	Possible contamination by food handlers, who were not investigated	Kobayashi *et al.* (2004)
Spain, 2002	40 (hospital staff and relatives)	nr	Salads and sandwiches	S and M	Hospital cafeteria	Cook employed in cafeteria worked while symptomatic. Hygiene deficiencies in kitchen	Sala *et al.* (2005)
US, 2002	>332	6	Wedding cakes with strawberry filling	S and M	Multiple weddings served by the same bakery	Fresh strawberries washed, sliced and hand-spread onto white chocolate mousse for cakes. Hygiene deficiencies in bakery; bakery employees ill	Friedman *et al.* (2005)
Japan, 2004	91	nr	Packed lunch boxes supplied by catering firm	D	8 different companies supplied by catering firm	Norovirus detected in samples from catering staff	Sakon *et al.* (2005)
Austria, 2005	~ 120	nr	Sandwiches supplied by catering company	M	Factory	Several staff of catering company were ill before this outbreak, one had worked while ill and prepared sandwiches without wearing gloves	Lederer *et al.* (2005)
US, 2005	>100	nr	Lettuce in sandwiches	D	3 catered events and community supplied with sandwiches by same restaurant	Food handler who sliced lettuce returned to work within few hours of vomiting. Heads of lettuce washed in sink in which employees washed their hands. Norovirus from food handler matched strains from customers	CDC (2006a)

[a] No deaths reported.
[b] M (microbiological): identification of an organism of the same type from cases and in the suspect vehicle or vehicle ingredient(s), or detection of toxin in faeces or food; D (descriptive): other evidence, usually descriptive, reported by local investigators as indicating the suspect vehicle or food; S (statistical): a significant statistical association between consumption of the suspect vehicle and being a case.
nr, not reported.

Table 2.14.9 Examples of outbreaks of foodborne disease caused by hepatitis A virus associated with shellfish.

Place, date	Cases	Number hospitalized	Deaths	Food implicated	Evidence[a]	Where food was provided	Comments and factors leading to outbreak	Reference
Shanghai, 1988	~ 290, 000	>132	32	Inadequately cooked clams	S and M	Community and 'eating out'	Contamination of harbour waters with untreated sewage and untreated effluent from fishing vessels	Halliday *et al.* (1991); Tang *et al.* (1991); Lees (2000)
US, 1988	61	17	0	Raw oysters	S and M	Restaurants and Oyster bars	Oysters were harvested illegally from unapproved waters that were subject to contamination with sewage	Desenclos *et al.* (1991)
Italy, 1996–97	886	nr	nr	Raw oysters	S	nr	Consumption of raw seafood	Malfait *et al.* (1996)
Australia, 1997	>400	~ 60	1	Oysters	S and M	Community	Sewage contamination of estuarine growing waters and unusually high rainfall	Conaty *et al.* (2000)
Spain, 1999	184	nr	nr	Frozen, imported coquina clams	S and M	Community	Contaminated growing waters	Sanchez *et al.* (2002)
Italy, 2004	>600	nr	nr	Shellfish	D	Community	nr	Boccia (2004)
US, 2005	39	nr	nr	Raw oysters	S, M	Restaurant	nr	Shieh *et al.* (2007)

[a] M (microbiological): identification of an organism of the same type from cases and in the suspect vehicle or vehicle ingredient(s), or detection of toxin in faeces or food; D (descriptive): other evidence, usually descriptive, reported by local investigators as indicating the suspect vehicle or food; S (statistical): a significant statistical association between consumption of the suspect vehicle and being a case.

nr, not reported.

Table 2.14.10 Examples of outbreaks of foodborne disease caused by hepatitis A virus associated with fruit and vegetables.

Place, date	Cases	Number hospitalized	Deaths	Food implicated	Evidence[a]	Where food was provided	Comments and factors leading to outbreak	Reference
Scotland, 1983	24	5	nr	Frozen raspberries	S	Hotel	Probable contamination in the field during picking	Reid and Robinson (1987)
US, 1988	202	nr	nr	Iceberg lettuce, imported	S	Restaurants	Possible contamination in the field or during transport	Rosenblum *et al.* (1990)
US, 1990	28	nr	0	Frozen strawberries	S	Elementary school, institution for developmentally disabled	Probable contamination during picking	Niu *et al.* (1992)
Finland, 1996	18	9	nr	Salad items, imported	S	Secondary school	Possibly associated with imported salad items contaminated in the field	Pebody *et al.* (1998)
Finland, 1996	12	3	nr	Salad items, imported	S	Bank employees	Possibly associated with imported salad items contaminated in the field	Pebody *et al.* (1998)
US, 1997	262	nr	nr	Frozen strawberries imported from Mexico	S	School lunches	Possible contamination during picking	Hutin *et al.* (1999)
Sweden, 2000–2001	16	nr	nr	Imported rocket salad	S	Community	Possible contamination in the field or during harvesting	Nygård *et al.* (2001)
New Zealand, 2002	27	nr	nr	Raw blueberries	S and M	Community	Opportunities for contamination in the field; open latrines for workers, one latrine near blueberry plants	Calder *et al.* (2003)
US, 2003 (September)	422	nr	nr	Green onions imported from Mexico	S	3 Restaurants	Possible contamination during harvesting or use of contaminated water for irrigation, processing or storage	CDC (2003b); Amon *et al.* (2005)
US, 2003 (October–November)	>600	>124	3	Green onions imported from Mexico	S	Restaurant	Possible contamination during harvesting or use of contaminated water for irrigation, processing or storage	CDC (2003b); Wheeler *et al.* (2005); Amon *et al.* (2005)
Egypt, 2004	331	>127	0	Orange juice	S	Hotel	Hygiene problems at supplier; juice not heat treated	Frank *et al.* (2007)

[a] M (microbiological): identification of an organism of the same type from cases and in the suspect vehicle or vehicle ingredient(s), or detection of toxin in faeces or food; D (descriptive): other evidence, usually descriptive, reported by local investigators as indicating the suspect vehicle or food; S (statistical): a significant statistical association between consumption of the suspect vehicle and being a case.

nr, not reported.

Table 2.14.11 Examples of outbreaks of foodborne disease caused by hepatitis A virus associated with food handlers.

Place, date	Cases	Number hospitalized	Deaths	Food implicated	Evidence[a]	Where food was provided	Comments and factors leading to outbreak	Reference
US, 1988	32	nr	nr	Ice slush drink bought from convenience store	S	Community	Food handler recovering from jaundice	CDC (1990)
US, 1998	32	nr	nr	Iced tea	S	Restaurant	Food handler IgM anti-HAV +, denied symptoms of HAV, IV drug user	CDC (1990)
US, 1990	110	nr	2	Salads	S	Restaurant	Food handler with hepatitis A	CDC (1993)
US, 1991	230	nr	nr	Sandwiches	S	Sandwich shops	Food handler with hepatitis A	CDC (1993)
US, 1994	64	4	nr	Sugar-glazed, baked goods	S	Retail buyers' club	Probable contamination by infected food handler, who sugar-glazed products after baking. Food handler had diarrhoea	Weltman *et al.* (1996)
US, 2001	43	nr	nr	Sandwiches	S	Restaurant	Food handler infected with hepatitis A and with a colostomy	CDC (2003a)
Italy, 2002	26	nr	nr	Sandwiches	M	Community	Symptomatic food handler	Chironna *et al.* (2004)
Germany, 2004	64	nr	nr	Pastries, filled doughnuts	S, M	Community	Two food handlers in bakery infected	Schenkel *et al.* (2006)

[a]M (microbiological): identification of an organism of the same type from cases and in the suspect vehicle or vehicle ingredient(s), or detection of toxin in faeces or food; D (descriptive): other evidence, usually descriptive, reported by local investigators as indicating the suspect vehicle or food; S (statistical): a significant statistical association between consumption of the suspect vehicle and being a case.
nr, not reported.

Table 2.14.12 Summary of EC and US standards for bacteria in live shellfish.

Shellfish treatment required	US FDA classification	Microbiological standard MPN per 100 mL water	EU classification	Microbiological standard MPN per 100 g shellfish
None required	Approved	GM <14 FC and 90th percentile <43 FC or GM <70 TC and 90% <230 TC	Category A	All samples, *E. coli* <230
Depuration or relaying	Restricted	GM <88 FC and 90th percentile <260 FC or GM <700 TC and 90th percentile <2300 TC	Category B	All samples, *E. coli* <4600
	–		Category C, prolonged relaying (>2 mo)	All samples, *E. coli* <46,000
Harvesting prohibited		Levels higher than above	Harvesting prohibited	Levels higher than above

Data from EC, 2004a, b, 2005; FDA/CFSAN, 2005b. Modified from Lees, 2000.
US specification for bivalve molluscs for direct consumption: from approved growing area.
EC specification for bivalve molluscs for direct consumption: from category A area, plus no detection of *Salmonella* in 25 g samples.
GM, geometric mean; TC, total coliform; FC, faecal coliform.

with cooked and raw oysters in the US in 1994–95 (Table 2.14.6), there was a 53% attack rate among persons who ate only cooked oysters (grilled, steamed, stewed or fried) compared with a 71% attack rate among persons who ate only raw oysters and a 3% attack rate in those who did not eat oysters. Domestic cooking of mussels in some recipes failed to give sufficient heat treatment to inactivate HAV (Croci *et al.*, 2005).

(c) *Prevent contamination of fruit and salad vegetables in the field and during harvesting, and wash thoroughly.* The use of sewage sludge, which is applied to land as a source of nutrients and organic matter, must be controlled in order to prevent contamination of food crops with pathogenic microorganisms. In the UK, the use of untreated sewage sludge on land used for food crops was banned in 1999, and a ban on its application to agricultural land used to grow non-food crops is intended (ADAS, 2001). The use of treated sewage sludge is regulated according to the 'Safe Sludge Matrix', which specifies the time interval that must apply between the use of a treated sludge and the date of harvest of specified crops (ADAS, 2001).

In some parts of the world control over the use and disposal of sewage may be lacking, and untreated human waste may be used as fertilizer for produce, with a consequent risk to consumers, particularly in areas where HAV is endemic. In the case of imports from such areas there is an onus on importers to ensure the microbiological safety of imported fresh fruit and salad crops (ACMSF, 1998), this requires a familiarity with the practices used during growth and harvesting. Contamination of river water, which may be used for irrigation of produce, may also result in contamination of crops.

Following the outbreak of NoV infection in Finland in 1998 associated with raspberries, suspicions of several similar outbreaks in Finland, and reports from other countries of viral infection associated with berries, a temporary ban on the serving of unheated dishes prepared from frozen berries was recommended for institutional kitchens in Finland (Ponka *et al.*, 1999).

(d) *Prevent transmission by food handlers.* Contamination by food handlers may be the most common route for contamination of food by viral pathogens (ACMSF, 1998; Fiore, 2004).

In England and Wales, it is required that food handlers who suffer illness due to gastrointestinal infection with NoV should not return to work until 48 hours after cessation of symptoms (DH, 1995, 1996). Workers in the Netherlands have suggested that making professional food handlers aware of their higher probability of being infected if a household member has gastroenteritis may be useful (de Wit *et al.*, 2003).

In the US, food employees who suffer illness due to norovirus infection and who work in a food establishment serving a highly susceptible population (e.g. in a hospital, nursing home, child or adult day-care centre, senior centre), should be excluded from work until 48 hours after the employee became asymptomatic (FDA/CFSAN, 2005a). An employee who is diagnosed with a norovirus infection and is asymptomatic, and who works in a similar establishment should also be excluded.

Shedding of NoV may continue for 2 weeks after symptoms have stopped (Parashar *et al.*, 1998; CDC, 2006c) but the extent to which viral shedding over 72 hours after recovery from symptoms signifies continued infectivity is not clear. Food handlers returning to work should be instructed that substantial numbers of norovirus may be shed for weeks after recovery from illness (Koopmans & Duizer, 2004), and strict hand washing after using the toilet and before handling food items is of major importance in preventing transmission of the virus.

In England and Wales, guidelines state 'Food handlers with hepatitis A should remain off work until seven days after symptoms have appeared, usually jaundice. Advice should be sought from relevant healthcare professionals' (DH, 1995, 1996). Symptomless contacts of a case of hepatitis A can continue food handling provided specified good hygienic practice is maintained. This may only provide limited protection against transmission of HAV by an infected handler, because high concentrations of HAV are shed in the stools of infected individuals for up to 3 weeks before development of symptoms and for a further 1–2 weeks after the onset of jaundice. In the US, food employees who work in a food establishment serving a highly susceptible population, and who are jaundiced or diagnosed with an infection from HAV with or without symptoms, or who have been exposed to HAV within the past 30 days, who are not immune to HAV, should be excluded from work (FDA/CFSAN, 2005a). They can be re-instated when more then 30 days have passed since the last day on which the food employee was exposed or a household contact of the food employee became jaundiced, or the food employee does not use a procedure that allows bare-hand contact with ready-to-eat food until at least 30 days after the potential exposure (FDA/CFSAN, 2005a).

Transfer of FCV (a model for NoV) and of HAV from contaminated fingers to foods has been demonstrated (Bidawid *et al.*, 2000, 2004). Good hand-washing is of major importance in prevention of transmission of viruses by food handlers, and appropriate facilities must be available. Washing fingers with water, or non-medicated soap and water or, to a lesser extent, with alcohol-based agents gave a marked reduction in the number of NoV and

HAV recovered from contaminated fingers and subsequently transferred to lettuce and ham. Proper hand washing and the use of disposable gloves by food handlers are advocated during preparation of ready-to-eat foods.

Vaccination against HAV is recommended for travellers from countries of low endemicity to areas of high endemicity and for certain at-risk groups (CDC, 1999; Crowcroft *et al.*, 2001; Craig & Schaffner, 2004). In institutions where hygiene may be difficult to maintain, such as nurseries for pre-school children, immunization against HAV may be justified (Crowcroft *et al.*, 2001).

In England and Wales, advice on management of close contacts of sporadic cases of HAV infection and outbreaks includes emphasis on good hygiene and in particular careful hand washing (Crowcroft *et al.*, 2001). For further prophylaxis, HAV vaccine (inactivated HAV) and/or Human Normal Immunoglobulin (HNIG) are recommended.

In the US, depending on the general rate of HAV infection in the area, in the event of an outbreak it is advised that programmes to vaccinate children or at-risk groups should be intensified (CDC, 1999). When outbreaks occur in settings such as day-care centres, hospitals, institutions and schools the administration of HNIG to persons in close contact with infected people is recommended. In the outbreak of HAV infection in the US in 2003, associated with imported green onions, which affected more than 600 people, HAV-infected staff at the restaurant involved may have been infectious during the incubation period. Consequently, more than 9000 persons who ate in the restaurant during the relevant period were provided with HNIG as a preventive treatment (CDC, 2003b). Since 1996, a programme has been implemented incrementally to immunize children in areas of the US with consistently high rates of HAV infection. Between 1997 and 2004, a marked reduction in the rate of HAV infection has occurred (CDC, 2006b). In 2006, routine HAV immunization of children aged at least 1 year old has been recommended with the aim of eventual elimination of indigenous transmission of HAV.

2.14.11 *Other viruses*

Several other viruses can be associated with food- and water-borne outbreaks (Cliver, 2001).

Hepatitis E virus (HEV) is a major cause of hepatitis in tropical and sub-tropical countries including parts of Asia, Africa and Central America where there is a lack of clean drinking water, and inadequate sanitation. Outbreaks in these countries are caused mainly by contaminated drinking water (WHO, 2002) and also by foodborne transmission. Subclinical infection is common (Smith, 2001). Infection can result in moderately severe jaundice that is usually self-limiting, but infection of pregnant women results in a high level of mortality of the women and their infants (Smith, 2001). Infection appears to result in long-term immunity to HEV. Humans are considered the natural hosts but animals may also serve as a reservoir. In Japan, infection has been linked to consumption of undercooked deer meat and wild boar meat (Li *et al.*, 2005). An outbreak in Japan in 2004 affected six people who ate pork liver in a restaurant where customers barbecued meat for themselves and many preferred the meat rare; one person (a man in his 60s) died of fulminant hepatitis (ProMED-mail, 2004). In Japan and in the UK strains of HEV isolated from humans infected locally were related closely to strains found in pigs (Banks *et al.*, 2004; Ijaz *et al.*, 2005), indicating a risk of transmission in undercooked pig liver and meat. In areas where HEV is endemic, contamination by sewage or by animal (particularly pig) manures of surface water used for irrigation of crops, and unsanitary conditions during harvesting, may result in contamination of fruit and vegetables. The increasing worldwide distribution of produce may result in an increased risk of transmission in developed countries (Seymour & Appleton, 2001; Smith, 2001). HEV may also be transmitted by shellfish (Lees, 2000).

Other viruses that may be transmitted on food are reviewed by Carter (2005). These include rotavirus which is considered mainly as a major cause of gastroenteritis in children. The extent to which this virus is foodborne is not clear (de Wit *et al.*, 2003); it has been associated with 13% of gastroenteritis outbreaks in aged-care facilities in Australia, whereas NoV was associated with 42% (Marshall *et al.*, 2003).

Summary of measures advised in healthcare settings

1. Shellfish should be obtained from approved growing areas and should be cooked thoroughly before consumption.
2. Salad vegetables and fruit that will not be cooked thoroughly should be obtained from areas where contamination by human waste/animal waste is prevented/as far as possible limited, and should be washed thoroughly before consumption. Uncooked salad vegetables and berry fruits should not be served to persons with compromised immune systems.
3. Meat, particularly pig meat, should be cooked thoroughly before serving.
4. Catering personnel should not handle food while suffering from symptoms of infection and, in the case of HAV, for more than 1 week after symptoms have appeared. Good hand-washing is essential at all times.

References

Advisory Committee on the Microbiological Safety of Food (ACMSF) (1998) *Report on Foodborne Viral Infections*. HMSO, London, pp. 107.

Agricultural Development Advisory Service (ADAS) (2001) *The Safe Sludge Matrix. Guidelines for the Application of Sewage Sludge to Agricultural Land*. Available from http://www.adas.co.uk/media_files/Publications/SSM.pdf. Accessed 13 March 2007.

Amon, J.J., Devasia, R., Xia, G. *et al.* (2005) Molecular epidemiology of foodborne hepatitis A outbreaks in the United States, 2003. *Journal of Infectious Diseases* **192**, 1323–1330.

Ang, L.H. (1997) An outbreak of viral gastroenteritis associated with eating raw oysters. *Communicable Disease and Public Health* **1**, 38–40.

Appleton, H. (2000) Control of foodborne viruses. *British Medical Bulletin* **56**, 125–133.

Banks, M., Bendall, R., Grierson, S. *et al.* (2004) Human and porcine hepatitis E virus strains, United Kingdom. *Emerging Infectious Diseases* **10**, 953–955.

Berg, D.E., Kohn, M.A., Farley, A. & McFarland, L.M. (2000) Multi-state outbreaks of acute gastroenteritis traced to fecally-contaminated oysters harvested in Louisiana. *Journal of Infectious Diseases* **181**(Suppl 2), S381–S386.

Bidawid, S., Farber, J. & Sattar, S. (2000) Contamination of foods by food handlers: experiments on hepatitis A virus transfer to food and its interruption. *Applied and Environmental Microbiology* **66**, 2795–2763.

Bidawid, S., Malik, N., Adegbunrin, O. *et al.* (2004) Norovirus cross-contamination during food handling and interruption of virus transfer by hand antisepsis: Experiments with feline calicivirus as a surrogate. *Journal of Food Protection* **67**, 103–109.

Boccia, D. (2004) Community outbreak of hepatitis A in southern Italy – Campania, January–May 2004. *Eurosurveillance Weekly* **8**(23), 040603. Available from http://www.eurosurveillance.org/ew/2004/040603.asp. Accessed 13 March 2007.

Boxman, I.L.A., Tilburg, J.J.H.C., te Loeke, N.A.J.M. *et al.* (2006) Detection of noroviruses in shellfish in the Netherlands. *International Journal of Food Microbiology* **108**, 391–396.

Calder, L., Simmons, G., Thornley, C. *et al.* (2003) An outbreak of hepatitis A associated with consumption of raw blueberries. *Epidemiology and Infection* **131**, 745–751.

Carter, M.J. (2005) Enterically infecting viruses: pathogenicity, transmission and significance for food and waterborne infection. *Journal of Applied Microbiology* **98**, 1364–1380.

Caul, E.O. (2000) Foodborne viruses. In: Lund, B.M., Baird-Parker, T.C. & Gould, G.W. (eds). *The Microbiological Safety and Quality of Food*. Aspen Publishers, Gaithersberg, MD, pp. 1457–1489.

Centers for Disease Control and Prevention (CDC) (1990) Epidemiologic notes and reports foodborne hepatitis A – Alaska, Florida, North Carolina, Washington. *Morbidity and Mortality Weekly Report* **39**(14), 228–232.

Centers for Disease Control and Prevention (CDC) (1993) Foodborne hepatitis A – Missouri, Wisconsin, and Alaska, 1990–1992. *Morbidity and Mortality Weekly Report* **42**(27), 526–529.

Centers for Disease Control and Prevention (CDC) (1999) Prevention of hepatitis A through active or passive immunization: recommendations of the Advisory Committee on Immunization Practices (ACIP). *Morbidity and Mortality Weekly Report* **48**(No. RR-12), 1–37.

Centers for Disease Control and Prevention (CDC) (2001) "Norwalk-Like Viruses" Public health consequences and outbreak management. *Morbidity and Mortality Weekly Report* **50**(RR-09), 1–18.

Centers for Disease Control and Prevention (CDC) (2003a) Foodborne transmission of hepatitis A – Massachusetts, 2001. *Morbitidy and Mortality Weekly Report* **52**(24), 565–567.

Centers for Disease Control and Prevention (CDC) (2003b) Hepatitis A outbreak associated with green onions at a restaurant – Monaco, Pennsylvania, 2003. *Morbidity and Mortality Weekly Report* **52**(47), 1155–1157.

Centers for Disease Control and Prevention (CDC) (2006a) Multistate outbreak of norovirus associated with a franchise restaurant – Kent County, Michigan, May 2005. *Morbidity and Mortality Weekly Report* **55**(14), 395–397.

Centers for Disease Control and Prevention (CDC) (2006b) Prevention of hepatitis A through active or passive immunization. *Morbidity and Mortality Weekly Report* **55**(RR07), 1–23.

Centers for Disease Control and Prevention (CDC) (2006c) Infectious Disease Information. Norovirus infection. Available from: http://www.cdc.gov. Accessed 9 July 2007.

Centers for Disease Control and Prevention (CDC) (2006d) Viral hepatitis. Hepatitis A. Available from: http://www.cdc.gov. Accessed 9 July 2007.

Chironna, M., Lopalco, P., Prato, R. *et al.* (2004) Outbreak of infection with hepatitis A virus (HAV) associated with a foodhandler and confirmed by sequence analysis reveals a new HAV genotype 1B variant. *Journal of Clinical Microbiology* **42**, 2825–2828.

Christensen, B.F., Lees, D., Wood, K.H. *et al.* (1998) Human enteric viruses in oysters causing a large outbreak of human food borne infection in 1996/97. *Journal of Shellfish Research* **17**, 1633–1635.

Cliver, D.O. (2001) Foodborne viruses. In: Doyle, M.P., Beuchat, L.R. & Montville, T.J. (eds). *Food Microbiology Fundamentals and Frontiers*, 2nd edn. ASM Press, Washington, DC, Chapter 24, pp. 501–511.

Conaty, S., Bird, P., Bell, G. *et al.* (2000) Hepatitis A in New South Wales, Australia, from consumption of oysters: the first reported outbreak. *Epidemiology and Infection* **124**, 121–130.

Cotterelle, B., Drougard, C., Rolland, J. *et al.* (2005) Outbreak of norovirus infection associated with the consumption of frozen raspberries, France, March 2005. *Eurosurveillance Weekly* **10**(4), E050428.1. Available from http://www.eurosurveillance.org/ew/2005/050428.asp. Accessed 14 March 2007.

Craig, A.S. & Schaffner, W. (2004) Prevention of hepatitis A with the hepatitis A vaccine. *New England Journal of Medicine* **350**, 476–481.

Croci, L., De Medici, D., Ciccozzi, M. *et al.* (2003) Contamination of mussels by hepatitis A virus: a public health problem in southern Italy. *Food Control* **14**, 559–563.

Croci, L., De Medici, D., Di Pasquale, S. & Toti, L. (2005) Resistance of hepatitis A virus in mussels subjected to different domestic cookings. *International Journal of Food Microbiology* **105**, 139–144.

Crowcroft, N.S., Walsh, B., Davison, K.L. & Gungabissoon, U. (2001) Guidelines for the control of hepatitis A virus infection. *Communicable Disease and Public Health* **4**, 213–227.

Crowcroft, N.S. (2003) Hepatitis A virus infections in injecting drug users. *Communicable Disease and Public Health* **6**, 82–84.

Daniels, N.A., Bergmire-Sweat, D.A., Schwab, K.J. *et al.* (2000) A foodborne outbreak of gastroenteritis associated with Norwalk-like viruses: first molecular traceback to deli sandwiches contaminated during preparation. *Journal of Infectious Diseases* **181**, 1467–1470.

Department of Health, UK (DH) (1995) *Food Handlers. Fitness to Work. Guidelines for Food Businesses, Enforcement Officers and Health Professionals*. Department of Health, London.

Department of Health, UK (DH) (1996) *Food Handlers. Fitness to Work. Guidelines for Food Business Managers*. Department of Health, London.

Desenclos, J.-C., Klontz, K.C., Wilder, M.H. *et al.* (1991) A multistate outbreak of hepatitis A caused by the consumption of raw oysters. *American Journal of Public Health* **81**, 1268–1272.

de Wit, M.A.S., Koopmans, M.P.G. & van Duynhoven, Y.T.H.P. (2003) Risk factors for norovirus, Sapporo-like virus, and group A rotavirus gastroenteritis. *Emerging Infectious Diseases* **9**, 1563–1570.

Dore, W.J., Mackie, M. & Lees, D.N. (2003) Levels of male-specific RNA bacteriophage in molluscan bivalve shellfish from commercial harvesting areas. *Letters in Applied Microbiology* **36**, 92–96.

Doyle, A., Barataud, D., Gallay, A. *et al.* (2004) Norovirus foodborne outbreaks associated with the consumption of oysters from the Etang de Thau, France, December 2002. *Eurosurveillance. European Communicable Disease Quarterly* **9**, 24–26.

EC (2004a) Commission Regulation (EC) No. 853/2004 of the European parliament and of the Council of 29 April 2004 laying down specific rules for food of animal origin. *Official Journal of the European Union* **L226/22**, 25 June 2004. Available from http://www.food.gov.uk/multimedia/pdfs/h2ojregulation.pdf. Accessed 13 March 2007.

EC (2004b) Commission Regulation (EC) No. 854/2004 of the European parliament and of the Council of 29 April 2004 laying down specific rules for the organization of official controls on products of animal origin intended for human consumption. *Official Journal of the European Union* **L226/83**, 25 June 2004. Available from http://www.food.gov.uk/multimedia/pdfs/h3ojregulation.pdf. Accessed 13 March 2007.

EC (2005) Commission Regulation (EC) No 2073/2005 of 15 November 2005 on microbiological criteria for foodstuffs. *Official Journal of the European Union* **L338/1**, 22 December 2005. Available from http://www.food.gov.uk/multimedia/pdfs/microbiolcriteria.pdf. Accessed 14 March 2007.

Estes, M.K., Prasad, B.V.V. & Atmar, R.L. (2006) Noroviruses everywhere: has anything changed? *Current Opinion in Infectious Diseases* **19**, 467–474.

Falkenhorst, G., Krusell, L., Lisby, M. *et al.* (2005) Imported frozen raspberries cause a series of norovirus outbreaks in Denmark, 2005. *Euro Surveillance* **10**(9), E050922.2. Available from http://www.eurosurveillance.org/ew/2005/050922.asp#2. Accessed 14 March 2007.

FDA (2006). *FDA News. FDA Investigating Norovirus Outbreak Linked to Oysters*. Available from http://www.fda.gov/bbs/topics/NEWS/2006/NEW01522.html. Accessed 14 March 2007.

FDA/CFSAN (2005a) *Food Code*. Available from http://www.cfsan.fda.gov. Accessed 14 March.

FDA/CFSAN (2005b) National Shellfish Sanitation Program.*Guide for the Control of Molluscan Shellfish 2005. Model Ordinance. IV. Shellstock Growing Areas*. Available from http://www.cfsan.fda.gov/~ear/nss3or04.html. Accessed 13 March 2007.

Fiore, A.E. (2004) Hepatitis A transmitted by food. *Clinical Infectious Diseases* **38**, 705–715.

Frank, C., Walter, J., Muehlen, M. *et al.* (2007) Major outbreak of hepatitis A associated with orange juice among tourists, Egypt, 2004. *Emerging Infectious Diseases* **13**, 156–158.

Friedman, D.S., Heisey-Grove, D., Argyros, F. *et al.* (2005) An outbreak of norovirus gastroenteritis associated with wedding cakes. *Epidemiology and Infection* **133**, 1057–1063.

Gallimore, C.I., Cubitt, D., du Plessis, N. & Gray, J.J. (2004) Asymptomatic and symptomatic excretion of noroviruses during a hospital outbreak of gastroenteritis. *Journal of Clinical Microbiology* **42**, 2271–2274.

Gallimore, C.I., Pipkin, C., Shrimptom, H. *et al.* (2005) Detection of multiple enteric virus strains within a foodborne outbreak of gastroenteritis: an indication of the source of contamination. *Epidemiology and Infection* **133**, 41–47.

Gaulin, C., Frigon, M., Poirier, D. & Fournier, C. (1999) Transmission of calicivirus by a foodhandler in the pre-symptomatic phase of illness. *Epidemiology and Infection* **123**, 475–478.

Gotz, H., de Jong, B., Lindback, J. *et al.* (2002) Epidemiological investigation of a food-borne gastroenteritis outbreak caused by Norwalk-like virus in 30 day-care centres. *Scandinavian Journal of Infectious Diseases* **34**, 115–121.

Guzewich, J. & Ross, M.P. (1999) *Evaluation of Risks Related to Microbiological Contamination of Ready-to-Eat Food by Food Preparation Workers and the Effectiveness of Interventions to Minimize Those Risks*. FDA/CFSAN. Available from http://vm.cfsan.fda.gov/~ear/rterisk.html. Accessed 14 March 2007.

Halliday, M.L., Kang, L.Y., Zhou, T.K. *et al.* (1991) An epidemic of hepatitis A attributable to the ingestion of raw clams in Shanghai, China. *Journal of Infectious Diseases* **164**, 852–859.

Health Protection Agency (HPA) (2004) *Statutory Notifications of Hepatitis A: England and Wales, by Region,* 1990–2003. Available from http://www.hpa.org.uk/infections/topics_az/hepatitis_a/data_not.htm. Accessed 14 March 2007.

Hernandez, F., Monge, R., Jimenez, C. & Taylor, L. (1997) Rotavirus and hepatitis A virus in market lettuce (*Latuca sativa*) in Costa Rica. *International Journal of Food Microbiology* **37**, 221–223.

Hjertqvist, M., Johansson, A., Svensson, A. *et al.* (2006) Four outbreaks of norovirus gastroenteritis after consuming raspberries, Sweden, June–August. *Euro Surveillance* **11**(9), EO60907.1. Available from http://www.eurosurveillance.org/ew/2006/060907.asp#1. Accessed 14 March 2007.

Hutin, Y.J., Pool, V., Cramer, E.H. *et al.* (1999) A multistate, foodborne outbreak of hepatitis A. *New England Journal of Medicine* **340**, 595–602.

Ijaz, S., Arnold, E., Banks, M. *et al.* (2005) Non-travel-associated hepatitis E in England and Wales: demographic, clinical, and molecular epidemiological characteristics. *Journal of Infectious Diseases* **192**, 1166–1172.

Jacobsen, K.H. & Koopman, J.S. (2004) Declining hepatitis A seroprevalence: a global review and analysis. *Epidemiology and Infection* **132**, 1005–1022.

Johansson, P.J.H., Torvén, M., Hammarlund, A.-C. *et al.* (2002) Food-borne outbreak of gastroenteritis associated with genogroup 1 calicivirus. *Journal of Clinical Microbiology* **40**, 494–798.

Kobayashi, S., Natori, K., Takeda, N. & Sakae, K. (2004) Immunomagnetic capture RT-PCR for detection of norovirus from foods implicated in a foodborne outbreak. *Microbiology and Immunology* **48**, 201–204.

Koopmans, M. & Duizer, E. (2004) Foodborne viruses: an emerging problem. *International Journal of Food Microbiology* **90**, 23–41.

Koopmans, M., Vennema, H., Heersma, H. *et al.* (2003) Early identification of common-source foodborne virus outbreaks in Europe. *Emerging Infectious Diseases* **9**, 1136–1142.

Koopmans, M., von Bonsdorff, C.-H., Vinje, J. *et al.* (2002) Foodborne viruses. *FEMS Microbiology Reviews* **26**, 187–205.

Lederer, I., Schmid, D., Pichler, A.-M. *et al.* (2005) Outbreak of norovirus infections associated with consuming food from a catering company, Austria, September 2005. *Euro Surveillance* **10**(10), E051020.7. Available from http://www.eurosurveillance.org/ew/2005/051020.asp#7. Accessed 14 March 2007.

Lees, D. (2000) Viruses and bivalve shellfish. *International Journal of Food Microbiology* **59**, 81–116.

Le Guyader, F.S., Mittelholzer, C., Haugarreau, L. *et al.* (2004) Detection of noroviruses in raspberries associated with a gastroenteritis outbreak. *International Journal of Food Microbiology* **97**, 179–186.

Le Guyader, F.S., Neill, F.H., Dubois, E. *et al.* (2003) A semiquantitative approach to estimate Norwalk-like virus contamination of oysters implicated in an outbreak. *International Journal of Food Microbiology* **87**, 107–112.

Li, T.-C., Chijiwa, K., Sera, N. *et al.* (2005) Hepatitis E virus from wild boar meat. *Emerging Infectious Diseases* **11**, 1958–1959.

Lo, S.V., Connolly, A.M., Palmer, S.R. *et al.* (1994) The role of the pre-symptomatic food handler in a common source outbreak of food-borne SRSV gastroenteritis in a group of hospitals. *Epidemiology and Infection* **113**, 513–521.

Lopalco, P.L., Malfait, P., Menniti-Ippolito, F. *et al.* (2005) Determinants of acquiring hepatitis A virus disease in a large Italian region in endemic and epidemic periods. *Journal of Viral Hepatitis* **12**, 315–321.

Lopman, B., Vennema, H., Kohli, E. *et al.* (2004a) Increase of viral gastroenteritis outbreaks in Europe and epidemic spread of new norovirus variant. *Lancet* **363**, 682–688.

Lopman, B.A., Adak, G.K., Reacher, M.H. & Brown, D.W.G. (2003a) Two epidemiologic patterns of Norovirus outbreaks: surveillance in England and Wales, 1992–2000. *Emerging Infectious Diseases* **9**, 71–77.

Lopman, B.A., Reacher, M.H., van Duijnhoven, Y. *et al.* (2003b) Viral gastroenteritis outbreaks in Europe, 1995–2000. *Emerging Infectious Diseases* **9**, 90–96.

Lopman, B.A., Reacher, M.H., Vipond, I.B. *et al.* (2004b) Clinical manifestation of norovirus gastroenteritis in health care settings. *Clinical Infectious Diseases* **39**, 318–324.

Malfait, P., Lopalco, P.L., Salmaso, S. *et al.* (1996) An outbreak of hepatitis A in Puglia, Italy. *Eurosurveillance Monthly* **1**(5), 33–35.

Marshall, J., Botes, J., Gorrie, G. *et al.* (2003) Rotavirus detection and characterisation in outbreaks of gastroenteritis in aged-care facilities. *Journal of Clinical Virology* **28**, 331–340.

Matin, N., Grant, A., Granerod, J. & Crowcroft, N. (2006) Hepatitis A surveillance in England – how many cases are not reported and does it really matter? *Epidemiology and Infection* **134**, 1299–1302.

Mattner, F., Sohr, D., Heim, A. *et al.* (2006) Risk groups for clinical complications of norovirus infections: an outbreak investigation. *Clinical Microbiology and Infection* **12**, 69–74.

McDonnell, S., Kirkland, K.B., Hlady, W.G. *et al.* (1997) Failure of cooking to prevent shellfish-associated viral gastroenteritis. *Archives of Internal Medicine* **157**, 111–116.

Millard, J., Appleton, H. & Parry, J.V. (1987) Studies on heat inactivation of hepatitis A virus with special reference to shellfish. *Epidemiology and Infection* **98**, 397–414.

Niu, M.T., Polish, L.B., Robertson, B.H. *et al.* (1992) Multistate outbreak of hepatitis A associated with frozen strawberries. *Journal of Infectious Diseases* **166**, 518–524.

Nygård, K., Andersson, Y., Lindkvist, Y. *et al.* (2001) Imported rocket salad partly responsible for increased incidence of hepatitis A cases in Sweden, 2000–2001. *Eurosurveillance Monthly* **6**(10), 151–153.

Parashar, U.D., Dow, L. & Fankhauser, R.L. (1998) An outbreak of viral gastroenteritis associated with consumption of sandwiches: implications for the control of transmission by food handlers. *Epidemiology and Infection* **121**, 615–621.

Patterson, T., Hutchings, P. & Palmer, S. (1993) Outbreaks of SRSV gastroenteritis at an international conference traced to food handled by a post-symptomatic caterer. *Epidemiology and Infection* **111**, 157–162.

Patterson, W., Haswell, P., Fryers, P.T. & Green, J. (1997) Outbreak of small round structured virus gastroenteritis arose after kitchen assistant vomited. *Communicable Disease Report Review* **7**, R101–R103.

Pebody, R.G., Leino, T., Ruutu, P. *et al.* (1998) Foodborne outbreaks of hepatitis A in a low-endemic country: an emerging problem? *Epidemiology and Infection* **120**, 55–59.

Ponka, A., Maunula, L., von Bonsdorff, C.-H. & Lyytikainen, O. (1999) An outbreak of calicivirus associated with consumption of frozen raspberries. *Epidemiology and Infection* **123**, 469–474.
ProMED-mail (2004) Hepatitis E virus, fatal – Japan (Hokkaido). ProMED-mail (2004); 28 November. Archive number 20041128.3183. Available from http://www.promedmail.org. Accessed 14 March 2007.
Reid, T.M.S. & Robinson, H.G. (1987) Frozen raspberries and hepatitis A. *Epidemiology and Infection* **98**, 109–112.
Rosenblum, L.S., Mirkin, I.R., Allen, D.T. *et al.* (1990) A multifocal outbreak of hepatitis A traced to commercially distributed lettuce. *American Journal of Public Health* **80**, 1075–1079.
Sakon, N., Yamazaki, K., Yoda, T. *et al.* (2005) A norovirus outbreak of gastroenteritis linked to packed lunches. *Japanese Journal of Infectious Diseases* **58**, 253.
Sala, M.R., Cardenosa, N., Arias, C. *et al.* (2005) An outbreak of food poisoning due to a genotype 1 norovirus. *Epidemiology and Infection* **133**, 187–191.
Sanchez, G., Pinto, R.M., Vanaclocha, H. & Bosch, A. (2002) Molecular characterization of hepatitis A virus isolates from a transcontinental shellfish-borne outbreak. *Journal of Clinical Microbiology* **40**, 4148–4155.
Schenkel, K., Bremer, V., Grabe, C. *et al.* (2006) Outbreak of hepatitis A in two federal states of Germany: bakery products as a vehicle of infection. *Epidemiology and Infection* **134**, 1292–1298.
Schwab, K.J., Neill, F.H., Fankhauser, R.L. *et al.* (2000) Development of methods to detect "Norwalk-like viruses" (NLVs) and hepatitis A virus in delicatessen foods: application to a food-borne NLV outbreak. *Applied and Environmental Microbiology* **66**, 213–218.
Seymour, I.J. & Appleton, H. (2001) Foodborne viruses and fresh produce. *Journal of Applied Microbiology* **91**, 759–773.
Shieh, Y.C., Khudyakov, Y.E., Xia, G. *et al.* (2007) Molecular confirmation of oysters as the vector for hepatitis A in a 2005 multistate outbreak. *Journal of Food Protection* **70**, 145–150.
Shieh, Y.-S.C., Monroe, S.S., Fankhauser, R.L. *et al.* (2000) Detection of Norwalk-like virus in shellfish implicated in illness. *Journal of Infectious Diseases* **181**(Suppl 2), S360–S366.
Slomka, M.J. & Appleton, H. (1998) Feline calicivirus as a model system for heat inactivation studies of small round structured viruses in shellfish. *Epidemiology and Infection* **121**, 401–407.
Smith, J.L. (2001) A review of hepatitis E virus. *Journal of Food Protection* **64**, 572–586.
Stafford, R., Strasin, D., Heymer, M. *et al.* (1997) An outbreak of Norwalk virus gastroenteritis following consumption of oysters. *Communicable Diseases Intelligence* **21**, 317–320.
Stevenson, P., McCann, R., Duthrie, R. *et al.* (1994) A hospital outbreak due to Norwalk virus. *Journal of Hospital Infection* **26**, 261–272.
Tang, Y.W., Wang, J.X., Xu, Z.Y. *et al.* (1991) A serologically confirmed, case-control study of a large outbreak of hepatitis A in China, associated with consumption of clams. *Epidemiology and Infection* **107**, 651–657.
Weltman, A.C., Bennett, N.M., Ackman, D.A. *et al.* (1996) An outbreak of hepatitis A associated with a bakery, New York, 1994: The 1968 'West Branch, Michigan' outbreak repeated. *Epidemiology and Infection* **117**, 333–341.
Wheeler, C., Vogt, T.M., Armstrong, G. *et al.* (2005) An outbreak of hepatitis A associated with green onions. *New England Journal of Medicine* **353**, 890–897.
Widdowson, M.-A., Sulka, A., Bulens, S.N. *et al.* (2005) Norovirus and foodborne disease, United States, 1991–2000. *Emerging Infectious Diseases* **11**, 95–102.
World Health Organization (WHO) (2000) Hepatitis A. WHO/CDS/CSR/EDC/2000.7. Available from http://www.who.int/csr/disease/hepatitis/HepatitisA_whocdscsredc2000_7.pdf. Accessed 14 March 2007.
World Health Organization (WHO) (2002) Hepatitis E virus. In: *Guidelines for Drinking Water Quality*, 2nd edn, Addendum. *Microbiological Agents in Drinking Water*. WHO, Geneva, pp. 23–25.

2.15 *Cryptosporidium* spp.

2.15.1 *Importance as a cause of foodborne disease*

Cryptosporidium was reported first as the cause of human disease in 1976, but it was not until 1982 that it aroused much attention, when this protozoan parasite was reported as the cause of severe disease in the US (Fayer, 2004). Techniques for detection and identification of the organism were developed in the early 1980s and large outbreaks associated with municipal water supplies were recognized in the US in the 1980s and early 1990s. Outbreaks and sporadic cases have now been reported worldwide. The most commonly identified outbreaks have been those associated with drinking water supplies or recreational water contact (Chapter 7), but outbreaks have also been reported associated with food, animal contact and person-to-person spread.

Cryptosporidium is associated with diarrhoeal illness in most areas of the world, and prevalence is highest in under-developed countries (Fayer, 2004). In the US it was estimated that *Cryptosporidium* infection affects 300,000 people annually, with ~10% of these cases foodborne (Mead *et al.*, 1999), although this is probably a substantial underestimate. In the UK in 2005 the reported incidence was 9.3 cases per 100,000 population, based on confirmed cases only (Semenza & Nichols, 2007).

2.15.2 *Characteristics of the organism*

Cryptosporidium species are found in a wide range of animals (Fayer, 2004; Hunter & Thompson, 2005). The species infecting man and other mammals has been named *C. parvum*, but it is now known that two distinct genotypes exist, genotype 1 which affects mainly humans, and genotype 2, which occurs mainly in animals but also affects humans. These two genotypes are now regarded as two separate species, genotype 1 named as *C. hominis* and genotype 2 as *C. parvum* (Hunter & Thompson, 2005).

The life cycle of *Cryptosporidium* occurs in the intestinal, mucosal epithelium of the infected host (Fig. 2.15.1). The infective stage is the mature, thick-walled oocyst, containing sporozoites, which is excreted from the host in faeces. After ingestion of the oocyst the sporozoites are released and penetrate epithelial cells of the gastrointestinal tract, where they are enclosed in a vacuole at the cell surface. In this vacuole morphological changes and asexual and sexual division take place resulting in the formation of thick-walled and thin-walled oocysts. Oocysts with a thick, two-layered wall are excreted in faeces and are responsible for transmission to other hosts; oocysts with a thin, single-layered membrane re-infect cells of the same host.

The thick-walled oocyst is a roughly spherical structure 4–6 μm diameter, which can be difficult to differentiate, microscopically, from other small particles such as yeasts, mould spores, algae and plant debris.

The organism does not multiply outside the host but oocysts can remain viable in cool, moist conditions in the environment for several months. Boiling in water inactivates oocysts, heating oocysts at 71.7°C for 5–15 seconds in water or milk gave a reduction of at least 1000-fold, freezing at −70°C or desiccation resulted in inactivation (Dawson, 2005). Oocysts are relatively resistant to disinfectants, particularly chlorine at the concentration used in

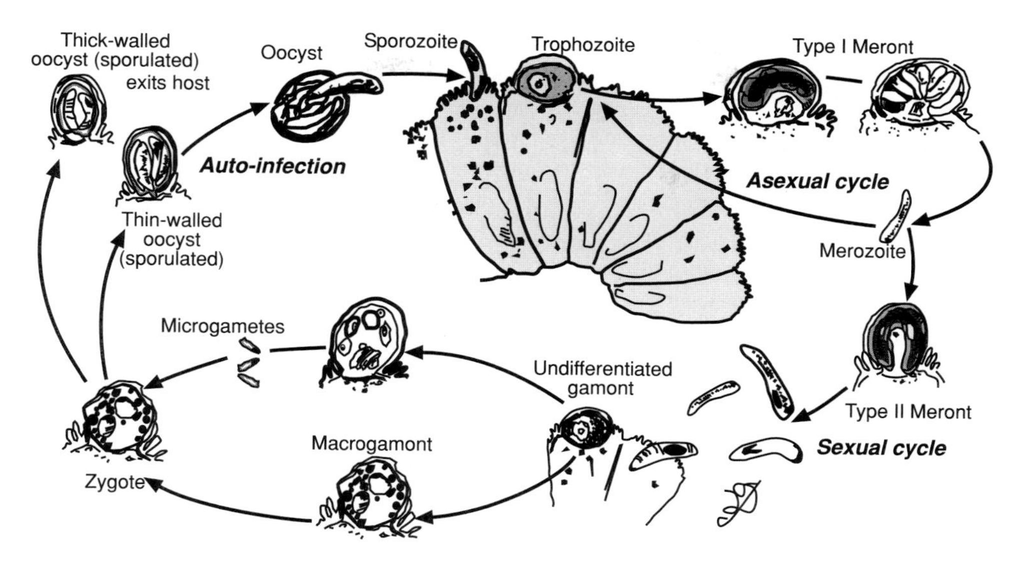

Fig. 2.15.1 Life cycle of *Cryptosporidium* in humans (From the US Centers for Disease Control and Prevention, Public Health Image Library, Content providers, CDC/Alexander J. da Silva/Melanie Moser)

municipal water supplies, but can be inactivated by ozone or by ultra violet irradiation (Betancourt & Rose, 2004; Erickson & Ortega, 2006).

Routine methods for growth of cryptosporidium in vitro are not yet available, but cell culture methods have been developed that are useful primarily for research (Arrowood, 2002). Isolates from clinical specimens can be stored unpreserved or in aqueous $K_2Cr_2O_7$ at 4°C.

Detection of *Cryptosporidium* in faeces and in environmental samples has relied on separation, concentration where necessary, and microscopy, including the use of immunofluorescent staining. PCR procedures have also been developed that should enable more specific characterization of the organisms (Gómez-Couso *et al.*, 2004; Sunnotel *et al.*, 2006).

2.15.3 *Disease in humans*

The dose required to cause infection may be as low as 1–10 oocysts. In normal, healthy people *Cryptosporidium* usually causes self-limiting diarrhoeal illness. The incubation period is usually 7–10 days (range 4–28 days) and acute symptoms last for 2–26 days, occasionally longer (Hunter *et al.*, 2004a). The main symptoms are watery diarrhoea, which may be profuse, abdominal pain and mild fever. Gastrointestinal symptoms may recur frequently after the initial acute phase of disease has resolved; *C. hominis* infection was reported more likely than *C. parvum* infection to be followed by systemic symptoms (Hunter *et al.*, 2004a).

In most immunocompetent patients illness is short term and recovery is complete, whereas in patients with AIDS the illness is commonly prolonged and life-threatening. In AIDS patients, cryptosporidia may cause atypical gastrointestinal disease, disease of the biliary tract and the pancreas, and contribute to respiratory tract disease. Illness in AIDS patients has been caused by at least nine other species of cryptosporidium in addition to *C. parvum* and *C. hominis* (Hunter & Nichols, 2002) and immunocompetent persons occasionally suffer illness caused by unusual species (Chalmers *et al.*, 2002). Further information is needed on the extent to which immunocompromised patients other than those with AIDS are at greater risk of cryptosporidiosis than immunocompetent persons.

In immunocompetent persons excretion of oocysts can start 3–5 days after ingestion of infective oocysts and can last for several weeks, but may be intermittent during later stages (Fayer, 2004). In immunocompromised persons whose symptoms last for several months, prolonged excretion may occur.

Acquired immunity to *Cryptosporidium* may reduce the risk of illness in immunocompromised persons (Frost *et al.*, 2005).

2.15.4 *Source and transmission*

Oocysts are infectious when excreted. Infection of a new host may be by ingestion or possibly by inhalation. Major risk factors for sporadic infection with *C. hominis* in England and Wales were foreign travel, and changing children's diapers, whereas for *C. parvum* infection touching farm animals was the main risk factor (Hunter *et al.*, 2004b).

Cryptosporidium can be present in effluent from sewage treatment plants, which passes into rivers and reservoirs, and manure and slurry particularly that from cattle is probably a major source of waterborne *Cryptosporidium* (Fayer, 2004). Contamination of water sources by livestock has been implicated as a major cause of sporadic infection in humans (Goh *et al.*, 2005) and of waterborne outbreaks (Caccio *et al.*, 2005).

Several studies have reported that consumption of unpasteurized milk or milk products is a risk factor for cryptosporidiosis (Smith, 1993; Gelletlie *et al.*, 1997). Calves are infected frequently, dairy cows can also be infected and inadequate udder hygiene may result in contamination of milk.

Vegetables and soft fruit may be contaminated with *Cryptosporidium* from irrigation water or as a result of the use of human or animal faeces as fertilizer (Moore *et al.*, 2007). *Cryptosporidium* oocysts have been detected on fresh vegetables collected from open markets in Costa Rica (Monge & Chinchilla, 1996), on 14.5% of vegetables from small markets in Peru (Ortega *et al.*, 1997), on lettuce, parsley, cilantro and blackberries in local agricultural markets in Costa Rica (Calvo *et al.*, 2004), and in low numbers on 9% of 149 samples of mung bean sprouts and 4% of 125 samples of lettuce in Norway (Robertson & Gjerde, 2001).

Oocysts of *Cryptosporidium* are liable to be present in estuarine water and in seawater near to sites of sewage outfall and agricultural runoff; oocysts of *Cryptosporidium* sp. and of *C. parvum* have been detected in oysters, clams, mussels and cockles from coastal waters of Europe and the US (Gómez-Couso *et al.*, 2003; Fayer, 2004; Li *et al.*, 2006). Consumption of raw molluscan shellfish is a potential risk factor for cryptosporidiosis (Butt *et al.*, 2004), and cooking mussels by steaming did not completely destroy infectivity of *C. parvum* (Gómez-Couso *et al.*, 2006). Although no reports have, as yet, implicated shellfish as a vehicle for *Cryptosporidium*, people who are immunocompromised are advised to avoid eating raw or undercooked shellfish.

The potential exists, therefore, for transmission of *Cryptosporidium* in drinking water, in unpasteurized milk products, on raw fruit and vegetables, on raw or undercooked molluscan shellfish and by food handlers.

On the basis of pulmonary symptoms in children and immune compromised persons with cryptosporidiosis it has been suggested that airborne transmission of *Cryptosporidium* occurs but this has not been proven.

2.15.5 *Examples of outbreaks*

The association of cryptosporidiosis with water supplies is discussed in Chapter 7. Examples of outbreaks associated with foods are shown in Table 2.15.1.

2.15.6 *Prevention and control*

Contamination of drinking water with *Cryptosporidium* oocysts is a particular risk if the water is obtained from an area where livestock is farmed (Goh *et al.*, 2005). Prevention of transmission of *Cryptosporidium* through drinking water requires protection of water sources from contamination with protozoa, effective water treatment and verification of the effectiveness of treatment (WHO, 2006). Because *Cryptosporidium* is very resistant to disinfection, it is difficult, even with a well-operated treatment system, to ensure that drinking water is free of this organism (Orlandi *et al.*, 2002; Betancourt & Rose, 2004; Dawson, 2005). Drinking water regulations in the UK require water companies to undertake risk assessment of water sources and, where there is judged to be a risk, to implement continuous monitoring of *Cryptosporidium* oocyst concentrations in treated water. The minimum standard required is an average of <1 oocyst per 10 L of water in any 24-hour period (Goh *et al.*, 2005). Because of the risk of cryptosporidiosis to highly immunosuppressed people, advice has been issued in the UK and the US to such people, in particular HIV-positive people with CD4 counts of less than $200/mm^3$, to boil their drinking water, from whatever source (Hunter & Nichols, 2002; Chapter 7). Most western countries have issued similar advice.

In the case of salad vegetables and fruits, there is a need to prevent contamination with manure from farm animals, contaminated irrigation water, and untreated sewage or treated sewage sludge (Dawson, 2005; Moore *et al.*, 2007).

Properly controlled pasteurization of milk will inactivate any oocysts that are present.

Summary of measures advised in healthcare settings

1. Drinking water should be from public water supplies that have undergone an effective treatment process. Patients who are severely immunosuppressed should be advised to use drinking water that has been boiled and cooled, whatever the source.
2. Avoid the use of raw vegetables and fruit that may have been contaminated via irrigation water; fruit and vegetables should be washed thoroughly.
3. Avoid the consumption of milk that has not been properly pasteurized and of products made with unpasteurized milk.
4. Avoid consumption of raw or undercooked shellfish.

Table 2.15.1 Examples of outbreaks of foodborne disease caused by *Cryptosporidium* spp.

Place, date	Cases	Food implicated	Evidence[a]	Where food was provided	Factors contributing to outbreak	Reference
Denmark, 1989	19 (18 with AIDS) (8 deaths)	Ice for cold drinks	D	Hospital	Psychotic, patient with AIDS had cryptosporidiosis and was incontinent; grossly negligent about basic hygiene, repeatedly hand-picked ice from ice-machine	Ravn *et al.* (1991)
US, 1993	160	Unpasteurized, fresh-pressed apple juice	S and M	School, agricultural fair	Apples were from trees on a pasture where cows had grazed recently. Some apples were collected from the ground and were later sprayed with water. Apple cider was not pasteurized	Millard *et al.* (1994)
UK, 1995	67	'Pasteurized' milk	S	School	Milk obtained from small-scale producer with an on-farm pasteurizer. Pasteurization plant had not been working properly and milk was not adequately heat treated	Djuretic *et al.* (1997); Gelletlie *et al.* (1997)
US, 1995	15	Chicken salad	D	Licenced day-care home	Salad required extensive handling during preparation; handler had changed diapers before preparing the salad	CDC (1996)
US, 1996	31	Unpasteurized, fresh-pressed apple juice	S	Community	Apples washed with well-water, dairy farm located nearby; juice not pasteurized	CDC (1997)
US, 1997	54	Salads containing uncooked green onions	D	Banquet	Green onions were served uncooked and not always washed before serving	CDC (1998)
US, 1998	88 + 60 possible cases	Meal with raw vegetables and fruits	S and M	University cafeteria	Food handler who was ill with copious watery diarrhea continued to work, cutting up vegetables and fruits that were served raw	Quiroz *et al.* (2000)
Australia, 2001	8	Unpasteurized milk	S and M	Community	Milk unpasteurized, probable faecally contaminated	Harper *et al.* (2002)
US, 2003	23 laboratory confirmed, 121 probable	Unpasteurized, ozonated apple cider	S and M	Community, school outings	'A few' dropped apples were used in the cider production. Juice was not pasteurized but was ozonated, this treatment has not been shown to give effective pathogen reduction	Blackburn *et al.* (2006)
Denmark, 2005	78	Raw carrot served in salad bar	S	Company canteen	Carrots kept in a bowl of water before serving, customers may have removed carrots from bowl using hands	Ethelberg *et al.* (2005)

[a] M (microbiological): identification of an organism of the same type from cases and in the suspect vehicle or vehicle ingredient(s), or detection of toxin in faeces or food; D (descriptive): other evidence, usually descriptive, reported by local investigators as indicating the suspect vehicle or food; S (statistical): a significant statistical association between consumption of the suspect vehicle and being a case.

5 Use good hygienic practices when handling food. Food handlers should not work in that capacity while suffering symptoms; good hand washing is essential and avoidance of bare hand contact with ready-to-eat foods is advisable.

References

Arrowood, M.J. (2002) In vitro cultivation of *Cryptosporidium* species. *Clinical Microbiology Reviews* **15**, 390–400.

Betancourt, W.Q. & Rose, J.B. (2004) Drinking water treatment processes for removal of *Cryptosporidium* and *Giardia*. *Veterinary Parasitology* **126**, 219–234.

Blackburn, B.G., Mazurek, J.M., Hlavsa, M. *et al.* (2006) Cryptosporidiosis associated with ozonated apple cider. *Emerging Infectious Diseases* **12**, 684–686.

Butt, A.A., Aldridge, K.E. & Sanders, C.V. (2004) Infections related to the ingestion of seafood. Part ii: parasitic infections and food safety. *Lancet Infectious Diseases* **4**, 294–300.

Caccio, S.M., Thompson, R.C.A., McLauchlin, J. & Smith, H.V. (2005) Unravelling *Cryptosporidium* and *Giardia* epidemiology. *Trends in Parasitology* **21**, 430–437.

Calvo, M., Carazo, M., Arias, M.L. *et al.* (2004) Prevalence of *Cyclospora* sp., *Cryptosporidium* sp., microsporidia and fecal coliform determination in fresh fruit and vegetables consumed in Costa Rica. *Archivos Latinamericanos de Nutricion* **54**, 428–432.

Centers for Disease Control and Prevention (CDC) (1996) Foodborne outbreak of diarrheal illness associated with *Cryptosporidium parvum* – Minnesota, 1995. *Morbidity and Mortality Weekly Report* **45**(36), 783–784.

Centers for Disease Control and Prevention (CDC) (1997) Outbreaks of *Escherichia coli* O157:H7 infection and cryptosporidiosis associated with drinking unpasteurized apple cider – Connecticut and New York, October 1996. *Morbidity and Mortality Weekly Report* **46**(01), 4–8.

Centers for Disease Control and Prevention (CDC) (1998) Foodborne outbreak of cryptosporidiosis – Spokane, Washington, 1997. *Morbidity and Mortality Weekly Report* **47**(27), 565–567.

Chalmers, R.M., Elwin, K., Thomas, A.L. & Joynson, D.H.M. (2002) Infection with unusual types of *Cryptosporidium* is not restricted to immunocompromised patients. *Journal of Infectious Diseases* **185**, 270–271.

Dawson, D. (2005) Foodborne protozoan parasites. *International Journal of Food Microbiology* **103**, 207–227.

Djuretic, T., Wall, P.G. & Nichols, G. (1997) General outbreaks of infectious intestinal disease associated with milk and dairy products in England and Wales: 1992 to 1996. *Communicable Disease Report* **7**, R41–R45.

Erickson, M.C. & Ortega, Y.R. (2006) Inactivation of protozoan parasites in food, water, and environmental systems. *Journal of Food Protection* **69**, 2786–2808.

Ethelberg, S., Lisby, M., Vertergaard, L.S. *et al.* (2005) Cryptosporidiosis associated with eating in a canteen, Denmark, August 2005. *Euro Surveillance* **10**(10), E051027.4. Available from http://www.eurosurveillance.org/ew/2005/051027.asp#4. Accessed 14 March 2007.

Fayer, R. (2004) *Cryptosporidium*: a water-borne zoonotic parasite. *Veterinary Parasitology* **126**, 37–56.

Frost, F.J., Tollestrup, K., Craun, G.F. *et al.* (2005) Protective immunity associated with a strong response to a *Cryptosporidium*-specific antigen group, in HIV-infected individuals. *Journal of Infectious Diseases* **192**, 618–621.

Gelletlie, R., Stuart, J., Soltanpoor, N. *et al.* (1997) Cryptosporidiosis associated with school milk. *Lancet* **350**, 1005–1006.

Goh, S., Reacher, M., Casemore, D.P. *et al.* (2005) Sporadic cryptosporidiosis decline after membrane filtration of public water supplies, England, 1996–2002. *Emerging Infectious Diseases* **11**, 251–259.

Gómez-Couso, H., Freire-Santos, F., Amar, C.F.L. *et al.* (2004) Detection of *Cryptosporidium* and *Giardia* in molluscan shellfish by multiplexed nested-PCR. *International Journal of Food Microbiology* **91**, 279–288.

Gómez-Couso, H., Freire-Santos, F., Martinez-Urtaza, J. *et al.* (2003) Contamination of bivalve molluscs by *Cryptosporidium* oocysts: the need for new quality control standards. *International Journal of Food Microbiology* **87**, 97–105.

Gómez-Couso, H., Mendez-Hermida, F., Castro-Hermida, J.A. & Ares-Mazas, E. (2006) Cooking mussels (*Mytilus galloprovincialis*) by steam does not destroy the infectivity of *Cryptosporidium parvum*. *Journal of Food Protection* **69**, 948–950.

Harper, C.M., Cowell, N.A., Adams, B.C. *et al.* (2002) Outbreak of *Cryptosporidium* linked to drinking unpasteurized milk. *Communicable Diseases Intelligence* **26**, 449–450.

Hunter, P.R., Hughes, S., Woodhouse, S. *et al.* (2004a) Health sequelae of human cryptosporidiosis in immunocompetent patients. *Clinical Infectious Diseases* **39**, 504–510.

Hunter, P.R., Hughes, S., Woodhouse, S. *et al.* (2004b) Sporadic cryptosporidiosis case-control study with genotyping. *Emerging Infectious Diseases* **10**, 1241–1249.

Hunter, P.R. & Nichols, G. (2002) Epidemiology and clinical features of *Cryptosporidium* infection in immunocompromised patients. *Clinical Microbiology Reviews* **15**, 145–154.

Hunter, P.R. & Thompson, R.C. (2005) The zoonotic transmission of *Giardia* and *Cryptosporidium*. *International Journal for Parasitology* **35**, 1181–1190.

Li, X., Guyot, K., Dei-Cas, E. *et al.* (2006) *Cryptosporidium* oocysts in mussels (*Mytilus edulis*) from Normandy (France). *International Journal of Food Microbiology* **108**, 321–325.

Mead, P.S., Slutsker, L., Dietz, V. *et al.* (1999) Foodborne illness and death in the United States. *Emerging Infectious Diseases* **5**, 607–625.

Millard, P.S., Gensheimer, K.F., Addiss, D.G. *et al.* (1994) An outbreak of cryptosporidiosis from fresh-pressed apple cider. *Journal of the American Medical Association* **272**, 1592–1596.

Monge, R. & Chinchilla, M. (1996) Presence of *Cryptosporidium* oocysts in fresh vegetables. *Journal of Food Protection* **59**, 202–203.

Moore, J.E., Millar, B.C., Kenny, F. *et al.* (2007) Detection of *Cryptosporidium parvum* in lettuce. *International Journal of Food Science and Technology* **42**, 385–393.

Orlandi, P.A., Chu, D.-M., Bier, J. & Jackson, G.J. (2002) Parasites and the food supply. *Food Technology* **56**, 72–81.

Ortega, Y.R., Roxas, C.R., Gilman, R.H. *et al.* (1997) Isolation of *Cryptosporidium parvum*, and *Cyclospora cayetanensis* from vegetables collected in markets of an endemic region in Peru. *American Journal of Tropical Medicine and Hygiene* **57**, 683–686.

Quiroz, E.S., Bern, C., MacArthur, J.R. *et al.* (2000) An outbreak of cryptosporidiosis linked to a food handler. *Journal of Infectious Diseases* **181**, 695–700.

Ravn, P., Lundgren, J.D., Kjaeldgaard, P. *et al.* (1991) Nosocomial outbreak of cryptosporidiosis in AIDS patients. *British Medical Journal* **302**, 277–280.

Robertson, L.J. & Gjerde, B. (2001) Occurrence of parasites on fruits and vegetables in Norway. *Journal of Food Protection* **64**, 1793–1798.

Semenza, J.C. & Nichols, G. (2007) Cryptosporidiosis surveillance and water-borne outbreaks in Europe. *Euro Surveillance* 12 (5). Available online: http://www.eurosurveillance.org/em/v12n05/1205-227.asp. Accessed 9 July 2007.

Smith, J.L. (1993) *Cryptosporidium* and *Giardia* as agents of foodborne disease. *Journal of Food Protection* **56**, 451–461.

Sunnotel, O., Lowery, C.J., Moore, J.E. *et al.* (2006) *Cryptosporidium*. *Letters in Applied Microbiology* **43**, 7–17.

World Health Organization (WHO) (2006) *Guidelines for Drinking-Water Quality.* First addendum to third edition. *Vol. 1: Recommendations*. WHO, Geneva.

2.16 *Cyclospora cayetanensis*

2.16.1 *Importance as a cause of foodborne disease*

Cyclospora species have been known since about 1881 as protozoans associated with animals. *Cyclospora cayetanensis* was recognized as a pathogen in humans in the early 1990s and named in 1994. Transmission of the organism to humans was linked initially with consumption of water, but foodborne transmission was suspected after outbreaks in 1995 and was established in outbreaks in the US and Canada in 1996, linked to imported raspberries (Herwaldt *et al.*, 1997).

Infection with *C. cayetanensis* is most frequent in certain tropical and subtropical countries including Nepal, Peru, Guatemala and Haiti. The prevalence of infection and development of symptoms are variable and seasonal. Disease affects visitors from Europe and North America, who have not been exposed previously to the organism, to countries where infection is common in the indigent population. The disease has probably been underreported in the past.

2.16.2 *Characteristics of the organism*

Cyclospora cayetanensis is a coccidian protozoan. It forms oocysts that are 8–10 μm in diameter and are excreted from infected humans in stools in an unsporulated state, which is not infectious (Herwaldt, 2000). In the environment sporulation occurs, this may take days or weeks and results in the formation of two internal sporocysts each containing two sporozoites (Fig. 2.16.1); the sporulated oocyst is infectious. In this respect oocysts of *C. cayetanensis* differ from those of *Cryptosporidium*, which are sporulated and infectious when excreted. The fact that the oocysts of *C. cayetanensis* require a period of time in the environment to sporulate and become infectious, decreases the chance of direct person-to-person spread.

Methods for culturing the organism are not available generally; the related protozoan *Eimeria acervulina*, which is found commonly in chickens and can be cultured, has been used as a model to investigate decontamination. Oocysts of *C. cayetanensis*, like those of *Cryptosporidium*, are probably highly resistant to the levels of chorine used in treatment of water supplies. After heating at 60°C for 1 hour or freezing at −20°C for 24 hours, oocysts could not be induced to sporulate, an indication that they were not viable (Sterling & Ortega, 1999).

It is not yet clear whether humans are the only sources of *C. cayetanensis*. Some non-human primates are infected with protozoa that are closely related to, but distinct from, *C. cayetanensis* (Eberhard *et al.*, 1999a) but attempts to establish infection with *C. cayetanensis* experimentally in animals and in humans have failed (Alfano-Sobsey *et al.*, 2004).

In early reports the organism was seen in the light microscope in acid-fast stained material from stools. In wet mounts viewed by ultraviolet fluorescence microscopy the oocysts of *C. cayetanensis* autofluoresce, in contrast with those of *Cryptosporidium*.

The numbers of oocysts present in stools of ill persons are often low, and concentration of the organism is needed before detection. Detection in water, berries and vegetables requires concentration from samples, fluorescence microscopy and PCR techniques (Shields & Olson, 2003; Orlandi *et al.*, 2004).

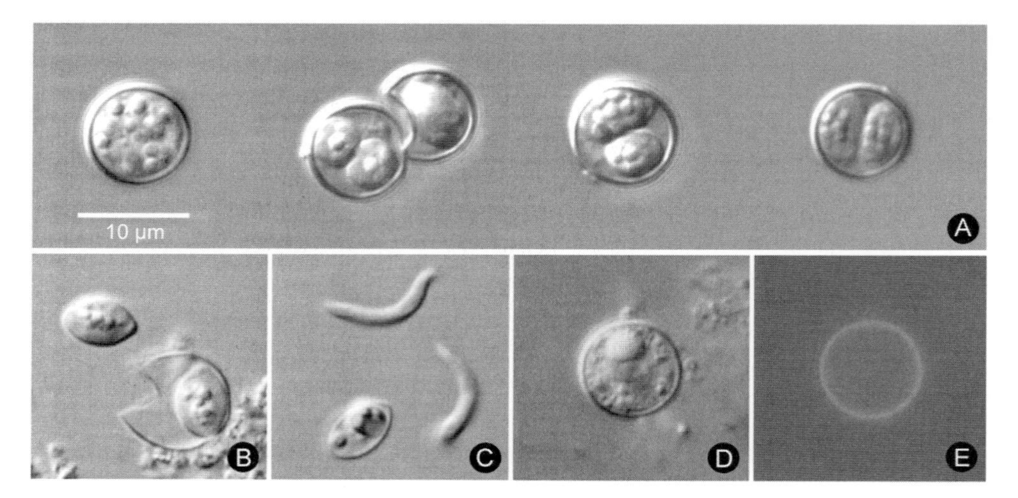

Fig. 2.16.1 Sporulation of *Cyclospora cayetanensis* and autofluorescence in wet preparations of stool specimens. A, from left to right shows the progression from an unsporulated oocyst, containing undifferentiated cytoplasm filled with refractile globules; to an oocyst containing two sporoblasts that is adjacent to another sporulating oocyst; to an oocyst containing two immature sporocysts; to an oocyst with more mature sporocysts. B, shows an oocyst that has been disrupted mechanically releasing one of its sporocysts. C, shows a free sporocyst and two sporozoites released from sporocysts. D, shows an unsporulated oocyst viewed by differential interference microscopy. E, shows the same oocyst viewed by ultraviolet fluorescence microscopy. (From Herwaldt (2000). Figure kindly supplied by Dr Michael J. Arrowood, Centers for Disease Control and Prevention, US, and published with his permission)

2.16.3 *Disease in humans*

In studies in Guatemala and Peru infection was most common in children aged between about 1.5 and 9 years and 2 to <4 years, respectively (Herwaldt, 2000). In children in developing countries infection is often asymptomatic (Herwaldt, 2000; Chacin-Bonilla *et al.*, 2001) or symptoms are relatively mild, whereas in non-immune adults living in these countries or in developed countries illness is generally more severe and long lasting. Following ingestion by humans, the incubation period ranges from 1 to 14 days, with an average of 1 week. The symptoms include relapsing and cyclical, watery diarrhoea (sometimes alternating with constipation) that lasts for an average of about 7 weeks in immunocompetent patients and can last for up about 4 months in AIDS patients. This can be accompanied by nausea, fever, fatigue and weight loss. After passage into the intestine, the oocyst ruptures releasing the sporocysts, which then release sporozoites. These invade epithelial cells of the small intestine, where they reproduce both asexually and sexually, resulting in changes in the intestinal mucosa (Ortega *et al.*, 1997a). Eventually gametes are formed that give rise to oocysts, which are passed out in the faeces.

Trimethoprim-sulphamethazole is the most effective treatment; no alternative treatment appears to be advised for persons who cannot tolerate these drugs as many antibiotics appear to be ineffective (CDC, 2004a), but there is some evidence that ciprofloxacin may be an alternative (Shields & Olson, 2003).

Guillain–Barré syndrome, reactive arthritis and disease of the gall bladder have been reported as complications of *Cyclospora* infection (Shields & Olson, 2003).

2.16.4 *Source and transmission*

Infection with *C. cayetanensis* occurs relatively frequently in Nepal, Peru, Guatemala, Indonesia and Haiti (Ortega *et al.*, 1993; Hoge *et al.*, 1995; Madico *et al.*, 1997; Bern *et al.*, 1999, 2002a, b; Eberhard *et al.*, 1999b) and the organism may be endemic in parts of Asia, Africa and South America (Shields & Olson, 2003). The prevalence of infection is seasonal, but this seasonality differs in different countries and the basis has not been explained (Lopez *et al.*, 2003). A high incidence of symptomatic infection has been reported in expatriate adults who have lived in Nepal or in West Java, Indonesia, for up to 2 years (Fryauff *et al.*, 1999; Shlim *et al.*, 1999) and a possible outbreak was reported among Dutch microbiologists attending a meeting in Indonesia (Blans *et al.*, 2005).

Several waterborne outbreaks of cyclosporiasis have been reported, and the organism may be more often waterborne than foodborne. It is transmitted to humans as a result of environmental factors and agricultural practices in the production of fruit and vegetables (Orlandi *et al.*, 2002). The organism is probably chlorine-resistant, like *Cryptosporidium* spp., but the oocysts should be removed more easily by conventional filtration, during water-treatment, than the smaller oocysts of *Cryptosporidium. Cyclospora cayetanensis* was reported in water supplies and water taps in Hanoi, suggesting that drinking water was an important source of the organism (Cam *et al.*, 2001).

Oocysts of *C. cayetanensis* have been detected on green vegetables in Nepal (CDC, 1991), on lettuce in Peru (Ortega *et al.*, 1997b) and Egypt (cited in Doller *et al.*, 2002), and in samples of wastewater from a lagoon in a shanty town in Lima; water from such lagoons was used for irrigation (Sturbaum *et al.*, 1998; Sterling & Ortega, 1999). Both sporulated and unsporulated oocysts were found in samples of river water in Guatemala (Bern *et al.*, 1999).

2.16.5 *Examples of outbreaks*

Examples of foodborne outbreaks of cyclosporiasis are shown in Table 2.16.1.

In Guatemala the peak time for human cyclosporiasis was reported to be May–July, overlapping with Guatemala's spring export season for raspberries (Bern *et al.*, 1999). The fruit may have become contaminated by exposure to contaminated water, possibly used when berries were sprayed with insecticides, fungicides and fertilizers.

Because of the difficulty in detecting low numbers of the organism, most of the evidence linking outbreaks to a food has been statistical. In the outbreak in the US in 1999, associated with basil included in salads, *Cyclospora* DNA was detected in the salad by an improved PCR technique, and one sporulated *Cyclospora* oocyst was detected by microscopy (Lopez *et al.*, 2001). In the outbreak in the US in 2000, *C. cayetanensis* was detected in the raspberry filling of a cake by PCR and sequencing of amplified products (Ho *et al.*, 2002).

2.16.6 *Prevention and control*

Illness caused by *C. cayetanensis* has probably been underdiagnosed and underreported in the past (Orlandi *et al.*, 2002). Increasing travel abroad is liable to result in illness in people not previously exposed to the organism. Unprocessed food, including fresh fruit and salad vegetables, is imported increasingly from parts of the world in which *Cyclospora*

Table 2.16.1 Examples of outbreaks of foodborne disease caused by *Cyclospora cayetanensis*.

Place, date	Cases	Food implicated	Evidence[a]	Where food was provided	Factors leading to outbreak	Reference
US, Canada 1996	1465	Raspberries imported from Guatemala	S	Events and community	Possible contamination on farms through use of contaminated water for spraying insecticides, fungicides or fertilizers on fruit	Herwaldt *et al.* (1997); Herwaldt (2000)
US, Canada, 1997	1012	Raspberries imported from Guatemala	S	Events and community	Possible contamination on farms through use of contaminated water for spraying insecticides, fungicides or fertilizers on fruit	CDC (1997a); Herwaldt *et al.* (1999); Herwaldt (2000)
US (Florida), 1997	>29	Mesclun (mix of baby lettuce and other salad greens), probably from Peru	D	Restaurants	nr	Herwaldt (2000)
US (Virginia, Washington, DC, Baltimore), 1997	341	Basil in pesto pasta salad	S	Sold by chain of gourmet food stores	Possibly contaminated by food handlers; basil handled several days before it was eaten, this time may have allowed sporulation	CDC (1997b); Herwaldt (2000)
US (Virginia), 1997	21	Fruit plate, probably including imported raspberries not from Guatemala	D	Inn	nr	Herwaldt (2000)
US (Florida), 1997	12	Mesclun, imported from Peru	D	Catered dinner	nr	Herwaldt (2000)
Canada (Ontario), 1998	315	Raspberries imported from Guatemala	S	13 events	nr	CDC (1998); Herwaldt (2000)
US (Georgia), 1998	17	Fruit salad, source not determined. Raspberries and blackberries from undetermined sources were among the fruits	D	Single establishment	nr	Herwaldt (2000)

Canada (Ontario), 1999	104	Dessert (berries) including fresh Guatemalan blackberries, frozen Chilean raspberries, fresh US strawberries	D	nr	nr	Herwaldt (2000)
US (Florida), 1999	94	Probably berry fruit, fruits served included foreign and domestic raspberries, imported blackberries, strawberries and blueberries	D	Convention	nr	Herwaldt (2000)
US (Missouri), 1999	64	Basil from Mexico or US	S and M	2 catered parties at a country club and a restaurant	Basil probably contaminated on farm, 2 possible sources were a Mexican farm and a US farm	Lopez *et al.* (2001); Herwaldt (2000)
US (Philadelphia), 2000	54	Wedding cake with cream filling including raspberries	S and M	Wedding reception	Possible sources of the raspberries were Guatemala, Mexico or US	Ho *et al.* (2002)
Germany, 2000	34	Lettuce or fresh green herbs	S	Restaurant	nr	Doller *et al.* (2002)
Canada, 2001	17	Raw Thai basil imported from the US	S	Community and restaurant	nr	Hoang *et al.* (2005)
US, 2004	96	Raw snow peas imported from Guatemala	S	Residential facility	nr	CDC (2004b)
US, 2005	592	Fresh basil	D	Restaurants and community	nr	Hammond (2005)

[a] M (microbiological): identification of an organism of the same type from cases and in the suspect vehicle or vehicle ingredient(s), or detection of toxin in faeces or food; D (descriptive): other evidence, usually descriptive, reported by local investigators as indicating the suspect vehicle or food; S (statistical): a significant statistical association between consumption of the suspect vehicle and being a case.

nr, not reported.

may be endemic, this is liable to increase the risk of foodborne infections. Good agricultural practice including avoidance of contact of the crop with contaminated surface water and human waste, and good hygienic practices during harvesting are required to prevent contamination.

The outbreaks of cyclosporiasis in North America have been associated with the import from Guatemala of fruit harvested in the first growing season (March–August). This resulted in a ban on the import into the US in 1998 of fruit produced during this growing season, whereas import of fruit from the second growing season was allowed. From 1999 to 2002, only Guatemalan farms that met specified hygiene standards were allowed to export fresh raspberries to the US during the spring season. Restrictions have also been applied in Canada. The UK has imported a relatively small quantity of raspberries from Guatemala, this import appears to be mainly in the second growing season.

The most important measure to prevent such infections is to prevent contamination of the produce, particularly in the field. The US FDA has worked with the Guatemalan Government and the Guatemalan Berry Commission to develop a strict version of Good Agricultural Practices (the Model Plan of Excellence) designed to reduce the risk of contamination of berry crops with *Cyclospora* (Orlandi *et al.*, 2002).

If berries have become contaminated it can be difficult to remove the organisms. *Eimeria acervulina* has been used as a model to investigate removal or inactivation of *C. cayetanensis* on raspberries (Lee & Lee, 2001). When berries were contaminated with 400–650 oocysts of *Eimeria*, washing in flowing water failed to remove the oocysts. More drastic treatments including heating at 80°C for 1 hour, freezing at −18°C, or irradiation at one kGy were required to inactivate oocysts on the raspberries.

Summary of measures advised in healthcare settings

1 Control the use of fruit and salad vegetables from regions where *C. cayetanensis* infection is endemic. In general, fruit and salad vegetables should be washed thoroughly before use.
2 Workers should not handle food while suffering from symptoms of infection.

References

Alfano-Sobsey, E.M., Eberhard, M.L., Seed, J.R. *et al.* (2004) Human challenge pilot study with *Cyclospora cayetanensis*. *Emerging Infectious Diseases* **10**, 726–728.

Bern, C., Hernandez, B., Lopez, M.B. *et al.* (1999) Epidemiologic studies of *Cyclospora cayetanensis* in Guatemala. *Emerging Infectious Diseases* **5**, 766–774.

Bern, C., Arrowood, M.J., Eberhard, M. & Maguire, J.H. (2002a) *Cyclospora* in Guatemala: further considerations. *Journal of Clinical Microbiology* **40**, 731.

Bern, C., Ortega, Y., Checkley, W. *et al.* (2002b) Epidemiologic differences between cyclosporiasis and cryptosporidiosis in Peruvian children. *Emerging Infectious Diseases* **8**, 581–585.

Blans, M.C.A., Ridwan, B.U., Verweij, J.J. *et al.* (2005) Cyclosporiasis outbreak, Indonesia. *Emerging Infectious Diseases* **11**, 1453–1455.

Cam, P.D., Sorel, N., Dan, L.C. *et al.* (2001) A new contribution to the epidemiological survey of *Cyclospora cayetanensis* in Hanoi water supplies (Viet-Nam); a 12-month longitudinal study. *Medecine et Maladies Infectieuses* **31**, 591–596.

Centers for Disease Control (CDC) (1991) Epidemiologic notes and reports outbreaks of diarrheal illness associated with cyanobacteria (Blue-green algae)-like bodies – Chicago and Nepal, 1989 and 1990. *Morbidity and Mortality Weekly Report* **40**(19), 325–327.

Centers for Disease Control and Prevention (CDC) (1997a) Update: Outbreaks of Cyclosporiasis – US and Canada 1997. *Morbidity and Mortality Weekly Report* **46**(23), 521–523.

Centers for Disease Control and Prevention (CDC) (1997b) Outbreak of Cyclosporiasis – Northern Virginia – Washington, D.C. – Baltimore, Maryland, Metropolitan area 1997. *Morbidity and Mortality Weekly Report* **46**(30), 689–691.

Centers for Disease Control and Prevention (CDC) (1998) Outbreak of cyclosporiasis – Ontario, Canada, May 1998. *Morbidity and Mortality Weekly Report* **47**(28), 806–809.

Centers for Disease Control and Prevention (CDC) (2004a) *Fact Sheet. Cyclospora Infection: Information for Healthcare Providers*. Available from http://www.cdc.gov/ncidod/dpd/parasites/cyclospora/2004_Cyclosporiasis_HCP.pdf. Accessed 16 March 2007.

Centers for Disease Control and Prevention (CDC) (2004b) Outbreak of cyclosporiasis associated with snow peas – Pennsylvania, 2004. *Morbidity and Mortality Weekly Report* **53**(37), 876–878.

Chacin-Bonilla, L., Estevez, J., Mosalve, F. & Quijada, L. (2001) *Cyclospora cayetanensis* infections among diarrheal patients from Venezuela. *American Journal of Tropical Medicine and Hygiene* **65**, 351–354.

Doller, P.C., Dietrich, K., Filipp, N. *et al.* (2002) Cyclosporiasis outbreak in Germany associated with the consumption of salad. *Emerging Infectious Diseases* **8**, 992–994.

Eberhard, M.L., da Silva, A.J., Lilley, B.G. *et al.* (1999a) Morphologic and molecular characterization of new *Cyclospora* species from Ethiopian monkeys: *C. cercopitheci* sp.n., *C. colobi* sp.n. and *C. papionis* sp.n. *Emerging Infectious Diseases* **5**, 651–658.

Eberhard, M.L., Nace, E.K., Freeman, A.R. *et al.* (1999b) *Cyclospora cayetanensis* infections in Haiti: a common occurrence in the absence of watery diarrhea. *American Journal of Tropical Medicine and Hygiene* **60**, 584–586.

Fryauff, D.J., Krippner, R., Prodjodipuro, P. *et al.* (1999) *Cyclospora cayetanensis* among expatriate and indigenous populations of West Java, Indonesia. *Emerging Infectious Diseases* **5**, 585–588.

Hammond, R. (2005) *Cyclospora* outbreak in Florida, 2005. Presented at the Institute of Medicine Forum on Microbial Threats. Washington, DC, 25–26 October 2005. Available from http://www.iom.edu/Object.File/master/30/719/Bodager%20Presentation.pdf. Accessed 16 March 2007.

Herwaldt, B.L. (2000) *Cyclospora cayetanensis*: a review focussing on the outbreaks of cyclosporiasis in the 1990s. *Clinical Infectious Diseases* **31**, 1040–1057.

Herwaldt, B.L., Ackers, M.-L. & the Cyclospora Working Group (1997) An outbreak in 1996 of cyclosporiasis associated with imported raspberries. *New England Journal of Medicine* **36**, 1548–1556.

Herwaldt, B.L., Beach, M.J. & Cyclospora Working Group (1999) The return of *Cyclospora* in 1997: another outbreak of cyclosporiasis in North America associated with imported raspberries. *Annals of Internal Medicine* **130**, 210–220.

Ho, A.Y., Lopez, A.S., Eberhart, M.G. *et al.* (2002) Outbreak of cyclosporiasis associated with imported raspberries, Philadelphia, Pennsylvania, 2000. *Emerging Infectious Diseases* **8**, 783–788.

Hoang, L.M.N., Fyfe, M., Ong, C. *et al.* (2005) Outbreak of cyclosporiasis in British Columbia associated with imported Thai basil. *Epidemiology and Infection* **133**, 23–27.

Hoge, C.W., Echeverria, P., Rajah, R. *et al.* (1995) Prevalence of *Cyclospora* species and other enteric pathogens among children less than 5 years of age in Nepal. *Journal of Clinical Microbiology* **33**, 3085–3060.

Lee, M.B. & Lee, E.H. (2001) Coccidial contamination of raspberries: Mock contamination with *Eimeria acervulina* as a model for decontamination treatment studies. *Journal of Food Protection* **64**, 1854–1857.

Lopez, A.S., Dodson, D.R., Arrowood, M.J. *et al.* (2001) Outbreak of cyclosporiasis associated with basil in Missouri in 1999. *Clinical Infectious Diseases* **32**, 1010–1017.

Lopez, A.S., Bendik, J.M., Alliance, J.Y. *et al.* (2003) Epidemiology of *Cyclospora cayetanensis* and other intestinal parasites in a community in Haiti. *Journal of Clinical Microbiology* **41**, 2047–2054.

Madico, G., McDonald, J., Gilman, R.H. *et al.* (1997) Epidemiology and treatment of *Cyclospora cayetanensis* infection in Peruvian children. *Clinical Infectious Diseases* **24**, 977–981.

Orlandi, P.A., Chu, D.-M.T., Bier, J.W. & Jackson, G.J. (2002) Parasites and the food supply. *Food Technology* **56**(4), 72–81.

Orlandi, P.A., Frazar, C., Cater, C. & Chu, D.-M.T. (2004) Detection of *Cyclospora* and *Cryptosporidium* from fresh produce: isolation and identification by polymerase chain reaction (PCR) and microscopic analysis. In: *Bacteriological Analytical Manual Online*. Chapter 19A. Available from http://www.cfsan.fda.gov/~ebam/bam-19a.html. Accessed 16 March 2007.

Ortega, Y.R., Sterling, C.R. & Gilman, R.H. (1993) *Cyclospora* species – a new protozoan pathogen of humans. *New England Journal of Medicine* **328**, 1308–1312.

Ortega, Y.R., Nagle, R., Gilman, R.H. *et al.* (1997a) Pathologic and clinical findings in patients with cyclosporiasis and a description of intracellular parasite life-cycle changes. *Journal of Infectious Diseases* **176**, 1584–1589.

Ortega, Y.R., Roxas, C.R., Gilman, R.H. *et al.* (1997b) Isolation of *Cryptosporidium parvum* and *Cyclospora cayetanensis* from vegetables collected in the markets of an endemic region in Peru. *American Journal of Tropical Medicine and Hygiene* **57**, 683–686.

Shields, J.M. & Olson, B.H. (2003) *Cyclospora cayetanensis*: a review of an emerging parasitic coccidian. *International Journal for Parasitology* **33**, 371–391.

Shlim, D.R., Hoge, C.W., Rajah, R. *et al.* (1999) Persistent high risk of diarrhea among foreigners in Nepal during the first 2 years of residence. *Clinical Infectious Diseases* **29**, 613–616.

Sterling, C.R. & Ortega, Y.R. (1999) *Cyclospora*: an enigma worth unravelling. *Emerging Infectious Diseases* **5**, 48–53.

Sturbaum, G.D., Ortega, Y.R., Gilman, R.H. *et al.* (1998) Detection of *Cyclospora cayetanensis* in wastewater. *Applied and Environmental Microbiology* **64**, 2284–2286.

2.17 *Giardia duodenalis* (syn. *G. intestinalis, G. lamblia*)

2.17.1 *Importance as a cause of foodborne disease*

Giardia duodenalis is the most commonly isolated intestinal protozoal parasite of humans in the world, and is especially prevalent in children in underdeveloped countries. Many of these infections are asymptomatic, but some children suffer chronic diarrhoea and weight loss (Adam, 1991). In 1998, the WHO estimated that 3000 million people lived in unsewered environments in developing countries and that the rate of giardiasis among them approached 30% (Upcroft & Upcroft, 2001).

Infection with *Giardia* is associated most frequently with consumption of contaminated water; several foodborne outbreaks have been reported and there is a risk of contamination, by irrigation water, of vegetables that are eaten raw. In 1999, *Giardia* was estimated as causing 2 million cases of illness annually in the US, 200,000 cases were estimated as foodborne accounting for 1.4% of total foodborne disease (Mead *et al.*, 1999). In England and Wales in 2005 the number of laboratory reports to the HPA was 2926 (HPA, 2007), equivalent to about 5.5 cases per 100,000 persons.

2.17.2 *Characteristics of the organism*

Giardias are flagellated protozoans with an anaerobic/microaerotolerant metabolism (Upcroft & Upcroft, 2001). Several species of *Giardia* have been distinguished and these are found in the intestinal tracts of primates, livestock, dogs, cats, rodents and other wild animals (Thompson, 2000; Hunter & Thompson, 2005). *Giardia duodenalis* is the only species reported to cause human illness; several 'assemblages' or genotypes (A–G) have been differentiated, of which assemblages A and B infect humans as well as other animals (Caccio *et al.*, 2005).

The infective form of the organism is the cyst, about 11–14 by 7–10 μm in size, which contains four nuclei and has a thick cyst wall that renders it resistant (Ortega & Adam, 1997). Following ingestion the organism reaches the stomach where excystation starts, and this is completed in the upper small intestine, releasing trophozoites each containing two nuclei (Fig. 2.17.1). The trophozoites attach to the membrane of small intestine epithelial cells, where they replicate by binary fission; usually they do not invade epithelial cells, but in certain conditions they can do so (Smith, 1993). Eventually some trophozoites become encysted and both nuclei divide, giving a mature cyst containing four nuclei (Adam, 1991). Both cysts and unencysted trophozoites are excreted, the resistant cysts survive but the trophozoites do not encyst outside the host, and eventually disintegrate. The cysts can remain infective for several months in moist and cool conditions.

Giardia does not multiply outside its host but in favourable conditions, such as in water at 4–10°C, cysts may remain viable for several months (Ortega & Adam, 1997; Lane & Lloyd, 2002). Some cysts may survive freezing for several days; cysts are inactivated by boiling but not by the concentrations of chlorine used in treatment of drinking water.

As in the case of *Cryptosporidium*, detection of *Giardia* in faeces and in environmental samples has relied on separation, concentration where necessary, and microscopy, including the use of immunofluorescent staining. PCR procedures have also been developed

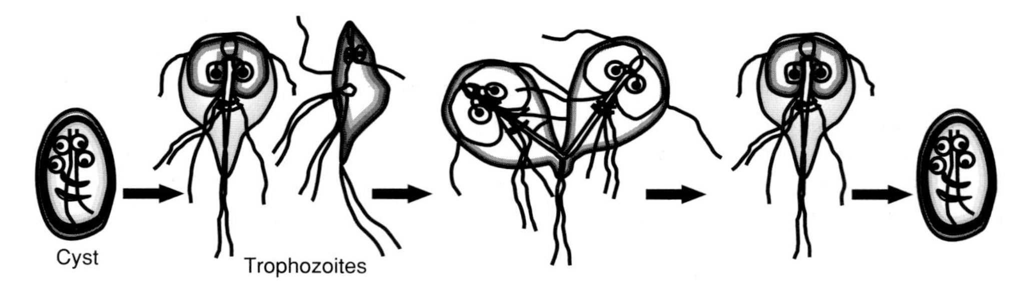

Fig. 2.17.1 Life cycle of *Giardia duodenalis* [*Giardia lamblia* (*intestinalis*)] (From the Centers for Disease Control and Prevention, Public Health Image Library. Content Providers: CDC/Alexander J. da Silva/Melanie Moser)

that should enable more specific characterization of the organisms (Gómez-Couso *et al.*, 2004).

2.17.3 *Disease in humans*

Between one and ten *Giardia* cysts may be sufficient to cause infection. The incubation period after ingestion is 1–2 weeks. Infection may be asymptomatic and individuals affected may be carriers who excrete cysts for several years and can disseminate the disease, thus asymptomatic infection is important epidemiologically (Smith, 1993). Symptoms include diarrhoea, nausea and stomach cramps (Ortega & Adam, 1997). In some cases the acute stage of illness can last for several months causing malabsorption of fat, sugars, vitamins and proteins (Smith, 1993), weight loss and debilitation. Lactose intolerance is common during the acute phase and may persist after this stage. The illness may resolve spontaneously, but often lasts for weeks or months if not treated. Persistent infection and diarrhoea may occur in certain immunocompromised persons (Upcroft & Upcroft, 2001). There is evidence of some antibody protection against re-infection (Ortega & Adam, 1997).

2.17.4 *Source and transmission*

Both humans and animals are a potential source of infection. The extent to which zoonotic transmission occurs is unclear at the time of writing. A proportion of dairy cattle have been reported to harbour *G. duodenalis* assemblage A, which is the most common genotype affecting humans (Thompson, 2000). Water supplies may be contaminated by human waste or by domestic livestock, particularly dairy cows. High numbers of cysts are removed by sewage treatment processes but some may still be present in sewage effluent; filtration is effective for final removal of cysts (Lane & Lloyd, 2002).

Major outbreaks of giardiasis have been caused by waterborne transmission (Rose & Slifco, 1999; Betancourt & Rose, 2004). Direct faecal–oral transmission can occur, particularly in child-care centres (Ortega & Adam, 1997); there is also a risk of transmission to humans from pet dogs and cats, although most strains found in pets are not from Assemblages A or B and so are not infectious to humans (Hunter & Thompson, 2005).

Infection can occur via food either as a result of contamination during production or by infected food handlers. There is a potential for the spread of the organism from contaminated human and animal faeces onto pastures, soil, crops and watercourses, and it occurs widely in water courses in many industrialized countries (Robertson & Gjerde, 2001). *Giardia* has been detected on fruits and vegetables grown in less-developed tropical and sub-tropical countries. In irrigation water, used for salad produce, from rivers, agricultural canals and a lake in South America and the US, *Giardia* cysts were detected in 50% of samples (Thurston-Enriquez *et al.*, 2002). Up to 17,000 cysts/100 mL were detected in river water in Costa Rica, up to 8900 cysts/100 mL in an agricultural canal in Mexico and up to 40/100 mL in agricultural canals in the US. In Norway, irrigation water was found to contain *Giardia* cysts and low numbers of cysts were found on 2% of 149 samples of mung bean sprouts, 2% of 125 samples of lettuce and 3% of 62 samples of strawberries (Robertson & Gjerde, 2001). Fruit and vegetables may be contaminated in the field or at harvest.

Giardia duodenalis cysts have been found in a high proportion of samples of mussels from estuaries on the Galician (northwest Spain) coast, including mussels from sites that conformed to the bacteriological criteria indicating safety for human consumption (Gómez-Couso *et al.*, 2005), and *Giardia* cysts were found in oysters harvested from sites in the Netherlands (Schets *et al.*, 2007). These reports indicate the potential for transmission in shellfish.

2.17.5 *Examples of outbreaks*

The main cause of reported foodborne outbreaks of giardiasis has been transmission by food handlers (Table 2.17.1).

2.17.6 *Prevention and control*

Prevention of transmission of *Giardia* through drinking water requires protection of water sources from contamination with protozoa, effective water treatment and verification of the effectiveness of treatment (WHO, 2006). The treatments necessary in production of drinking water to remove or inactivate *Cryptosporidium* oocysts will also be effective against *Giardia* cysts (Betancourt & Rose, 2004).

To prevent foodborne infection it is important to prevent contamination of crops in the field from human and animal waste, sewage sludge and contaminated irrigation water. There is a need to prevent contamination of water used for shellfish production, and shellfish should be cooked thoroughly before consumption. Water used in processing, production and preparation of food must be free from contamination. Food handlers should not work while suffering from symptoms of infection; good handwashing is essential as workers may be asymptomatic carriers or may have contact with infants or persons who are sources of infection.

Summary of measures advised in healthcare settings

1 Drinking water should be from water supplies that have undergone an effective treatment process.

Table 2.17.1 Examples of outbreaks of foodborne disease caused by *Giardia duodenalis*.

Place, date	Cases	Food implicated	Evidence[a]	Where food was provided	Factors contributing to outbreak	Reference
US, 1979	29 (employees at school)	Home-canned salmon	S	School	Cans opened and juice drained by a person caring for a 1-year-old child who was excreting *Giardia* cysts	Osterholm *et al.* (1981)
US (Minnesota), 1986	88 (35 nursing home residents, 15 children, 38 staff)	Sandwiches	S	Nursing home/child-care centre	Infection started in day-care centre. Mother of toddler in centre worked in food service area, developed giardiasis, infected other kitchen workers. Transmission to residents, other kitchen workers via uncooked food. Person-to-person infection of residents by contact with toddlers	White *et al.* (1989)
US, 1986	9	Fruit salad	S	Private home	The person who prepared the fruit salad had a diapered child and a pet rabbit at home who both excreted *Giardia lamblia*	Porter *et al.* (1990)
US, 1985	13	Cold noodle salad	D	Picnic	Person who prepared noodle salad developed symptoms one day later. She may have been excreting *Giardia* cysts on the day she prepared salad, which she mixed with bare hands	Petersen *et al.* (1988)
Thailand (Phuket), 1988	7 (tourists from Italy) also infected with *Entamoeba histolytica*	Drinks with ice; ice cream; raw fruit in ice	S	Luxury hotel	*Giardia* and *E. histolytica* were probably endemic in the area. Possibly water contaminated with *Giardia* was used	De Lalla *et al.* (1992)
US, 1989	21	Lettuce, onions, tomatoes, tea/coffee	S	Church dinner	Possible contamination of water supply	CDC (1989)
US, 1990	27	Ice	D	Restaurant, daylong meeting	Of employees serving ice, one was asymptomatic but excreting cysts, one had a child excreting cysts	Quick *et al.* (1992)
US, 1990	26	Sliced raw vegetables	S	Company cafeteria	Infected, asymptomatic food handler prepared salad vegetables with bare hands	Mintz *et al.* (1993)

[a]D (descriptive): other evidence, usually descriptive, reported by local investigators as indicating the suspect vehicle or food; S (statistical): a significant statistical association between consumption of the suspect vehicle and being a case.
nr, not reported.

2 Avoid the use of raw vegetables and fruit that may have been exposed to contaminated irrigation water. Fruit and salad vegetables should be washed thoroughly
3 Avoid consumption of milk that has not been properly pasteurized and of products made with unpasteurized milk.
4 Avoid consumption of raw or undercooked shellfish.
5 Ensure good hygienic practices by those handling food. Food handlers should not work in that capacity while suffering symptoms; good hand washing is essential and avoidance of bare hand contact with ready-to-eat foods is advisable.

References

Adam, R.D. (1991) The biology of *Giardia* spp. *Microbiological Reviews* **55**, 706–732.

Betancourt, W.Q. & Rose, J.B. (2004) Drinking water treatment processes for removal of *Cryptosporidium* and *Giardia*. *Veterinary Parasitology* **126**, 219–234.

Caccio, S.M., Thompson, R.C.A., McLauchlin, J. & Smith, H.V. (2005) Unravelling *Cryptosporidium* and *Giardia* epidemiology. *Trends in Parasitology* **21**, 430–437.

Centers for Disease Control and Prevention (CDC) (1989) Epidemiologic notes and reports Common-source outbreak of giardiasis – New Mexico. *Morbidity and Mortality Weekly Reports* **38**(23), 405–407.

De Lalla, F., Rinaldi, E., Santoro, D. *et al.* (1992) Outbreak of *Entamoeba histolytica* and *Giardia lamblia* infections in travellers returning from the tropics. *Infection* **20**, 78–83.

Gómez-Couso, H., Freire-Santos, F., Amar, C.F.L. *et al.* (2004) Detection of *Cryptosporidium* and *Giardia* in molluscan shellfish by multiplexed nested-PCR. *International Journal of Food Microbiology* **91**, 279–288.

Gómez-Couso, H., Méndez-Hermida, F., Castro-Hermida, J.A. & Ares-Mazás, E. (2005) Occurrence of *Giardia* cysts in mussels (*Mytilus galloprovincialis*) destined for human consumption. *Journal of Food Protection* **68**, 1702–1705.

Health Protection Agency (HPA) (2007) Topics A-Z. *Giardia*. Available from http://www.hpa.org.uk. Accessed 10 July 2007.

Hunter, P.R. & Thompson, R.C.A. (2005) The zoonotic transmission of *Giardia* and *Cryptosporidium*. *International Journal for Parasitology* **35**, 1181–1190.

Lane, S. & Lloyd, D. (2002) Current trends in research into the waterborne parasite *Giardia*. *Critical Reviews in Microbiology* **28**, 123–147.

Mead, P.S., Slutsker, L., Dietz, V. *et al.* (1999) Food-related illness and death in the United States. *Emerging Infectious Diseases* **5**, 607–625.

Mintz, E.D., Huson-Wragg, M., Mshar, P. *et al.* (1993) Foodborne giardiasis in a corporate office setting. *Journal of Infectious Diseases* **167**, 250–253.

Ortega, Y.R. & Adam, R.D. (1997) *Giardia*: overview and update. *Clinical Infectious Diseases* **25**, 545–550.

Osterholm, M.T., Forfang, J.C., Ristinen, T.R. *et al.* (1981) An outbreak of foodborne giardiasis. *New England Journal of Medicine* **304**, 24–28.

Petersen, L.R., Cartter, M.L. & Hadler, J.L. (1988) A food-borne outbreak of *Giardia lamblia*. *Journal of Infectious Diseases* **157**, 846–848.

Porter, J.D.H., Gaffney, C., Heymann, D. & Parkin, W. (1990) Food-borne outbreak of *Giardia lamblia*. *American Journal of Public Health* **80**, 1259–1260.

Quick, R., Paugh, K., Addiss, D. *et al.* (1992) Restaurant-associated outbreak of giardiasis. *Journal of Infectious Diseases* **166**, 673–676.

Robertson, L.J. & Gjerde, B. (2001) Occurrence of parasites on fruits and vegetables in Norway. *Journal of Food Protection* **64**, 1793–1798.

Rose, J.B. & Slifco, T.R. (1999) *Giardia*, *Cryptosporidium* and *Cyclospora* and their impact on foods: a review. *Journal of Food Protection* **62**, 1059–1070.
Schets, F.M., van den Berg, H.H.J.L., Engels, G.B. *et al.* (2007) *Cryptosporidium* and *Giardia* in commercial and non-commercial oysters (*Crassostrea gigas*) and water from the Oosterschelde, the Netherlands. *International Journal of Food Microbiology* **113**, 189–194.
Smith, J.L. (1993) *Cryptosporidium* and *Giardia* as agents of foodborne disease. *Journal of Food Protection* **56**, 451–461.
Thompson, R.C.A. (2000) Giardiasis as a re-emerging infectious disease and its zoonotic potential. *International Journal for Parasitology* **30**, 1259–1267.
Thurston-Enriquez, J.A., Watt, P., Dowd, S.E. *et al.* (2002) Detection of protozoal parasites and microsporidia in irrigation waters used for crop production. *Journal of Food Protection* **65**, 378–382.
Upcroft, P. & Upcroft, J.A. (2001) Drug targets and mechanisms of resistance in the anaerobic protozoa. *Clinical Microbiology Reviews* **14**, 150–164.
White, K.E., Hedberg, C.W., Edmonson, L.M. *et al.* (1989) An outbreak of giardiasis in a nursing home with evidence for multiple routes of transmission. *Journal of Infectious Diseases* **16**, 298–303.
World Health Organization (WHO) (2006) *Guidelines for Drinking-Water Quality.* First addendum to third edition. *Vol. 1: Recommendations*. WHO, Geneva.

2.18 *Toxoplasma gondii*

2.18.1 Importance as a cause of foodborne disease

Infection with *Toxoplasma gondii* has probably occurred in nearly one third of the human population (Tenter *et al.*, 2000). It is estimated that in the UK and the US, 16–40% of people become infected, while in Central and South America and continental Europe the proportion infected is 50–80% (Dubey, 2004). In immunocompetent people infection is usually asymptomatic, but immunocompromised individuals are liable to develop severe disease, and infection of pregnant women can result in severe disease in their infants.

In the US, it was estimated that in 50% of cases of illnesses due to *T. gondii* the organism was foodborne and that it was responsible for 21% of food-related deaths (Mead *et al.*, 1999). Work in the Netherlands indicated that toxoplasmosis causes the highest disease burden among seven foodborne pathogens studied, which included *Campylobacter* spp., *Salmonella* spp. and norovirus (Kemmeren *et al.*, 2006).

2.18.2 Characteristics of the organism

Toxoplasma gondii is a protozoan parasite of animals and humans and is the only known species of toxoplasma. The genus name was introduced in 1909 and toxoplasmosis was recognized in humans in the late 1930s (Tenter *et al.*, 2000; Percival *et al.*, 2004). Vertical transmission in humans was recognized in 1942 and in 1953–54 it was suggested that transmission to humans may occur in undercooked meat (pork).

The only known definitive hosts are felids, whereas intermediate hosts include pigs, sheep, goats and a range of animals, including poultry (Dubey *et al.*, 2003); severe disease involving abortion and neonatal death is caused particularly in sheep and goats (Dubey, 2004).

The life cycle is shown in Figure 2.18.1. Cats are infected after ingestion of raw or undercooked meat with tissue cysts containing bradyzoites (Dubey, 2004). In the stomach the tissue wall is digested by proteolytic enzymes, and bradyzoites are released in the small intestine. Here some penetrate the lamina propria, multiply as tachyzoites, and may disseminate to extraintestinal tissues. Other bradyzoites penetrate epithelial cells of the small intestine and give rise to numerous generations of asexual types followed by sexual forms and production of oocysts. Mature oocysts are released into the lumen of the intestine by rupture of intestinal cells and are shed in the faeces. Cats can also become infected after ingesting tachyzoites or oocysts. *T. gondii* persists in the intestinal and extraintestinal tissue of cats for at least several months and possibly for the life of the cat. After primary infection of cats with tissue cysts, oocysts or tachyzoites, a period of days elapses after which a proportion of cats shed oocysts in their faeces for between about 7 and 20 days (Tenter *et al.*, 2000). Further periods of shedding may occur, the causes of which are largely unknown.

In fresh cat faeces the oocysts are roughly spherical, 10 × 12 μm size, unsporulated and not infective. Sporulation occurs within 1–5 days, depending on aeration and temperature; the sporulated oocysts are infective.

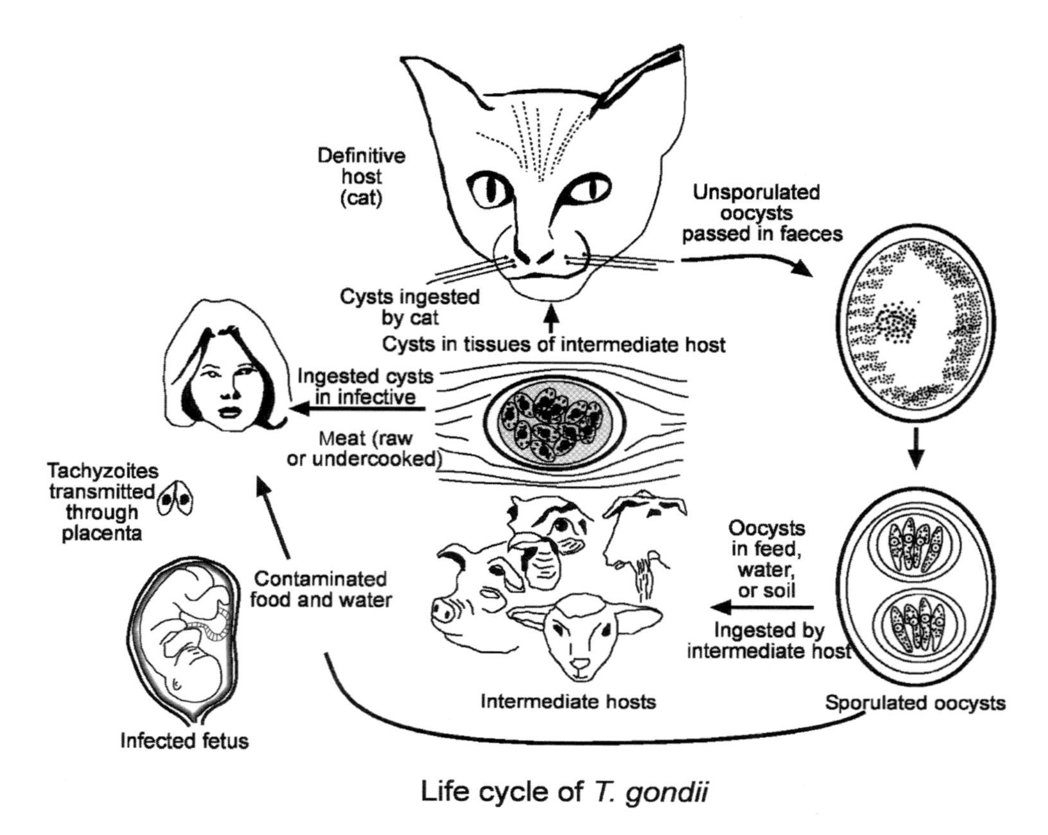

Fig. 2.18.1 Life cycle of *T. gondii*. (Reprinted from Dubey, 2004. Copyright (2004) with permission from Elsevier B.V.)

Hosts including other cats, other animals and humans can be infected with *T. gondii* by ingesting meat of infected animals, containing tissue cysts, or food or drink contaminated with sporulated oocysts, or by transplacental transmission. After ingestion, bradyzoites released from tissue cysts or sporozoites released from oocysts penetrate intestinal tissues and give rise to tachyzoites, which multiply locally and are disseminated through the body in the blood or lymph system. Tachyzoites can then give rise to bradyzoites and tissue cysts in muscles, neural tissue and viscera, in some animals these tissue cysts may persist for the life of the host (Tenter *et al.*, 2000). Infection in animals is generally asymptomatic, but infection during pregnancy can result in transfer of tachyzoites via the placenta to infect the fetus, causing abortion in animals such as sheep and goats, as well as abortion or the birth of a severely ill infant in humans.

Three genotypes (I, II and III) have been distinguished, that differ in virulence in mice (Boothroyd & Grigg, 2002).

Sporulated oocysts can survive in moist soil for months or years, and in water at temperatures between 4 and 22°C for over 1 year (Dubey, 2004). Most tissue cysts are killed by cooling to −13°C or lower, and tissue cysts in meat are killed by heating to an internal temperature of 67°C; both tissue cysts and oocysts are readily killed by irradiation (Dubey, 2004).

Detection of oocysts in cat faeces by concentration and microscopy is difficult and serological prevalence is used to measure infection. Techniques for detection of oocysts in soil and water were discussed by Dubey (2004). Detection of tissue cysts in meat has involved animal feeding tests (Dubey *et al.*, 2005), and methods based on PCR of several DNA sequences have been developed (Jauregui *et al.*, 2001; Aspinall *et al.*, 2002).

2.18.3 *Disease in humans*

Serological evidence shows that a high proportion of the population has been infected with *T. gondii*. In 90% of cases the acute infections are asymptomatic; where symptoms do occur, the most common is lymphadenitis, which may be accompanied by fever, malaise, fatigue, muscle pain and headache (Dubey, 2004). Following acute infection most people become chronically infected, although this is nearly always asymptomatic.

In rare cases in immunocompetent individuals severe ocular infection occurs (Boothroyd & Grigg, 2002; Jones *et al.*, 2006) which has been associated with type I strains (Switaj *et al.*, 2006). The organism is stated to be the most common cause of retinal infection throughout the world, leading eventually to 24% of affected eyes becoming legally blind (Kijlstra *et al.*, 2004). In an outbreak in Canada in 1995 that was linked to oocyst contamination of a municipal water supply, 20 of 100 patients with acute toxoplasmosis showed ocular symptoms (Dubey, 2004).

Following infection of pregnant women, transplacental infection may cause abortion or congenital infection. Congenital infections cause symptoms ranging from mild, slightly diminished vision to mental retardation or blindness in infants, and severe symptoms including chorioretinitis and hydrocephalus, convulsions and intracerebral calcification may occur (Dubey, 2004; Percival *et al.*, 2004). The severity of infection in the infant depends on the stage of pregnancy at the time of infection. Infection acquired by the mother in the first and second trimester is liable to lead to more severe toxoplasmosis in the infant or to abortion/stillbirth, while infection in the third trimester usually results in subclinical infection of the infant (Kemmeren *et al.*, 2006).

Many cases of symptomatic toxoplasmosis in immunocompromised persons result from re-activation of earlier, latent infection; other cases follow new infection of seronegative persons (Smith, 1993). Reactivation has been reported in a patient receiving anti-TNF treatment, and other medication, for rheumatoid arthritis (Young & McGwire, 2005). In immunocompromised persons infection can result in encephalitis, and in patients with AIDS the infection is a major cause of death (Percival *et al.*, 2004).

In the US, most of the deaths caused by toxoplasmosis are due to congenital infection in neonates and disease in patients with AIDS or who are otherwise immunocompromised (Smith, 1993).

2.18.4 *Source and transmission*

Cats are the primary source of *T. gondii*, and infected cats can excrete millions of oocysts which may be ingested by other animals. The organism may be transmitted between intermediate hosts as tissue cysts indefinitely, even in the absence of primary host (Tenter *et al.*, 2000). Infection of humans can occur, therefore, by the flowing routes:

1 Handling cats, cat litter and contaminated soil and transfer on the hands, or by inhalation of oocysts from litter (Smith, 1993).
2 Consumption of raw or undercooked meat containing tissue cysts.
3 Consumption of water contaminated with oocysts (Dubey, 2004; Percival *et al.*, 2004).
4 Consumption of raw vegetables or fruit that have been contaminated with oocysts.
5 Consumption of raw milk containing tachyzoites.

Studies in Italy and Norway have indicated that consumption of raw or undercooked meat containing tissue cysts was the main route of infection in pregnant women (cited in Aspinall *et al.*, 2002).

Meat-producing animals, particularly pigs, sheep and goats, are liable to be infected. A study of 71 meat samples (mainly pork and lamb) from shops in the UK, using PCR reactions, indicated the presence of *T. gondii* in 27 samples (38%) (Aspinall *et al.*, 2002). It was considered probable that a significant proportion of the organisms detected by PCR were viable, although this was not proven. This work suggested that if the products were not frozen, and were consumed undercooked, these types of meat would pose a significant risk of infection. In a study to detect the presence of viable *T. gondii* in 6282 samples (2094 each of beef, chicken and pork) from 698 stores in many areas of the US, the organism was found in pork samples but not in beef or chicken (Dubey *et al.*, 2005). The prevalence was very low, but the risk of infection from ingestion of undercooked meat was stressed.

Some measures can be taken to reduce the incidence of *T. gondii* in food animals (Smith, 1993; Tenter *et al.*, 2000; Kijlstra *et al.*, 2004); however, the move from indoor to outdoor farming of pigs may lead to a re-emergence of *Toxoplasma* infections, and require further on-farm prevention strategies (Kijlstra *et al.*, 2004).

2.18.5 *Examples of outbreaks*

Outbreaks of toxoplasmosis are rarely detected because most individuals affected show few or no symptoms. For example, an outbreak in Canada in 1995, attributed to contamination of a municipal water system, was estimated to have affected between 2894 and 7718 individuals of whom 100 were detected with symptoms (Bowie *et al.*, 1997). Many incidents have been reported that were linked to consumption of raw meat (including lamb, beef and venison) or raw goat milk (Smith, 1993; Tenter *et al.*, 2000). An outbreak of toxoplasmic chorioretinitis in Korea affected three people who had eaten raw boar meat, and in a further outbreak five soldiers developed lymphadenopathy after consumption of raw pork liver and meat (Choi *et al.*, 1997).

2.18.6 *Main methods of prevention and control*

Reduction of the risk of infection from drinking water depends in part on the potential for felid faeces to enter water systems (Percival *et al.*, 2004); in the production of safe drinking water, control should include prevention of contamination of source water by wild or domesticated cats (WHO, 2006). The oocysts show high resistance to chlorine, but are larger than those of *Cryptosporidium* and treatments that remove or inactivate *Cryptosporidium* will probably be effective against *Toxoplasma*.

In general, vulnerable persons in particular, including pregnant women, should as far as possible avoid contact with cat litter and wear gloves when gardening or handling soil that may have been contaminated by cats.

When preparing raw meat, any cutting boards, sinks, knives and other utensils that might have touched the raw meat should be washed thoroughly to avoid cross-contamination of other foods.

Hands should be washed thoroughly after handling raw meat.

Meat should be cooked thoroughly, to an internal temperature of at least 71°C (160°F).

Consumption of unpasteurized milk, particularly goat milk, should be avoided.

Salad vegetables and fruits should be washed thoroughly before consumption.

Summary of measures advised in healthcare settings

1. Drinking water should be from public water supplies that have undergone an effective treatment process.
2. Obtain foods from a reputable supplier.
3. Prevent cross-contamination from raw meat.
4. Meats should be cooked thoroughly to an internal temperature of at least 71°C.
5. Unpasteurized milk should be avoided.
6. Salad vegetables and fruits should be washed thoroughly before serving.

References

Aspinall, T.V., Marlee, D., Hyde, J.E. & Sims, P.F.G. (2002) Prevalence of *Toxoplasma gondii* in commercial meat products as monitored by polymerase chain reaction – food for thought. *International Journal for Parasitology* **32**, 1193–1199.

Boothroyd, J.C. & Grigg, M.E. (2002) Population biology of *Toxoplasma gondii* and its relevance to human infection: do different strains cause different disease. *Current Opinion in Microbiology* **5**, 438–442.

Bowie, W.R., King, A.S., Werker, D.H. *et al.* (1997) Outbreak of toxoplasmosis associated with municipal drinking water. *Lancet* **350**, 173–177.

Choi, W.-Y., Nam, H.-W., Kwak, N.-H. *et al.* (1997) Foodborne outbreaks of human toxoplasmosis. *Journal of Infectious Diseases* **175**, 1280–1282.

Dubey, J.P. (2004) Toxoplasmosis – a waterborne zoonosis. *Veterinary Parasitology* **126**, 57–72.

Dubey, J.P., Hill, D.E., Jones, J.L. *et al.* (2005) Prevalence of viable *Toxoplasma gondii* in beef, chicken and pork from retail meat stores in the United States: risk assessment to consumers. *Journal of Parasitology* **91**, 1082–1093.

Dubey, J.P., Navarro, I.T., Graham, D.H. *et al.* (2003) Characterization of *Toxoplasma gondii* isolates from free-range chickens from Parana, Brazil. *Veterinary Parasitology* **117**, 229–234.

Jauregui, L.H., Higgins, J., Zarlenga, D. *et al.* (2001) Development of a real-time PCR assay for detection of *Toxoplasma gondii* in pig and mouse tissues. *Journal of Clinical Microbiology* **39**, 2065–2071.

Jones, J.L., Muccioli, C., Belfort, R., Jr. *et al.* (2006) Recently acquired *Toxoplasma gondii* infection, Brazil. *Emerging Infectious Diseases* **12**, 582–587.

Kemmeren, J.M., Mangen, M.-J.J., Duynhoven, Y.T.H.P. & Havelaar, A.H. (2006) *Priority Setting of Foodborne Pathogens*. RIVM Report 330080001/2006. Available from http://www.rivm.nl/bibliotheek/rapporten/330080001.pdf. Accessed 20 March 2007.

Kijlstra, A., Eissen, O.A., Cornelissen, J. *et al.* (2004) *Toxoplasma gondii* infection in animal-friendly pig production systems. *Investigative Ophthalmology and Visual Science* **45**, 3165–3169.

Mead, P.S., Slutsker, L., Dietz, V. *et al.* (1999) Food-related illness and death in the United States. *Emerging Infectious Diseases* **5**, 607–625.

Percival, S., Embry, M., Hunter, P.R. *et al.* (2004) *Toxoplasma gondii*. In: *Microbiology of Waterborne Diseases. Microbiological Aspects and Risks.* Academic Press, London, pp. 325–336.

Smith, J.L. (1993) Documented outbreaks of toxoplasmosis: transmission of *Toxoplasma gondii* to humans. *Journal of Food Protection* **56**, 630–639.

Switaj, K., Master, A., Borkowski, K. *et al.* (2006) Association of ocular toxoplasmosis with type I *Toxoplasma gondii* strains: direct genotyping from peripheral blood samples. *Journal of Clinical Microbiology* **44**, 4262–4264.

Tenter, A.M., Heckeroth, A.J. & Weiss, L.M. (2000). *Toxoplasma gondii*: from animals to humans. *International Journal of Parasitology* **30**, 1217–1258.

World Health Organization (WHO) (2006) *Guidelines for Drinking-Water Quality.* First addendum to third edition. *Vol. 1: Recommendations*. WHO, Geneva.

Young, J.D. & McGwire, B.S. (2005) Infliximab and reactivation of cerebral toxoplasmosis. *New England Journal of Medicine* **353**, 1530–1531; discussion 1531.

Summaries

Table 2A Association of food groups and pathogens.

Food group	Main human pathogens
Beef	*Salmonella* spp., Shiga toxin-producing *Escherichia coli*, *Clostridium perfringens*
Poultry	*Campylobacter jejuni*, *Salmonella* spp., *Listeria monocytogenes*, *Clostridium perfringens*
Pork	*Staphylococcus aureus*, *Yersinia enterocolitica*, *Salmonella* spp., *Clostridium perfringens*, *Toxoplasma gondii*, *Trichinella*
Lamb	*Toxoplasma gondii*, *Clostridium perfringens*
Cooked, refrigerated, RTE meats	*Listeria monocytogenes*
Fermented meats	*Salmonella* spp., *Yersinia enterocolitica*, Shiga toxin-producing *Escherichia coli*
Dairy products	*Listeria monocytogenes*, Shiga toxin-producing *Escherichia coli*, *Salmonella* spp., *Campylobacter jejuni*, *Brucella* spp., *Staphylococcus aureus*, *Clostridium botulinum*, *Toxoplasma gondii* (*goat, sheep milk*)
Eggs	*Salmonella* spp.
Fresh produce (Salad vegetables and fruits)	*Salmonella* spp., *Shigella* spp., Shiga toxin-producing *Escherichia coli*, *Listeria monocytogenes*, *Yersinia pseudotuberculosis*, norovirus, hepatitis A virus, *Cryptosporidium* spp. *Cyclospora cayetanensis*, *Giardia duodenalis*, *Toxoplasma gondii*
Fin fish	Histamine poisoning (scombroid)[a], ciguatera poisoning[a], helminth parasites, *Clostridium botulinum*
RTE smoked fish	*Listeria monocytogenes*
Shellfish	*Vibrio* spp., norovirus, hepatitis A virus, *Shigella* spp., *Cryptosporidium* spp., marine toxins[a]
Cooked meat, poultry, vegetables and starchy foods	*Bacillus cereus*, *Clostridium perfringens*
Cooked rice	*Bacillus cereus*
Underprocessed canned foods; cooked chilled foods (REPFEDS); vegetables/herbs in oil	*Clostridium botulinum*
Honey given to infants under 1 yr old	*Clostridium botulinum*
Powdered infant formula	*Enterobacter sakazakii*, *Salmonella* spp.

Modified from Doyle *et al.* (2001).

[a] Centres for Disease Control and Prevention (CDC) (2005) *Marine Toxins*. Available from: http://www.cdc.gov/ncidod/dbmd/diseaseinfo/marinetoxins_g.htm. Accessed 23 March 2007.

RTE, ready-to-eat; REPFEDS, refrigerated, processed foods of extended durability.

Potentially hazardous food (PHF) (time/temperature control for safety food [TCS]) means a food that requires time/temperature control for safety, to limit pathogenic microorganism growth or toxin production (FDA/CFSAN, 2005). Many of these foods are termed 'high-risk foods' which have been defined as ready-to-eat foods which, under favourable conditions, support the multiplication of pathogenic bacteria and are intended for consumption without treatment, which would destroy such organisms (Sprenger, 2005).

Some examples of high-risk foods were given by Department of Health (UK) (2006), with guidance on the special precautions required for care homes. The guidance states that:

- pasteurized eggs should be used for all uncooked or lightly cooked dishes
- vulnerable people should avoid pâté and soft-ripened cheeses and should only eat cook-chill meals and ready-to-eat chicken if they have been reheated until they are piping hot
- only milk and milk-based products that have been pasteurized should be offered for consumption by residents
- all meat and poultry should be thoroughly cooked until the juices run clear before being served. Piping hot meat is safest; the use of food thermometers is recommended
- shellfish, especially if eaten raw or undercooked, is a high-risk food. If served to vulnerable residents, particular care should be taken to ensure proper preparation, cooking and handling of fresh, tinned and frozen shellfish
- where possible, fruit and vegetables should have skin removed provided this does not mean excessive manual handling. Leafy vegetables, such as lettuce, and fruit should be washed thoroughly in running water. All fruit and vegetables should be purchased from a reputable supplier
- food brought in by visitors for residents should preferably be of a low-risk nature, food listed are fruit, biscuits, chocolates, pre-packaged fruit drinks.

References

Department of Health, UK (2006) *Infection Control Guidance for Care Homes*. Available from: http://www.dh.gov.uk/assetRoot/04/13/63/84/04136384.pdf. Accessed 13 February 2007.

Doyle, M.P., Beuchat, L.R. & Montville, T.J. (eds) (2001) *Food Microbiology, Fundamentals and Frontiers*, 2nd ed. American Society for Microbiology Press, Washington, DC.

Sprenger, R.A. (2005) *Hygiene for Management*, 12th edn. Highfield Co UK Ltd, Doncaster.

Table 2B Microbiological hazards in production of foods, and control measures in healthcare settings.

Hazard	Associated foods	Comments and control measures
Bacteria		
Bacillus cereus (infection by strains forming diarrhoeal toxin or intoxication caused by heat-stable, preformed toxin). Spores survive cooking	Cooked meat, poultry, starchy foods such as rice and potatoes, puddings, soup, vegetables	After cooking thoroughly, either keep above 63°C (145°F) or cool promptly to <7–8°C (44.6–46.4°F) (ideally <4°C, <39°F). Reconstituted, dried milk and powdered foods should be used promptly or kept at <4–5°C (39–41°F). Clean and disinfect equipment to avoid build-up of spores
Brucella spp.	In regions where brucellosis occurs in animals, unpasteurized milk from sheep, goats and cows and cheese made with unpasteurized milk	Avoid consumption of unpasteurized milk and unpasteurized milk products
Campylobacter jejuni/ *C. coli*	Poultry meat; unpasteurized, inadequately pasteurized and recontaminated pasteurized milk	Obtain poultry meat from approved suppliers[a]. Avoid cross-contamination from raw meat to other foods. Cook meat thoroughly. Avoid unpasteurized milk. Apply employee health policy[b]. Good handwashing
Clostridium botulinum		
Intoxication caused by preformed, heat-labile toxin. Spores survive cooking	Under-processed, canned foods (mainly home-prepared); minimally heated, vacuum and modified atmosphere packed, chilled foods (prepared, chilled foods); 'sous vide' foods; vegetables/herbs in oil	Obtain canned foods and other preserved foods from approved suppliers. Store minimally heated, prepared, chilled foods (pasteurized chilled foods) at < 8°C (46.4°F) for a limited time (<10 days) unless other factors are present that inhibit *C. botulinum*
Infant botulism (infection by spores and toxin formation in intestine)	Honey	Do not give honey to infants less than 1 yr old
Clostridium perfringens. Spores survive cooking	Cooked meat, poultry, vegetables	Cooked foods should be eaten immediately, or kept above 63°C (146°F) for a short time, or be cooled rapidly, to prevent growth from spores, maintained below 4–5°C (39–41°F) and heated to >72°C (162°F) before serving

(*cont.*)

Table 2B *(continued)*

Hazard	Associated foods	Comments and control measures
Enterobacter sakazaki	Powdered infant formula. (This is not a sterile product)	Reconstitute formula in sterile conditions, using hot water (>70°C, >158°F), cool rapidly, use immediately ('hang time' <4 h)[c], discard any remainder
Escherichia coli O157:H7 and other Shiga toxin-producing *E. coli* (STEC)	Undercooked meat; cross-contamination of cooked meats, unpasteurized, inadequately pasteurized, and recontaminated pasteurized milk, cheese made with unpasteurized milk, unpasteurized fruit juice, raw seed sprouts, raw vegetables	Obtain foods from approved suppliers. Cook meats thoroughly. Prevent cross-contamination from raw to cooked meat. Avoid unpasteurized milk and milk products and unpasteurized fruit juice. Avoid raw seed sprouts. Raw vegetables and fruits should be washed thoroughly. Apply employee health policy. Good handwashing. No bare-hand contact with RTE food.
E. coli other than STEC	Raw vegetables exposed to contaminated water. Contamination by food handlers. Information on foods incomplete	As for STEC
Listeria monocytogenes	Hot dogs; sliced, processed meats; refrigerated pâtés; soft cheeses particularly those made with unpasteurized milk; cold-smoked fish; refrigerated, ready-to-eat foods	Obtain foods from approved suppliers. Cook meats and fish thoroughly. Store PHFs at <5°C (<41°F). Observe 'Use-by' date. Avoid unpasteurized milk products; soft-ripened cheeses; blue-veined cheeses; refrigerated pâté; refrigerated cooked meats, refrigerated, cold-smoked fish
Salmonella spp. non-typhoid/paratyphoid	Eggs, meat and poultry, seafood, raw seed sprouts, raw vegetables and fruits, unpasteurized milk, unpasteurized juice	Obtain foods from approved suppliers. Cook eggs thoroughly. Cook foods thoroughly. Raw vegetables and fruits should be washed thoroughly. Avoid unpasteurized milk and milk products. Avoid raw seed sprouts. Apply employee health policy. Good handwashing. No bare-hand contact with RTE foods
Salmonella Typhi, *S.* Paratyphi A, B and C	Fruit and vegetables exposed to contaminated water. Contamination of food by human case or occasionally by asymptomatic carrier	Obtain foods from approved suppliers. Avoid fruit and vegetables that may have been contaminated in the field or at harvest. Raw vegetables and fruits should be washed thoroughly. Apply employee health policy. Good handwashing

Shigella spp.	Raw vegetables and herbs, seafood, other foods contaminated by infected workers via faecal–oral route	Obtain foods from approved suppliers. Raw vegetables and fruit should be washed thoroughly. Obtain shellfish from approved growing areas. Cook food, particularly seafood, thoroughly. Apply employee health policy. Good handwashing. No bare-hand contact with RTE foods
Staphylococcus aureus (intoxication caused by preformed, heat-stable toxin)	Foods contaminated and subject to time/temperature abuse before cooking. Cooked foods touched by bare hands after cooking, and subsequently time/temperature abused. Other RTE foods touched by bare hands and inadequately refrigerated	Obtain foods from approved suppliers. Avoid temperature abuse of foods before cooking. Store PHFs at $<5°C$ ($<41°F$). Avoid unpasteurized milk and milk products. Apply employee health policy. Good handwashing. No bare-hand contact with RTE foods
Vibrio cholerae	Raw vegetables, fish and shellfish	Avoid vegetables and fruit that may have been exposed to contaminated irrigation water. Raw vegetables and fruit should be washed thoroughly. Obtain shellfish from approved growing areas and cook thoroughly. Prevent cross-contamination. Apply employee health policy. Good handwashing. No bare-hand contact with RTE foods
Vibrio parahaemolyticus, *V. vulnificus*	Seafood, shellfish	Obtain foods from approved suppliers. Obtain shellfish from approved growing areas and cook thoroughly. Prevent cross-contamination
Yersinia enterocolitica, *Y. pseudotuberculosis*	Pork chitterlings. Foods contaminated from pigs. Raw vegetables	Avoid use of pork chitterlings. Cook meats, particularly pork, thoroughly. Obtain salad vegetables from approved suppliers and ensure protection from wild animals during growth and storage. Raw vegetables and fruit should be washed thoroughly
Viruses		
Hepatitis A and E	Shellfish; fresh fruit and salad vegetables from endemic regions; RTE foods contaminated by food handler, pig meat	Obtain foods from approved suppliers. Obtain shellfish from approved growing areas and cook thoroughly. Raw vegetables and fruit should be washed thoroughly. Cook meat (especially pig meat) thoroughly. Apply employee health policy. Good handwashing. No bare-hand contact with RTE foods

(*cont.*)

Table 2B *(continued)*

Hazard	Associated foods	Comments and control measures
Noroviruses	Shellfish; fresh fruit and salad vegetables; RTE foods contaminated by food handler	Obtain foods from approved suppliers. Obtain shellfish from approved growing areas and cook thoroughly. Raw vegetables and fruit should be washed thoroughly. Apply employee health policy. Good handwashing. No bare-hand contact with RTE foods
Protozoa		
Cryptosporidium spp.	Raw fruit and vegetables contaminated via irrigation water; unpasteurized milk and juice; shellfish; RTE foods contaminated by food handler	Use foods from approved supplier. Raw vegetables and fruit should be washed thoroughly. Avoid unpasteurized milk and fruit juice. Obtain shellfish from approved growing areas and cook thoroughly. Apply employee health policy. Good handwashing. No bare-hand contact with RTE foods
Cyclospora cayetanensis	Fresh fruit and salad vegetables from endemic regions	Use foods from approved suppliers. Avoid vegetables and fruit that may have been exposed to contamination in the field. Raw vegetables and fruit should be washed thoroughly
Giardia duodenalis	Raw fruit and vegetables contaminated via irrigation water; unpasteurized milk; RTE foods contaminated by food handler	Use foods from approved supplier. Raw vegetables and fruit should be washed thoroughly. Apply employee health policy. Good handwashing. No bare-hand contact with RTE foods
Toxoplasma gondii	Raw and undercooked meat, (especially pork, lamb, venison); unpasteurized milk of goats and sheep; unwashed fruit and vegetables	Use foods from approved suppliers. Cook meat thoroughly. Wash fruit and vegetables thoroughly. Avoid unpasteurized milk

Additional controls are needed for vulnerable persons requiring a low microbial diet, see Chapters 1 and 5. Temperatures are based mainly on UK and EFSA specifications and are for guidance. Control measures in relation to drinking water are given by WHO (2006) and in Chapter 7. Please see Notes on Table 2B on the next page.

2.19 Notes on Table 2B

[a] In order to assess the acceptability of a supplier, hygiene officers may be employed to inspect the premises to ensure that they are able to provide a satisfactory product that consistently meets the agreed specification (Sprenger, 2005). If this is impractical copies of the food safety policy, the HACCP plan, the prerequisite programmes and the monitoring records can be requested together with specifications and assessment on delivery. Assessment of a supplier may also be based on evidence of: (1) good hygienic practice, (2) HACCP plan, (3) process designed and validated to meet a food safety objective (FSO), (4) process criteria in HACCP plan, (5) product criteria including organoleptic, chemical, physical and biological specifications and (6) Complete and accurate records (ICMSF, 2002). The standards required by the UK NHS for manufacturers and suppliers of foods are given in NHS (2001).

[b] Employee health policy, in the UK, Department of Health (1995, 1996), Salmon *et al.* (2004); in the US, FDA/CFSAN (2005).

[c] Hang time: the length of time for which a formula is at room temperature in the feeding bag and accompanying lines during enteral feeding.

RTE food, ready-to-eat food; PHF, potentially hazardous food (time/temperature control for safety food [TCS]).

References

Department of Health, UK (1995) *Food Handlers. Fitness to Work. Guidelines for Food Businesses, Enforcement Officers and Health Professionals*. Department of Health, London. Available from the Food Standards Agency, UK.

Department of Health, UK (1996) *Food Handlers. Fitness to Work. Guidelines for Food Business Managers*. Department of Health, London. Available from the Food Standards Agency, UK.

FDA/CFSAN (2005) *Food Code*. Available from http://www.cfsan.fda.gov.

ICMSF (2002) Parameters that can be used to assess the acceptability of a supplier. In: *Microorganisms in Foods. Vol. 7: Microbiological Testing in Food Safety Management*. Kluwer Academic/Plenum Publishers, New York, pp. 87.

National Health Service (NHS) (2001) NHS *Code of Practice for the Manufacture, Distribution and Supply of Food Ingredients and Food-Related Products*. Available from http://www.pasa.doh.gov.uk/food/docs/code_of_practice_2001.pdf. Accessed 22 March 2007.

Salmon, R. & Working Group (2004) Preventing person-to-person spread following gastrointestinal infections: guidelines for public health physicians and environmental health officers. *Communicable Disease and Public Health* 7, 362–384.

Sprenger, R.A. (2005) *Hygiene for Management*. 12th edn., Highfield Co UK Ltd, Doncaster.

3 The Surveillance of Foodborne Disease and its Application in Healthcare Settings

John M. Cowden and Jacqui Reilly

3.1 Introduction

Patients and staff in healthcare settings are also members of the public at large. Consequently, the surveillance systems for diseases that operate in the general population apply to patients and staff in healthcare settings. In addition, there are surveillance systems specific to foodborne disease in the general population and to disease in general in the healthcare setting. Although none of the surveillance systems in the healthcare setting are specifically designed to capture information on the occurrence of potentially foodborne disease, many do identify foodborne disease in addition to other conditions.

As well as formal surveillance systems, informal mechanisms exist in the general population, but more especially in the healthcare setting, which, although they may fall short of the accepted definitions of surveillance, facilitate the early detection of single cases and clusters of disease, including those that are potentially foodborne.

This chapter will address surveillance systems relevant to foodborne disease that operate in the general population (and as such apply to both within and outside healthcare settings) and systems that are targeted specifically at healthcare settings (and as such capture information on both non-foodborne and foodborne disease).

Within the healthcare setting, as well as systems for detecting infectious diseases, including foodborne disease, healthcare staff including ward managers, care-home managers, domestic service managers and infection control teams (ICTs) have responsibility for ensuring that food hygiene regulations are complied with. The measurement of this compliance is a form of surveillance. There are also systems within and outside the healthcare setting that seek information on the microbiological status of food and the standards of food premises, which could be deemed surveillance systems. It is beyond the scope of this chapter to address these non-human systems outside the healthcare setting, but the role of audit in kitchen and food inspections in healthcare settings is described.

3.2 The principles and practice of public health surveillance in the general population in relation to foodborne disease

3.2.1 *The scope of surveillance*

Surveillance has been defined as the:

> ongoing systematic collection, collation, analysis, and interpretation of data; and the dissemination of information to those who need to know in order that action be taken. (WHO, 2007)

A more comprehensive definition is:

> The ongoing systematic collection, analysis, and interpretation of health data essential to the planning, implementation, and evaluation of public health practice, closely integrated with the timely dissemination of these data to those who need to know. The final link in the surveillance chain is the application of these data to prevention and control. A surveillance system includes a functional capacity for data collection, analysis, and dissemination linked to public health programs. (CDC, 1988)

One of the most important aspects of both these definitions is that surveillance is not merely monitoring. The most important difference between monitoring and surveillance is that the latter is designed to inform appropriate action.

The immediate objectives of infectious disease surveillance include:

- the identification of trends
- the detection of outbreaks
- the detection of individual cases or severe illness
- the evaluation of interventions and control programmes
- the prediction of future trends and outbreaks
- the generation of hypotheses about transmission and prevention for further study

The ultimate aim of the surveillance of any infectious disease is to describe the extent and causes of disease transmission sufficiently well to inform evidence-based strategies for disease prevention and control.

Prevention and control of foodborne disease can be achieved by:

- immediate intervention, e.g. the removal of a contaminated foodstuff from the market
- the identification and promotion of good practice, e.g. encouraging, through health promotion, hand washing or the thorough cooking of raw meats
- the formulation and implementation of rational policy, e.g. the use of HACCP (Hazard Analysis, Critical Control Point) systems during food production and preparation

All these activities depend upon a knowledge of the causes of cases and outbreaks of foodborne disease, which in turn depends upon their investigation, and which in turn depends upon the identification of cases and outbreaks through surveillance. The identification of cases and outbreaks, and the synthesis of the information gleaned from their investigation in order to inform evidence-based policy therefore depends ultimately upon appropriate surveillance.

3.2.2 *Categories of surveillance systems*

Surveillance systems can be categorized as:

- *Voluntary or mandatory*: While the difference between voluntary and mandatory may seem self evident, it is often not clear-cut. There is a spectrum between systems backed by a legal obligation at one extreme, and systems with no formal documentation or agreement between the reporter of data and the collector at the other. At one end of this spectrum a surveillance system for which there was a statutory or legal obligation to comply would obviously be mandatory, as would be a system guaranteed by a legally binding contract. At the other end of the spectrum, however, a system where there was no written protocol or declared obligation on the part of the data reporter to report is obviously voluntary. Between these two extremes lie such managerial arrangements as 'service level agreements' where a non-legal obligation is accepted by both reporter and collector and 'organizational imperatives' where management require reporting from employees. In this chapter, 'mandatory' is used to mean both legal and managerial obligation to participate.
- *Active or passive*: A surveillance system is described as active if the person (often a microbiologist, clinician or nurse in healthcare settings) responsible for identifying a condition is asked at regular intervals (e.g. every day, week, or month) to report whether they have identified cases fulfilling a specified case definition or not. They are prompted, therefore, to report not only the presence of cases, but their absence. If they are merely asked to report, unprompted, the cases they identify, then the system is described as passive. Most surveillance systems for infectious disease are passive. It should be noted that this does not mean that a passive system requires little effort. A passive system may require the commitment of considerable resources by those reporting and those collecting data.
- *Comprehensive or sentinel*: If all potential reporters are required or invited, either actively or passively, to report all cases they identify, then that system is designated comprehensive. If, however, only a subset of reporters are asked to report all the cases they identify, or if all reporters are asked to report a subset (e.g. one in ten cases, or all cases every Monday, or all cases every alternate day) then that surveillance system would be described as sentinel. The subset of the population selected for sentinel surveillance should be representative of the population under surveillance if it is intended to give a picture of the population as a whole. Other strategies are possible however: the sentinel population might be a particularly susceptible or vulnerable group (analogous to the miner's canary) whose experience is informative, but not typical.

- *Based on individual or aggregate data*: Reporters may be asked to give details of every case they identify, or merely to report the numbers of cases. The former system is an individual data system, the latter an aggregate data system. It may be that reporters participate in an individual data system locally, but submit aggregate data to their regional or national surveillance centre.
- *National, regional or local*: Ideally national, regional and local surveillance systems should be integrated so that no more data are sought nationally than are needed regionally, and no more data are sought regionally than are needed locally. To ask of those people responsible for collecting data more than they need for their own purposes is to risk collecting incomplete and unreliable information. This is especially true of voluntary surveillance systems. While the overall aims of surveillance at each of these three levels may be similar, the emphases vary: at national level there is likely to be more emphasis on the identification and prediction of national trends and outbreaks, and the evaluation of national interventions and control programmes, whereas at the local level, local trends, outbreaks and individual cases of severe illness are likely to be more important, as are the evaluation of local interventions and initiatives, and the generation of hypotheses about local problems of transmission and prevention. Furthermore, although all surveillance systems should adhere to the WHO definition, being, for example 'ongoing and systematic', local systems may be less formally structured than those at either national or regional level.

As a rule, active, comprehensive, individual data systems provide the best quality data, but they may be inappropriate because of their expense. There is a widespread view that mandatory systems achieve better compliance than voluntary ones, but the enthusiastic commitment of those involved is a much better guarantor of success.

3.2.3 *Examples of surveillance systems*

Various methods are used at national, regional and local levels in the surveillance of diseases. These capture information on many different conditions including infectious diseases, and within that, infectious intestinal disease (IID), and within that, potentially foodborne disease. Those systems particularly relevant to potentially foodborne disease include:

- death certification
- clinical syndrome surveillance
- laboratory reports
- outbreak surveillance

In addition to routine surveillance systems, valuable information can be gleaned from special studies (including outbreak investigations) and surveys. Special studies and surveys do not, strictly, constitute surveillance as they are not 'ongoing'. The distinction is not a clear one, however. Regularly repeated surveys may be deemed to be surveillance.

Death certification is carried out in most countries and is usually required by law. Statutory death certification is a mandatory, passive, comprehensive surveillance system based on individual data. Data are collected locally and are collated locally, regionally and nationally.

Clinicians complete a death certificate when a person under their care dies. In the UK, the certificate indicates the primary cause of death, contributory factors, the deceased's name, address, date of birth and gender. In practice, as most foodborne disease in the developed world is non-fatal, death certification is a very poor measure of the burden of disease.

Clinical syndrome surveillance includes statutory notifications, hospital discharge diagnoses and primary care diagnoses. Probably the most important of these in the UK and elsewhere is the statutory notification. Statutory notification is a mandatory, passive, comprehensive system based on individual data. Data are collected locally and are collated locally, regionally and nationally.

All doctors in clinical practice in the UK have a legal duty to notify the proper officer of the local authority (in England and Wales) or the Chief Administrative Medical Officer (in Scotland) of cases, or suspected cases, of certain infectious diseases, or food poisoning. The diseases, although named, are not defined in law. For many notifiable diseases, the absence of a specific definition is not a problem, as the diseases are pathogen-specific (e.g. cholera, typhoid, paratyphoid) and in any case, the system is intended to be sensitive; i.e., it is deemed more important to identify all case that do occur, even at the expense of wrongly including as cases some people who are not. Food poisoning, however, is not pathogen-specific: indeed it is not even syndrome-specific, and what should be notified as 'food poisoning' is therefore far from clear. In 1992, the UK Advisory Committee on the Microbiological Safety of Food (ACMSF) recommended to all doctors in the UK the definition of food poisoning already in use in Scotland (ACMSF, 1993). This definition had been adopted previously by the World Health Organization (WHO, 1984).

The definition of food poisoning is:

> any disease of an infectious or toxic nature, caused by or thought to be caused, by the consumption of food or water.

The definition therefore includes:

- all food- and waterborne illness regardless of the presenting symptoms and signs; it thus includes not only acute illnesses characterized by diarrhoea or vomiting, but also illnesses presenting with manifestations not related to the gastrointestinal tract, such as scombrotoxin poisoning, paralytic shellfish poisoning, botulism and listeriosis
- illness caused by the ingestion of toxic organic and inorganic chemicals

The definition excludes:

- illness due to known allergies and food intolerances
- food- and waterborne disease notifiable in its own right, such as dysentery or typhoid

This definition has numerous drawbacks not least that:

- it cannot be applied with any degree of precision to a single case because it is virtually impossible to deduce the means of acquisition of an infection or intoxication in a single case, and
- as food poisoning is not pathogen specific, and potentially foodborne pathogens can be acquired by a non-foodborne route the diagnosis cannot be confirmed by laboratory tests

It is doubtful, however, that many clinicians are aware of the definition, although no surveys have been published to confirm this. It is more likely, although again it has never been investigated, that clinicians are notifying either cases already confirmed in the laboratory as being infected with a potentially food- or waterborne pathogen, or cases of presumed infectious diarrhoea. If clinicians are notifying cases in whom potentially food- and waterborne pathogens have already been confirmed, then their notifications must be slower than the laboratory reporting system, and are also probably less sensitive and specific. They are, in effect, a haphazardly selected subset of laboratory reports with the name of the organism removed. For these and other reasons, the system of statutory notification of food poisoning in the UK has severe limitations for the interpretation of long-term trends, and the identification of outbreaks (Cowden, 2000). Food poisoning statistics are available for England and Wales (HPA, 2006a) and for Scotland (NHS National Services Scotland, 2006). In some countries, hospital discharge diagnosis data can provide a valuable measure of the burden of disease requiring hospital admission. In Scotland, Scottish Morbidity Returns (SMRs) include such data according to the International Classification of Disease code 10 (ICD10) on all hospital discharges. This system is a passive, comprehensive system based on individual data and collated locally, regionally and nationally. Nationally, the data are available from the ISD website (NHS National Services Scotland, 2006). Similar comprehensive or sentinel systems operate in other countries.

A clinical surveillance system is operated by the Royal College of General Practitioners (RCGP) at the primary care level in England and Wales (RCPG, 2006). This system is a voluntary, passive, sentinel system based on individual data. Data are collected routinely from a network of approximately 73 practices, well distributed across England and Wales. The practices provide a tabular summary of the number of patients seen each week categorized by gender, age group and disease or disease group, including 'infectious intestinal disease'. The data are collated and published by the RCGP Research Unit to provide incidence and prevalence rates of diseases. It should be noted that GPs are not provided with a case definition for IID, and there is no way of identifying which of the cases they report were suspected to result from food poisoning.

Laboratory reports are the mainstay of the surveillance of food- and waterborne disease in the UK. The Health Protection Agency (HPA) in England and Wales, and Health Protection Scotland (HPS) in Scotland invite clinical laboratories in their respective countries to report to them laboratory-confirmed infections. Both systems are passive comprehensive systems based on individual data. Whether the system is mandatory or voluntary is open to debate. The HPA's predecessor, the Public Health Laboratory Service, deemed reporting to be an 'organizational imperative' for its laboratories, and in Scotland the reporting of certain identifications (for example methicillin-resistant *Staphylococcus aureus* from blood culture) is required by the Scottish Executive Health Department. Both these requirements fall short of legal obligation, and the sanctions applicable in the event of non-compliance are unclear. On balance therefore, laboratory reporting is probably best regarded as voluntary.

Laboratory reports are collected locally, and collated and published regionally and nationally. Often at local level, laboratory reports of potentially foodborne pathogens prompt further investigation by local authority Environmental Health Officers (EHOs), or health authority public health Infection Control Nurses (ICNs). Practices vary in different authorities. There is little consistency in what cases prompt further investigation (for example, in some areas, cases of campylobacteriosis are followed up, and in others they are not), or the method of follow up (for example, in some areas, a postal questionnaire is administered, in others personal interviews are carried out), or the content of the supplementary information sought (while most authorities have standard data collection forms, there is no uniformity between authorities across the country). The supplementary information collected by these local follow-ups is not, as a rule, submitted nationally.

In the UK, the data held nationally include:

- the source (initial laboratory at which the specimen was examined) and reference laboratory (for the subset of pathogens whose identity is confirmed or further characterized)
- a case identifier (name or laboratory identification number)
- date of birth or age
- sex
- date of onset of symptoms
- specimen type
- some clinical and epidemiological features
- the name of the organism or toxin and the tests used to identify it

The quality and quantity of information in laboratory reports is dependent on the information included in the initial laboratory request form which accompanies the specimen to the laboratory. Reports are not updated and therefore data on outcomes, including case fatality, are incomplete.

The numbers of reports of potentially foodborne and other pathogens from England and Wales and Scotland are published on their websites (HPA, 2006b; HPS, 2006).

Laboratory surveillance has the advantage of specificity: If a pathogen is reported, that pathogen was undoubtedly present. However, the system has the disadvantage of not being sensitive. Many cases will not consult their doctor, or have a stool specimen taken.

The issue of by what factor the number of reports of a particular pathogen must be multiplied by to give the true number occurring in the population was addressed by a large study in England in the early 1990s (FSA, 2000). This study estimated that, to obtain the true number of infections in the community, the number of reports of salmonella in the national laboratory reporting system in England should be multiplied by 3.8, those of campylobacter by 10.3, those of small round structured virus (now termed norovirus) by 315. The study did not, however, address cases in closed populations such as prisons, boarding schools, or hospitals. It is certain that the ascertainment of infection in these settings, especially the healthcare setting, is greater than in the population at large.

Neither did the study make any attempt to quantify how many of these infections were likely to be foodborne. The proportion of cases of infection with commonly foodborne pathogens in the general population has been estimated (Adak *et al.*, 2002), but it is improbable that these proportions are applicable within the healthcare setting where person-to-person spread is more likely than in the general population, and where (hopefully) standards of food hygiene are superior. Neither of these studies, therefore, have any utility in estimating the true level of foodborne disease in the healthcare setting.

As well as the problem of being unable to identify all the cases of infection occurring in the population or the source of infection, laboratory surveillance has the disadvantage of unrepresentativeness. Clinicians have no standard set of criteria to determine from whom they request a faecal specimen. Again, there is little or no published information on this issue, but anecdotally, specimens are more likely to be requested by clinicians when cases:

- have recently returned from abroad
- have severe illness
- report other associated cases
- are at the extremes of age

It should be stressed that, in the absence of a specific policy, these criteria are not applied with any consistency. The consequence of this is that laboratory surveillance cannot be relied upon to give accurate information on the proportion of cases in any of these categories in the population. The relationship between the proportion of true cases in any category and the cases reported in any category will be governed, not only by their numbers in the population, but also by the unknown and haphazard sampling bias of clinicians.

Despite these limitations, the laboratory surveillance system in the UK has identified numerous foodborne outbreaks over recent decades, for example outbreaks of *Salmonella* infection associated with imported foods such as Italian chocolate bars in 1982 (Gill *et al.*, 1983), German salami sticks in 1988 (Cowden *et al.*, 1989), Irish cheese in 1989 (Maguire *et al.*, 1992), Israeli snacks in 1995 (Killelea *et al.*, 1996) and more recently Spanish eggs (FSA, 2004). The utility of laboratory surveillance in the UK is exemplified by the fact that every one of these outbreaks was detected either much later, or not at all, in the countries from which the foods were imported.

A further testament to the success of laboratory surveillance is that in recent years a number of international laboratory surveillance networks have been established in Europe and further afield. One of the most successful of these has been 'Enternet' (HPA, 2006c), an international laboratory surveillance system for salmonella and verocytotoxin producing *Escherichia coli*. In addition, regular reports are issued by the WHO surveillance programme for control of foodborne infections and intoxications in Europe (WHO, 2004).

Outbreak surveillance is a separate activity from outbreak identification or management. Outbreak identification is a function of clinical syndrome surveillance or laboratory reports. Once an outbreak has been identified, it becomes a surveillance event in its own right. Whether or not a case is part of an outbreak is a characteristic of that case, but an outbreak has characteristics of its own, e.g.:

- When did it start?
- How long did it last?
- How many cases were there?
- How many people were at risk?
- What was the main mode of transmission?
- If food- or waterborne, what vehicle was suspected?
- What was the evidence upon which this suspicion was based?

Reports of outbreak investigations that appear in the peer-reviewed literature are likely to contain this information and more. Published reports tend, however, to be of outbreaks which are unusual in some way (either particularly large, or due to a surprising vehicle of infection). They, therefore, constitute a biased subset of the whole. Rational, evidence-based food policy must be based on the results of the investigations of typical outbreaks, not those sufficiently noteworthy to warrant publication. A surveillance system for all outbreaks is therefore necessary to achieve a more representative picture of the vehicles and sources of infection in foodborne outbreaks, as well as their size, severity, length and the factors deemed to have caused them. A new programme for the surveillance of general outbreaks of IID was set up in England and Wales in 1992, and a similar programme was initiated in Scotland in 1996.

No surveillance system is ideal. Epidemiologists should be aware of the limitation of the surveillance systems available to them. The surveillance of foodborne disease is complicated by the fact that the means of acquisition of infection can rarely if ever be inferred with complete confidence. Most of the systems of value in the monitoring of foodborne disease are directed at the surveillance of potentially foodborne pathogens. They all inevitably suffer, to a greater or lesser extent, from deficiencies, for example under-ascertainment and imprecise (or non-existent) case definitions.

It is clear from the above that surveillance of any infectious disease, including foodborne disease, in the healthcare setting cannot rely solely upon systems that cover the population as a whole. The remainder of this chapter addresses surveillance systems designed specifically for the healthcare setting, with particular reference to foodborne disease.

3.3 The principles and practice of public health surveillance in the healthcare setting in relation to foodborne disease

3.3.1 *The history of surveillance in the healthcare setting*

In 1859, Florence Nightingale said, 'Hospitals should do the sick no harm' (Nightingale, 1859). She can be credited with the development of the first system of healthcare-associated infection (HAI) surveillance. During the Crimean war, through the collection and collation of data, she established that soldiers in the hospital were more at risk of dying than those on the front line, as a result of HAI. However, there is little evidence to suggest that any HAI surveillance activity commenced in other healthcare settings until a much later date.

In the late 1960s, epidemiologists in the US noticed that feedback of information about staphylococcal infection epidemics in hospitals could change the behaviour of the physicians, nurses and other personnel in such a way as to reduce infection rates (Raven & Haley, 1980). A large multi-centre study, the 'SENIC study', in the 1970s (Haley *et al.*, 1985) suggested that four components were required to reduce nosocomial infection:

- surveillance
- control
- an ICN to collect data and
- a physician actively involved

Hospitals that had all of these elements could reduce HAI by 32%, over a 5-year period (Haley *et al.*, 1985).

It can be seen therefore that the role of the ICT is central to the success of surveillance in healthcare settings. Infection control is an essential service in healthcare settings. Most hospitals in the UK have an ICT, which consists of an infection control doctor and ICN and support staff, which may also include surveillance nurses. The priorities and strategy for infection control at the individual healthcare setting level are developed by the infection control committee, which usually has representation from the ICT, clinical staff, services, management and public health.

In the time since Haley's study, components of HAI surveillance programmes, both within the US and UK, have been introduced in response to clinical and political pressures. With changing hospital environments, patterns of care and new infection risks, it was not known what proportion of nosocomial infections were preventable, if (and by how much) infection control programmes reduced the incidence of nosocomial infections, or if they did, which particular components were responsible for achieving this.

3.3.2 *Examples of surveillance programmes of HAI*

As a result of the published HAI prevalence studies (Emmerson *et al.*, 1993, 1996) and the SENIC study (Haley *et al.*, 1985), many countries, including the UK, have recognized the importance of HAI as an outcome indicator and have established targeted incidence surveillance programmes for HAI.

The programmes have been set up on a country-by-country basis and in the majority of cases have involved the adoption of the US Centre for Disease Control and Prevention (CDC) definitions for HAI (CDC, 1999). These definitions were developed as part of the National Nosocomial Infection Surveillance (NNIS) programme, which was the first national programme of targeted HAI surveillance to be established worldwide. Since this time, the NNIS definitions of HAI have been accepted internationally.

Most current HAI programmes are incidence-based prospective studies, either organism-, speciality- or intervention-specific, with the aim of:

- Promoting the concept of surveillance for HAI prevention and control by offering hospitals an efficient and effective tool for data collection and analysis, as well as technical and scientific assistance in its implementation.
- Allowing each hospital to compare its own incidence figures with those of other hospitals, and thereby evaluate its prevention and control activities.
- Obtaining a national perspective of the incidence of HAIs, trends over time, sites, risk factors, patient outcomes, major pathogens and antimicrobial resistance.

A recent development has been the Hospitals in Europe Link for Infection Control through Surveillance (HELICS) project – a collaborative project aiming to achieve standardization in surveillance methods across Europe, as well as to establish networks to share protocols, organizational methods, informatics tools, analytic approaches and feedback. The common targeted surveillance programmes adopted by HELICS, countries in the UK and further afield have little relevance to foodborne HAI. This is due to the fact that it is considered not to be a common HAI problem. Data on foodborne HAI is sourced mainly from reports on outbreaks.

Table 3.1 Common types of healthcare-associated infection (HAI) and their relative proportions: results from UK HAI Prevalence Survey, 1994.

HAI type	Number of infections	Prevalence (%) (95% CI)
Urinary tract	894	2.41 (2.26–2.57)
Lower respiratory tract	882	2.38 (2.22–2.53)
Surgical wound	413	1.11 (1.0–1.22)
Skin	370	1.00 (0.90–1.10)
Total HAI, number of patients affected[a]	3353	9.03 (8.74–9.32)

[a]Number of patients surveyed, 37,111
Data from Emmerson *et al.*, 1996.

3.3.3 *The burden of HAI*

During the 1970s and 1980s, many countries undertook national prevalence surveillance studies to determine the extent and common types of HAI. Prevalence surveys are a snap shot of existing HAI on a single day (point prevalence) or over several days (period prevalence). These studies found that between 6 and 14% of patients acquired infections after admission to hospital (Meers *et al.*, 1981; Haley, 1985). The most recent national prevalence study in the UK, carried out in 1993, demonstrated broadly similar HAI rates to those published previously, with an overall rate of 9% (Emmerson *et al.*, 1996). Four main groups of HAI were identified: urinary tract (23.2%), surgical wound (10.7%), lower respiratory tract (22.9%) and skin (9.8%) (Table 3.1). A small proportion (5%) of HAI reported by the study was gastrointestinal HAI, although the study did not identify how much of this was potentially foodborne. However, prevalence surveys, by their very nature, are limited snap shots of the HAI burden and incidence studies give a better source of information on the burden of HAI and provide more useful surveillance data for evaluation of the impact of any interventions on subsequent HAI rates.

Since the SENIC study was published (Haley *et al.*, 1985), there has been a steady promotion of the benefits of targeted incidence-based surveillance over hospital-wide prevalence-based surveillance. Incidence surveillance is continual monitoring of all patients for new HAI. Targeted surveillance focuses preventive effort and resources on high-risk patient groups (for example surgical patients), units (for example Intensive Care Units), or infection sites (e.g. blood stream infections). It has the potential to yield more meaningful data as case finding is more accurate if targeted in a specific area, and risk adjustment is more feasible for targeted units (Gaynes *et al.*, 2001).

To link surveillance more effectively to prevention of HAI and reduce the financial and staff requirements of hospital-wide surveillance, Haley has proposed the system of surveillance by objectives, with hospitals focusing on their priority HAI problems based on morbidity, mortality and cost, and developing a specific surveillance and control strategy directed at reducing HAI.

One of the most important HAI incidence studies carried out in recent years is that of the Public Health Laboratory Service (PHLS) (Plowman *et al.*, 2001). This study of the socioeconomic burden of HAI in a single hospital in England was the first to carry out hospital wide ($n = 4000$), incidence surveillance over a full year (1994–1995), establishing

Table 3.2 The burden of healthcare-associated infection by type, in a single hospital in 1994–1995.

Type of HAI	Number of infections	Incidence (%) (95% CI)	Additional cost attributable to HAI (£)	Additional LoS[a] attributable to HAI (days)
Urinary tract	107	2.7 (2.2–3.2)	1327	6.1
Lower respiratory tract	48	1.2 (0.9–1.6)	2398	12.5
Surgical wound	38	1.0 (0.7–1.3)	1618	6.5
Skin	25	0.6 (0.4–0.9)	1790	12
Bloodstream	4	0.1 (0.0–0.3)	5397	1.9
Other	30	0.8 (0.5–1.1)	2263	13.4
Multiple	57	1.4 (1.1–1.9)	9152	37.8
Total HAI[b]	309	7.8 (7.0–8.6)	3154	14.1

[a] LoS, length of stay.
[b] Number of patients surveyed, 3980.
Data from Plowman *et al.*, 2001.

the burden by HAI type. The study indicated an overall HAI rate of 7.8% (95% CI; 7.0–8.6%), with each HAI costing £3154, on average, to treat. The findings from this study were applied, by the authors, to NHS-wide data and they estimated that the cost to the hospital sector is £930.62 million (95% CI; £780.26–£1080.97 million) per annum. It is not known what proportion of the HAI was potentially foodborne. Table 3.2 summarizes the burden of HAI by type of HAI found by the PHLS study.

HAI rates from surveillance studies are often compared between hospitals, countries and over time, but comparisons of crude infection rates should be made with due caution. Rates may be affected by factors such as differences in numerator or denominator definitions, surveillance methods with different sensitivities and specificities for case detection and different intensities of surveillance activities.

Some patients are at greater risk than others for acquiring HAIs due to the presence of certain risk factors, which alter their susceptibility to infection. The risk of HAI is dependent on the dose of bacteria, their potential virulence, and the resistance of the host patient (Altemeier & Culbertson, 1965), given by the equation:

$$\frac{\text{Dose of bacterial contamination} \times \text{Virulence}}{\text{Host resistance}}$$

Published research suggests that a number of variables fit into this equation and may affect the risk of developing HAI. Many of these are intrinsic in the patient population, i.e. extremes of age, obesity, malnutrition, co-morbidity such as diabetes, malignancy or immunosuppression (Ayliffe & Babb, 1995; Chapter 5). However, several identified influences on HAI are extrinsic, i.e. patient care practice-based and as such may be addressed by adopting evidence-based practice.

Thus, crude HAI rates should not be compared without considering the underlying intrinsic risk of infection, however further research is needed on risk factors for HAI and how these data might be used for comparative purposes. This is important if 'risk' of HAI is to be effectively communicated to clinicians, politicians and the wider public at large.

3.3.4 *Surveillance of potentially foodborne disease in the healthcare setting*

The targeted approach to the surveillance of HAI has the advantage of offering flexibility for healthcare institutions' own identified priorities. However, a potential limitation of the targeted approach is that cases or outbreaks of infection of conditions that have not been targeted may be missed. Haley (1992) recommends that in healthcare settings ICTs should train ward staff to be alert for, and report, clustering of any infection. The ICT should then investigate such clusters. It is therefore clear that the relationship between the clinical staff in healthcare settings and the ICT is very important. Indeed, reporting of possible problems early in their development, in order that intervention can be taken to control the problem, is dependent upon this relationship.

Haley's exhortation falls short of the definitions of surveillance presented earlier, and no formal surveillance system for the detection of potential outbreaks specific to healthcare settings is, to our knowledge, described in the literature. In many hospitals there is, however, an alert-organism information system in which the laboratory serving the hospital informs the ICT when a patient is confirmed as being infected or colonized with previously specified organisms. The hospital's patient administration system then normally allows the ICT to track the patient's movements through the hospital.

The detection of cases and outbreaks of any infection, including of foodborne disease, in the healthcare setting is therefore the result of both formal surveillance programmes for the detection of single cases, as described, and the continued vigilance on the part of all healthcare staff, especially the ICT, for the occurrence of outbreaks. Once detected, outbreaks become surveillance events in their own right, and should be investigated appropriately, both to inform appropriate control measures, and provide evidence to help prevent future outbreaks.

The development of HAI outbreak surveillance is a relatively new concept. Outbreaks of HAI vary widely with respect to the organism(s) involved, the numbers and types of patients affected, the severity and consequences of the resulting morbidity, and the nature of the infection control measures implemented. No national data are routinely available on the numbers and types of outbreaks of HAI that occur in many countries, including the UK, and on the impact and implications for the health services.

Most of the published literature on HAI outbreaks reports single incidents.

A review of outbreaks of potentially foodborne disease in hospitals in England and Wales over a 10-year period Joseph and Palmer (1989) showed that of 248 outbreaks, foodborne infection accounted for 57 (24%) outbreaks and presumed person-to-person spread for 70 (30%). No information was available for 107 (46%) of the outbreaks. Frequently, person-to-person transmission is established by the exclusion of foodborne infection.

Because of the importance of data from outbreaks, Wales and Scotland have both recently developed programmes for HAI outbreak surveillance. In Wales, reporting of HAI outbreaks is mandatory as a result of a health department directive and they have developed a web-based data management system to this end. ICTs enter data directly and are able to use the national database in an interactive way to interrogate these data.

In Scotland, the programme for the surveillance of HAI outbreaks (the Scottish HAI Outbreak Reporting System – 'SHORS') is a development of the system for the surveillance of general outbreaks of IID ('ObSurv') described earlier. Outbreaks of IID in the healthcare setting are, therefore, captured in both surveillance systems. Preliminary results from SHORS have identified that most reports of outbreaks of HAI are of norovirus infection, just as ObSurv shows that most outbreaks of norovirus infection are in healthcare settings.

HAI in relation to potentially foodborne disease may result from inappropriate patient care practices, the impact of which on outcome is likely to be greater in the healthcare setting than the general population, because the patient population may be particularly susceptible to such infections, due to age or co-morbidities. In addition, immunocompromised patients, including infants or patients undergoing chemotherapy, are likely to develop more serious disease and the infection may result in death. The principles of food hygiene should therefore be applied with particular rigour to prevent infection in healthcare settings. As food hygiene, preparation and storage practice is governed by regulations, audit of compliance with these is an important part of infection control.

3.4 Audit of food hygiene in the healthcare setting

Healthcare staff, including ward managers, care-home managers, domestic service managers and ICTs have responsibility for ensuring that food hygiene regulations are complied with in the healthcare setting. This was highlighted by the UK government's Department of Health, which has stated that:

> All catering facilities in healthcare settings will comply with current food safety legislation. This will include catering management, food handlers and premises from where food is sourced, stored, prepared or served. (DH, 2003)

Compliance with this legislation is achieved through audit. Audit forms part of the clinical governance structure of healthcare settings and is defined as:

> the systematic and critical analysis of the quality of clinical care. This includes the procedures used for diagnosis and treatment, the associated use of resources and the effect of care on the outcome and quality of life for the patient. (DH, 1989)

or put more simply asking ourselves 'are we doing what we should be doing?' and doing something about it if we are not.

Audit, like surveillance, provides information for action but differs from surveillance in that its focus is measuring practice against an agreed standard, and usually occurs over a defined period of time, rather than being ongoing. The practice of audit is focused on identified areas of high risk with regard to potentially foodborne disease. In healthcare settings, these can be areas including food preparation, cooking food, food storage, kitchens, fridges, microwave ovens, crockery and cutlery, personal hygiene, meal delivery systems, enteral feeds and hand hygiene.

It follows that while potentially foodborne HAI can be identified through targeted surveillance or untargeted vigilance, it can actually be prevented through audit: as such all three activities are important components of infection control in healthcare institutions. Control of all HAI, including foodborne infections, is important because of the recognized burden of the disease, both in terms of morbidity and mortality, and its subsequent cost. Encouraged by the move towards clinically effective (and cost effective) care provision, the development of performance indicators based on the incidence of HAI has progressed rapidly in recent years. As such, the resulting data from national HAI surveillance programmes in the UK are utilized within the performance assessment framework of the NHS in Scotland and as part of the Health Commission's star rating assessment for hospitals in England.

3.5 Conclusion

The surveillance of foodborne disease, and the surveillance of HAI (including potentially foodborne disease) in the UK and elsewhere is far from ideal. HAI surveillance programmes, both within and outwith the UK, have developed empirically and will continue to do so. Priority areas will continue to be identified through reviewing the evidence from published studies. The most recent systematic review of the literature (Harbarth *et al.*, 2003) has confirmed that the potential for prevention in HAI is most likely in the targeted areas of:

- surgical site infection
- ventilator-associated pneumonia
- central line-associated bacteraemias and
- catheter-associated urinary tract infections

Foodborne HAI is not specified, and it is not known what proportion of these infections are foodborne or result from secondary spread from person-to-person following primary foodborne infection. Indeed, the prevalence of HAI in general has not been measured in the UK for many years. There is consequently a need for another national prevalence study of HAI within the UK to ascertain the relative proportions and types of HAI commonly occurring in hospitals at a given time, as it has now been 10 years since the last one. Other European countries, such as Germany, are advocating this approach for identifying future targeting activities for surveillance. It would seem wise to include the identification of potentially foodborne disease within the remit of any future prevalence studies.

Further work is needed also on the impact of outbreaks of HAI. This has been recognized both in Wales, where the surveillance of HAI outbreaks is now a mandatory requirement in all healthcare settings, and in Scotland where a new web-based surveillance programme was piloted in the summer of 2005. The purpose of both these programmes is to collect data on the numbers and types of outbreaks of HAI, their aetiology where known, and the outcomes of interventions aimed at their control. Such information would provide a knowledge-base of outbreaks of HAIs, the organisms responsible, methods of transmission, factors that contribute to outbreaks and effective control interventions. Where similar problems appear in different locations the experience gained in an earlier outbreak, along with new information, may reveal answers to hitherto unanswered questions and/or assist in the implementation of effective control measures. This is particularly important as ICTs in the hospital setting are burdened with the impact these infections have on their time and the subsequent costs associated with the management of these.

Extension of such national surveillance to all healthcare premises would be extremely valuable since even less is known about HAI in community healthcare premises, e.g. GP premises, care homes than in hospitals.

The challenge for governments and other key stakeholders interested in establishing national surveillance systems is to establish a system with a standardized approach, flexible enough to be relevant to all healthcare settings of all sizes at an individual level and for comparison between healthcare settings, producing meaningful data for feedback in order that quality improvement may be addressed.

References

Adak, G., Long, S.M. & O'Brien, S. (2002) Trends in indigenous food-borne disease and deaths England and Wales: 1992–2002. *Gut* **51**, 832–841.

Advisory Committee on the Microbiological Safety of Food (ACMSF) (1993) *Report of Progress 1990–1992*. HMSO, London.

Altemeier, W.A. & Culbertson, W.R. (1965) Surgical infection. In: Moyer, C.A., Rhoads, J.E., Allen, J.G. & Harkins, H.N. (eds). *Surgery, Principles and Practice*. Lippincott, Philadelphia, pp. 185–218.

Ayliffe, G.A.J. & Babb, J.R. (1995) *Hospital Acquired Infections*. Science Press Ltd, London.

Centers for Disease Control and Prevention (CDC) (1988) *CDC Surveillance Update*. CDC, Atlanta, GA.

Centers for Disease Control and Prevention (CDC) (1999) *National Nosocomial Infections Surveillance (NNIS) Manual*. CDC, Atlanta, GA.

Cowden, J.M. (2000) Food poisoning notification: time for a rethink. *Health Bulletin* **58**, 328–331.

Cowden, J.M., O'Mahony, M., Bartlett, C.L.R. *et al.* (1989) National outbreak of *Salmonella typhimurium* DT124 caused by contaminated salami sticks. *Epidemiology and Infection* **103**, 219–225.

Department of Health (DH) (1989) *Working for Patients*. DH, London, UK.

Department of Health (DH) (2003) *Winning Ways: Working Together to Reduce Healthcare Associated Infection in England.* Report from the Chief Medical Officer. DH, London, UK.

Emmerson, A.M., Ayliffe, G.A.J., Casewell, M.W. *et al.* (1993) National prevalence survey of hospital acquired infections: definitions. *Journal of Hospital Infection* **24**, 69–76.

Emmerson, A.M., Enstone, J.E., Griffin, M. *et al.* (1996) The second national prevalence survey of infection in hospitals: overview of the results. *Journal of Hospital Infection* **32**, 175–190.

Food Standards Agency (FSA) (2000) *A Report of the Study of Infectious Intestinal Disease in England.* Food Standards Agency, London.

Food Standards Agency (FSA) (2004) *Action Stepped up on Salmonella Outbreaks*. Available from http://www.food.gov.uk/news/newsarchive/2004/oct/spanisheggs. Accessed 2 March 2007.

Gaynes, R., Richards, C., Edwards, J. *et al.* (2001) Feeding back surveillance data to prevent hospital-acquired infections. *Emerging Infectious Diseases* **7**, 295–298.

Gill, O.N., Bartlett, C.L.R., Sockett, P.N. *et al.* (1983) Outbreak of *Salmonella napoli* infection caused by contaminated chocolate bars. *Lancet* **1**, 574–577.

Haley, R.W. (1985) Surveillance by objective: a new priority-directed approach to the control of nosocomial infections. *American Journal of Infection Control* **13**, 78–89.

Haley, R.W. (1992) Cost benefit analysis of infection control programmes. In: Brachman, P.S. & Bennett, J.V. (eds). *Hospital Infections.* Little, Brown (Medical Division), Boston, pp. 507–532.

Haley, R.W., Culver, D.H., White, J.W. *et al.* (1985) The efficacy of infection surveillance and control programmes in preventing nosocomial infections in US hospitals. *American Journal of Hospital Epidemiology* **121**, 182–205.

Harbarth, S., Sax, H. & Gastmeier, P. (2003) The preventable proportion of nosocomial infections: an overview of published reports. *Journal of Hospital Infection* **54**, 258–266.

Health Protection Agency (HPA) (2006a) *NOIDS Food Poisoning*. Available from http://www.hpa.org.uk/infections/topics_az/noids/food_poisoning.htm. Accessed 2 March 2007.

Health Protection Agency (HPA) (2006b) *Gastrointestinal Disease*. Available from http://www.hpa.org.uk/infections/topics_az/gastro/menu.htm. Accessed 2 March 2007.

Health Protection Agency (HPA) (2006c) *International Surveillance Network for the Enteric Infections Salmonella and VTEC O157*. Available from http://www.hpa.org.uk/hpa/inter/enter-net_menu.htm. Accessed 19 February 2007.

Health Protection Scotland (HPS) (2006) *Gastrointestinal and Zoonoses*. Available from http://www.hps.scot.nhs.uk/. Accessed 2 March 2007.

Joseph, C.A. & Palmer, S.R. (1989) Outbreaks of salmonella infection in hospitals in England and Wales 1978–87. *British Medical Journal* **298**, 1161–1163.

Killelea, D., Ward, L.R., Roberts, D. *et al.* (1996) International epidemiological and microbiological study of outbreak of *Salmonella agona* infection from a ready to eat savoury snack I England and the United States. *British Medical Journal* **313**, 1105–1107.

Maguire, H., Cowden, J., Jacob, M. *et al.* (1992) An outbreak of *Salmonella dublin* infection in England and Wales associated with soft unpasteurised cows' milk cheese. *Epidemiology and Infection* **109**, 389–396.

Meers, P.D., Ayliffe, G.A.J., Emmerson, A.M. *et al.* (1981) Report of the national survey of infection in hospitals – 1980. *Journal of Hospital Infection* **2** (Suppl), 1–53.

NHS National Services Scotland (2006) *Infectious Diseases*. Available from http://www.isdscotland.org/isd/info3.jsp?pContentID=1523&p_applic=CCC&p_service=Content.show&. Accessed 2 March 2007.

Nightingale, F. (1859) *Notes on Nursing: What It Is and What It Is Not*. Harrison, London.

Plowman, R., Graves, N., Griffin, M.A.S. *et al.* (2001) The rate and cost of hospital-acquired infections occurring in patients admitted to selected specialities of a district general hospital in England and the national burden imposed. *Journal of Hospital Infection* **47**, 198–209.

Raven, B.H. & Haley, R.W. (1980) Social influence in a medical context. In: Bickman, L. (ed). *Applied Social Psychology Annual*, Vol. 1. Sage Publications, London, pp. 255–277.

Royal College of General Practitioners (RCPG) (2006) *The Birmingham Research Unit*. Available from http://www.rcgp.org.uk/bru/index.asp. Accessed 2 March 2007.

World Health Organization (WHO) (1984) *WHO Surveillance Programme for Control of Food-Borne Infections and Intoxications in Europe: Third Report 1982*. Institute of Veterinary Medicine – Robert von Ostertag Institute, Berlin.

World Health Organization (2004) *WHO Surveillance Programme for Control of Infections and Intoxications in Europe*. Available from http://www.euro.who.int/foodsafety/Surveillance/20020903_3. Accessed 2 March 2007.

World Health Organization (WHO) (2007) *Integrated Disease Surveillance and Response*. Available from http://www.who.int/countries/eth/areas/surveillance/en/. Accessed 2 March 2007.

4 Foodborne Disease Outbreaks in Healthcare Settings

Sarah J. O'Brien

4.1 Introduction

Ensuring a continuous supply of food in healthcare settings is an important and complicated business. Twenty-four hours a day, 7 days a week, patients, visitors and staff must be fed. A nutritious diet is essential for patient treatment and recovery. Therefore, food must be safe, of good quality, wholesome, and served at times that are convenient and appropriate, i.e. not only at conventional mealtimes. In 2001, the Audit Commission published a review of hospital catering. At that time, 71% of Trusts had in-house catering departments whilst the remainder had contracted out their catering services. The audit considered five areas: identifying and meeting patients' nutritional needs, quality of service provision and the relationship between quality and cost, the amount spent on catering and the variation between Trusts, the management and control of costs and the potential savings from reducing food being wasted. Quality was judged in terms of patient satisfaction. Ironically, microbiological food safety seems not to have been considered (Audit Commission, 2001). However, in a survey of 1660 hospital kitchens in England and Wales, inspectors found that 1119 were below acceptable sanitary standards and in 153 the infractions were sufficiently serious to justify prosecution (Gellert *et al.*, 1989). In England alone, it was reported by NHS Estates in 2005 that the National Health Service spends some £500 million annually, providing in the region of 300 million meals for patients. In this chapter, we will consider the consequences when food safety breaks down.

4.2 The significance of foodborne outbreaks

Foodborne disease outbreaks are by their nature unexpected. At best they are inconvenient; at worst they are life-threatening. Outbreaks of foodborne disease in hospital cause considerable disruption to services for patients and staff. They constitute avoidable causes of illness and death in a vulnerable population (Joseph & Palmer, 1989; Wall *et al.*, 1996; Meakins *et al.*, 2003).

Pinning down a food vehicle in hospital outbreaks can be problematic:

- By the time an outbreak comes to light it might be difficult to establish if contaminated food was likely to have been the source. The time taken to establish that an outbreak has occurred may be lengthened depending on the level of contamination in food (Guallar *et al.*, 2004), the distribution of patients across an institution, or the clinical diagnostic protocols used within hospitals to investigate patients with diarrhoea (Bruins *et al.*, 2003).
- Obtaining an accurate food history from elderly or infirm patients can be challenging. Although printed hospital menus can be helpful, they are not kept for long and a patient offered certain food items might not actually eat them. Experience has shown that mobile patients may also supplement their diet with food from hospital canteens or shops, or with food brought in by relatives.
- Person-to-person transmission might hide the fact that the initial insult was foodborne infection.
- Contaminated food might have been eaten or discarded.

4.2.1 *Reviews of outbreaks in hospitals: what do we learn?*

The significance of foodborne disease outbreaks in hospitals has been reviewed by several authors and their findings are summarized in Table 4.1. In a review of 50 hospital-based outbreaks of food poisoning in Scotland between 1973 and 1977, at least 1530 people eating hospital-prepared food were involved (Sharp *et al.*, 1979). Thirty-one episodes were associated with *Clostridium perfringens*, 11 were due to foodborne salmonella infection, three to enterotoxigenic *Staphylococcus aureus*, and in five incidents the aetiology was not determined.

In the UK since 1968, the number of reported salmonella outbreaks in hospitals has fallen from, on average, 53 per year (Abbott *et al.*, 1980) to 6 per year (Meakins *et al.*, 2003). The reasons for this fall are likely to be multi-factorial and probably reflect better control of cross-infection on the one hand, coupled with improvements in food hygiene on the other.

One of the notable features of hospital outbreaks is that psychiatric wards and elderly-care wards (including psychogeriatric wards) are the units that tend to be most badly affected (Sharp *et al.*, 1979; Joseph & Palmer, 1989; Wall *et al.*, 1996; Meakins *et al.*, 2003). These patient groups are, perhaps, especially vulnerable and preventing cross-infection following a foodborne disease outbreak can be a particular challenge.

Although foodborne disease outbreaks in hospitals in the UK appear to be becoming less common, the mortality risk associated with these outbreaks is still high. In the review

Table 4.1 Recent reviews of outbreaks of foodborne disease in hospitals.

Country	Period of review	Nature of review	Number of outbreaks	Cases	Deaths	Proportion diagnosed as foodborne (*N*)	Reference
UK	1973–1977	Food poisoning outbreaks	50	1530	–	Not applicable	Sharp *et al.*, 1979
UK	1968–1977	Salmonella outbreaks	522	2152	42	32% (24)[a]	Abbott *et al.*, 1980
UK	1980–1982	Prospective survey of salmonella outbreaks	55	Average 8 cases per outbreak	–	1.1% (6)	Palmer & Rowe, 1983
UK	1978–1987	Salmonella outbreaks	248	3000+	110	24% (57)	Joseph & Palmer, 1989
UK	1992–1994	Salmonella outbreaks	22	260	15	36% (8)	Wall *et al.*, 1996
Australia	1995–2000	All foodborne disease outbreaks	3	64	3	Not applicable	Dalton *et al.*, 2004
Republic of Ireland	1998–1999	All outbreaks of infectious intestinal disease	6	Not stated	2	Not stated	Bonner *et al.*, 2001
UK	1992–2000	All outbreaks of infectious intestinal disease	1396	29,507	82	1.8% (25)	Meakins *et al.*, 2003

[a]Note that in only 76 outbreaks was the potential source recorded.
–, not reported.

Table 4.2 General outbreaks of foodborne illness in humans, in hospitals, England and Wales, quarterly reports 1996–2004.

Report date	Local authority	Organism	Number of ill	Cases positive	Suspect vehicle	Evidence
Jan–Mar/96	Northumberland	*Salmonella* Hadar PT2	NA	NA	NA	NA
Jan–Mar/96	East Dyfed	*Clostridium perfringens*	NA	NA	NA	NA
Jan–Mar/96	Manchester	*C. perfringens*	5	4	None	M
Jan–Mar/96	Stoke on Trent	*C. perfringens*	12	5	None	M
Oct–Dec/96	Gwent	*S.* Enteritidis PT6	NA	NA	NA	NA
Oct–Dec/96	Scunthorpe	*S.* Typhimurium DT104	2	2	None	–
Oct–Dec/96	Thamesdown	*S.* Typhimurium DT104	4	4	None	–
Apr–June/97	Newcastle	*S.* Typhimurium DT104	36	12	None	–
Apr–June/97	East Kent	*S.* Typhimurium DT104	NA	NA	NA	NA
Apr–June/97	Taunton	*C. perfringens*	14	4	Cottage pie	M
Jul–Sept/97	Westminster	*S.* Bovis-morbificans	10	9	None	–
Jul–Sept/97	Bro Taf	*S.* Enteritidis PT4	NA	NA	NA	NA
Oct–Dec/97	Lambeth	*S.* Enteritidis PT4	2	2	Chicken and eggs	D
Oct–Dec/97	Wandsworth	*S.* Enteritidis PT4	2	2	None	–
Oct–Dec/97	Liverpool	*S.* Newport	10	10	None	–
Jan–Mar/98	Chelmsford	*S.* Enteritidis PT4	4	2	None	–
Jan–Mar/98	Ealing, Hammersmith and Hounslow	*S.* Hadar PT2	NA	NA	NA	NA
Apr–June/98	Basildon	*S.* Agona RDNC	6	6	None	–
Jul–Sept/98	Kings Lynn	*S.* Enteritidis PT4	27	23	Egg sandwiches	M
Jul–Sept/98	Derby	*S.* Enteritidis	3	3	None	–

Jul–Sept/98	Manchester	*S.* Typhimurium DT104	3	3	None	–
Jul–Sept/98	Aylesbury Vale	*S.* Typhimurium DT104	10	3	None	–
Oct–Dec/98	Camden	*S.* Enteritidis PT4	5	5	None	–
Oct–Dec/98	Surrey Heath	*C. perfringens*	100	2	Beef sandwiches	
Jan–Mar/98	Southampton	*S.* Montevideo	7	7	Banana custard	D
Jan–Mar/98	Bournemouth	*S.* Typhimurium DT193	3	3	None	–
Apr–June/99	Shropshire	*S.* Enteritidis PT4	NA	NA	NA	NA
July–Sept/99	Blackburn	*S.* Heidelberg	12	11	None	–
Apr–June/00	Chelmsford	*S.* Enteritidis PT4	6	6	None	–
Apr–June/00	West Surrey	*S.* Enteritidis PT4	NA	NA	NA	NA
Oct–Dec/00	Plymouth	Norovirus	106	6	Salad and rolls	D
Oct–Dec/01	Vale of Glamorgan	*S.* Typhimurium DT104B	4	4	None	–
Apr–June/02	Lincoln	*S.* typhimurium DT104	3	3	None	–
Oct–Dec/02	Lambeth, Lewisham and Southwark	*S.* Enteritidis PT1	31	31	None	–
Oct–Dec/02	Lambeth, Lewisham and Southwark	*S.* Enteritidis PT6A	29	29	Imported shell egg	M, S
Jul–Sep/03	Suffolk	*S.*Enteridis PT8	12	12	Made-up drinks	–
Oct–Dec/04	Salford	*S.*Virchow PT8	8	7	Chicken	D

Source: Health Protection Agency Centre for Infections. *Note*: M (microbiological): identification of an organism of the same type from cases and in the suspect food or food ingredient, or detection of toxin in faeces or food; D (descriptive): other evidence reported by local investigators as indicating the suspect food; S (statistical): a significant statistical association between consumption of the suspect food and being a case.

NA, not available; –, not reported.

by Meakins *et al.* (2003), the authors noted that there was a threefold increase in the risk of death in outbreaks of foodborne disease in hospitals compared with outbreaks in other settings. Similarly, in Australia, Dalton and colleagues (2004) reviewed 214 foodborne disease outbreaks and noted that 35% of deaths occurred in aged-care and hospital settings, despite the fact that only 5% of foodborne outbreaks and fewer than 3% of foodborne disease outbreak cases were reported in these settings. The case fatality rate was ten times higher in foodborne outbreaks in aged-care and hospital settings than in other foodborne outbreaks.

In the review by Wall *et al.* (1996), all eight foodborne outbreaks of salmonellosis during a 3-year period were due to *Salmonella* Enteritidis phage type 4 (PT4), although in only two of these outbreaks were specific food vehicles implicated. These were chicken and a range of sandwiches in one outbreak and egg sandwiches in the other. Shell eggs were also implicated in 4 of 25 foodborne disease outbreaks in the review by Meakins *et al.* (2003).

One of the disappointing characteristics that Table 4.2 highlights is the frequent failure to identify a suspected food vehicle in hospital outbreaks that are judged to have been foodborne. In the review by Abbott *et al.* (1980) of hospital outbreaks of gastrointestinal infection, in only 76 of 522 (36%) was the source of infection recorded; 24 outbreaks (32%) were considered to be foodborne. More than 20 years later, only 1.8% of outbreaks of gastrointestinal disease in hospitals in England and Wales were thought to be foodborne (Meakins *et al.*, 2003); many of the outbreaks in which a food vehicle was not identified were small (Table 4.2). Investigation policies and the availability of resources to investigate outbreaks will have a bearing on the outcome. In eight outbreaks no information was forthcoming at all, which might also reflect compliance with the surveillance system. It is possible, therefore, that the true burden of foodborne outbreaks is greater than is reported to national surveillance systems.

4.2.2 *Foodborne disease outbreaks in hospitals: how much do they cost?*

One way of judging the significance of foodborne disease outbreaks in hospitals is to look at the economic consequences, yet few investigators have done so. Those that have attempted to define precisely the economic impact of a hospital outbreak have demonstrated that the costs are substantial. In Australia in 1996, 52 people were affected in an outbreak of *S.* Typhimurium traced back to contamination of sandwiches prepared in the hospital kitchen (McCall *et al.*, 1999). Spearing and colleagues (2000) estimated the direct costs incurred by the hospital associated with this outbreak. The medical costs amounted to some AU$67,000. The costs of the investigation were in the region of AU$37,000 whilst the costs of lost productivity were over AU$15,000. Taking into account various miscellaneous items (photocopying, external catering, telephone calls etc.), the costs of this one outbreak were substantial at over AU$120,000 (nearly US$95,000). Indirect costs were not included in the calculations but are likely to have been considerable, comprising the expenses incurred by individuals affected in the outbreak and their relatives (suffering, loss of life and lost productivity), and the costs elsewhere in the healthcare system and in the community.

In a hospital outbreak of salmonellosis in Scotland, both direct and indirect costs were determined (Yule *et al.*, 1988). In this instance, the economic impact of the outbreak was estimated to be between £200,000 and £900,000.

In an outbreak of campylobacteriosis amongst 31 members of hospital staff in the north of England, 46 working days were lost at an estimated cost of £3000 (Murphy *et al.*, 1995), whilst the identifiable costs of an outbreak of salmonellosis that affected 22 patients and seven staff in the south of England in 1993 were estimated to be in the region of £33,000 (Dryden *et al.*, 1994).

What emerges from studies looking at the cost of outbreaks is that few have offered a comprehensive account. Most have attempted to calculate the costs to the hospital but few have gone beyond the healthcare sector or considered the longer-term sequelae for those caught up in outbreaks. There is a tendency to forget that, at least in a proportion of victims, the acute episode can lead to longer-term problems like reactive arthritis and irritable bowel syndrome. Therefore, the costs presented above are mostly minimum estimates, the real costs of the outbreaks having been much greater.

4.3 Foodborne disease outbreaks in hospitals: what goes wrong?

Outbreak reports in the literature highlight a number of different failures and these are summarized in the following sections.

4.3.1 *Outbreaks due to contaminated raw ingredients*

One of the hazards for the catering manager is bringing into the hospital kitchen foods that are contaminated at source, particularly if those foods are not going to undergo further cooking (Table 4.3). Salad items are an obvious example. A variety of organisms have been implicated including *Listeria monocytogenes*, Shiga toxin-producing *Escherichia coli* (STEC) O157 and norovirus.

Contamination of lettuce with STEC O157 has been implicated in two outbreaks. The first was in Canada in 1995 when 21 people were affected in an outbreak at a community hospital. In a case–control study consumption of green salad was associated with becoming unwell. During the course of environmental investigations, it was noted that the iceberg lettuce used to make the green salad had been heavily contaminated with soil (Preston *et al.*, 1997). In the first foodborne outbreak of STEC O157 to be recognized in Sweden, in 1999, contaminated lettuce was judged to be the most likely source for 11 cases of infection amongst nursing staff in a children's hospital (Welinder-Olsson *et al.*, 2004). Lettuce from a wholesaler who supplied the lettuce served at a buffet was heavily contaminated with a mixture of Gram-negative bacteria. Although STEC were not identified in a suspension of lettuce, using polymerase chain reaction (PCR), taking into account all the evidence, lettuce was the most likely source. Contaminated lettuce or other salad items are now well-recognized sources of foodborne outbreaks of STEC O157 (Wachtel *et al.*, 2002).

Contaminated salad vegetables have also been implicated in outbreaks of infection with norovirus and *L. monocytogenes*. In a large outbreak of norovirus amongst hospital staff in India in 1999, only those who had eaten salad sandwiches were ill (Girish *et al.*, 2002). In a widespread outbreak affecting 23 patients in eight hospitals in Boston, Massachusetts in 1979, the majority of patients infected with *L. monocytogenes* serotype 4b had gastrointestinal symptoms commencing at the same time as their fever (Ho *et al.*, 1986). In a

Table 4.3 Outbreaks due to contaminated raw ingredients.

Country of report and year of outbreak	Causative organism	Setting	Number of cases (number at risk, where stated)	Number of deaths	Implicated food vehicle	Evidence implicating food vehicle	Reference
US, 1973	*S.*Typhimurium	Hospital	32 – 18 patients, 14 staff	–	Egg nog	Epidemiological and microbiological studies	Steere *et al.*, 1975
US, 1979	*Listeria monocytogenes*	Eight hospitals in Boston	23	–	Tuna fish, chicken salad and cheese	Anecdotally all three dishes had common ingredients – namely raw celery, tomatoes and lettuce	Ho *et al.*, 1986
US, 1987	*S.* Enteritidis	Municipal hospital	404	9	Hospital-made mayonnaise containing raw eggs	Epidemiological study plus macaroni salad and raw eggs culture positive	Telzak *et al.*, 1990
Canada, 1995	STEC O157	Community hospital	21 – 8 patients, 10 staff, 3 volunteers (360)	–	Green salad	Case–control study	Preston *et al.*, 1997
Denmark, 1995	*S.* Enteritidis PT6	Hospital	35	–	? Eggs	Circumstantial	Skibsted *et al.*, 1998
Finland, 1999	*Listeria monocytogenes*	Tertiary care hospital	25 patients	6	Pasteurized butter	Matched case–control study and isolation of an indistinguishable organism from butter samples and dairy environment	Lyytikäinen *et al.*, 2000
India, 1999	Norovirus	Nurses' hostel of a Civil hospital	130	–	Salad sandwiches	Descriptive study	Girish *et al.*, 2002
Sweden, 1999	STEC O157	Children's hospital	11 (250)	–	Lettuce	Descriptive study	Welinder-Olsson *et al.*, 2004

–, not reported.

case–control study, food preferences amongst the cases included tuna fish, chicken salad and cheese. Common ingredients served with all these dishes were raw celery, lettuce and tomatoes and these were thought to be the source of the bacteria although the evidence for this was circumstantial.

A very unusual food vehicle was reported from Finland by Lyytikäinen and colleagues (2000). In 1999, an outbreak of infection with *L. monocytogenes* serotype 3a affecting 25 severely immunosuppressed patients was traced to contaminated, pasteurized butter. The butter supplied to the hospital in 7 g packs was manufactured in a continuous butter maker. The outbreak strain was isolated from 13 butter samples from the hospital kitchen and from several batches of butter packs of varying sizes from the dairy and a wholesale store, and from points along the production line. In the majority of butter samples, the numbers of *L. monocytogenes* were less than 100 colony-forming units per gram (CFU/g), but one pooled sample of 7 g butter packs contained 11,000 CFU/g. The butter was probably contaminated after pasteurization, during the packaging process. Using data generated from this outbreak, Maijala *et al.* (2001) were able to estimate the exposure and attack rate for the hospital population based on the delivery, consumption and microbiological studies of the implicated butter. They considered two different hypotheses. In the first a relatively high dose of *L. monocytogenes* could have occurred occasionally in the 7 g butter packs, in which case the dose in one meal that caused a case of listeriosis would have been 7.7×10^4 CFU. Alternatively, prolonged daily consumption of contaminated butter could have led to the cases. If this were the correct scenario then the dose would have been somewhere in the range 1.4×10^1 to 3.1×10^5 CFU/day. This latter hypothesis implies that low levels of *L. monocytogenes* in food can be hazardous to susceptible populations, especially if exposure occurs over a prolonged period.

In an outbreak of *S.* Typhimurium in the US in 1973, affecting 32 people, 'egg nog' made with raw egg was implicated (Steere *et al.*, 1975). Raw shell eggs were used to make the 'egg nog', which was used as part of a liquid diet for patients recovering after surgery. There was a highly statistically significant association between 'egg nog' consumption and becoming a case of *S.* Typhimurium. Cases included members of the catering staff who had tasted the 'egg nog'. As part of the investigation samples of chicken faeces and environmental swabs from the farm supplying the eggs yielded the outbreak strain. Control measures included changing the supply eggs and purchasing 'egg nog' as a commercially produced pasteurized product. This outbreak demonstrated another consequence of foodborne disease outbreaks in hospitals – namely, person-to-person spread following the initial foodborne insult. Eight members of staff and six patients acquired infection via secondary spread after the original foodborne source had been removed.

Raw eggs as ingredients of contaminated food vehicles have become a recurring theme over the years. Telzak and colleagues (1990) described a very large nosocomial outbreak of *S.* Enteritidis at a municipal facility in New York in 1987. More than 400 patients were affected and nine died when raw shell eggs used to make mayonnaise were contaminated with *S.* Enteritidis PT8. During detailed enquiries, 300 out of 985 grade A eggs were tested individually for the presence of *S.* Enteritidis and the remainder were tested in pooled samples. Five hundred and fifty ovaries from hens from the same farm corporation that had supplied the eggs were also examined. Five pooled samples of raw shell egg and 383 ovary samples were positive for *S.* Enteritidis, demonstrating convincingly that raw shell eggs were most likely to have been the source of the outbreak.

More tenuous evidence linking raw eggs with an outbreak in a hospital comes from Denmark (Skibsted *et al.*, 1998). As part of an exercise to differentiate better outbreak cases from sporadic cases of *S.* Enteritidis, Skibsted *et al.* used various microbiological methods (RAPD, PFGE and phage-typing) to group isolates obtained from hospital patients and cases in the surrounding community. Thirty-five patients who had developed symptoms of salmonellosis more than 48 hours after admission to hospital were infected with *S.* Enteritidis PT6 of three RAPD types (A, B and C). The investigators hypothesized that all three strains might have been introduced into the hospital from shell eggs delivered from a single producer. Contaminated shell eggs have also been implicated in a recent outbreak of foodborne *S.* Enteritidis PT6a in the UK (Table 4.2). In its food specifications, the NHS Purchasing and Supply Agency (PASA) specified that shell eggs should be produced in accordance with the Lion Quality Code of Practice (responsibility for specifications for supply of food to the NHS passed from the PASA to the NHS Supply Chain in autumn of 2006). Where shell eggs have been implicated recently in hospital outbreaks they have not conformed to this standard.

4.3.2 *Outbreaks due to commercially prepared, contaminated ingredients*

Commercially prepared, contaminated ingredients can also cause problems for hospital caterers (Table 4.4). In an outbreak involving *S.* Chester, *S.* Tennessee and *S.* Habana in Vermont consumption of roast beef and cold cuts was associated with illness amongst 43 hospital staff (Spitalny *et al.*, 1984). An unopened package of pre-cooked roast beef from a commercial supplier was found to be contaminated with *S.* Chester, *S.* Tennessee and *S.* Livingston. Other cold cuts were thought to have become cross-contaminated in the hospital kitchen from a meat slicer. In this outbreak, two of three patients affected were thought to have been infected by hospital staff.

In a similar incident in the UK, 17 patients were affected when pre-cooked, vacuum-packed pork, which was contaminated with *C. perfringens* type A, was supplied by a local meat producer. A hospital kitchen inspection was satisfactory and hygiene practices within the hospital kitchen were found to be adequate. However, when the meat production company was visited, it was established that there was a key failure in the production process. Blast chilling was not used for cuts of meat that were, in any event, too large. Meat at 68°C was cooled to 58°C in a home-made water shower designed for cooling meat. After a further 48 hours in a refrigerator at 2°C the meat temperature was still 28°C. After this outbreak the hospital in question stopped purchasing ready-cooked joints, changed its meat supplier and installed a blast chiller (Regan *et al.*, 1995).

Staphylococcal food poisoning occurred amongst hospital staff in New York after consuming contaminated canned mushrooms imported from China (Levine *et al.*, 1996). Forty-eight people became unwell and two unopened tins of mushrooms from the implicated batch contained staphylococcal enterotoxin A. In total there were four confirmed staphylococcal outbreaks across the US and a further three suspected, associated with canned mushrooms from China. Investigations led to the Food and Drug Administration prohibiting the importation of cans from China, destined for institutional caterers. This was thought to be the first time staphylococcal food poisoning was associated with canned food.

Table 4.4 Outbreaks due to commercially prepared contaminated ingredients.

Country of report and year of outbreak	Causative organism	Setting	Number of cases (number at risk, where stated)	Number of deaths	Implicated food vehicle	Evidence implicating food vehicle	Reference
US, 1981	*S.* Chester, Tennessee and Habana	Hospital	43 staff (292)	–	Roast beef or cold cuts	Case–control study plus *S.* Chester, Tennessee and Livingston isolated from unopened package of precooked roast beef	Spitalny *et al.*, 1984
US, 1989	Staphylococcal enterotoxin A	Hospital cafeteria	48 staff	–	Canned mushrooms	Cohort study plus Staphylococcal enterotoxin A found in unopened cans from three plants	Levine *et al.*, 1996
England, 1989	*C. perfringens*	District General Hospital	17 (44)	–	Roast pork	Case–control study plus heavy contamination of pork with *C. perfringens* type A	Regan *et al.*, 1995
Wales, 2001	*S.* Indiana	Acute hospital	17	–	Egg sandwiches	Postulated failure of pasteurization of a batch of egg roll	Mason *et al.*, 2001

–, not reported.

In an outbreak of *S.* Indiana infection in south Wales, 17 people (patients, relatives and staff) were ill following consumption of egg mayonnaise sandwiches (Mason *et al.*, 2001). Epidemiological evidence implicated the egg sandwiches; inadequate pasteurization of the batch of egg roll used to prepare the sandwiches was considered to be a cause of the outbreak.

4.3.3 *Outbreaks due to inadequate cooking, handling and/or storage of food in hospital kitchens*

Food handling failures within the hospital kitchen environment frequently cause problems (Table 4.5). Temperature abuse and storage problems have contributed to a number of outbreaks and it is not surprising that both *C. perfringens* and *Salmonella* sp. feature in this table.

In one of the largest foodborne outbreaks in a hospital ever recorded in the UK, 379 patients were affected by *C. perfringens* (Thomas *et al.*, 1977). This amounted to a third of the patients in the hospital. One frail man died. Those at greatest risk were on a minced diet. A detailed investigation revealed a catalogue of problems including decrepit infrastructure, too few staff (especially trained staff), dirty pans and utensils, poor organization of food in refrigerators, and mincers and choppers used for both raw and cooked meats (and containing raw meat residues). In an inspection by an environmental health officer 2 months prior to the outbreak urgent recommendations to upgrade the hospital's catering facilities had been recorded. Unfortunately, it would appear that they had not been acted upon. The preparation of the minced diet was faulty in a number of ways – joints of meat were too large to be cooked and cooled properly, lengthy delays occurred between the end of the cooking time for the meat and the beginning of refrigeration, and cooked meat was minced in machines used for raw meat that had not been properly cleaned in between. Contaminated minced beef was also implicated in a large outbreak of *C. perfringens* type A in a psychiatric hospital in England (Pollock & Whitty, 1990), whilst poor preparation of a tuna salad caused a *C. perfringens* outbreak in the US when large quantities of ingredients had been cooled inadequately (Khatib *et al.*, 1994).

Major disruption to services in an orthopaedic and trauma hospital in the UK occurred following an outbreak of *S.* Typhimurium DT 197 infection affecting staff (White, 1986). In total 104 staff out of a total complement of 954 reported illness, which was laboratory-confirmed in 82 individuals. The scale of the outbreak, and the fact that it involved so many key personnel, meant that exclusion guidelines had to be modified in order to keep services running. Whilst food handlers were still required to stay away from work until three consecutive stool samples were negative for salmonella, other staff were allowed to return to work after 48 hours symptom-free, provided that they paid scrupulous attention to their personal hygiene, did not take patients' temperatures, administer food, drink or oral medication to the patients or attend to patients' oral hygiene. Despite this relaxation in hospital policy, no secondary cases occurred amongst patients. The change in standard infection control procedures was backed by evidence from a study of convalescent salmonella excreters (Pether & Scott, 1982) and meant that the only accident and emergency department serving a third of a million people was able to stay open. The food vehicle implicated in the outbreak was tartare sauce made in the hospital kitchen using raw shell eggs. The mayonnaise base for the tartare sauce was made with 60 egg yolks. Although never proven, the eggs were suspected to be the source of the organism.

Table 4.5 Outbreaks due to inadequate cooking, handling and/or storage of food.

Country of report and year of outbreak (if stated)	Causative organism	Setting	Number of cases	Number of deaths	Implicated food vehicle	Evidence implicating food vehicle	Reference
England, 1976	*C. perfringens*	Teaching hospital	379	1	Minced ham dish	Cohort study plus evidence of faulty cooking and storage of ham	Thomas *et al.*, 1977
US, 1985	*Bacillus cereus*	Chronic disease hospital	28	–	Turkey loaf	Epidemiological study plus isolation of *B. cereus* from turkey loaf	Giannella & Brasile, 1979
England, 1980	*S.* Muenchen	Children's hospital	28	3	None identified	*S.* Muenchen isolated from a refrigerated cupboard in one kitchen	Kumarasinghe *et al.*, 1982
England, 1985	*S.* Typhimurium DT 197	Hospital canteen	82 staff	–	Tartare sauce	*S.* Typhimurium isolated in low numbers from samples of tartare sauce retrieved from the kitchen	White, 1986
US	*C. perfringens*	Teaching hospital	52 staff	–	Tuna salad	?	Khatib *et al.*, 1994
England, 1989	*C. perfringens* type A	Psychiatric hospital	50 patients	2	Minced beef meal	Epidemiological study	Pollock & Whitty, 1990
Wales, 1990	*S.* Enteritidis phage type 4	Hospital for mentally handicapped people	109 – 101 patients, 8 staff	–	Beef rissoles	Cohort study plus organism isolated from shell eggs in hospital kitchen	Evans *et al.*, 1996
Mexico, 1994	*S.* Enteritidis	Hospital cafeteria	97 staff	–	Egg-covered meat plate	Case–control study	Molina-Gamboa *et al.*, 1997
France, 1999	*S.* Paratyphi B	Hospital	8	0	Minced beef	*S.* Paratyphi B isolated from a batch of raw product	Haeghebaert *et al.*, 2001
France, 1999–2000	*S.* Typhimurium	Hospital	29	3	Hamburgers	Case-crossover study plus isolation of S. Typhimurium from two batches of hamburgers	Haeghebaert *et al.*, 2001
The Netherlands, 2001	*S.* Enteritidis phage type 6	Hospital and nursing home served by the same hospital kitchen	82	5	Bavaroise and strawberry custard	Cohort study	Bruins *et al.*, 2003

–, not reported.

In recent years, the epidemiology of foodborne salmonella outbreaks in hospitals has been dominated by *S.* Enteritidis and outbreaks in many countries have involved dishes made with contaminated raw shell eggs. A large outbreak of *S.* Enteritidis PT4 occurred in a hospital for mentally handicapped people in Wales, affecting 101 inhabitants and eight members of staff, when beef rissoles were inadequately cooked (Evans *et al.*, 1996). Shell eggs from the hospital kitchen, used as an ingredient in the rissoles, were culture-positive for *S.* Enteritidis PT4. Ninety-seven people who regularly ate in a hospital cafeteria in Mexico became infected with *S.* Enteritidis following a breakfast meal (Molina-Gamboa *et al.*, 1997). An egg-covered meat plate was implicated in an epidemiological investigation and the eggs were found to have been inadequately cooked. In the Netherlands, a strawberry bavaroise and a bavaroise-custard dessert made with raw shell eggs caused an outbreak of *S.* Enteritidis PT6 infection amongst patients and staff at a large multi-site hospital and a nursing home served by the same hospital kitchen (Bruins *et al.*, 2003). The desserts had been made by using raw egg yolks as well as pasteurized egg white. This outbreak, in which five frail people died, led to changes in the law. As well as banning the sale of *Salmonella*-contaminated eggs for direct consumption by consumers, food hygiene codes for health institutions and the hotel and catering industries were revised to ban the use of raw shell eggs in the preparation of dishes for consumption that do not undergo sufficient cooking.

Improper cooking of minced beef products led to two unusual nosocomial outbreaks in France (Haeghebaert *et al.*, 2001). In the first outbreak eight patients who had been in-patients for several weeks in four different departments of the same hospital developed an infection with *S.* Paratyphi B; there were no deaths. *Salmonella* Paratyphi B was isolated from samples of frozen pre-cooked hamburgers from the hospital kitchen. The traceback exercise led to a minced beef producer that had undergone an official inspection earlier that year. During that inspection *S.* Paratyphi B had been identified in the batch of raw product. Although the producer had been required to use the meat in thoroughly cooked products such as Bolognese sauce he had, instead, distributed the batch as frozen hamburgers that were insufficiently pre-cooked. The strains isolated from the patients, from hamburgers taken from the hospital, and from the official inspection of the minced beef producer all showed the same DNA macro-restriction profile. All hamburgers in the implicated batch were recalled.

The second outbreak involved 35 cases of *S.* Typhimurium infection, of which 29 (83%) were nosocomial cases distributed across six care institutions (hospitals or retirement homes) in the same locality. Twenty-five cases were individuals who had been in-patients for several weeks when the outbreak started, and four cases were health-workers from the same care institutions. Three patients with severe underlying disease died. The results of a preliminary investigation suggested a common foodborne source of contamination. In a case-crossover study, consumption of hamburgers was significantly associated with occurrence of the illness. Trace-back investigations showed that the frozen hamburgers supplied to the institutions came from a single producer. *Salmonella* Typhimurium was cultured from two batches of hamburgers from the affected organizations. Isolates of *S.* Typhimurium from the patients and the food all belonged to lysotype, definitive type 104 and exhibited resistance phenotype ACSSuT (resistance to amoxicilline, chloramphenicol, streptomycin, sulphonamides and tetracycline). They also had the same DNA macrorestriction profile. Two implicated batches were recalled nationwide.

Foodborne outbreaks of *Bacillus cereus* in hospitals are rarely reported, despite the organism being a well-known cause of food poisoning. Twenty-eight patients were involved in an outbreak of *B. cereus* infection in the US traced to contaminated turkey loaf served at an evening meal (Giannella & Brasile, 1979). The organism was cultured from symptomatic patients and from the turkey loaf.

It is not always easy to pin down a food vehicle in a hospital-based outbreak, as was demonstrated at an investigation at a children's hospital in London (Kumarasinghe *et al.*, 1982). The organism involved was a strain of *S.* Muenchen exhibiting resistance to multiple antibiotics. Twenty-eight of 393 children were found to be excreting the organism and three children died, although in only one child was salmonella infection thought to have contributed directly to death. During investigations *S.* Muenchen was isolated from a refrigerated cupboard in which both raw and cooked food, including chicken carcases, were stored on the same shelves. The kitchen was used to prepare meals for children on special diets. No food vehicle was ever identified and the investigators surmised that the outbreak was propagated by person-to-person spread. They were unable to rule out food as the original source of the outbreak. The outbreak caused considerable disruption to children's services, which did not get back to normal for several months because of the necessary deferment of waiting list admissions and surgical procedures during the outbreak itself. Emergency admissions were suspended for 5 days and led to strain on hospital services elsewhere that had to cope with the diverted emergencies.

Similar difficulties in discovery of a food vehicle were encountered in the investigation of an outbreak of *S.* Poona infection in the US (Stone *et al.*, 1993). In 1991, during a national outbreak of *S.* Poona infection traced to contaminated cantaloupe melon, a small outbreak of the same salmonella strain occurred on a neonatal intensive care unit (NICU). The outbreak started with two adult patients. One of the adult cases acquired infection in the community, but the second case might have been hospital-acquired. The second case was an insulin-dependent diabetic in premature labour; following her delivery she developed fever and diarrhoea and *S.* Poona infection was confirmed. She was re-admitted to hospital shortly after discharge with severe diarrhoea and septicaemia. Her infant, who had remained on the NICU, was stool-culture positive for *S.* Poona but never developed symptoms. However, two other infants became symptomatic and culture-positive. The only connection linking the two adult cases was the dietary porters but no common food item emerged. As regards infection in the infants one possibility is silent maternal transfer of salmonellosis from the diabetic mother to her baby. Thereafter the organism was probably passed amongst the infants via the staff.

In a mixed outbreak of *S.* Brandenburg and *S.* Corvallis infection in Japan in 2001 the specific food-handling problem was never identified (Hamada & Tsuji, 2001). Seven out of 315 patients developed symptoms after eating a lunch, at which three of the dishes were subsequently found to be contaminated with *S.* Brandenburg and one with *S.* Corvallis. How the four dishes became contaminated, however, was not determined.

4.3.4 *Outbreaks due to cross-contamination of food in hospital kitchens*

Cross-contamination of food items during preparation is a common fault identified in foodborne disease outbreaks. In a review of foodborne disease outbreaks in hospitals, Meakins *et al.* (2003) found that in 8 of 25 outbreaks cross-contamination probably occurred. Cross-contamination was also the food-handling fault identified in three of eight foodborne outbreaks of salmonella infection in hospitals reviewed by Wall *et al.* (1996). Table 4.6 contains a summary of outbreaks in which the principal food-handling fault was considered to be cross-contamination. There was one fatality in an outbreak of *S.* Typhimurium infection in a hospital in Brisbane affecting 52 people (McCall *et al.*, 1999; Spearing *et al.*, 2000). The incriminated food vehicle was sandwiches prepared in the hospital kitchen; during the course of investigations cleaning of food preparation equipment was found to be inadequate.

An outbreak of *S.* Enteritidis infection in a university hospital in Germany was caused, probably, by cross-contamination of a vanilla pudding from raw turkey that was being prepared in an area of the kitchen adjacent to the pudding preparation area (Kistemann *et al.*, 2000). Within the hospital campus nine buildings were affected in the outbreak – patients in 26 out of 76 wards were involved. An environmental inspection revealed a number of inadequacies in kitchen design and equipment. Turkey was thawed and cooked in close proximity to where vanilla pudding (made from milk and custard powder) was being cooled. Juices from the thawing turkey might have been introduced into the bottom of the pudding containers and/or pudding either directly or on the hands of the caterers. What ensued was a large outbreak affecting patients, staff and nursery children.

Cross-contamination from turkey was also implicated in an outbreak of *S.* Senftenberg infection in the US (L'Ecuyer *et al.*, 1996). The outbreak lasted for more than a year and 22 individuals were affected. Food prepared by the hospital kitchen was found to be the cause of the outbreak and in an epidemiological study consumption of lettuce, cauliflower, cottage cheese and cold turkey were associated with becoming a case. It is thought that kitchen equipment might have become contaminated with *S.* Senftenberg from the turkey, thus enabling the transfer of organisms to other foods. In an outbreak with a similarly low attack rate nine immunosuppressed patients died when food provided by the hospital kitchen became infected with *S.* Enteritidis (Guallar *et al.*, 2004). Clinical and food isolates were indistinguishable using PFGE. It is unclear precisely how the food became contaminated, but reinforcing hygiene and cleaning procedures aborted the outbreak.

Molecular fingerprinting of organisms proved invaluable in tracking an outbreak of *S.* Infantis to its source in Minnesota (Johnson *et al.*, 2001). Modified repetitive element PCR-typing (rep-PCR) was used in the investigation of clinical and environmental isolates traced to a hospital cafeteria steam table. One of the cutting boards on the steam table, which was otherwise being operated properly, yielded *S.* Infantis. The outbreak isolates, both clinical and environmental, were all grouped by re-PCR. All the cutting boards were replaced and the cleaning regimen was revised, following which no further cases occurred. In retrospect it was concluded that the routine use of rep-PCR might have led to slightly earlier detection of the outbreak.

No account of foodborne outbreaks in hospitals would be complete without mentioning the events that occurred at Stanley Royd Hospital in Wakefield, UK, in 1984 (Department of Health and Social Security, 1986). Three hundred and fifty-five patients and 106 members of staff were affected in an outbreak of *S.* Typhimurium DT 49. Nineteen patients died. This was the most serious foodborne outbreak in a hospital to date in the UK and its aftermath led to major reform of the public health system. The public inquiry, which ensued identified serious failings throughout the system.

Table 4.6 Outbreaks due to cross-contamination of food in hospital kitchens.

Country of report and year of outbreak	Causative organism	Setting	Number of cases	Number of deaths	Implicated food vehicle	Evidence implicating food vehicle	Reference
England, 1984	*S.* Typhimurium DT 49	Psychiatric hospital	461 – 355 patients, 106 staff	19	Roast beef	Descriptive study plus microbiological confirmation	Department of Health and Social Security, 1986
US, 1993	*S.* Senftenberg	University hospital	22 – 18 patients, 4 staff	–	Lettuce, cauliflower, cottage cheese, deli turkey	Case–control study	L'Ecuyer *et al.*, 1996
Australia, 1996	*S.* Typhimurium	Tertiary care hospital	52 – 17 patients, 3 neonates, 5 spouses of maternity patients, 27 staff	1	Sandwiches	Epidemiological study plus microbiological confirmation	McCall *et al.*, 1999; Spearing *et al.*, 2000
US, 1996	*S.* Infantis	Hospital cafeteria	14 – 12 outpatients, 1 staff, 1 visitor	–	Food served from the steam table	Cutting board from steam table culture positive	Johnson *et al.*, 2001
Germany, 1995	*S.* Enteritidis	University hospital	102 – 44 patients, 31 staff, 26 nursery school children	–	Vanilla pudding	Temporo-spatial analysis using a geographical information system	Kistemann *et al.*, 2000
Spain, 1998	*S.* Enteritidis	Tertiary care hospital	22	9	Unspecified meal	Microbiological	Guallar *et al.*, 2004

–, not reported.

The outbreak at Stanley Royd began over the August Bank Holiday weekend in 1984. At first there was little warning of the impending scale of the outbreak or the speed with which it would proceed to claim lives. What started with a few cases of diarrhoea at breakfast time on Sunday had, around 2 hours later, escalated to affect 36 patients on eight wards. By the end of the first day 94 patients were ill and 1 had died. On the second day the case count jumped to 153 and nursing and medical staff became overwhelmed by the rapidly evolving situation. Nursing staff also began to develop symptoms. A little over a week later 461 people had been involved and 19 patients were dead. The outbreak affected 45% of the patient complement and 11% of the hospital's staff. It was caused by contaminated roast beef that had been served cold on Saturday evening. Exactly how the meat became contaminated is still unclear. It is thought that it became contaminated after cooking when it was left at room temperature in a storeroom where chickens were being defrosted. Chicken was also being handled in the kitchen on the day that the cold roast beef was served to patients. Microbiological testing of food confirmed that the epidemic strain, *S.* Typhimurium DT 49, originated from chicken rather than from beef, lending weight to the fact that cross-contamination must have occurred. The fact that the ambient temperature was high would have allowed salmonella to grow on the beef, which underwent no further cooking prior to being served.

The public inquiry that followed the outbreak reached some damning conclusions (DHSS, 1986). As well as identifying failures that occurred in the hospital kitchens immediately prior to the outbreak, the inquiry uncovered serial neglect to act on recommendations to improve the infrastructure at Stanley Royd. The kitchens were originally built in 1865 and had remained virtually unaltered since that time. Six years before the outbreak began a major refurbishment of the kitchens had been recommended costing around £155,000, the kitchens having been described as a 'culinary disaster area' (Pennington, 2003). What appears to have followed was 6 years of prevarication during which, inevitably, the costs of the proposed improvements escalated. By the time agreement to build a new hospital kitchen (then at a cost of around £600,000) had been reached on 23 July 1984, it was far too late – the outbreak occurred just a month later.

Managers were also admonished. At one end of the scale, the hospital catering manager and her assistant were criticized for being largely absent from the kitchen, particularly after normal working hours and at weekends, and for their poor supervision of hygiene and catering practices. At the other end of the scale regional managers, responsible for the organization of health services within their region, were criticized for an apparently laissez-faire approach to the problem of Stanley Royd's kitchen in the first instance and, secondly, their response to the outbreak itself (or rather the lack of it).

The repercussions of the outbreak at Stanley Royd continued to reverberate through the public health system. The events of the summer of 1984 in Wakefield, coupled with an outbreak of Legionnaires' disease in Stafford in 1985, highlighted a decline in the medical expertise available to investigate and control communicable diseases. In 1986, at the behest of the Secretary of State for Social Services, Sir Donald Acheson, then Chief Medical Officer, chaired the 'Committee of Inquiry into the Future Development of the Public Health Function' (Acheson, 1988). Amongst the Group's wide-ranging recommendations were that a clearly identified medical practitioner should be responsible for necessary action on communicable disease and infection control. The Consultant in Communicable Disease

Control was born. Despite the recommendations of the Acheson Committee, however, relationships between infection control personnel in the community and in hospitals have sometimes continued to be a little uneasy.

4.3.5 *Outbreaks due to contaminated food prepared outside the hospital*

The increasing drive to make the hospital environment feel more like home, especially for the elderly, brings with it unforeseen risks and consumption of food prepared outside the hospital by patients' relatives and friends, or even by staff outside or within the hospital, has been implicated in outbreaks (Table 4.7).

In a large foodborne outbreak of Shiga toxin-producing *E. coli* O157 in a general hospital in Scotland, 37 people on four continuing-care wards for the elderly were infected, 16 were patients and 11 were members of staff (O'Brien *et al.*, 2001). There were additional cases in the community at the same time as the hospital outbreak. In an epidemiological investigation home-baked, cream-filled cakes, brought into the hospital for a concert party, were implicated as the contaminated food. One of the remarkable features of this outbreak was that there were no deaths. This contrasts with many other outbreaks caused by STEC involving the elderly, when the mortality rate has been very high (Carter *et al.*, 1987; Levine *et al.*, 1991; Kohli *et al.*, 1994). In an outbreak affecting patients at a psychiatric hospital in Scotland, and which led to a Fatal Accident Enquiry, food brought into the hospital and eaten by the index case was thought to be the source. This included meat sandwiches, cold meat and potted meat (Kohli *et al.*, 1994).

As well as food prepared in the domestic environment, food prepared in factory settings and brought into hospital can cause problems. In the UK in recent years there has been a doubling in cases of *L. monocytogenes* in the non-pregnancy-associated category (Fig. 4.1). Several clusters of cases have been detected and at least three have been linked to sandwiches prepared off the premises (Table 4.8). In an outbreak in the north east of England affecting four patients, one of whom died, raises interesting issues in terms of food safety responsibilities (Graham *et al.*, 2002). The sandwiches were prepared by an external caterer and sold through the hospital shop; the same strain of *L. monocytogenes* was isolated from two of the patients, sandwiches bought in the hospital and the sandwich manufacturing environment. However, inspection of the hospital kitchens by an infection control team accompanied by an environmental health officer (sanitarian) would not have prevented this outbreak. Hospital shops are often run by voluntary organizations and might not be regarded as food premises.

Two cases of listeriosis occurred in Wales in May 2003 amongst cancer patients attending the same oncology outpatient clinic. Epidemiological investigations implicated consumption of either ham or tuna sandwiches from a hospital canteen and there was evidence of poor temperature control at the point of sale to the patients. Food testing and environmental investigations at the factory that produced the sandwiches revealed frequent, low level contamination by *L. monocytogenes*. Isolates from the two cases were indistinguishable from isolates from 49 sandwiches and 11 environmental sites within the sandwich manufacturing environment (Gillespie *et al.*, 2005).

Four pregnancy-associated cases of listeriosis occurred in South West England in October 2003 (HPA, 2003). The cases had all eaten sandwiches from the same hospital retailer. The

Table 4.7 Outbreaks due to contaminated food prepared outside the hospital.

Country of report and year of outbreak	Causative organism	Setting	Number of cases (number at risk, where stated)	Number of deaths	Implicated food vehicle	Evidence implicating food vehicle	Reference
US, 1985	*Bacillus cereus*	Hospital cafeteria	160 (291)	–	Rice and chicken dishes	Isolation of *B. cereus* in food items	Baddour *et al.*, 1986
Scotland, 1990	STEC O157	Psychogeriatric wards of a large psychiatric hospital	11 – 8 patients and 3 staff	4 (all patients)	Food brought into hospital	Anecdotal	Kohli *et al.*, 1994
Republic of Ireland, 1996	*S.* Enteritidis phage type 4	Psychiatric hospital	65 – 36 patients, 29 staff	–	Chocolate mousse cakes made with unpasteurized eggs	Case–control study	Grein *et al.*, 1997
Scotland, 1997	STEC O157 phage type 8	Continuing care wards for the elderly	37 – 16 patients and 11 staff (106)	0	Home-baked cream-filled cakes	Cohort study	O'Brien *et al.*, 2001
Northern England, 1999	*Listeria monocyto-genes*	District general hospital	4 patients	1	Sandwiches bought in the hospital shop	Indistinguishable organism found in sandwich sample and environmental samples from caterer's premises	Graham *et al.*, 2002

–, not reported.

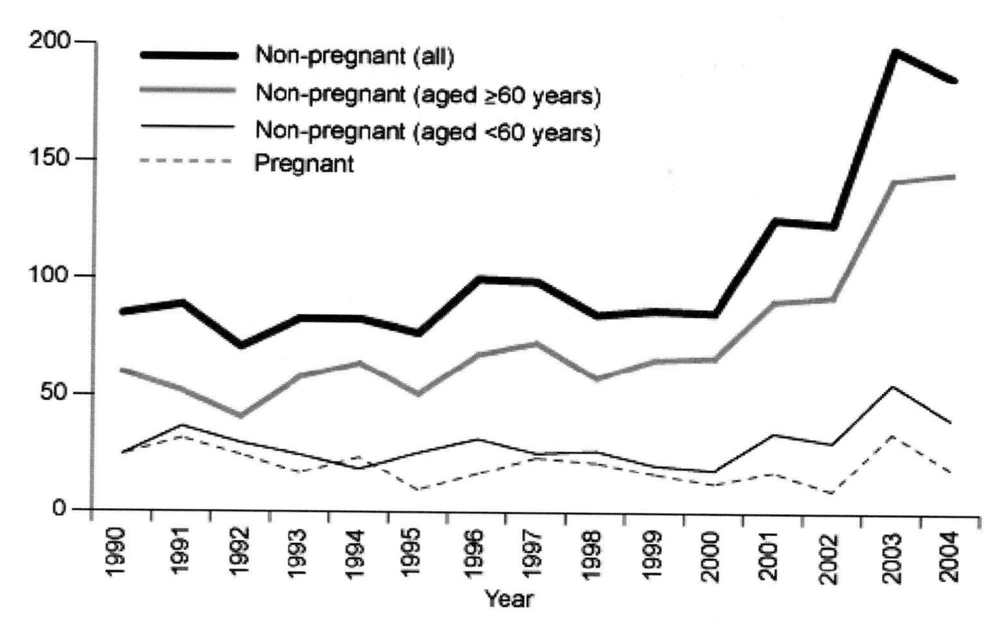

Fig. 4.1 Listeriosis in England and Wales, 1990–2004. (*Source*: Health Protection Agency Centre for Infection)

outbreak strain was subsequently recovered from the sandwich manufacturer (one sandwich and one environmental site). Two other *L. monocytogenes* strains were also recovered from either one sandwich or three other environmental sites in the factory. All sandwiches were contaminated at low levels.

The final sandwich-associated cluster occurred in two patients attending the same renal unit in South East England in September 2004. The patient isolates of *L. monocytogenes* were different types but epidemiological investigations implicated the consumption of hospital sandwiches. The organism was isolated from sandwiches from the hospital, sandwiches from the factory and from the factory environment. Out of a total of 117 isolates recovered from sandwiches and the factory, multiple *L. monocytogenes* strains were identified. Two

Table 4.8 Clusters of human listeriosis in which hospital food was implicated, England and Wales 1993–2004.

Year	Region	Number of cases (deaths)	Pregnancy associated	*L. monocytogenes* type (Serovar)	Vehicles of infection
1999	NE England	4 (1)	0	4b	Hospital sandwiches
2003	S Wales	2	0	1/2a	Hospital sandwiches
2003	SW England	4	4	1/2a	Hospital sandwiches
2004	SE England	2	0	4b	Hospital sandwiches

Source: Health Protection Agency Centre for Infections.

were indistinguishable from those recovered from blood cultures taken from one of the patients (Gillespie *et al*., 2005).

Outside caterers can cause problems as well. One hundred and sixty staff were affected by *Bacillus cereus* when outside caterers prepared meals for the hospital cafeteria (Table 4.7) (Baddour *et al*., 1986). The organism was cultured from chicken and rice dishes served at each of two implicated meals. Eighty-four members of staff were sufficiently unwell to attend the emergency room, and although 63 submitted stool specimens, none were culture-confirmed cases since none of the stools were examined for *B. cereus*. No one thought to tell laboratory staff, who did not routinely test stool specimens for this enteric pathogen, that *B. cereus* infection was suspected.

Outside catering was also responsible for a large outbreak of *S*. Enteritidis PT4 infection at a psychiatric hospital in Dublin (Grein *et al*., 1997). In a case–control study there was a 28-fold increase in risk of illness associated with eating a chocolate mousse cake dessert. The epidemic strain was isolated from cases and from a sample of leftover chocolate mousse cake. The implicated dessert had been made by a local bakery using raw shell eggs. Once delivered to the hospital the desserts were stored for a long period of time at room temperature because of inadequate cold storage space. This will have allowed growth of *S*. Enteritidis within the food.

4.3.6 *Outbreaks due to food prepared by hospital staff other than catering staff*

Introducing food from uncontrolled conditions has implications for staff as well as patients (Table 4.9). A large outbreak of *S*. Enteritidis occurred amongst hospital staff who consumed a Chinese meal prepared by a nurse at home (Metz *et al*., 2001). In this instance, no cross-infection occurred but what the outbreak illustrates is the disruption to services that can occur. In total 15 people were ill and excluded from work for the duration of their illness plus the time taken for stool cultures to become negative for *S*. Enteritidis. For most members

Table 4.9 Outbreaks due to food prepared by hospital staff other than catering staff.

Country of report and year of outbreak	Causative organism	Setting	Number of cases (number at risk, where stated)	Number of deaths	Implicated food vehicle	Evidence implicating food vehicle	Reference
England, nr	Campylobacter	Hospital	12 staff members (31)	–	Chicken dish served at a departmental party	Cohort study	Murphy *et al*., 1995
Germany, 1998	*Salmonella* Enteritidis	Surgical Unit and Anaesthetics Department	17 members of staff (21)	–	Chinese meal made at home by a member of staff	Cohort study	Metz *et al*., 2001

–, not reported.

of staff the time taken to stop excreting the organism was less than 5 days although one member of staff continued to excrete *S*. Enteritidis for an extended period. The outbreak illustrates once again that although there are stringent requirements for the production and preparation of food in hospital kitchens, bringing in food from the domestic environment bypasses all these safeguards. Educating healthcare workers of the risks inherent in food produced under uncontrolled conditions is, therefore, needed.

Reported outbreaks of campylobacter infection are notoriously rare (Pebody *et al.*, 1997). However, 12 of 31 members of staff became infected with *Campylobacter* sp. at a hospital departmental party in the north of England (Murphy *et al.*, 1995). Most of the food was prepared by members of staff (not catering staff) at home, with the remaining items being bought from local shops and prepared the night before the party. There was a statistically significant sixfold increase in risk of illness amongst those who ate chicken pieces. These had been purchased from a local retail outlet by a member of staff who failed to realize that they should have been cooked prior to being served. As a result the chicken pieces were consumed without any further heat treatment. This outbreak also served as a reminder that acute infection can be followed by chronic symptoms because one of the people affected subsequently developed erythema nodosum and reactive arthritis.

4.3.7 *Outbreaks due to contaminated enteral feeds, powdered infant formula and human breast milk*

Although fairly uncommon, foodborne disease outbreaks in hospital arising from contamination of enteral feeds, powdered infant formula or human breast milk also pose potential risks for the patients (Table 4.10). These outbreaks serve as a reminder to think widely in considering potential food vehicles/sources in any investigation of nosocomial gastroenteritis. In an outbreak of *S*. Saintpaul infection in a children's hospital in Seattle, 11 children infected with the outbreak strain were much more likely to have received infant formula mixed by the hospital than controls (Bornemann *et al.*, 2002). This outbreak reinforced the need for strict adherence to good hygiene in the preparation and delivery of formula.

Infant formula contaminated with *Citrobacter freundii* cause an outbreak on an NICU in Germany (Thurm & Gericke, 1994). Thirty-eight strains of the bacteria isolated during the course of clinical and environmental investigations were sub-typed using allozyme, whole-cell protein and resistance pattern analysis, and clinical and food isolates were found to be indistinguishable from each other. Not all the children infected displayed clinical disease.

Contaminated infant formula has also been implicated in several outbreaks of *Enterobacter sakazakii* infection worldwide. The clinical presentation in neonates in these outbreaks has mainly been of meningitis or bacteraemia and the mortality tends to be high. Contaminated dried infant formula has been implicated in outbreaks of *E. sakazakii* in the Netherlands (Muytjens *et al.*, 1983), Iceland (Biering *et al.*, 1989), the US (Simmons *et al.*, 1989; Clark *et al.*, 1990), Israel (Block *et al.*, 2002) and Belgium (van Acker *et al.*, 2001). In the latter outbreak the clinical presentation was of necrotising enterocolitis, the first time that this has been reported as the predominant clinical presentation associated with the isolation of *E. sakazakii* in patients and powdered infant formula milk.

Table 4.10 Outbreaks due to contaminated enteral feeds, powdered infant formula or human breast milk.

Country of report and year of outbreak	Causative organism	Setting	Number of cases (number at risk)	Number of deaths	Implicated food vehicle	Evidence implicating food vehicle	Reference
US, 1977	*Salmonella* Kottbus	Neonatal Intensive Care Unit (NICU)	7 (22)	–	Human breast milk	Case–control study plus organism isolated from human milk	Ryder *et al.*, 1977
The Netherlands, 1977–1981	*E. sakazakii*		8	6	Prepared infant formula	Organism isolated from prepared milk formula, dish brush and stirring spoon	Muytjens *et al.*, 1983
US, 1988	*E. sakazakii*	NICU	4 (20)	–	Infant formula	Plasmid profile and MLEE profiles of clinical and food isolates indistinguishable	Simmons *et al.*, 1989
Iceland, 1986–1987	*E. sakazakii*		3	1	Milk powder	Organism isolated at low concentration	Biering *et al.*, 1989
Germany, 1991	*Citrobacter freundii*	NICU	38	–	Infant formula	Organism isolated from milk powder	Thurm & Gericke, 1994
Belgium, 1998	*Enterobacter sakazakii*	NICU, University hospital	12 (50)	2	Powdered milk formula	*E. sakazakii* isolated from prepare formula and from several unopened cans of a single batch	Van Acker *et al.*, 2001
Brazil, 1999	*S.* Enteritidis	General hospital	8	3	Enteral feed	Diet contained lyophilized egg albumin – though to be source of organism	Matsuoka *et al.*, 2004
Israel, 1999–2000	*E. sakazakii*	University hospital	2	0	Prepared infant formula	Prepared formula and kitchen blender positive (organism not isolated from infant formula powder)	Block *et al.*, 2002
US, 2001	*S.* Saintpaul	University-affiliated children's hospital	11 (208)	–	Infant formula	Case–control study	Bornemann *et al.*, 2002
US, 2001	*E. sakazakii*	NICU	10 – 1 confirmed, 2 suspected, 7 colonized (49)	1	Commercial powdered formula	Opened and unopened formula positive – food isolates indistinguishable from clinical isolates using PFGE	CDC (2002)

–, not reported

In 2002, the Centers for Disease Control and Prevention (CDC) (2002) issued interim recommendations for preparation of powdered infant formula in the NICU settings that included:

1. Selection of formula products on the basis of nutritional needs; alternatives to powdered forms should be chosen when possible.
2. Preparation of powdered formula by trained personnel using aseptic techniques in a designated preparation room.
3. Following the manufacturer's instructions; product should be refrigerated immediately and discarded if not used within 24 hours after preparation.
4. An administration time for continuous enteral feeding not exceeding 4 hours.
5. Availability of written hospital guidelines in the event of a manufacturer product recall, including notification of healthcare providers, a system for reporting and follow-up of specific formula products used, and retention of recall records.

Proper handling and use of infant formula is an important patient safety issue (CDC, 2002). Powdered formulas are used commonly in hospitals but clinicians need to remember that they are not sterile products. This means that they may contain opportunistic bacterial pathogens, including *E. sakazakii*, thus posing a hazard in certain circumstances. The risk of infection relates to the number of organisms present in the product, unhygienic handling during and after preparation and host factors such as prematurity, low birth weight or immunosuppression.

Another hazard posed by enteral nutrition was highlighted in an outbreak amongst patients in a hospital in Brazil (Matsuoka *et al.*, 2004). Eight cases of *S.* Enteritidis diarrhoea occurred in patients aged between 17 and 79 years. All were receiving enteral nutrition at the time that they developed symptoms. *Salmonella* Enteritidis was isolated from one sample of the enteral feed and, using molecular typing methods, the organisms were found to be indistinguishable from those of seven of the patients. During the investigation it emerged that the formulation contained lyophilized egg albumin, thought by the investigators to be the probable source of *S.* Enteritidis. The severity of this outbreak was evident from the fact that two patients developed septicaemia and three patients died.

Finally, a highly unusual outbreak occurred in an NICU in the US (Ryder *et al.*, 1977). Seven of 22 neonates were infected with *S.* Kottbus and a thorough investigation implicated only consumption of human milk, from a single donor, which was subsequently found to be contaminated with *S.* Kottbus. The risks of infection associated with human breast milk are more usually considered in terms of three main viral agents, namely cytomegalovirus, Human Immunodeficiency Virus and human T cell leukaemia virus I (HTLV-I), and although maternal bacterial infections are rarely transmitted to their infants through breast milk (Lawrence & Lawrence, 2004), the possibility should not be overlooked in any investigation.

4.3.8 *Outbreaks due to infected food handlers*

Table 4.11 summarizes outbreaks in which infected food handlers have been identified as the origin of the event. Outbreaks involving a range of organisms have been described, but perhaps the most unusual is Methicillin-resistant *Staphylococcus aureus* (MRSA). Control and prevention of MRSA in hospitals in the UK is a major government priority and mandatory surveillance of MRSA bacteraemia was introduced in 2001. Regarded as a cross-infection

Table 4.11 Outbreaks caused by infected food handlers.

Country of report and year of outbreak	Causative organism	Setting	Number of cases (number at risk, where stated)	Number of deaths	Implicated food vehicle	Evidence implicating food vehicle	Reference
US, 1973	Hepatitis A	General hospital	66 staff – 44 clinical and 22 subclinical cases	–	Sandwiches	Epidemiological study	Meyers *et al.*, 1975
England	Norovirus	Two separate hospitals	?	?	Chicken sandwiches	?	Pether & Caul, 1983
US, 1983	*Shigella dysenteriae* Type 2	US Naval hospital	107 (1490)	–	Salad bar	Case–control study	CDC, 1983
US, 1987	*S.* Typhimurium	Hospital cafeteria	27 staff	–	Salad items	Cohort study	Opal *et al.*, 1989
Jordan, 1989	*S.* Enteritidis	Tertiary care university hospital	183 – 26 patients, 150 staff, 7 visitors	–	Mashed potato	Cohort study plus positive stool cultures from 11 of 61 kitchen staff	Khuri-Bulos *et al.*, 1994
England, 1991	Norovirus	Hospital for the elderly	164 – 95 patients, 69 staff	–	Sandwiches	Three separate cohort studies	Stevenson *et al.*, 1994
Wales	Norovirus	Four hospitals served by one central kitchen	195 – 81 patients, 114 staff	–	Turkey salad sandwiches, tuna salad	Cohort study amongst staff; case–control study amongst patients	Lo *et al.*, 1994
England, 1993	*S.* Enteritidis phage type 4	Two hospitals with shared catering facilities	29 – 22 patients, 7 staff	–	Meals prepared by one carrier	Case–control study	Dryden *et al.*, 1994
The Netherlands, 1992/1993	Methicillin-resistant *Staphylococcus aureus*	University hospital	21	5	Banana	Isolation of MRSA from banana and from throat swabs of a dietary worker	Kluytmans *et al.*, 1995
Wales	None confirmed but norovirus suspected	Hospital for patients with learning difficulties	80 – 47 patients and 33 staff (460)	–	Party food prepared by the hospital kitchen but supplemented by food brought in from outside	Cohort study	Fone *et al.*, 1999
England, 1994	*S.* Virchow phage type 26	Private hospital	11 – 1 patient, 9 staff, 1 food handler's child	–	Turkey sandwiches	Cohort study	Maguire *et al.*, 2000
Canada, 2002	STEC O157 phage type 32 PFGE profile (0756)	Long-term psychiatric care centre	64 – 38 employees (203), 26 patients (80)	2	Sandwiches and salads	Case–control study	Bolduc *et al.*, 2004
Spain, 2002	Norovirus	Hospital cafeteria	40 – all staff	–	Sandwiches and salads	Epidemiological study and M (RT-PCR)	Sala *et al.*, 2005

–, not reported.

issue, it is unlikely that foodborne transmission is ever considered in the investigation of upsurges of MRSA in hospitals. However, Kluytmans and colleagues (1995) provide an important lesson in 'thinking outside the box' when investigating such outbreaks. The incident started in the haematology unit of a university hospital in the Netherlands, where the index case was severely neutropenic because of treatment for leukaemia. The patient died from overwhelming sepsis. MRSA was spread from the haematology unit to a surgical unit, probably via a colonized nurse who was transferred between the two units. In total 27 patients (including the index patient) became infected or colonized with MRSA, 21 of whom developed clinical disease, and four surgical patients also died. The index case appeared to have become infected from eating banana peeled by a dietary worker who worked on the haematology unit. This was picked up by routine bacterial screening of food prepared for neutropenic patients, when MRSA was found in a sample of peeled banana. Indistinguishable organisms were cultured from the sample of peeled banana, the throat swabs from the dietary worker and samples from the haematology patient who died. This was the first report of foodborne transmission of MRSA in the literature.

Infected food handlers have been implicated as the potential source of outbreaks of norovirus in the UK (Pether & Caul, 1983; Lo *et al.*, 1994; Stevenson *et al.*, 1994) and in Spain (Sala *et al.*, 2005). In the outbreak described by Pether and Caul (1983) a food handler who was incubating the infection was thought to have been the cause. In the light of more recent data concerning possible pre-symptomatic shedding of norovirus (Parashar *et al.*, 1998) this explanation seems reasonable. This was also thought to be the reason for a large outbreak of norovirus infection linked with the consumption of hospital sandwiches in North West England, when 95 patients and 69 staff were made ill (Stevenson *et al.*, 1994). Perhaps the most compelling evidence comes from a hospital outbreak in Spain, in which 40 members of staff were affected following consumption of contaminated sandwiches and salads (Sala *et al.*, 2005). The head cook, who was the index case, continued to work after developing symptoms of diarrhoea and vomiting and was also the person who usually prepared the sandwiches and salads that were implicated as food vehicles in the outbreak.

A symptomatic food handler was found to have continued to work during the investigation of an outbreak of *S.* Typhimurium in the US (Opal *et al.*, 1989). In an epidemiological investigation eating individually wrapped salads was associated with developing illness. One of three symptomatic food handlers was found to have been working whilst suffering from acute diarrhoea and had been preparing individually wrapped salads at the time that the outbreak occurred. As a result of the outbreak the sick leave policy of the hospital at the centre of the outbreak was changed so that symptomatic staff were paid from the first day of absence because of illness (as opposed to after the first day of illness, which had previously been the case).

Asymptomatic food handlers have also been implicated in foodborne outbreaks of salmonellosis in hospitals. In an outbreak in Jordan (Khuri-Bulos *et al.*, 1994) massive contamination of mashed potato with *S.* Enteritidis was thought to have occurred during preparation by an asymptomatic member of the catering staff. During the outbreak investigation 11 of 61 kitchen staff were found to be excreting the organism, including the member of staff who mashed the potatoes.

An asymptomatic salmonella carrier was also thought to be the origin of an outbreak of *S.* Enteritidis PT4 in the south of England (Dryden *et al.*, 1994). The index case was a patient who had been in hospital for 6 weeks. In total 29 patients and staff were affected and control measure instituted early in the outbreak seemed to have little effect on its course. Cases were widely dispersed in time and place and the only feature they had in common was consumption of food from the hospital kitchen. A case–control study failed to reveal specific food vehicles and extensive sampling of raw and cooked food and the kitchen environment was all negative. However, more than 12% of food handlers were found to be excreting *S.* Enteritidis PT4. Following a detailed examination of food histories and work rotas and a re-analysis of the case–control study using these data, an association between eating food prepared by one particular stool-culture-positive chef and becoming a case of *S.* Enteritidis PT4 emerged. Coupled with evidence of imperfect hand hygiene, and the fact that the chef in question had previously been warned about touching cooked food, the investigators concluded that the chef was the source of the outbreak. Intermittent contamination of meals could explain the pattern of cases.

In a slightly more unusual situation Maguire and colleagues (2000) concluded a catering assistant might have contaminated turkey sandwiches, served to hospital patients, with *S.* Virchow PT 26. The outbreak involved two strains of *S.* Virchow PT 26, one fully sensitive to antimicrobial drugs and one resistant to sulphonamides and trimethoprim (R-type SuTm). In a cohort study there was a 13-fold increase in risk of becoming a case after consuming turkey sandwiches and a sixfold increase in risk following consumption of bacon sandwiches. A food handler who became symptomatic after the first hospital case had a child who displayed symptoms a few days before this case. However, it transpired that mother and child were infected with the two different strains of *S.* Virchow. An alternative explanation, therefore, is that the turkey used in sandwich preparation was contaminated with more than one strain of *S.* Virchow PT 26 and was thus the source of the outbreak.

In a very serious outbreak in Canada an infected food handler was thought to be the source of STEC O157 infection amongst patients and staff at a psychiatric hospital (Bolduc *et al.*, 2004). The first case in the outbreak was the individual involved in vegetable preparation in the hospital kitchen. Despite being symptomatic the individual had continued to work, though it is not clear why this was so. Subsequently 109 possible, probable and confirmed cases of STEC O157 PT 32 were identified. In a case–control study preparation of, or consumption, of salads or sandwiches were associated with becoming a case. Preparation or consumption of salads conferred a 12-fold increase in risk, whilst there was a ninefold increase in risk after preparing or eating sandwiches. The symptomatic food handler had almost exclusive use of the vegetable preparation facilities, which comprised a sink and a wooden board. The vegetables were rinsed, chopped and placed in a bowl prior to being used in salads, sandwiches and other meals. Lending further weight to the food handler as the source of this outbreak was the fact that most of the cases occurred during the time that the food handler was excreting STEC O157 and that there were no cases of STEC O157 PT 32 linked to other places provided with vegetables by the wholesaler that supplied the hospital.

In the UK, transmission of hepatitis A through food from infected food handlers incubating the illness is rare. The last foodborne hospital outbreak documented in the literature occurred in 1973 when a total of 66 hospital staff became infected after eating sandwiches (Meyers *et al.*, 1975). Transmission was thought to have occurred because the sandwiches were prepared by two subclinically infected cafeteria workers.

Table 4.12 Foodborne outbreaks in hospital nurseries or day care centres.

Country of report and year of outbreak	Causative organism	Setting	Number of cases (number at risk, where stated)	Number of deaths	Implicated food vehicle	Evidence implicating food vehicle	Reference
England, 1994	STEC O157 phage type 49	Hospital nursery	29 (31)	–	Not determined	–	Cheasty *et al.*, 1998

–, not reported.

Finally, an unusual cause of a foodborne disease outbreak in hospital is *Shigella dysenteriae*. In 1983, an outbreak of *S. dysenteriae* type 2 occurred at a US naval hospital affecting 95 clinical staff, three patients, four visitors and five food handlers (CDC, 1983). One hundred and three of 107 known symptomatic individuals and 102 controls matched by job category were interviewed, and food-specific histories were obtained. Eating food prepared in the staff cafeteria was significantly associated with illness. Patients were significantly more likely than controls to have eaten raw vegetables from the salad bar although no single salad item or dressing was specifically implicated. No samples of salad from the days in question were available for culture, and no cultures taken from available food items were positive. Five of 63 food handlers had compatible illnesses with onset at the same time as the other cases. Although no particular individual was identified, it was noted that gastrointestinal illnesses was a common cause of absence from work among food handlers during the 3 weeks leading up to the outbreak.

4.3.9 *Outbreaks in hospital nurseries or hospital-associated day-care centres*

One of the unintended consequences of providing childcare for children belonging to hospital staff is that they may be caught up in a hospital outbreak (Table 4.12). This happened in the north west of England in 1994 (Cheasty *et al.*, 1998). An outbreak of STEC O157 phage type 49 was detected when two children attending the hospital nursery developed haemolytic uraemic syndrome. A further 13 children and five members of the nursery staff were also found to have had diarrhoeal symptoms. Ten children, three nursery staff and 11 kitchen staff were found to have serum antibodies to the lipopolysaccharide of *E. coli* O157. The kitchen staff all denied symptoms. Food for the nursery was supplied by the hospital kitchen. No specific food vehicle was ever identified and, fortunately, the infection did not spread more widely in the hospital. The reason why so many kitchen staff were antibody-positive in the absence of symptoms could not be explained.

4.4 Lessons learned from foodborne disease outbreaks in hospitals

Several common themes emerge amongst the outbreaks described in this chapter and are summarized below:

4.4.1 *Hospital outbreaks are serious, causing avoidable deaths in patients*

By and large the organisms that cause foodborne disease in the community are also the organisms that cause foodborne disease in hospital. However, an irrefutable fact is that outbreaks in hospitals can have serious consequences for patients. While healthy individuals might shrug off a bout of food poisoning, which is normally self-limiting, debilitated patients may suffer much more serious sequelae. Patients may be predisposed to infection, or superinfection, through surgical interventions or medical therapy such as immunosuppressive agents, including steroids. There are many examples showing that children and the elderly tend to be the worst affected and deaths in hospital outbreaks of foodborne disease are recorded regularly. These deaths are preventable.

4.4.2 *Unusual organisms can be foodborne and affect vulnerable patients*

In certain circumstances unusual organisms like *Enterobacter sakazakii* and *Citrobacter freundii* can cause clinical disease. Infection in vulnerable neonates is an important reminder that infant formula is not sterile and extra care needs to be taken with its preparation, and in particular with its subsequent use.

4.4.3 *Hospital staff, including clinical staff, are frequently affected causing major disruption to services*

As well as having severe consequences for patients, many of the outbreaks described in this chapter have affected hospital staff. Large numbers of staff have been made unwell, leading to major problems in continuing to provide a hospital service. Ward closures, and sometimes even closure of whole hospitals, produces strains on clinical teams in the affected institution, but also in those who have to cope with extra work as a result of their neighbour's closure. Where specialist institutions are affected it can take many months for services to return to normal. The opportunity costs of resources forgone because of preventable infection are high. A bed occupied by someone (patient or staff member) affected unnecessarily by infection deprives someone else of treatment (Roberts, 2000). Thousands of working days (and bed days) are lost and costs can run into several hundreds of thousands of pounds.

4.4.4 *Food handlers can cause outbreaks, as well as be victims of them*

Received wisdom is that it is difficult to pinpoint food handlers as the cause of foodborne outbreaks since finding that food handlers are culture-positive for an outbreak strain may simply mean that they are victims of events, rather than the causes of them. Yet, there is fairly compelling evidence that the food handler is both victim and cause. So, why do food handlers who are symptomatic continue to work? One reason is almost certainly to do with remuneration. In certain parts of the world sick leave is not an entitlement. The result is that if catering staff take time off sick they do not get paid, so the incentive is to turn up for work, even though they pose an infection risk. The Department of Health (1996) issued authoritative guidelines for the food industry on food handlers' fitness to work. In addition, in 2005, a group of occupational physicians with expertise in food manufacturing and retail developed a code of practice for food handlers using a risk assessment-based approach (Smith *et al.*, 2005). There has been wide consultation on these guidelines, which present a set of practical, minimum standards for food handlers in food businesses of all sizes.

4.4.5 *Inadequate kitchen staffing levels may lead to outbreaks*

One of the recurring themes in outbreak investigations is inadequate levels of staffing. This was certainly the case at Stanley Royd where bad working practices included adding cleaning duties to those of the existing kitchen staff as part of their bonus scheme, rather than employing cleaning staff. The cleaning practices were, frankly, dreadful. For example, a shortage of cleaning cloths meant re-using dirty cloths that had been soaked in soapy water, and squeegees that had been used on the floor were also used to wipe the metal-topped tables. The kitchen was understaffed to start with and this was compounded by very high sickness–absence rates. Thus, supervisory staff spent most of their time cooking and food was prepared at times that suited the staff, rather than just prior to consumption by the patients. On top of all this, food safety training had not taken place since 1982 (Pennington, 2003).

Following an outbreak of *C. perfringens* in a long-stay hospital in London, Pollock and Whitty (1990) explored the reasons for poor food handling amongst the catering staff. They looked at staffing rotas, staff vacancies and interviewed the catering staff. They found that poor morale could be explained by a number of issues including long-term problems with recruitment leading to staff shortages. This meant that existing staff had to work double shifts and were often working for weeks at a time without a break. Low wages, lack of support from managers and the fact that the hospital had a poor workplace image all compounded problems with recruiting and retaining ancillary staff. It is, perhaps, ironic that very high standards are required amongst ancillary workers to prevent outbreaks of foodborne disease in hospitals, yet their remuneration does not necessarily reflect this.

Inadequate staffing levels, including the availability of too few trained staff, were also highlighted during the investigation of one of the largest foodborne disease outbreaks in a hospital in the UK (Thomas *et al.*, 1977). Consequently, supervision was poor, particularly at the weekends and in the early morning.

4.4.6 *High-risk foods continue to be served to high-risk patients*

Perhaps one of the most remarkable features of foodborne disease outbreaks in hospital settings is the fact that high-risk foods are served to high-risk patients. The two clearest examples of this are serving sandwiches and foods made with raw shell eggs to patients. In the outbreaks reviewed in this chapter the single most frequently identified food vehicle is the sandwich. This is not necessarily surprising. Sandwiches are handled a lot during their preparation giving ample opportunity for cross-contamination to occur. They often contain salad items, which can be contaminated at source, and which do not undergo any further cooking.

There can be no excuse for not knowing that raw shell eggs can be contaminated with *S.* Enteritidis, yet despite countless warnings and messages of advice, hospitals in the UK continue to serve foods prepared with raw or undercooked shell eggs to patients in hospital.

As recently as 2002, shell eggs imported from Spain were found during the investigation of an outbreak of *S*. Enteritidis PT6a in a hospital in London. The outbreak ceased once the use of raw shell eggs was discontinued (O'Brien & Ward, 2002). It is nearly 20 years since the Department of Health (1988) issued advice that raw shell eggs should not be used in recipes for patients in vulnerable groups. The use of pasteurized liquid egg was recommended. This advice has not been rescinded – indeed it has been re-iterated on several occasions by the UK Food Standards Agency. Although the introduction of new quality assurance programmes in UK egg production (Section 2.9.6) has substantially reduced salmonella contamination of shell eggs, even those bearing the 'Lion Mark' cannot be guaranteed to be salmonella-free. The situation is compounded by the fact that, recently, the epidemiology of *S*. Enteritidis in the UK has been driven by contamination of shell eggs bought from elsewhere within the EU (HPA, 2004). For the avoidance of doubt, Government guidance is still that raw shell eggs should not be used when catering for people in vulnerable groups, and patients in hospital fall into this category. In the Netherlands they have gone further, suggesting the introduction of legislation to prevent the sale of *Salmonella*-contaminated eggs for direct consumption (Bruins *et al*., 2003).

4.4.7 *Most outbreaks could have been avoided by adhering to good personal and food hygiene and HACCP*

Most investigators conclude that better attention to hygiene, both personal and food hygiene, and adherence to HACCP principles would have prevented the outbreaks that they have investigated. Over the years that outbreak investigations have been published the same issues arise time and again – poor hygiene, inadequate cooking, handling and storage of food and failure to ensure that the food supplied to the hospital in the first place was of appropriate quality. It is worth reflecting at this point that the outbreaks that have been published are likely to be unusual. Firstly, they are the ones in which an investigation has, by and large, been successful. There will have been many more instances where the source of a hospital outbreak has remained elusive. Secondly, they are sufficiently original to warrant inclusion in a peer-reviewed journal where new discoveries tend to be favoured over well-known problems. Thus, what appears in the peer-reviewed literature only tells part of the story (O'Brien *et al*., 2006). Developing prevention strategies depends on knowing what commonly occurs, and therefore needs to be remedied, as well as what is unusual, and surveillance of outbreaks is therefore a crucial part of intelligence gathering. A further challenge is to understand properly why food handlers make the same mistakes over and over again. More targeted training programmes might be needed.

4.5 Continuing challenges

4.5.1 *A changing environment*

Hospitals are continuing to evolve. At one time the thought that most major hospital trusts in the UK would contain shopping malls that included food outlets would have not have been contemplated. Today these are commonplace. Hospitals are using greater quantities and more varied ingredients to meet the demands placed upon them. Some, as witnessed in this chapter, use outside caterers to supplement their own production. This means the hospital catering manager needs to be able to assure him/herself that the standards that are imposed on the hospital catering side apply equally to those outside caterers supplying food for patients and staff. Anyone proposing to prepare food for hospital patients should be familiar with the NHS (2001) Code of Practice for the Manufacture, Distribution and Supply of Food Ingredients and Food-Related Products (see also Chapter 6).

The Health and Social Care Act 2003 heralds a fundamental change in the way hospital services will be delivered in the future, establishing NHS foundation trusts as independent public benefit corporations and will be modelled on co-operative and mutual traditions. The move to create foundation trusts, outside the control of the NHS, means that hospitals are likely to seek to expand services into the community. Their provision of food services might also expand.

4.5.2 *New foodborne diseases*

A new threat to public health has emerged in the shape of extended-spectrum β-lactamase producing (ESBL) Enterobacteriaceae. Since 2003, new highly resistant strains of *E. coli* have become widespread in England and parts of Northern Ireland. They are important nosocomial pathogens and *E. coli* producing a specific family of ESBL (the CTX-M enzymes) are emerging worldwide (Rodriguez-Bano *et al.*, 2006). These strains were not recorded in the UK prior to 2000 but have spread rapidly since 2003, causing urinary tract infections (UTIs) in hospital patients as well as those treated in the community. Extended-spectrum beta-lactamase resistance in *Salmonella* sp. has also been reported in isolates from Dutch poultry, poultry meat and hospitalized humans (Hasman *et al.*, 2005). Perhaps future investigations of ESBL organism outbreaks in hospitals should include consideration of foodborne transmission.

4.5.3 *An aging population*

More people are being cared for in their old age in the community as the elderly population grows and the nature of healthcare changes. Nursing homes are admitting increasingly frail individuals with ever more complex medical and nursing needs. Safe food preparation in nursing and residential homes is therefore becoming increasingly important. As if to emphasize the vulnerability of elderly people, two nursing home residents died in an outbreak of *S.* Enteritidis PT5a after an outbreak at a party in a residential home (Mansell *et al.*, 1998). Twenty-five people were ill and infection was associated with a variety of foods served at the party. The investigators' task in identifying contaminated food vehicles was complicated by poor recall among the residents, who found it difficult to complete questionnaires.

In a review of outbreaks in residential institutions in England and Wales, Ryan *et al.* (1997) reported that 58 of 282 outbreaks were foodborne. The resident populations in these institutions were the elderly (95%), people with learning difficulties (3%), children's home, a home for those with physical disabilities and a single mothers' hostel (1% each). The predominant foodborne pathogens were *Clostridium perfringens* (40%), *Salmonella* sp. (43%) and norovirus (10%). In a review of nursing home outbreaks in the US, Levine *et al.* (1991) reported a total of 115 foodborne outbreaks in a 13-year period. Nearly 5000 people were made ill and 51 died. The predominant pathogens were *Salmonella* sp. (52%) and *Staphylococcus aureus* (23%). Eighty-one per cent of deaths occurred in people with *Salmonella* sp. with *S.* Enteritidis infection being the predominant cause. Contaminated shell eggs were a leading cause of *S.* Enteritidis outbreaks during the surveillance period.

Raw shell eggs were implicated in an outbreak of *S.* Typhimurium DT135 in an elderly care home in south Australia (Tribe *et al.*, 2002). Sixteen residents and two members of staff became unwell after eating meat-based potato pie and rice pudding. The outbreak strain was recovered from samples of both dishes. It transpired that raw shell eggs had been whisked into the warm rice pudding immediately prior to serving. Raw egg had also been used in the potato topping of the pie. In a trace back exercise the outbreak strain was also recovered from samples of chicken manure at the farm that had supplied the eggs to the nursing home. Although three nursing home residents were admitted to hospital, fortunately no one died.

Finally, Smerdon and colleagues (2001) noted, in their review of outbreaks linked to red meat in England and Wales, that a large proportion of *C. perfringens* and *Salmonella* sp. outbreaks occurred in residential care homes for the elderly. Good food hygiene, therefore, continues to be essential.

4.6 Conclusion

Foodborne disease outbreaks in hospitals are both avoidable and costly. A hazard analysis-based approach to food safety management and adherence to the '4Cs' – avoiding cross-contamination, ensuring cleanliness, chilling food properly and cooking food properly should all help to prevent foodborne outbreaks. After all, patients are especially vulnerable and some pay the ultimate price – with their lives!

References

Abbott, J.D., Hepner, E.D. & Clifford, C. (1980) Salmonella infections in hospitals. A report from the Public Health Laboratory Service Salmonella Subcommittee. *Journal of Hospital Infection* **1**, 307–314.

Acheson, D. (1988) *Public Health in England. The Report of the Committee of Inquiry into the Future Development of the Public Health Function.* HMSO, London (Cm 289)

Audit Commission (2001) *Acute Hospital Portfolio: Review of National Findings – Catering. London: Audit Commission Publications.* Available from http://www.healthcarecommission.org.uk/_db/_documents/04000237.pdf. Accessed 2 March 2007.

Baddour, L.M., Gaia, S.M., Griffin, R. & Hudson, R. (1986) A hospital cafeteria-related food-borne outbreak due to *Bacillus cereus*: unique features. *Infection Control* **7**, 462–465.

Biering, G., Karlsson, S., Clark, N.C. *et al.* (1989) Three cases of neonatal meningitis caused by *Enterobacter sakazakii* in powdered milk. *Journal of Clinical Microbiology* **27**, 2054–2056.

Block, C., Peleg, O., Minster, N. *et al.* (2002) Cluster of neonatal infections in Jerusalem due to unusual biochemical variant of *Enterobacter sakazakii. European Journal of Clinical Microbiology and Infectious Disease* **21**, 613–616.

Bolduc, D., Srour, L.F., Sweet, L. *et al.* (2004) Severe outbreak of *Escherichia coli* O157:H7 in health care institutions in Charlottetown, Prince Edward Island, Fall 2002. *Canada Communicable Disease Report* **30**, 81–88.

Bonner, C., Foley, B., Wall, P. & Fitzgerald, M. (2001) Analysis of outbreaks of infectious intestinal disease in Ireland: 1998 and 1999. *Irish Medical Journal* **94**, 140 and 142–144.

Bornemann, R., Zerr, D.M., Heath, J. *et al.* (2002) An outbreak of *Salmonella* serotype Saintpaul in a children's hospital. *Infection Control and Hospital Epidemiology* **23**, 671–676.

Bruins, M.J., Fernandes, T.M.A., Ruijs, G.J.H.M. *et al.* (2003) Detection of a nosocomial outbreak of salmonellosis may be delayed by application of a protocol for rejection of stool cultures. *Journal of Hospital Infection* **54**, 93–98.

Carter, A.O., Borczyk, A.A., Carlson, J.A. *et al.* (1987) A severe outbreak of *Escherichia coli* O157:H7-associated hemorrhagic colitis in a nursing home. *New England Journal of Medicine* **317**, 1496–1500.

Centers for Disease Control and Prevention (CDC) (1983) Hospital-associated outbreak of *Shigella dysenteriae* type 2 – Maryland. *Morbidity and Mortality Weekly Report* **32**(19), 250–252.

Centers for Disease Control and Prevention (CDC) (2002) *Enterobacter sakazakii* infections associated with the use of powdered infant formula – Tennessee, 2001. *Morbidity and Mortality Weekly Report* **51**(14), 298–300.

Cheasty, T., Robertson, R., Chart, H. *et al.* (1998) The use of serodiagnosis in the retrospective investigation of a nursery outbreak associated with *Escherichia coli* O157:H7. *Journal of Clinical Pathology* **51**, 498–501.

Clark, N.C., Hill, B.C. & O'Hara, C.M. (1990) Epidemiologic typing of *Enterobacter sakazakii* in two neonatal nosocomial outbreaks. *Diagnostic Microbiology and Infectious Disease* **13**, 467–472.

Dalton, C.B., Gregory, J. & Kirk, M.D. (2004) Foodborne disease outbreaks in Australia, 1995–2000. *Communicable Disease Intelligence* **28**, 211–224.

Department of Health (1988) *Raw Shell Eggs*. Department of Health, London (EL/88/136).

Department of Health (1995). *Food Handlers: Fitness to Work: Guidelines for Food Businesses, Enforcement Officers and Health Professionals.* Department of Health, London.

Department of Health and Social Security (DHSS) (1986) *The Report of the Committee of Inquiry into an Outbreak of Food Poisoning at Stanley Royd Hospital.* HMSO, London (Cmnd 9716).

Dryden, M.S., Keyworthm, N., Gabb, R. & Stein, K. (1994) Asymptomatic foodhandlers as the source of nosocomial salmonellosis. *Journal of Hospital Infection* **28**, 195–208.

Evans, M.R., Hutchings, P.G., Ribeiro, C.D. & Westmoreland, D. (1996) A hospital outbreak of salmonella food poisoning due to inadequate deep-fat frying. *Epidemiology and Infection* **116**, 155–160.

Fone, D.L., Lane, W. & Salmon, R.L. (1999) Investigation of an outbreak of gastroenteritis at a hospital for patients with learning difficulties. *Communicable Disease and Public Health* **2**, 35–38.

Gellert, G.A., Tormey, M., Rodriguez, G. *et al.* (1989) Food-borne disease in hospitals: prevention in a changing food service environment. *American Journal of Infection Control* **17**, 136–140.

Giannella, R.A. & Brasile, L. (1979) A hospital food-borne outbreak of diarrhoea caused by *Bacillus cereus*: clinical, epidemiologic and microbiologic studies. *Journal of Infectious Diseases* **139**, 366–370.

Gillespie, I., McLauchlin, J., Adak, B. *et al.* (2005) *Changing Pattern of Human Listeriosis in England and Wales, 1993–2004.* Health Protection Agency, London. Available from http://www.food.gov.uk/multimedia/pdfs/acm753.pdf. Accessed 2 March 2007.

Girish, R., Broor, S., Dar, L. & Ghosh, D. (2002) Foodborne outbreak caused by a Norwalk-like virus in India. *Journal of Medical Virology* **6**, 603–607.

Graham, J.C., Lanser, S., Bignardi, G. *et al.* (2002) Hospital-acquired Listeriosis. *Journal of Hospital Infection* **51**, 136–139.

Grein, T., O'Flanagan, D., McCarthy, T. & Prendergast, T. (1997) An outbreak of *Salmonella enteritidis* food poisoning in a psychiatric hospital in Dublin, Ireland. *Eurosurveillance Monthly* **2**(11), 84–86.

Guallar, C., Ariza, J., Dominguez, M.A. *et al.* (2004) An insidious nosocomial outbreak due to *Salmonella enteritidis*. *Infection Control and Hospital Epidemiology* **25**, 10–15.

Haeghebaert, S., Duche, L., Gilles, C. *et al.* (2001) Minced beef and human salmonellosis: review of the investigation of three outbreaks in France. *Eurosurveillance Monthly* **6**(2), 21–26.

Hamada, K. & Tsuji, H. (2001) *Salmonella brandenburg* and *S. corvallis* involved in a food poisoning outbreak in a hospital in Hyogo prefecture. *Japanese Journal of Infectious Diseases* **54**, 195–196.

Hasman, H., Mevius, D., Veldman, K. *et al.* (2005) β-Lactamases among extended-spectrum β-lactamase (ESBL)-resistant Salmonella from poultry, poultry products and human patients in The Netherlands. *Journal of Antimicrobial Chemotherapy* **56**, 115–121.

Health Protection Agency (HPA) (2003) *Cluster of Pregnancy Associated Listeria Cases in the Swindon Area. Communicable Disease Report CDR Weekly* **13**(50). Available from http://www.hpa.org.uk/cdr/archives/2003/cdr5003.pdf. Accessed 2 March 2007.

Health Protection Agency (HPA) (2004) *Salmonella* Enteritidis non phage-type 4 infections in England and Wales: 200 to 2004 – report from a multi-agency national outbreak control team. *Communicable Disease Report CDR Weekly* **14**(42): 14102004. Available from http://www.hpa.org.uk/cdr/archives/2004/cdr4204.pdf. Accessed 2 March 2007.

Ho, J.L., Shands, K.N. & Friedland, G. (1986) An outbreak of type 4b *Listeria monocytogenes* infection involving patients from eight Boston hospitals. *Archives of Internal Medicine* **146**, 520–524.

Johnson, J.R., Clabots, C., Azar, M. *et al.* (2001) Molecular analysis of a hospital cafeteria-associated Salmonellosis outbreak using modified repetitive element PCR fingerprinting. *Journal of Clinical Microbiology* **39**, 3452–3460.

Joseph, C.A. & Palmer, S.R. (1989) Outbreaks of Salmonella infection in hospitals in England and Wales 1978–87. *British Medical Journal* **298**(661), 1161–1164.

Khatib, R., Naber, M., Shellum, N. *et al.* (1994) A common source outbreak of gastroenteritis in a teaching hospital. *Infection Control and Hospital Epidemiology* **15**, 534–535.

Khuri-Bulos, N.A., Abu Khalaf, M., Shehabi, A. & Shami, K. (1994) Foodhandler-associated Salmonella outbreak in a university hospital despite routine surveillance cultures of kitchen employees. *Infection Control and Hospital Epidemio*logy **15**, 311–314.

Kistemann, T., Dangendorf, F., Krizek, L. *et al.* (2000) GIS-supported investigation of a nosocomial *Salmonella* outbreak. *International Journal of Hygiene and Environmental Health* **203**, 117–126.

Kluytmans, J., Van Leeuwen, W., Goessens, W. *et al.* (1995) Food-initiated outbreak of methicillin-resistant *Staphylococcus aureus* analyzed by pheno- and genotyping. *Journal of Clinical Microbiology* **33**, 1121–1128.

Kohli, H.S., Chaudhuri, A.K. & Todd, W.T. (1994) A severe outbreak of *E. coli* 0157 in two psychogeriatric wards. *Journal of Public Health Medicine* **16**, 11–15.

Kumarasinghe, G., Hamilton, W.J., Gould, J.D. *et al.* (1982) An outbreak of *Salmonella muenchen* infection in a specialist paediatric hospital. *Journal of Hospital Infection* **3**, 341–344.

Lawrence, R.M. & Lawrence, R.A. (2004) Breast milk and infection. *Clinical Perinatology* **31**, 501–528.

L'Ecuyer, P.B., Diego, J., Murphy, D. *et al.* (1996) Nosocomial outbreak of gastroenteritis due to *Salmonella senftenberg*. *Clinical Infectious Diseases* **23**, 734–742.

Levine, W.C., Bennett, R.W., Choi, Y. *et al.* (1996) Staphylococcal food poisoning caused by imported canned mushrooms. *Journal of Infectious Diseases* **173**, 1263–1267.

Levine, W.C., Smart, J.F., Archer, D.L. *et al.* (1991) Foodborne disease outbreaks in nursing homes, 1975 through 1987. *Journal of the American Medical Association* **266**, 2105–2109.

Lo, S.V., Connolly, A.M., Palmer, S.R. *et al.* (1994) The role of the pre-symptomatic food handler in a common source outbreak of food-borne SRSV gastroenteritis in a group of hospitals. *Epidemiology and Infection* **113**, 513–521.

Lyytikäinen, O., Autio, R., Maijala, R. *et al.* (2000) An outbreak of *Listeria monocytogenes* serotype 3a infections from butter in Finland. *Journal of Infectious Diseases* **181**, 1838–1841.

Maguire, H., Pharoah, P., Walsh, B. *et al.* (2000) Hospital outbreak of *Salmonella virchow* possibly associated with a food handler. *Journal of Hospital Infection* **44**, 261–266.

Maijala, R., Lyytikainen, O., Autio, T. *et al.* (2001) Exposure of *Listeria monocytogenes* within an epidemic caused by butter in Finland. *International Journal of Food Microbiology* **70**, 97–109.

Mansell, A.L., Sen, S., Sufi, F. & McCallum, A. (1998) An outbreak of *Salmonella enteritidis* phage type 5a infection in a residential home for elderly people. *Communicable Disease and Public Health* **1**, 172–175.

Mason, B.W., Williams, N., Salmon, R.L. *et al.* (2001) Outbreak of *Salmonella* Indiana associated with egg mayonnaise sandwiches at an acute NHS hospital. *Communicable Disease and Public Health* **4**, 300–304.

Matsuoka, D.M., Costa, S.F., Mangini, C. *et al.* (2004) A nosocomial outbreak of *Salmonella enteritidis* associated with lyophilized enteral nutrition. *Journal of Hospital Infection* **58**, 122–127.

McCall, B., McCormack, J.C., Stafford, R. & Towner, C. (1999) An outbreak of *Salmonella typimurium* at a Teaching hospital. *Infection Control and Hospital Epidemiology* **20**, 55–56.

Meakins, S.M., Adak, G.K., Lopman, B.A. & O'Brien, S.J. (2003) General outbreaks of infectious intestinal disease (IID) in hospitals, England and Wales, 1992–2000. *Journal of Hospital Infection* **53**, 1–5.

Metz, R., Jahn, B., Kohnen, W. *et al.* (2001) Outbreak of *Salmonella enteritidis* gastrointestinal infections among medical staff due to contaminated food prepared outside the hospital. *Journal of Hospital Infection* **48**, 324–325.

Meyers, J.D., Romm, F.J., Tihen, W.S. & Bryan, J.A. (1975) Food-borne hepatitis A in a general hospital. Epidemiologic study of an outbreak attributed to sandwiches. *Journal of the American Medical Association* **231**, 1049–1053.

Molina-Gamboa, J.D., Ponce-de-Rosales, S., Guerrero-Almeida, M.L. *et al.* (1997) Salmonella gastroenteritis outbreak among workers from a tertiary care hospital in Mexico City. *Revista de Investigacion Clinica* **49**, 349–353.

Murphy, O., Gray, J., Gordon, S. & Bint, A.J. (1995) An outbreak of Campylobacter food poisoning in a health care setting. *Journal of Hospital Infection* **30**, 225–228.

Muytjens, H.L., Zanen, H.C., Sonderkamp, H.J. *et al.* (1983) Analysis of eight cases of neonatal meningitis and sepsis due to *Enterobacter sakazakii*. *Journal of Clinical Microbiology* **18**, 115–120.

National Health Service (NHS) (2001) Code of Practice for the manufacture, distribution and supply of food ingredients and food-related products. Available from http://www.pasa.doh.gov.uk/food/docs/code_of_practice_2001.pdf . Accessed 15 June 2007.

O'Brien, S. & Ward, L. (2002) Nosocomial outbreak of *Salmonella* Enteritidis PT 6a (Nx, Cp_L) in the United Kingdom. *Eurosurveillance Weekly* **6**(43), 021024. Available from http://www.eurosurveillance.org/ew/2002/021024.asp#6. Accessed 2 March 2007.

O'Brien, S.J., Gillespie, I.A., Sivanesan, M.A. *et al.* (2006) Publication bias in foodborne outbreaks of infectious intestinal disease and its implications for evidence-based food policy. England and Wales 1992–2003. *Epidemiology and Infection* **134**, 667–674.

O'Brien, S.J., Murdoch, P.S., Riley, A.H. *et al.* (2001) A foodborne outbreak of Vero cytotoxin-producing *Escherichia coli* 0157:H- phage type 8 in hospital. *Journal of Hospital Infection* **49**, 167–172.

Opal, S.M., Mayer, K.H., Roland, F. *et al.* (1989) Investigation of a food-borne outbreak of salmonellosis among hospital employees. *American Journal of Infection Control* **17**, 141–147.

Palmer, S.R. & Rowe, B. (1983) Investigation of outbreaks of salmonella in hospitals. *British Medical Journal* **287**, 891–893.

Parashar, U.D., Dow, L., Fankhauser, R.L. *et al.* (1998) An outbreak of viral gastroenteritis associated with consumption of sandwiches: implications for the control of transmission by food handlers. *Epidemiology and Infection* **121**, 615–621.

Pebody, R.G., Ryan, M.J. & Wall, P.G. (1997) Outbreaks of campylobacter infection: rare events for a common pathogen. *Communicable Disease Report CDR Review* **7**(3), R33–R37.

Pennington, T.H. (2003) *When Food Kills*. Oxford University Press, Oxford.

Pether, J.V. & Caul, E.O. (1983) An outbreak of food-borne gastroenteritis in two hospitals associated with a Norwalk-like virus. *Journal of Hygiene (Lond)* **91**, 343–350.

Pether, J.V. & Scott, R.J. (1982) Salmonella carriers; are they dangerous? A study to identify finger contamination with Salmonellae by convalescent carriers. *Journal of Infection* **5**, 81–88.

Pollock, A.M. & Whitty, P.M. (1990) Crisis in our hospital kitchens: ancillary staffing levels during an outbreak of food poisoning in a long stay hospital. *British Medical Journal* **300**(6721), 383–385.

Preston, M., Borcyzk, A., Davidson, R. *et al.* (1997) Hospital-associated outbreak of *Escherichia coli* O157:H7 associated with a rare phage type – Ontario. *Canada Communicable Disease Report*, **23-05**, 33–37.

Regan, C.M., Syed, Q. & Tunstall, P.J. (1995) A hospital outbreak of *Clostridium perfringens* food-poisoning – implications for food hygiene review in hospitals. *Journal of Hospital Infection* **29**, 69–73.

Roberts, J.A. (2000) Economic aspects of food-borne outbreaks and their control. *British Medical Bulletin* **56**, 133–141.

Rodriguez-Bano, J., Navarro, M.D., Romero, L. *et al.* (2006) Clinical and molecular epidemiology of extended-spectrum beta-lactamase-producing *Escherichia coli* as a cause of nosocomial infection or colonization: implications for control. *Clinical Infectious Diseases* **42**, 37–45.

Ryan, M.J., Wall, P.G., Adak, G.K. *et al.* (1997) Outbreaks of infectious intestinal disease in residential institutions in England and Wales 1992–1994. *Journal of Infection* **34**, 49–54.

Ryder, R.W., Crosby-Ritchie, A., McDonough, B. & Hall, W.J., III (1977) Human milk contaminated with *Salmonella kottbus*. A cause of nosocomial illness in infants. *Journal of the American Medical Association* **238**, 1533–1534.

Sala, M.R., Cardenosa, N., Arias, C. *et al.* (2005) An outbreak of food poisoning due to a genogroup I norovirus. *Epidemiology and Infection* **133**, 187–191.

Sharp, J.C., Collier, P.W. & Gilbert, R.J. (1979) Food poisoning in hospitals in Scotland. *Journal of Hygiene (Lond)* **83**, 231–236.

Simmons, B.P., Gelfand, M.S., Haas, M. *et al.* (1989) *Enterobacter sakazakii* infections in neonates associated with intrinsic contamination of a powdered infant formula. *Infection Control and Hospital Epidemiology* **10**, 398–401.

Skibsted, U., Baggesen, D.L., Dessau, R. & Lisby, G. (1998) Randon amplification of polymorphic DNA (RAPD), pulsed-field gel electrophoresis (PFGE) and phage-typing in the analysis of a hospital outbreak of *Salmonella enteritidis*. *Journal of Hospital Infection* **38**, 207–216.

Smerdon, W.J., Adak, G.K., O'Brien, S.J. *et al.* (2001) General outbreaks of infectious intestinal disease linked with red meat, England and Wales, 1992–1999. *Communicable Disease and Public Health* **4**, 259–267.

Smith, T.A., Kanas, R.P., McCoubrey, I.A. & Belton, M.E. (2005) Code of practice for food handler activities. *Occupational Medicine (Lond)* **55**, 369–370.

Spearing, N.M., Jenen, A., McCall, B.J. *et al.* (2000) Direct costs associated with a nosocomial outbreak of *Salmonella* infection: an ounce of prevention is worth a pound of cure. *American Journal of Infection Control* **28**, 54–57.

Spitalny, K.C., Okowitz, E.N. & Vogt, R.L. (1984) Salmonellosis outbreak at a Vermont hospital. *Southern Medical Journal* **77**, 168–172.

Steere, A.C., Hall, W.J., III, Wells, J.G. *et al.* (1975) Person-to-person spread of *Salmonella typhimurium* after a hospital common-source outbreak. *Lancet* **8**, 319–322.

Stevenson, P., McCann, R., Duthie, R. *et al.* (1994) A hospital outbreak due to Norwalk virus. *Journal of Hospital Infection* **26**, 261–272.

Stone, A., Shaffer, M. & Sautter, R.L. (1993) *Salmonella* Poona infection and surveillance in a neonatal nursery. *American Journal of Infection Control* **21**, 270–273.

Telzak, E.E., Budnick, L.D., Greenbeerg, M.S. *et al.* (1990) A nosocomial outbreak of *Salmonella enteritidis* infection due to the consumption of raw eggs. *New England Journal of Medicine* **9**, 394–397.

Thomas, M., Noah, N.D., Male, G.E. *et al.* (1977) Hospital outbreak of *Clostridium perfringens* food-poisoning. *Lancet* **309** (8020), 1046–1048.

Thurm, V. & Gericke, B. (1994) Identification of infant food as a vehicle in a nosocomial outbreak of *Citrobacter freundii*: epidemiological subtyping by allozyme, whole-cell protein and antibiotic resistance. *Journal of Applied Bacteriology* **76**, 553–558.

Tribe, I.G., Cowell, D., Cameron, P. & Cameron, S. (2002) An outbreak of *Salmonella* Typhimurium phage type 135 infection linked to the consumption of raw shell eggs in an aged care facility. *Communicable Disease Intelligence* **26**, 38–39.

Van Acker, J., De Smet, F., Muyldermans, G. *et al.* (2001) Outbreak of necrotizing enterocolitis associated with *Enterobacter sakazakii* in powdered milk formula. *Journal of Clinical Microbiology* **39**, 293–297.

Wachtel, M.R., Whitehead, L.C. & Mandrell, R.E. (2002) Association of *Escherichia coli* O157:H7 with preharvest leaf lettuce upon exposure to contaminated irrigation water. *Journal of Food Protection* **65**, 18–25.

Wall, P.G., Ryan, M.J., Ward, L.R. & Rowe, B. (1996) Outbreaks of salmonellosis in hospitals in England and Wales: 1992–1994. *Journal of Hospital Infection* **33**, 181–190.

Welinder-Olsson, C., Stenqvist, K., Badenfors, M. *et al.* (2004) EHEC outbreak among staff at a children's hospital – use of PCR for verocytotoxin detection and PFGE for epidemiological investigation. *Epidemiology and Infection* **132**, 43–49.

White, P.M. (1986) Food poisoning in a hospital staff canteen. *Journal of Infection* **13**, 195–198.

Yule, B.F., Macleod, A.F., Sharp, J.C. & Forbes, G.I. (1988) Costing of a hospital-based outbreak of poultry-borne salmonellosis. *Epidemiology and Infection* **100**, 35–42.

5 Vulnerable Populations and their Susceptibility to Foodborne Disease

David W.K. Acheson and Lisa F. Lubin

5.1 Introduction

The general consequences of foodborne illness have been described in other chapters. In a healthy adult, the symptoms typically resolve spontaneously without the need for significant medical intervention and without long-term consequence. However, certain subsets of the population are at a greater risk of acquiring foodborne infections and have a greater propensity to develop serious complications.

One of the most difficult elements is actually defining 'vulnerable populations', and in subsequent sections of this chapter we will address how one may approach this, in view of the fact that there are varying degrees of vulnerability to the same pathogen and there may be much greater vulnerability to a given pathogen in one vulnerable population than in another vulnerable population. The degree to which an individual is considered vulnerable and the pathogens from which that person has increased risk will vary depending on a wide range of factors. Some of these factors relate to the properties of the pathogen in question, others relate to the immune status of the individual concerned, and yet others are determined by how the food is stored and prepared. Some of the physiological defences against foodborne pathogens will also be addressed.

Finally, there is also a critical need to communicate the most appropriate risk-based information to the patient. Such information may provide the patient with specific advice regarding certain types of food to avoid, or offer recommendations on food preparation and storage that will reduce the likelihood of the food containing harmful pathogens. The final portion of this chapter will offer some insight into these types of questions.

5.2 Physiological defences

Susceptibility to foodborne infection is dependent on a variety of factors. These factors largely relate to the status of an individual's 'defence systems' in regard to both preventing and mitigating foodborne illness. The 'defence system' has a variety of components that serve to safeguard the human host from outside attack, and there are a number of barriers present in the normal host that offer protection from infection with orally acquired agents. These barriers include the physical barriers and components of the gastrointestinal tract that will either block or destroy foodborne agents (Table 5.1). In addition, the host's immune system has the ability to incapacitate, and protect the host from, infective agents. In the context of foodborne pathogens, many of those defence systems pertain to the gastrointestinal tract. Table 5.1 summarizes many of the components that are part of the multifaceted and complex nature of the mucosal immune response.

5.2.1 *The intestinal mucosal barrier*

The intestinal mucosal barrier has an enormous surface area, approximately 400 m^2, and comprises a variety of cellular and non-cellular elements (Maury *et al.*, 1995). Structurally, the barrier is formed by an epithelial cell lining with a complex array of agents on its luminal surface, and by organized lymphoid tissues designed to assist in the protective function against harmful foreign antigens (Table 5.1). Although the epithelial barrier consists of only a single layer of columnar cells, it serves the balanced function of providing a physical deterrent as well as providing the portal for uptake of important nutrients. Epithelial cells in the intestine have microvilli on their apical surfaces with a filamentous brush border glycocalyx at the tips (Maury *et al.*, 1995). This cellular anatomy helps to prevent penetration by foreign antigens, and these cells simultaneously express Major Histocompatability Complex Class II receptors (MHC II) to facilitate antigen presentation to immune cells as needed. Moreover, cells of this villous epithelium produce a variety of functional molecules, such as defensins (Ayabe *et al.*, 2000), trefoil factors, and mucins, which help to further protect the human host (Kindon *et al.*, 1995). Mucins are the principal components of mucus, which lines the surface epithelium throughout the intestinal tract (Lamont, 1992). Microbes of all types (i.e. bacteria, viruses and protozoa) become trapped in the mucus layer and are expelled from the intestine by peristalsis. Proteolytic enzymes are another important component of mucus. Proteolytic enzymes not only facilitate digestion of polypeptides, but also alter/diminish the immunogenic properties of these peptides; small peptides (less than 8–10 amino acids) are poor immunogens (Mayer, 2003). Defensins fall into two major groups (alpha and beta). Alpha defensins have both antimicrobial and antiviral properties as well as numerous other properties such as the regulation of complement activation, degranulation of mast cells and certain immunoadjuvant effects in animal models. Beta defensins are also antibacterial and induce a number of other immune modulator functions (Oppenheim

Table 5.1 A summary of the various components that are important in the make up of the gastrointestinal mucosal barrier.

Mucosal defensive factor	Function of the defensive factor
Epithelium: glycocalyx, villi	Innate immune response Antigen presentation Block penetration of ingested antigens
Defensins	Antimicrobial peptides
Trefoil factors (3)	Protection from a variety of deleterious agents (bacterial toxins, chemicals and drugs); provide restitution after mucosal injury
Mucus/mucins	Block penetration of ingested antigens
Proteases: pepsins, pancreatic enzymes	Breakdown of ingested antigens
Secretory-IgA (s-IgA)	Block adhesion to epithelial surface by pathogens
Gastric acid (pH)	Breakdown of ingested antigens
Bile acids	Breakdown of ingested antigens
Intestinal peristalsis	Block penetration of ingested antigens
Indigenous microflora (IMF) (Rolfe, 1997)	Competitive inhibition: *Direct*: competition for essential nutrients and bacterial receptor sites; creation of restrictive physiologic environments; secretion of antibiotic-like substances *Indirect*: chemical modification of bile salts and dietary fats, induction of protective Ig responses, stimulation of peristalsis
GALT-associated	
IgA, *IgG, *IgM (serum)	Clear antigens penetrating GI barrier/systemic immunity Assist in opsonization and phagocytosis of antigens *Complement activation
Lymphoid follicles Lamina propria	Clear antigens penetrating GI barrier
(LP) Intraepithelial cells (IEL)	Innate and acquired immune responses
Mesenteric lymph nodes (MLN)	Phagocytosis and antigen presentation

Modified from Acheson & Luccioli, 2004.

et al., 2003). The trefoil factor (TFF) family comprises the gastric peptides pS2/TFF1, spasmolytic peptide (SP)/TFF2 and the intestinal trefoil factor (ITF)/TFF3. Their fundamental action is to promote epithelial-cell restitution within the gastrointestinal tract. TFFs are secreted abundantly onto the mucosal surface by mucus-secreting cells and their expression is rapidly and coordinately upregulated at the margins of mucosal injury. TTFs then serve to augment barrier function or may act intracellularly effecting transcriptional and signaling events (Taupin & Podolsky, 2003).

Interspersed with the villous epithelial monolayer is the follicle-associated epithelium (FAE), which overlies a vast network of organized mucosa-associated lymphoid tissue (MALT) (Kraehenbuhl & Neutra, 2000). This specialized epithelium is distributed throughout the intestinal tract as part of the gut-associated lymphoid tissue (GALT) and is found in a more organized fashion in areas reflecting a high presence of foreign materials and microorganisms, such as the Peyer's patches in the distal small intestine, the palatine tonsils and the pharyngeal mucosa of Waldeyer's ring and the appendix (Kraehenbuhl & Neutra, 2000). The GALT also contains important regulatory cells of the mucosal immune system such as lymphocytes and phagocytes. Lymphocytes organize and mount rapid, selective and potent immune responses against harmful foreign pathogens, and phagocytes play a role in the sampling, presentation and destruction of pathogens. The anatomy of this network facilitates a close association between these cells and antigens or microbes present in the lumen (Kraehenbuhl & Neutra, 2000; Mowat, 2003). These interactions are also important in differentiating between harmful or beneficial food antigens and/or pathogens through a process involving oral tolerance (Mayer, 2003; Mowat, 2003).

The GALT is comprised of four distinct lymphoid compartments: Peyer's patches and other lymphoid follicles associated with the FAE; lamina propria (LP); intraepithelial lymphocytes (IELs); and mesenteric lymph nodes (MLN) (Cazac & Roes, 2000). Lymphoid follicles are characterized by aggregates of immature B cells and CD4 (+) helper T cells sitting within specific pockets of the M cell (Kraehenbuhl & Neutra, 2000) and resemble lymph nodes without the afferent lymphatics. Thus, these lymphoid aggregates come into contact solely with antigens from the gut lumen and serve as inductive sites for intestinal immune responses (Brandtzaeg, 2002). Another unique feature of these lymphoid tissues is the propensity for IgA production. The production of immunoglobulin typically requires T cell help, and specific interactions between the CD4 (+) T cells and dendritic cells (DCs) found within the dome surrounding these follicles lead to secretion of transforming growth factor-β (TGF-β) which favours B cell class switching to IgA (Cazac & Roes, 2000). B cells then migrate out of the follicles into the surrounding mucosa and release secretory IgA (s-IgA) into the gut lumen, where it functions mainly to bind pathogens and prevent their attachment to epithelial cells (Gaskins, 1997). IgA may also remove antigens from the subepithelial space by binding free antigens while it is being transported from the basolateral surface to the apical epithelial cell surface and secreted into the gut lumen (Robinson *et al.*, 2001). In these capacities, IgA has a role in the tolerance function of the gut as well (Mowat, 2003).

The LP represents the basement membrane layer of cells residing below the FAE and around lymphoid follicles. The LP contains an enormous number of terminally differentiated B cells, probably more than any other organ in the body, as well as T cells, DCs, macrophages, mast cells and polymorphonuclear leukocytes (Brandtzaeg, 2002). Although antigen presentation also occurs, this compartment provides important effector sites for preventing the entry and systemic spread of pathogens across the intestinal cell barrier and for destroying pathogens when invasion occurs. Also, in cooperation with the underlying MLN and vascular endothelium, they serve as important sites for cellular expansion and differentiation within GALT (Brandtzaeg, 2002).

The LP lymphocytes are largely IgA secreting plasma cells of the B cell lineage and memory T effector cells. T cell effector functions include two types of helper T cells (T_H), mostly of the CD4 (+) lineage, based on their pattern of cytokine secretion. This T cell population also includes a significant percentage of CD8 (+) cells which have cytotoxic functions, but which may also elaborate effector responses. Macrophages, DCs and epithelial cells sample and engulf antigens to present to T cells (Ayabe *et al.*, 2000). The differentiation into cytokine-producing cells is dependent on specific cellular interactions with antigen and on the cytokine milieu (Paul & Seder, 1994). T_H1 cell responses occur in environments where IL-12 is released by these antigen-presenting cells (APCs). IL-12 is a known inducer of natural killer (NK) cell cytotoxic function (Shiloh & Nathan, 2000) and induces T cells to produce interferon gamma (IFN-γ). IFN-γ is important for cell-mediated responses and for killing of intracellular pathogens by phagocytes. In particular, IFN-γ upregulates production within macrophages of reactive oxygen intermediates (ROI) and reactive nitrogen intermediates (RNI) which have important antimicrobial properties (Shiloh & Nathan, 2000). Other important cytokines in this cascade are IL-1 and TNF-α, which activate macrophages, and IL-8, which recruits other phagocytes such as neutrophils. T_H2 cells traditionally respond to multicellular pathogens and/or are produced abnormally in response to exogenous environmental proteins. They secrete a cytokine repertoire of IL-4, IL-5 and IL-13 to induce B cell activation and differentiation and recruitment of eosinophils and mast cells to tissues (Paul & Seder, 1994). A third effector cell type (T_H3), referred to as T regulatory cells, has also been identified and responds to immunosuppressive cytokines in the monitoring of oral tolerance (McGuirk & Mills, 2002; Mayer, 2003; Mowat, 2003).

Populations of IELs, predominantly cytotoxic T cells, inhabit the interdigitating spaces between epithelial cells above the basement membrane. NK-like lymphocytes have also been detected in these spaces (Leon *et al.*, 2003). To date, IELs' role in immune defence remains to be fully elucidated. However, because of their location and the observation that most cells are cytotoxic, IELs are believed to play an important role in innate defence and tumor surveillance in the gut. Recent immunohistological studies in mice have elucidated regional and quantitative differences within the small intestine between IELs and lamina propria lymphocytes (LPLs) (Tamura *et al.*, 2003). These studies also highlighted differences between the T cell receptor (TCR) $\gamma\delta$ and TCR $\alpha\beta$ lymphocyte populations. The functional significance of the unusually high percentage of $\gamma\delta$ T cell populations, both at baseline and following certain infections, in the gut is not fully understood (Ziegler *et al.*, 1994; Gaskins, 1997). Mice deficient in $\gamma\delta$ T cells generally have greater difficulty with immune regulatory function than with ability to clear intracellular pathogens (Born *et al.*, 1999). Also, the observation that $\gamma\delta$ T cells, as opposed to $\alpha\beta$ T cells, respond directly to certain bacterial antigens without antigen presentation suggests that these cells also are involved in innate defence against pathogens (Williams, 1998).

The MLN and the vascular endothelium also serve important functions in the GALT. After ingestion of foreign antigen, DCs and macrophages often migrate to MLN and present antigen to MLN T cells. These interactions aid MLN B cell differentiation into primarily IgA-producing plasma cells and induce a systemic response to the foreign antigen. The plasma cells are released into the bloodstream to then eventually 'home' back through the gut vascular endothelium to populate the LP, where they are needed to release IgA into the gut lumen. Gut endothelial cells, which release and recognize these homing factors, thus serve the important functions as 'gatekeepers' for recruitment of immune cells to the infected or damaged mucosal sites (Brandtzaeg, 2002). Furthermore, killing and removal of pathogens also occurs in the MLN. Pathogens able to bypass immune defences and reach these targets have an optimal chance to spread systemically through the bloodstream or lymphatic system.

5.2.2 *Breakdown of mucosal barrier function*

The mucosal barrier helps maintain symbiosis between microbes residing in the gut and the host animal. The integrity of this barrier is regulated by a complex network of physical, physiologic, and immune factors which includes dietary influences, the host environment (which can be modified by age, external factors such as antibiotics, and immune competency) and the indigenous microbial flora (IMF) of the gut (Rolfe, 1997). Modification and/or breakdown of these factors leads to ineffective clearance or degradation of harmful ingested antigens and/or disruption of regulatory cell function resulting in mucosal damage, increased gut permeability and overgrowth of harmful pathogens which may result in disease.

Diet not only introduces carcinogens or toxins, which may be directly toxic to the gut mucosa, but also provides nutrients that modify the physiologic environment and allow growth of certain bacteria in the gut. For example, infant diets comprised solely of breast milk are rich in acetic/propionic acids which induce an acidic environment, while diets comprised of infant formula have increased iso-(butyric, valeric, caproic) acids favouring higher pH. Lower pH appears to inhibit growth of certain *Escherichia coli* and *Vibrio cholera* strains, and this has been postulated as a reason for the decreased incidence of intestinal symptoms or diseases in breast-fed compared to formula-fed infants. Also, starvation leads to a gut environment with unconjugated bile acids and decreased short-chain fatty acid concentrations which tend to inhibit anaerobic bacterial growth while favouring growth of pathogenic coliform bacteria, such as *E. coli* (Tannock, 1997).

Age or stage of gut development may dictate which pathogens colonize the gut mucosa and contribute to disease. Infants tend to have a higher prevalence of gastrointestinal disease than adults, especially from enteric viruses. Developmental immaturity of the infant gut plays a factor. Newborns typically lack IgA and IgM in exocrine secretions until a few months of age, and s-IgA is found in low levels in the saliva and gut following birth. These low antibody levels may provide an inefficient barrier to microbes. Moreover, porous villous membranes, low basal acid output and immature proteolyic activity in the gut are additional factors that lead to altered clearance of potentially harmful microbes. Infants are further colonized by a unique repertoire of bacteria compared to adults, and this difference has been suggested to make the infant more susceptible to botulism following ingestion of *Clostridium botulinum* spores (Sugiyama & Mills, 1978; Rolfe, 1997). At the same time, up to 90% of infants are colonized with *Clostridium difficile*, but rarely develop disease from this organism (Rolfe, 1997).

Antibiotics are beneficial in treating illness by certain enteric pathogens. However, they may also contribute to disease by eliminating from the IMF certain microbes that inhibit or suppress the growth of other pathogenic microbes in the gut. Moreover, prolonged use of antibiotics may favour colonization by antibiotic-resistant strains of bacteria, which can pose serious consequences if these strains become pathogenic. A prime example of the suppressive effect of antibiotics is infection with *C. difficile*, leading to pseudomembranous colitis (Rolfe, 1997). Although spontaneous cases of *C. difficile* colitis occur, most cases appear with use of antibiotics. Antibiotics that preferentially eliminate anaerobes (i.e. Clindamycin) rather than facultative organisms tend to cause higher prevalence of infection. Also, it may take up to 30 days for the suppressive effects of these antibiotics to wear off (Berg, 1999). *C. difficile* is non-invasive and mediates a neutrophil-mediated colitis through elaboration of cellular toxins. One of these toxins, Toxin A, has been shown to directly interact with enteric neurons, and subsequent release of neuropeptides may contribute to gut hypermotility as well as inflammatory responses (Xia *et al.*, 2000). Disease states and/or pathogens are kept in check by competent cells or pathways of the immune system. Immune competency can be altered by drugs (e.g. steroids), age (i.e. infants and elderly individuals), acquired disease states (e.g. infections, malignancies), stressors (e.g. trauma, surgery) and germline immunodeficiency. The end result is naïve, delayed or defective IgA, phagocytic, or T cell effector immune responses leading to ineffective neutralization of pathogens and increased susceptibility to disease. IgA represents a first line of defence by blocking the association of certain non-indigenous microbes, namely *Salmonella*, *V. cholera* and enteric viruses, with the gut epithelium (Gaskins, 1997), and therefore reducing penetration of these pathogens into the LP. Phagocytes and T cells offer a second line of defence against cellular invasive organisms and help eliminate pathogens from cellular tissues. Also, interactions between these cells and epithelial cells help elaborate cytokines or other factors which support the symbiotic relationship with the IMF. Thus, immune deficiency leads not only to acute microbial disease, but also favours chronic persistence of these pathogens in diseased tissues because of ineffective measures to clear offending pathogens, contributing to malignancies, inflammatory disorders, or other systemic illness (Mowat, 2003).

5.2.3 *Translocation of pathogens*

Breakdown of the mucosal barrier network also facilitates the invasion or entry through host cells by pathogenic organisms, some of which would not otherwise normally cause disease. Although previously described for IMF, the term 'translocation' refers to the process of migration of viable microbes from the lumen of the gastrointestinal tract to extraintestinal locations such as the MLN, liver, spleen and bloodstream (Berg, 1999). Factors that favour translocation include bacterial overgrowth, immune deficiencies and mucosal injury with loss of barrier integrity. Evidence of increased translocation of pathogens has been observed in a variety of disease states (Berg, 1999) and has been shown associated with cases of postoperative sepsis (O'Boyle *et al.*, 1998) and with spontaneous bacterial peritonitis in cirrhotic rats (Guarner *et al.*, 1997). The ability of a pathogen to translocate efficiently through cells, coupled with mechanisms to avoid immune detection or destruction, appears to provide a survival advantage for a pathogen and may account for the systemic spread of certain organisms.

Bacterial overgrowth can occur from a variety of factors (as described above) and facilitates translocation. For some organisms, such as *E. coli*, the potential for translocation is directly related to levels of organisms. Other microbes, such as *Pseudomonas* species and gram-negative enterobacteria, are more efficient at translocating across cells than other bacteria in the gut such as obligate anaerobes (e.g. *Bacteroides*). This may explain why these former organisms are more frequent causes of opportunistic infections and sepsis in immunocompromised hosts. This ability for translocation is independent of their ability to adhere to epithelial cells or of host genetic factors (Berg, 1999), but may be facilitated by defective host immune responses and mucosal injury. Translocation routinely occurs across epithelial cell walls. However, ingested agents (e.g. castor oil) and clinical conditions (e.g.

shock) can cause direct damage to tight junctions between epithelial cells, and this may then allow microbes to utilize other portals of entry into the mucosa (Berg, 1999).

The mechanisms to protect the human host from ingested pathogens are numerous and complex, and the material discussed above is a brief summary of some of the mucosal factors involved. There are, of course, a myriad of systemic immune responses that will also serve to protect the host or, when compromised, increase the risks of disease. It is beyond the scope of this chapter to discuss the systemic immune response and readers are referred to other texts that address that component of the host defence network.

5.3 Vulnerable populations

The simple answer to the question of who constitutes a vulnerable population can be answered by saying it consists of the young, the elderly and the immunocompromised. However, at a deeper level, there are many other questions that emerge, such as are these populations all equally susceptible? How do you define young and how do you define elderly? Are the immunocompromised all equally susceptible to the same range of specific agents? What constitutes the definition of increased vulnerability? Does increased susceptibility mean the individual has a more prolonged or severe illness compared with a non-susceptible person? Is there a great likelihood of long-term complications or death in the susceptible person? Is there a greater chance of becoming ill following exposure? There are no clear answers to many of these questions, however the subsequent sections will delve into some of them.

An increased susceptibility to infection may not be generalized to all pathogens and may even result in a situation in which microbial agents that will not make a 'normal' person sick will result in illness in a susceptible person. While there are no clear answers to these questions, there have been a number of studies that have examined the variable susceptibility of a variety of populations to specific pathogens, for example to *Listeria monocytogenes*.

The populations most at risk for foodborne illness and subsequent death are collectively characterized by suppressed immune function, whether because of age (the very young or the very old), reproductive status (pregnancy), pharmacologic therapy (chemotherapy or organ transplantation), or disease (human immunodeficiency virus infection) (Smith, 1997). Together, these groups represent approximately 20% of the American population (Gerber *et al.*, 1996; Smith, 1997). Although each of these high-risk groups is immune suppressed or compromised, they differ in the length of time that immune function is affected and in the source of the physiologic insult on the immune system (Kendall *et al.*, 2003). As discussed above, there are many conditions that may result in a person developing increased susceptibility to infectious agents in food. Table 5.2 summarizes the main groups. There are two principal types of immune deficiency, congenital and acquired. Congenital immune deficiencies are due to disorders in which a portion of the immune system is either not present or not functioning properly. Acquired immune deficiency is due to some secondary event, which may be natural, age-related or due to some medical treatment or medical condition (Table 5.2). There are many different types of congenital (or primary) immune deficiency. Some congenital immune deficiencies lead to increased bacterial or viral infections that can

Table 5.2 Examples of conditions that may result in increased susceptibility to foodborne infections.

Condition
Congenital or primary
Myeloperoxidase deficiency
Wiskott–Aldrich syndrome
Chronic granulomatous disease
Hyper-IgM syndromes
Leukocyte adhesion deficiency
Chediak–Higashi syndrome
Severe combined immunodeficiency
Acquired or secondary
Age
Young
Elderly
Temporary
Pregnancy
Chronic disease
End-stage renal disease
Cirrhosis
Diabetes mellitus
Inflammatory bowel disease
HIV/AIDS
Malignancy
Solid tumours
Leukaemia's etc.
Iatrogenic (medication induced)
Reduced gastric acidity
Immunosuppressive medication

be treated with antibiotics or administration of passive antibodies, while others eliminate adaptive immunity and are fatal within the first year of life. Before the advent of antibiotics, all immune deficiencies were fatal, as mutant genes responsible for inherited deficiencies are generally recessive. However, immune deficiencies are associated with recurrent infections of many different types and not just foodborne infections. While it is not possible to go into a detailed discussion of specific immune deficiencies in this chapter, there are several categories that require specific mention.

5.4 Age-related immune deficiency

5.4.1 *Elderly*

The population in the United States and in many other countries throughout the world is aging. Life expectancy has increased during the past century, from 47 years for Americans born in 1900 to 77 years for those born in 2001 (National Center for Health Statistics, 2003). Not only are people in the United States living longer, but the proportion of the population that is age 65 years and older is also growing; a trend that will continue to increase as baby

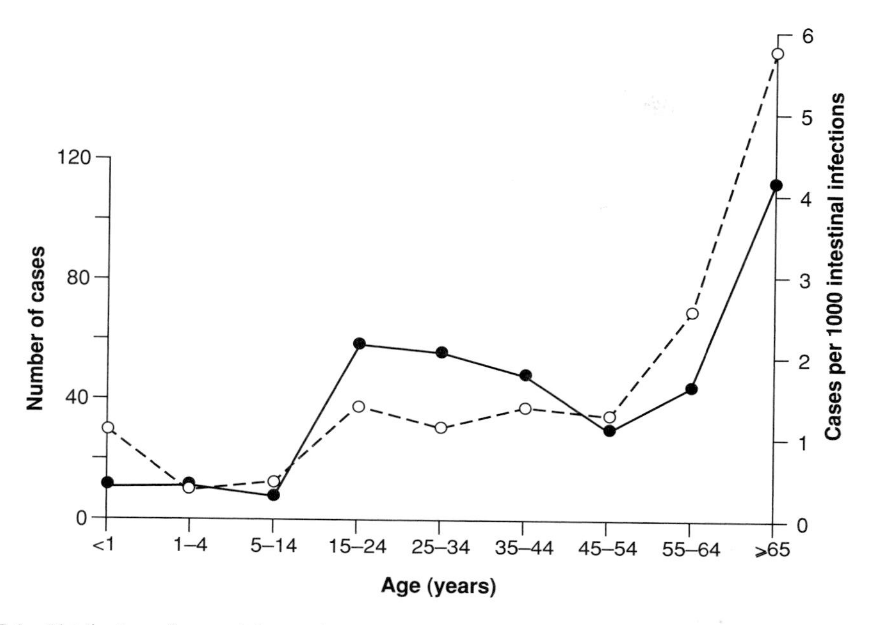

Fig. 5.1 Distribution of campylobacter bacteraemia cases by age, England and Wales, 1981–1991. Solid line number of cases ($n = 374$); dashed line cases per 1000 intestinal infections. (Reprinted with permission from Skirrow *et al.*, 1993. Copyright Cambridge University Press)

boomers (i.e. born between 1946 and 1964) reach age 65. Since 1900, the population of the United States has tripled; however, the number of older adults has increased 11-fold, from 3.1 million in 1900 to 35 million in 2000. By 2030, when all of the baby boomers have reached age 65, the number of older Americans is expected to reach 71 million, or roughly 20% of the population (CDC, 2003).

Infectious diseases are a major problem in the elderly as a result of decreases in humoral and cellular immunity, age-related changes in the gastrointestinal tract, malnutrition, lack of exercise, entry into nursing homes and excessive use of antibiotics (Meyers, 1989; Smith, 1998). As a result, outbreaks of foodborne-induced gastroenteritis can be devastating in nursing homes resulting in a significantly higher morbidity and mortality than the general population (Meyers, 1989). Mortality from diarrhoea shows a binomial curve with the greatest risk of mortality occurring among the very young and very old. In the United States, a majority of diarrhoeal deaths occur among individuals older than 74 years of age (51%) followed by adults 55–74 years of age (27%) and those younger than 5 years (11%). These deaths have a clear winter peak that may suggest a rotavirus aetiology (Lew *et al.*, 1991). This trend is illustrated for *Campylobacter* and *Salmonella* in Figures 5.1 and 5.2, which demonstrate the increase in bacteraemia from these pathogens with increasing age.

Most epidemiological studies concerning a specific agent in the elderly are focused on nursing homes, since the impact can be more easily observed with a confined group of individuals. Case fatality rates for specific enteric pathogens are frequently markedly greater in this group than the general population (Gerber *et al.*, 1996). One documented outbreak

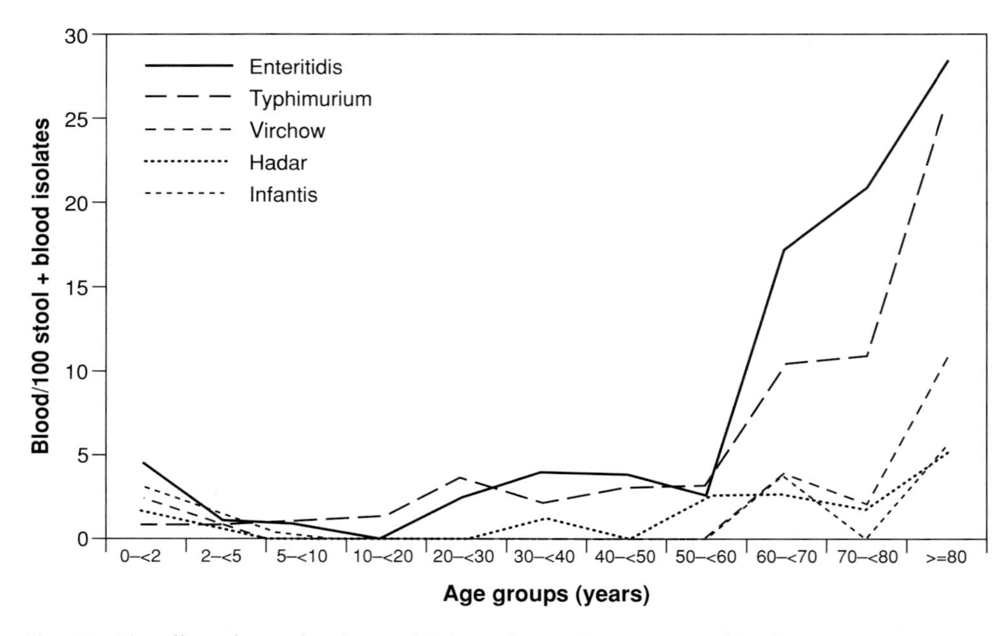

Fig. 5.2 The effect of age of patient and *Salmonella enterica* serotype on blood invasiveness ratio, estimated as number of blood isolates per 100 stool + blood isolates. (Reprinted with permission from Weinberger *et al.*, 2004. Copyright Cambridge University Press)

of rotavirus in a nursing home was characterized by high attack rates (66%), with few if any asymptomatic cases (Halvorsrud & Orstavik, 1980). While the number of days of illness was within the range observed for other age groups (1–5 days), the convalescence was prolonged for some individuals and the mortality rate was 1%. While many cases of gastrointestinal distress may be short-lived, secondary long-term complications may arise that are life-threatening and require hospitalization. Gordon *et al.* (1990) reported a mortality rate of 1.3% among a retirement community during a foodborne outbreak of the Snow Mountain agent, a norovirus. The authors noted that several of the residents sustained serious injuries from falling because of near-syncopal episodes due to dehydration from the gastroenteritis. The elderly would be expected to be more prone to such injuries than younger adults because of greater illness severity. Outbreaks of noroviruses and enteric adenoviruses have also been reported in nursing homes and geriatric hospital wards (Kaplan *et al.*, 1982; Reid *et al.*, 1990, Chapter 2.14), but these may not necessarily be transmitted via food. Higher attack rates occurred among the residents, as well as a more severe or protracted illness when compared to the staff (Oshiro *et al.*, 1981; Pether & Caul, 1983; Mattner *et al.*, 2006). Although death is rare in cases of norovirus infection (<1 per 1000 cases), those deaths that do occur are in previously hospitalized patients or residents of elderly care institutions (Lopman *et al.*, 2003).

Hepatitis A virus usually causes a mild and often asymptomatic infection in children. However, in adults the virus typically produces clinical illness that can lead to death (Ledner *et al.*, 1985). Waterborne outbreaks are often characterized by high attack rates with all or most of the infected individuals exhibiting clinical illness (Bowen & McCarthy, 1983). In an

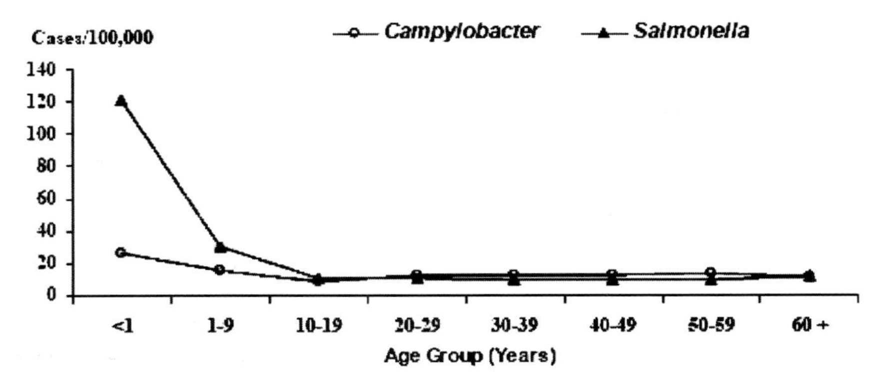

Fig. 5.3 Incidence of *Campylobacter* and *Salmonella* infections by age group, FoodNet sites 2004. (CDC, 2006a)

8-year review of hepatitis A cases from England, Wales, and Ireland, the case fatality rate of hepatitis A for patients of less than 55 years of age was 0.02–0.03%, 0.9% for patients 55–64 years of age and 1.5% for older patients. The median age of those dying from hepatitis A was reported as over 60 in the United Kingdom (Gust, 1990). The elderly also experience higher mortality rates from enteric bacterial gastroenteritis. The overall case fatality ratio for foodborne outbreaks in nursing homes from 1975 to 1987 was 1.0%, compared to 0.1% for outbreaks at other locations (Levine *et al.*, 1991). For domestically acquired cases of typhoid in the United States, the case fatality rate is higher among individuals 55 years or older than in younger persons (Ryan *et al.*, 1989).

5.4.2 *Children*

Children and youth deserve added attention because the risks of some foodborne illnesses, such as salmonellosis, are relatively higher for children than for any other demographic group. Children are at higher risks due a less-developed, more permeable gut tissue, less gastrointestinal reserve capacity, lower body weight (i.e. it takes a smaller quantity of pathogens to make them sick), potential for rapid dehydration and limited recognition of thirst (Gottschlich, 1993; Buzby, 2001). The impact of age on the likelihood of acquiring a foodborne infection is well illustrated from 2004 FoodNet data for *Salmonella* and *Camplylobacter*. For children less than 1 year of age, the incidence of *Salmonella* infection was 121.57 per 100,000 individuals and the incidence of *Campylobacter* infection was 26.98 per 100,000 individuals, substantially higher than for other age groups (Fig. 5.3) (CDC, 2006a).

5.5 Acquired immunodeficiency syndrome (AIDS)

Infections in the immunocompromised host constitute a relatively new and severe problem magnified by the current AIDS epidemic and by the escalation in organ and tissue transplantation. In 2005, the estimated number of AIDS diagnoses in the United States and

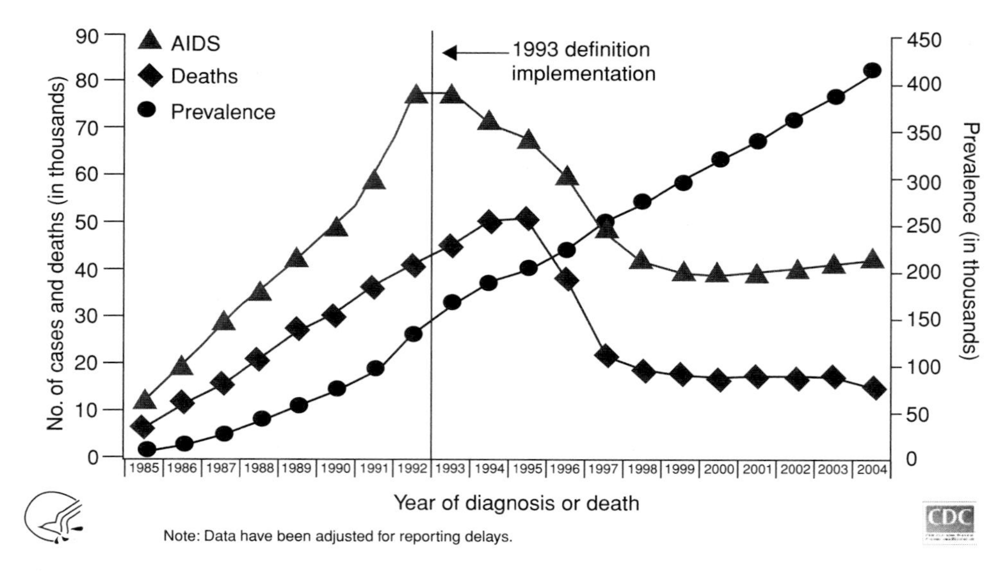

Fig. 5.4 Incidence, prevalence and deaths among persons with AIDS, 1985–2004 in the United States. (CDC, 2006c)

dependent areas was 45,669. Of these, 44,198 were in the 50 states and District of Columbia and 1096 were in the dependent areas. In the 50 states and District of Columbia, adult and adolescent AIDS cases totaled 44,140, with 32,430 cases in males, 11,710 cases in females and 58 cases estimated in children under age 13. The cumulative estimated number of AIDS diagnoses through 2005 in the United States is 988,376. Of these, 956,666 were in the 50 states and District of Columbia and 30,523 were in dependent areas. In the 50 states and District of Columbia, adult and adolescent AIDS cases totaled 947,585 with 764,763 cases in males, 182,822 cases in females and 9078 cases were estimated in children under age 13 (CDC, 2005). Figure 5.4 illustrates how the trends in AIDS have altered over time and continue to rise.

Enteric pathogens are among the many agents that take advantage of an impaired or destroyed immune system to develop persistent and generalized infections in the immunocompromised host. Such infections are difficult to treat, tend to be long-term, add to the burden of the debilitation in the patient, and can result in a significantly higher mortality than in immunocompetent persons (Gerber *et al.*, 1996).

The impact of the AIDS epidemic has increased the number of diarrhoeal deaths in the 25- to 54-year-old age group (Lew *et al.*, 1991). Enteric diseases are among the most common and devastating problems that affect persons with AIDS. The majority of AIDS patients (50–90%) suffer from chronic diarrhoeal illnesses, the effects of which can be fatal. Many studies have shown that the rates of diarrhoeal disease among HIV-infected persons in developing countries are higher than the rates in developed countries, probably reflecting more frequent exposure to enteric pathogens by contaminated food and water (Janoff &

Smith, 1988). Adenoviruses and rotaviruses are the most common enteric viruses isolated in the stools of AIDS infected persons. A comprehensive study of Australian men showed that 54% of diarrhoeal illnesses in AIDS patients were caused by viruses and that 37% of the viral diarrhoeas were adenovirus-related (Cunningham *et al.*, 1988). Overall, it is estimated that 12% of AIDS patients with clinical symptoms suffer from adenovirus infections and that 45% of these cases result in death within 2 months (Hierholzer, 1992). The other enteric viruses do not appear to be a significant problem in AIDS-associated gastroenteritis (Gouandjika-Vasilache *et al.*, 2005; Rodriguez-Guillen *et al.*, 2005; Asturias *et al.*, 2006). One study from Venezuela found an increased carriage of caliciviruses other than norovirus in HIV positive children but not adults, and virus carriage was not associated with presence of diarrhoeal symptoms (Rodriguez-Guillen *et al.*, 2005). In addition, enteric bacterial infections are more severe in AIDS patients. For example, patients with *Salmonella*, *Shigella* and *Campylobacter* infection often develop bacteraemia (Baine *et al.*, 1982). *Cryptosporidium* is also a serious problem among AIDS patients (Hunter & Nichols 2002). A severe and protracted diarrhoea results, with fluid losses of several litres per day in some cases. Symptoms may persist for months, resulting in severe weight loss and mortality. Mortality rates of 50% have been reported for this organism (Clifford *et al.*, 1990). *Giardia* does not cause more frequent or more severe infections in AIDS patients than in otherwise healthy individuals (Smith & Janoff, 2002).

5.6 Iatrogenic immune suppression

Cancer patients undergo intensive chemotherapy with cytotoxic and immunosuppressive drugs and often radiation treatment in an attempt to destroy the neoplastic growth. These measures also attack the immune system, leaving the patient with little defence against opportunistic pathogens (Gerber *et al.*, 1996). For example, in cancer immunosuppressed patients the fatality rate for adenovirus infection was 53% (Hierholzer, 1992).

Bone marrow transplantation is an effective therapy in patients with severe aplastic anaemia or acute leukaemia. However, because of a weakened immune system, patients are especially susceptible to a variety of infections, including enteric infections. The mortality rate among allogenic bone marrow transplant patients with enteric viral infection (i.e. rotavirus, adenovirus and coxsackievirus) was observed to be 59% in one study (Yolken *et al.*, 1982). Of eight patients with rotavirus infections, five individuals died. The case fatality rates for adenoviruses for bone marrow patients range from 53 to 69% depending on subgenus (Hierholzer, 1992). Coxsackie Al, which seldom appears to cause diarrhoea in healthy individuals, resulted in the deaths of six of seven bone marrow patients in one outbreak (Townsend *et al.*, 1982). In a study from South India, there was significantly higher mortality ($p < 0.01$) in patients who had undergone an allogenic bone marrow transplant with symptomatic or asymptomatic gastrointestinal infections caused by bacteria than in patients with parasitic or viral infections or without enteric infections (Kang *et al.*, 2002). In contrast to many of the other enteric viruses, Norwalk virus (norovirus) and hepatitis A virus do not appear to be associated with a greater severity or chronic illness in the immunocompromised. Bone marrow transplants are one of the most extreme examples of immunosuppresion; however, there are other instances in which there may be significant iatrogenic immunosuppression. For example, one study demonstrated neutropenia in patients receiving anti-tumour necrosis factor treatment for rheumatoid arthritis (Rajakulendran *et al.*, 2006).

While the examples of iatrogenic situations discussed above clearly lead to major immune suppression, there are other causes leading to a less severe immune suppression which are, in fact, much more common. For example, diminished gastric acidity will increase a person's risk to enteric infection. Decreased gastric acidity may occur as a result of age or from ingestion of gastric acid-lowing agents that are available over the counter and are very commonly used throughout the world.

5.7 Pregnancy-related immune suppression

During pregnancy, alteration of cellular immune function leads to increased susceptibility to intracellular infections, most notably toxoplasmosis and listeriosis (Smith, 1999). It is difficult to obtain an accurate assessment of the number of pregnancies in which either toxoplasmosis or listeriosis result in morbidity or mortality. However, studies in France estimate that close to 25% of listeriosis cases are maternal–fetal infections (Goulet *et al.*, 2006). In the United States, the annual incidence of *Listeria* infections decreased 32% (confidence interval = 16–45%), from 1996 to 2005 (CDC, 2006b). However, this trend has not been seen in some European countries. For example, in Germany for the years 2001–2005, the number of pregnancy-associated listeriosis cases (including neonates) showed some fluctuation but no clear trend, while non-pregnancy-associated listeriosis (excluding neonates) increased dramatically during this period (Koch & Stark, 2006).

Women may also be at an increased risk during pregnancy from enteric viruses and may act as a source of infection for neonates. In the decade up to 1992, at least 30 outbreaks of hepatitis E have been documented in 17 countries due to contaminated water (Craske, 1992) and foodborne outbreaks have been suspected (Caredda *et al.*, 1986). Although outbreaks of hepatitis E have not been reported in the United States, cases do occur among tourists returning from developing countries. Waterborne outbreaks have at times involved thousands of individuals. Overall, the case fatality ratios have ranged from 1 to 2% during outbreaks, which is significantly higher than that for the hepatitis A virus. For pregnant women, however, the ratio is generally between 10 and 20% and can be as high as 40% (Gust & Purcell, 1987; Craske, 1992).

Infection during pregnancy may also result in the transmission of infection from mother to child in utero, during birth, or shortly thereafter with a potentially serious outcome (Gerber *et al.*, 1996). This is especially true with *Listeria monocytogenes*. Neonates are uniquely susceptible to enterovirus infections and this group of viruses can cause severe disease and death when infection occurs within the first 10–14 days of life. Enteroviruses are often acquired from the fecal–oral route, and one of the most significant viruses in this group with regard to children is coxsackie B. Acquisition of coxsackie B infections early in life is the most significant risk factor leading to fatal disease. The most fatal cases caused by this virus are probably transmitted transplacentally at term (Kaplan *et al.*, 1983). Among 41 documented cases of fatal infection in infants, 24 of their mothers had symptomatic illness consisting of fever, symptoms of upper respiratory tract involvement, pleurodynia or meningitis. Symptomatic infection occurred between 10 days antepartum and 5 days postpartum.

Table 5.3 Estimates of the risk of serious illness from *Listeria monocytogenese* in different susceptible populations relative to the general population.

Condition	Relative susceptibility
Transplant	2584
Cancer – blood	1384
AIDS	865
Dialysis	476
Cancer – pulmonary	229
Cancer – gastrointestinal/liver	211
Non-cancer liver disease	143
Cancer – bladder and prostate	112
Cancer – gynaecological	66
Diabetes – insulin dependent	30
Diabetes – non-insulin dependent	25
Alcoholism	18
Perinatals	14
Over 65 years of age	7.5
Over 60 years of age	2.6

Data from WHO, 2004.

The observed case fatality rate for coxsackievirus in a New England county in the United States was almost 13%, with a morbidity rate of 50.2 per 100,000 live births (Gerber *et al.*, 1996). An average case fatality of 3.4% was also observed in 16 documented outbreaks of echovirus in newborn nurseries. In two outbreaks of coxsackie B virus in nurseries, the infant mortality rate from myocarditis ranged from 50 to 60% (Modlin & Kinney, 1987). Stillbirth late in pregnancy has also been reported for the echoviruses and coxsackie B viruses. More recently, coxsackie B viruses have been implicated as a potentially important causative agent in spontaneous abortions (Frisk & Diderholm, 1992). Anomalies (e.g. urogenital system defects, heart defects and digestive malformation) in children born to mothers infected with coxsackie B viruses have been suggestive in several studies (Modlin & Kinney, 1987).

5.8 Factors contributing to foodborne infection

While it may not be possible to protect all immunocompromised individuals from all food-borne infections, there are a number of steps that can be undertaken to minimize the potential risk. At the outset, it is important to recognize that certain groups are at much greater risk than others. This is well illustrated in the context of listeriosis, in which the likelihood of developing illness varies in relation to a variety of underlying conditions (Table 5.3). For example, there is a 2584 times greater risk of a transplant patient becoming sick from listeriosis compared to an individual under the age of 65 with no underlying medical conditions (WHO, 2004).

According to the Council for Agricultural Science and Technology (CAST), a majority of foodborne illnesses can be attributed to improper food handling behaviours (CAST, 1994). Leading causal behaviours are failure to (1) hold and cool foods appropriately, (2) practice proper personal hygiene, (3) prevent cross-contamination, (4) cook to proper internal temperatures and (5) procure food from safe sources.

Another critical but often unrecognized food safety factor is the temperature at which potentially hazardous foods are received by food services. Rural stores and foodservices are particularly at risk for poor delivery temperatures. In 2005, the United States Food and Drug Administration (FDA) issued the 2005 edition of the Food Code, which contains the latest science-based information on food safety for retail and food service industries. This edition provided temperature standards by which potentially hazardous foods are to be received, as well as the best temperature standards for preparation, holding and storage of foods. The temperature range for the 'danger zone' identified in the food code is from 41°F (5°C) to 135°F (57°C) (FDA, 2005a).

5.9 Basic food handling practices

Irrespective of the type of immune deficiency a person may have, and even if their immune systems are normal, there are four basic rules that should be followed to prevent foodborne illness. The four steps include appropriate cleaning, proper separation of foods, adequate cooking and maintaining foods at safe temperatures. A summary of the critical issues for each can be found in the subsequent sections. For those interested in more information, it is suggested that the reader visit the website U.S. Gateway to Government Food Safety Information (2007). This website is supported by a variety of Federal, state and industry partners and contains links to a great deal of information, often in many different languages.

5.9.1 *Clean: wash hands and surfaces often*

Wash hands and surfaces often with hot, soapy water (for at least 20 s) before and after handling food and after using the bathroom, changing diapers and handling pets. Always wash hands, cutting boards, counter tops, dishes, utensils and spills in the refrigerator with hot, soapy water after they come in contact with raw foods and ready-to-eat foods (e.g. hot dogs, luncheon meats, cold cuts, fermented and dry sausage, and other deli-style meat and poultry products). Wash hands after use of the bathroom or after handling animals or animal waste. Thorough washing helps eliminate any bacteria that might get on hands or other surfaces from food before it is reheated. Consider using paper towels to clean up kitchen surfaces. If you use cloth towels wash them often in the hot cycle of your washing machine.

5.9.2 *Separate: do not cross-contaminate*

Cross-contamination is how bacteria can be spread. Ready-to-eat foods and raw meat, poultry, and seafood can contain dangerous bacteria. As a result, keep these foods and their juices separate from vegetables, fruits, breads and other foods that are already prepared for eating. For example, separate raw meat, poultry, seafood and eggs from other foods in your grocery shopping cart, grocery bags and in your refrigerator. Use one cutting board for fresh produce and a separate one for raw meat, poultry and seafood. Never place cooked food on a plate that previously held raw meat, poultry, seafood or eggs.

5.9.3 *Cook: cook to proper temperatures*

Food is safely cooked when it reaches a high enough internal temperature to kill the harmful bacteria that cause foodborne illness. It is important to use a food thermometer to measure the internal temperature of cooked foods. When cooking at home, keep hot foods hot (140°F/60°C or above) and cold foods cold (40°F/4°C or below). Harmful bacteria can grow rapidly in the danger zone between these temperatures (Hayes *et al.*, 2003). These temperatures are slightly different from those identified in the food code and offer an added degree of safety in a home environment, which may not be as controlled as a food service or retail establishment. The following recommendations in relation to cooking are especially important to avoid infection with *L. monocytogenes*. Reheat until steaming hot the following types of ready-to-eat foods: hot dogs, luncheon meats, cold cuts, fermented and dry sausage and other deli-style meat and poultry products. Thoroughly reheating food can help kill any bacteria that might be present. If you cannot reheat these foods, do not eat them.

We all enjoy the benefits of using the microwave oven for cooking and reheating foods in minutes, even seconds. However, microwaves often cook food unevenly, thus creating hot and cold spots in the food. Bacteria can survive in the cold spots. This uneven cooking occurs because the microwaves bounce around the oven irregularly. Microwaves also heat food elements like fats, sugars and liquids more quickly than carbohydrates and proteins. Extra care must be taken to even out the cooking so that harmful bacteria are destroyed.

5.9.4 *Chill: refrigerate promptly*

Refrigerate food quickly because cold temperatures keep most harmful bacteria from multiplying. Do not over-stuff the refrigerator. Cold air must circulate to help keep food safe. Keeping a constant refrigerator temperature of 40°F (4°C) or below is one of the most effective ways to reduce the risk of foodborne illness. Use an appliance thermometer to be sure the temperature is correct. The freezer temperature should be 0°F (−18°C) or below.

Perishables, prepared food and leftovers should be refrigerated or frozen within 2 hours, and large amounts of leftovers should be divided into shallow containers will allow for quick cooling in the refrigerator. At family outings or barbecues, use a cooler to keep perishable foods cold. Always use ice or cold packs and fill the cooler with food. A full cooler will maintain its cold temperatures longer than one that is partially filled. Foods must be kept at a safe temperature during thawing. The safe ways to defrost food are in the refrigerator, in cold water, and in the microwave oven. Food thawed in cold water or in the microwave oven should be cooked immediately.

5.10 Food handling practices for vulnerable populations

Food safety guidelines are fundamental in controlling foodborne illnesses in all populations. While general education is essential in reducing cases and outbreaks of foodborne illnesses among all consumers, it is of particular importance for high-risk populations, either because

Table 5.4 Consumer food handling behaviours of special importance to pregnant women, infants and young children.

Behaviour[a]	Pathogen
Pregnant women, infants and young children	
Avoid soft cheeses, cold-smoked fish and cold deli salads (e.g. chicken salad, tuna salad and shrimp salad)	*Listeria monocytogenes*
Avoid hot dogs and lunchmeats that have not been reheated to steaming hot or 165°F	*Listeria monocytogenes*
Use cheese and yogurt made from pasteurized milk	*Salmonella* species
Avoid eating foods containing raw eggs/cook eggs until both the yolk and white are firm	*Salmonella* Enteritidis
Pregnant women only	
Do not clean cat litter boxes if pregnant	*Toxoplasma gondii*
Use plastic gloves when cleaning cat litter boxes	*Toxoplasma gondii*
Do not handle pets when preparing foods	*Toxoplasma gondii*
Keep pets out of food preparation areas	*Toxoplasma gondii*
Infants and young children only	
Drink only pasteurized milk and fruit juices	*Escherichia coli* O157:H7; *Listeria monocytogenes*; *Campylobacter jejuni*; *Yersinia enterocolitica*; *Salmonella* species
Avoid eating raw vegetable sprouts	*Escherichia coli* O157:H7
Wash knives, cutting boards and food preparation surfaces with hot water and soap after contact with raw poultry, meat and seafood	*Salmonella* species; *Campylobacter jejuni*; *Yersinia enterocolitica*; *Listeria monocytogenes*; *Toxoplasma gondii*; *Salmonella* Enteritidis; *Escherichia coli* O157:H7; *Vibrio* species, *Shigella* species
Thoroughly rinse fresh fruits and vegetables under running water before eating	*Escherichia coli* O157:H7
Use water from a safe water supply for drinking and food preparation	*Shigella species*; Norovirus; *Yersinia enterocolitica*; *Escherichia coli* O157:H7

[a] Behaviours that 80% of a national panel of food safety experts ($n = 28$) rated as being of special importance to pregnant women and/or infants and young children, with those rated as important to both groups presented first. Reprinted from *Journal of the American Dietetic Association* **103**, Kendall, P., Medeiros, L.C., Hillers,V., Chen, G. & DiMascola, S. Food handling behaviors of special importance for pregnant women, infants and young children, the elderly, and immune-compromised people pp. 1646–1649, Copyright Elsevier (2003).

of food consumption practices or greater vulnerability to specific pathogens. There are specific sets of advice that should be targeted to certain high-risk groups, for example pregnant women or those who are liable to become neutropenic, including persons who have had a bone marrow transplant or undergone chemotherapy or radiotherapy for cancer.

5.10.1 *Pregnant women, infants and young children*

There are certain behaviours that are of special importance to pregnant women, infants and young children. A list of these behaviours and pathogens associated with the behaviours are presented in Table 5.4. The behaviours most important to both pregnant women and children are associated with *L. monocytogenes* or *Salmonella* species. *Listeria monocytogenes* has a unique predilection for pregnant women, with an estimated 17-fold increase in incidence compared to other healthy adults (Mylonakis *et al.*, 2002). Pregnant women account for 27% of all cases and 60% of cases among people ages 10–40 (Lorber, 1997). In the United Kingdom, the proportion of all reported cases of listeriosis that were pregnancy-associated was 41.1% during the years 1983–1989, 19.4% for the years 1990–1999 and 11.4% for the years 2000–2005 (HPA, 2007). This decline probably reflects a strong public health message directed at pregnant women. Although pregnant women are no more susceptible to *Salmonella* species than the normal population, bacteraemia resulting from salmonellosis can lead to intrauterine sepsis (Scialli & Rarick, 1992). Neonates and infants under 1 year of age are also highly susceptible to infection from *L. monocytogenes* and *Salmonella* species. The 2001, crude incidence rates in the United States for these two pathogens were higher among infants under age 1 than among any other age group (CDC, 2002).

The behaviours most important to pregnant women involve avoiding cat or other pet feces and are associated with *Toxoplasma gondii*. Alteration of cell-mediated immunity, which is critical during pregnancy to prevent rejection of the fetus, increases susceptibility to *T. gondii* (Szekeres-Bartho, 1993). An estimated 1.5 million *T. gondii* infections occur each year in the United States, and most of those infected, including pregnant women, are asymptomatic (Todd *et al.*, 1989). However, when a woman not previously exposed to *T. gondii* is infected with the parasite during pregnancy, there is a 30–40% chance that the fetus will be infected (Smith, 1999). Behaviours important in reducing the risk of toxoplasmosis include avoidance of cleaning cat litter boxes or using plastic gloves while cleaning cat litter boxes (cats shed an infective form of the parasite in their feces), not handling pets when preparing foods, cooking meats adequately and washing vegetables well (Kapperud *et al.*, 1996).

The behaviours of special importance for infants and young children are primarily associated with pathogens for which the 2001 crude incidence rates in the United States were highest for infants under age 1 (i.e. *Salmonella* species, *Campylobacter jejuni*, *Yersinia enterocolitica* and *L. monocytogenes*) or for children ages 1–4 (i.e. *Shigella* species and *E. coli* O157:H7 (CDC, 2002). The most important behaviour for infants and young children is to drink only pasteurized milk and fruit juices. Milk and fruit juices are important foods for young children, and thus it is critical that they be served in a safe manner (CAST, 1994; Beuchat, 1996).

Despite the clear need for pregnant women to pay attention to what they eat, a recent publication from Ogunmodede *et al.* (2005) indicates there is more work to be done. In this study, pregnant women were recruited to complete a questionnaire to assess their knowledge of listeriosis and its prevention. One study was a national survey of 403 women from throughout the United States, and the other survey was limited to 286 Minnesota residents. The data showed that 74 of 403 respondents (18%) had some knowledge of listeriosis in the multi-state survey, compared with 43 of 286 (15%) respondents to the Minnesota survey. The majority of respondents reported hearing about listeriosis from a medical professional. In the multi-state survey, 33% of respondents knew listeriosis could be prevented by not eating delicatessen meats, compared with 17% in the Minnesota survey ($p = 0.01$). Similarly, 31% of respondents to the multi-state survey compared with 19% of Minnesota survey respondents knew listeriosis could be prevented by avoiding unpasteurized dairy products ($p = 0.05$). As for preventive behaviours, 18% of multi-state and 23% of Minnesota respondents reported avoiding delicatessen meats and ready-to-eat foods during pregnancy, whereas 86% and 88%, respectively, avoided unpasteurized dairy products. Clearly, there is a need for further education of this particular group.

5.10.2 *Elderly and immunocompromised*

A number of specific behaviours are also considered to be of special importance to elderly people and/or to individuals immunocompromised because of conditions such as HIV/AIDS or pharmacologic therapy (e.g. chemotherapy and organ transplantation patients) (Table 5.5). Important behaviours include avoiding raw or undercooked seafood, cooking shellfish and fish properly, and obtaining shellfish from approved sources. These behaviours are associated with reducing the risk of *Vibrio* species infection (FDA/CFSAN, 2005b) and norovirus (FDA/CFSAN, 2006). In 2001, the highest incidence rate for *Vibrio* species infection was seen in people ages 65–74 years (0.7 cases per 100,000 persons) (CDC, 2002). Other important behaviours include avoiding raw foods (e.g. sprouts and eggs) and minimally processed foods eaten without further heating (e.g. soft cheeses, cold-smoked fish and cold deli meat and salads). From 1996 to 1998, raw vegetable sprouts were associated with more than half of all outbreaks in California that were multi-state in nature (Mohle-Boetani *et al.*, 2001). Eggs are a known source of *Salmonella enteritidis* (CDC, 2000), yet are commonly consumed by these high-risk groups because they are relatively inexpensive, easy to prepare, and generally well-liked. Altekruse *et al.* (1999) found that people older than 60 years of age were much more likely to report consumption of undercooked eggs than any other risky behaviour and reported more frequent consumption of undercooked eggs than did people less than 60 years of age.

5.10.3 *Bone marrow transplant patients*

One of the recent trends in relation to protecting individuals undergoing bone marrow transplantation is the 'low microbial diet'. A low microbial diet is essentially a diet that avoids the types of foods that are more likely to contain foodborne pathogens. An example of the types of foods to eat and the types to avoid are shown in Table 5.6. However, it is important to remember that even when starting with a food that is free of microbial pathogens, proper handling of that food is the only way to ensure that it remains safe.

While this concept is one that extends to a variety of high-risk groups, those with bone marrow transplants are likely to follow this diet to a more extreme level. Health establishments may have various approaches to achieve a low microbial diet, including preparing food in a separate kitchen using aseptic techniques, providing a diet consisting of well-cooked foods or foods containing a minimum number of pathogen-forming units, or simply focusing on safe food handling guidelines and avoiding foods associated with foodborne illness. Alternatively, some establishments just modify the standard diet by excluding fresh fruits and vegetables. A survey by French *et al.* (2001) on the use of low microbial diets for pediatric bone marrow transplantation patients at hospitals in Canada and the northwestern United States indicated that five out of seven hospitals responding to the survey do provide

Table 5.5 Consumer food handling behaviours of special importance to elderly and immune compromised individuals.

Behaviour[a]	Pathogen
Elderly and immune compromised individuals	
Avoid soft cheeses, cold-smoked fish and cold deli salads (e.g. chicken salad, tuna salad and shrimp salad)	*Listeria monocytogenes*
Avoid hot dogs and lunchmeats that have not been reheated to steaming hot or 165°F	*Listeria monocytogenes*
Store eggs and poultry in the refrigerator	*Salmonella* Enteritidis
Avoid raw or partially cooked eggs, foods containing raw eggs. Cook eggs until both the yolk and white are firm. Use a thermometer to make sure that foods containing eggs are cooked to 160°F	*Salmonella* Enteritidis
Cook shellfish until the shell opens and the flesh is fully cooked; cook fish until flesh is opaque and flakes easily with a fork	Norovirus
Obtain shellfish from approved sources	Norovirus; *Vibrio* species
Avoid eating raw or undercooked seafood/ shellfish (clams, oysters, scallops and mussels). Cook fish and shellfish until it is opaque; fish should flake easily with a fork. When eating out, order foods that have been thoroughly cooked and make sure they are served piping hot	*Vibrio* species
Avoid eating raw vegetable sprouts	*Escherichia coli* O157:H7
Thoroughly rinse fresh fruits and vegetables under running water before eating	*Escherichia coli* O157:H7
Drink only pasteurized milk and fruit juices Use cheese and yogurt made from pasteurized milk	*Escherichia coli* O157:H7; *Listeria monocytogenes*; *Salmonella* species; *Campylobacter jejuni*; *Yersinia enterocolitica*; *Cryptosporidium parvum*
Wash knives, cutting boards and food preparation surfaces with hot water and soap after contact with raw poultry, meat and seafood	*Salmonella* species; *Campylobacter jejuni*; *Yersinia enterocolitica*; *Listeria monocytogenes*; *Toxoplasma gondii*; *Salmonella* Enteritidis; *Escherichia coli* O157:H7; *Vibrio* species; *Shigella* species

[a] Behaviours that 80% of a national panel of food safety experts ($n = 28$) rated as being of special importance to the elderly and/or immunocompromised individuals, with those rated as important to both groups presented first. Reprinted from *Journal of the American Dietetic Association* **103**, Kendall, P., Medeiros, L.C., Hillers,V., Chen, G. & DiMascola, S. Food handling behaviors of special importance for pregnant women, infants and young children, the elderly, and immune-compromised people, pp. 1646–1649, Copyright Elsevier (2003).

some type of a low microbial diet to this population to reduce the potential risk posed by food pathogens. Various guidelines were also used to determine when to initiate and discontinue the low microbial diet. These guidelines included criteria such as a specific day relative to transplantation and a patient's absolute neutrophil count. A more recent study by Larson and Nirenberg (2004) examined the effectiveness of selected nursing interventions

Table 5.6 Example of the breakdown of the types of products that are allowed and those that should be excluded as part of a low microbial diet.

	Allowed	Excluded
Dairy products	• low-fat milk • nonfat milk • chocolate milk • non-dairy cream • canned eggnog • hospital milkshakes or milkshakes made with evaporated or condensed milk • all pasteurized yogurts and hard cheeses	• soft-serve ice cream • raw milk • non-pasteurized dairy products • moldy cheeses and soft cheeses (e.g. bleu and brie)
Vegetables	• cooked fresh or frozen vegetables • canned vegetables except sauerkraut • canned vegetable juices	• all raw or uncooked vegetables • all salads • sauerkraut • stir-fry vegetables • potato skins
Fruit and juices	• any canned or stewed fruit • any pasteurized canned or bottled fruit juice	• all fresh fruit • non-pasteurized fruit juices • raisins and other dried fruit, unless part of a baked food such as a raisin oatmeal cookie
Breads, cereals and other starches	• any white or whole grain breads or rolls • muffins • bagels • biscuits • crackers and melba toast • pretzels • sweet rolls and donuts • pancakes, waffles and French toast • any cooked or dry cereals • white or sweet potatoes (without skin) • yams, potato chips • macaroni, noodles and spaghetti • rice	• sweet rolls with custard or cream filling • unprocessed bran, unless part of a baked or cooked food
Meat and protein products	• well done cooked beef, veal, ham, pork, lamb, chicken, turkey and fish • fresh or pasteurized eggs thoroughly cooked, any style • casseroles or stews using allowed foods • macaroni and cheese, peanut butter • Roasted nuts	• raw fish and shellfish • raw or rare meats • raw or soft-cooked eggs including 'over-easy' fried eggs • stir-fried foods such as Chinese entrees • raw nuts
Soup	• any cooked hot homemade, canned frozen or dehydrated soup	• cold soups such as Gazpacho
Fats	• margarine and butter • vegetable oils • crisp bacon • cooked gravies • white sauces • oil and vinegar type dressings	• whipped topping • Avocado • bleu cheese • Roquefort cheese dressing
Desserts	• plain cakes • cookies • custard pudding • gelatin desserts • commercially made hard ice cream, ices and sherbet	• non-commercial ice cream • all other desserts made with foods in the exclude list (see also Breads, Cereals and Starches)
Sweets and spices	• salt spices, herbs and seasonings may be used in food only during the cooking process • sugar • jam, jelly and preserves • honey, syrup and molasses • candy • chocolate and cocoa	• added pepper, spices, herbs and seasonings to food after they have been cooked • shredded coconut
Beverages	• coffee, caffeinated and decaffeinated • tea • cocoa • carbonated beverages lemonade (reconstituted from powder with sterile water) • canned or bottled beverages if pasteurized	• beverages made with frozen concentrate or reconstituted with non-sterile water

Table modified from UCSF (2006) University of San Francisco Bone Marrow Transplant information on dietary concerns during bone marrow transplant.

Box 5.1 Suggestions from the American Cancer Society for ways to reduce the risk from foodborne pathogens in high-risk populations.

- Avoid raw milk or milk products; any milk or milk product that has not been pasteurized, including cheese and yogurt made from unpasteurized milk
- Avoid raw or undercooked meat, fish, chicken, eggs, tofu
- Do not eat cold-smoked fish
- Do not eat hot dogs, deli meats, processed meats (unless they have been cooked again just before eating)
- Avoid any food that contains mold (for example, blue cheese, including that in salad dressings)
- Avoid uncooked vegetables and fruits
- Avoid uncooked grain products
- Avoid unwashed salad greens
- Do not eat vegetable sprouts (alfalfa, bean and others)
- Avoid fruit and vegetable juices that have not been pasteurized
- Avoid raw honey (honey that has not been pasteurized)
- Do not eat raw nuts or nuts roasted in their shells
- Do not drink beer that has not been pasteurized (home brewed and some microbrewery beers); also avoid brewer's yeast
- Avoid any outdated food
- Do not eat any cooked food left at room temperature for 2 h or more
- Avoid any food that has been handled or prepared with unwashed hands

Note: Talk with your doctor about any dietary concerns you may have, or ask to talk with a registered dietician.
Data from American Cancer Society, 2005.

used in hospitals to prevent healthcare-associated infections in neutropenic patients with cancer. One of the interventions examined was low microbial diets, and the authors concluded that few studies have demonstrated the effectiveness of low microbial food and water in reducing infections in neutropenic patients with cancer. However, they felt that this type of approach was prudent and reasonable, but that further studies were needed to determine if it was effective in preventing infection. A review by Dadd *et al.* (2003) indicated that a variety of protective isolation measures such as positive-pressure single rooms, low microbial diets and strict hand washing should be used only for children requiring allogeneic transplants. Children undergoing autologous transplants should not require the same level of protective isolation. A review by Hayes-Lattin *et al.* (2005) also recommended a low microbial diet for recipients of allogeneic stem cell transplantation until all immunosuppressive drugs are discontinued. Patients are also discouraged from drinking well water from private wells or public wells in small communities where bacterial contamination is tested less than twice daily. Similar advice is provided by the American Cancer Society (2005) (Box 5.1)

5.11 Conclusions

The immune system is a highly complex and integrated mechanism that can readily be compromised, thus increasing the risk of a person acquiring a foodborne infection. With improved medical therapies and the lengthening of the life span, the number of individuals with compromised immune systems is increasing and will continue to do so into the future. There is a significant amount of research and information available on how and why food-borne pathogens cause illness and how to prevent it from occurring. In theory, all foodborne illness is preventable and everyone who handles food from the grower to the consumer has a responsibility to ensure its safety while in their custody. Despite this, foodborne illnesses and deaths continue to occur. All consumers, including the immunocompromised, have a right to assume the food they buy is safe; however, certain types of food, such as raw meat are to be expected to have foodborne pathogens present. Therefore, proper handling of foods in healthcare settings and in the home is also critical, and health professionals have a responsibility to advise their patients on behaviours to minimize the likelihood of foodborne infection. Protecting the immunocompromised from foodborne infection is a team effort that includes the manufacturer, the health professional and the patient. There are many educational tools available, and many being developed, to assist the busy health professional with this mission.

References

Acheson, D.W.K. & Luccioli, S. (2004) Host enteropathogen interactions in the gut. *Best Practices and Research, Clinical Gastroenterology* **18**, 387–404.

Altekruse, S.F., Yang, S., Timbo, B.B. & Angulo, F.J. (1999) A multi-state survey of consumer food-handling and food-consumption practices. *American Journal of Preventive Medicine* **16**, 216–221.

American Cancer Society (2005) *Infections in People with Cancer*. Available from http://www.cancer.org/docroot/ETO/content/ETO_1_2X_Infections_in_people_with_cancer.asp. Accessed 2 March 2007.

Asturias, E.J., Grazioso, C.F., Luna-Fineman, S. *et al.* (2006) Poliovirus excretion in Guatemalan adults and children with HIV infection and children with cancer. *Biologicals* **34**, 109–112.

Ayabe, T., Satchell, D.P., Wilson, C.L. *et al.* (2000) Secretion of microbicidal alpha-defensins by intestinal Paneth cells in response to bacteria. *Nature Immunology* **1**, 113–118.

Baine, W.B., Gangarosa, E.J. & Bennett, J.V. (1982) Institutional salmonellosis. *Journal of Infectious Diseases* **128**, 357–682.

Berg, R.D. (1999) Bacterial translocation from the gastrointestinal tract. *Advances in Experimental Medicine and Biology* **473**, 11–30.

Beuchat, L.R. (1996) Pathogenic microorganisms associated with fresh produce. *Journal of Food Protection* **59**, 204–216.

Born, W., Cady, C., Jones-Carson, J. *et al.* (1999) Immunoregulatory functions of gamma delta T cells. *Advances in Immunology* **71**, 77–144.

Bowen, G.S. & McCarthy, M.A. (1983) Hepatitis A associated with a hardware store water fountain and a contaminated well in Lancaster County, Pennsylvania 1980. *American Journal of Epidemiology* **117**, 695–705.

Brandtzaeg, P.E. (2002) Current understanding of gastrointestinal immunoregulation and its relation to food allergy. *Annals of the New York Academy of Sciences* **964**, 13–45.

Buzby, J.C. (2001) Children and microbial foodborne illness. *Food Review* **24**, 32–37.

Caredda, F., Antinori, S., Re, T., Pastecchia, T. & Moroni, M. (1986) Acute non-B hepatitis after typhoid fever. *British Medical Journal* **292**, 1492.

Cazac, B.B. & Roes, J. (2000) TGF-beta receptor controls B cell responsiveness and induction of IgA in vivo. *Immunity* **13**, 443–451.

Centers for Disease Control and Prevention (CDC) (2000) Outbreaks of *Salmonella* serotype Enteritidis infection associated with eating raw or undercooked shell eggs – United States, 1996–1998. *Morbidity and Mortality Weekly Report* **49**(4), 74–79.

Centers for Disease Control and Prevention (CDC) (2002) Preliminary FoodNet data on the incidence of foodborne illnesses – selected sites, United States, 2001. *Morbidity and Mortality Weekly Report* **51**(15), 325–329.

Centers for Disease Control and Prevention (CDC) (2003) Public health and aging: trends in aging – United States and worldwide. *Morbidity and Mortality Weekly Reports* **52**(6), 101–106.

Centers for Disease Control and Prevention (CDC) (2005) HIV/AIDS basic statistics. *HIV/AIDS Surveillance Report: HIV Infection and AIDS in the United States*, 2005. Available from http://www.cdc.gov/hiv/topics/surveillance/basic.htm#aidscases. Accessed 2 March 2007.

Centers for Disease Control and Prevention (CDC) (2006a) *FoodNet Surveillance Report for 2004.* Available from http://www.cdc.gov/foodnet/annual/2004/Report.pdf. Accessed 2 March 2007.

Centers for Disease Control and Prevention (CDC) (2006b) Preliminary FoodNet data on the incidence of infection with pathogens transmitted commonly through food – 10 States, United States, 2005. *Morbidity and Mortality Weekly Report* **55**(14), 392–395.

Centers for Disease Control and Prevention (CDC) (2006c) *AIDS Surveillance – Trends 1985–2004.* Available from http://www.cdc.gov/hiv/topics/surveillance/resources/slides/trends/index.htm. Accessed 2 March 2007.

Clifford, C.P., Crook, D.W.M., Conlon, C.P. *et al.* (1990) Impact of waterborne outbreak of cryptosporidiosis on AIDS and renal transplant patients. *Lancet* **1**, 1455–1456.

Council for Agricultural Science and Technology (CAST) (1994) *Foodborne Pathogens: Risks and Consequences.* Task Force Report No. 122. Ames, IA. Available from CAST, 4420 West Lincoln Way, Ames, IA 50014-3447.

Craske, J. (1992) Hepatitis C and non-A non-B hepatitis revisited: Hepatitis E, F and G. *Journal of Infection* **25**, 243–250.

Cunningham, A.L., Grohman, G.S., Hatkness, J. *et al.* (1988) Gastrointestinal viral infections in homosexual men who were symptomatic and seropositive for immunodeficiency virus. *Journal of Infectious Diseases* **158**, 386–391.

Dadd, G., McMinn, P. & Monterosso, L. (2003) Protective isolation in hemopoietic stem cell transplants: a review of the literature and single institution experience. *Journal of Pediatric Oncology Nursing* **20**, 293–300.

FDA/CFSAN (2005a) *Food Code*. Available from http://www.cfsan.fda.gov. Accessed 2 March 2007.

FDA/CFSAN (2005b) *Foodborne Pathogenic Microorganisms and Natural Toxins Handbook: Vibrio Parahaemolyticus.* Available from http://www.cfsan.fda.gov/~mow/chap9.html. Accessed 2 March 2007.

FDA/CFSAN (2006) *Foodborne Pathogenic Microorganisms and Natural Toxins Handbook: The Norwalk Virus Family.* Available from http://vm.cfsan.fda.gov/~mow/chap34.html. Accessed 2 March 2007.

French, M.R., Levy-Milne, R. & Zibrik, D. (2001) A survey of the use of low microbial diets in pediatric bone marrow transplant programs. *Journal of the American Dietetic Association* **101**, 1194–1198.

Frisk, G. & Diderholm, H. (1992) Increased frequency of coxsackie B virus IgM in women with spontaneous abortion. *Journal of Infection* **24**, 141–145.

Gaskins, H.R. (1997) Immunological aspects of host/Microbiota interactions at the intestinal epithelium. In: Mackie, R.I. & White, B.A. (eds). *Gastrointestinal Microbiology.* Chapman & Hall, New York, pp. 537–587.

Gerber, C.P., Rose, J.B. & Haas, C.N. (1996) Sensitive populations: who is at the greatest risk? *International Journal of Food Microbiology* **30**, 112–123.

Gordon, S.M., Oshiro, L.S., Jarvis, W.R. *et al.* (1990) Foodborne Snow Mountain agent gastroenteritis with secondary person-to-person spread in a retirement community. *American Journal of Epidemiology* **131**, 702–710.

Gottschlich, M.M. (1993) Nutrition in the burned pediatric patient. In: Queen, P.M. & Lang, C.E. (eds). *Handbook of Pediatric Nutrition*. Aspen Publishers, Gaithersburg, MD, pp. 537.

Gouandjika-Vasilache, I., Akoua-Koffi, C., Begaud, E. & Dosseh, A. (2005) No evidence of prolonged enterovirus excretion in HIV-seropositive patients. *Tropical Medicine & International Health* **10**, 743–747.

Goulet, V., Jacquet, C., Martin, P. *et al.* (2006) Surveillance of human listeriosis in France, 2001–2003. *Euro Surveillance* **11**(6), 79–81.

Guarner, C., Runyon, B.A., Young, S. *et al.* (1997) Intestinal bacterial overgrowth and bacterial translocation in cirrhotic rats with ascites. *Journal of Hepatology* **26**, 1372–1378.

Gust, I. (1990) Design of hepatitis A vaccines. *British Medical Bulletin* **46**, 319–328.

Gust, I. & Purcell, R.H. (1987) Report of a workshop: waterborne non-A non-B hepatitis. *Journal of Infectious Diseases* **156**, 630–635.

Halvorsrud, J. & Orstavik, I. (1980) An epidemic of rotavirus-associated gastroenteritis in nursing home for the elderly. *Scandinavian Journal of Infectious Diseases* **12**, 161–164.

Hayes, C., Elliot, E., Krales, E. & Downer, G. (2003) Food and water safety for persons infected with Human Immunodeficiency Virus. *Clinical Infectious Diseases* **36**(Suppl 2), S106–S109.

Hayes-Lattin, B., Leis, J.F. & Maziarz, R.T. (2005) Isolation in the allogeneic transplant environment: how protective is it? *Bone Marrow Transplantation* **36**, 373–381.

Health Protection Agency (HPA) (2007) *Listeria moncytogenes Human Cases in Residents of England and Wales 1983–2005*. Available from http://www.hpa.org.uk/infections/topics_az/listeria/data_ew.htm. Accessed 2 March 2007.

Hierholzer, J.C. (1992) Adenoviruses in the immunocompromised host. *Clinical Microbiology Reviews* **5**, 262–274.

Hunter, P.R. & Nichols, G. (2002) The epidemiology and clinical features of cryptosporidium infection in immune-compromised patients. *Clinical Microbiology Reviews* **15**, 145–154.

Janoff, E.D. & Smith, P.D. (1988) Perspectives on gastrointestinal infections in AIDS. *Gastroenterology Clinics of North America* **17**, 451–463.

Kang, G., Srivastava, A., Pulimood, A.B. *et al.* (2002) Etiology of diarrhea in patients undergoing allogeneic bone marrow transplantation in South India. *Transplantation* **73**, 1247–1251.

Kaplan, J.E., Schonberger, L.B., Varano, G. *et al.* (1982) An outbreak of acute nonbacterial gastroenteritis in a nursing home. *American Journal of Epidemiology* **116**, 940–948.

Kaplan, M.H., Klein, S.W., McPhee, J. & Harper, R.G. (1983) Group B coxsackievirus infections in infants younger than three months of age: a serious childhood illness. *Reviews of Infectious Diseases* **5**, 1019–1032.

Kapperud, G., Jenum, P.A., Stray-Pedersen, B. *et al.* (1996) Risk factors for *Toxoplasma gondii* infection in pregnancy: results of a prospective case–control study in Norway. *American Journal of Epidemiology* **144**, 405–412.

Kendall, P., Medeiros, L.C., Hillers, V. *et al.* (2003) Food handling behaviors of special importance for pregnant women, infants and young children, the elderly, and immune-compromised people. *Journal of the American Dietetic Association* **103**, 1646–1649.

Kindon, H., Pothoulakis, C., Thim, L. *et al.* (1995) Trefoil peptide protection of intestinal epithelial barrier function: cooperative interaction with mucin glycoprotein. *Gastroenterology* **109**, 516–523.

Koch, J. & Stark, K. (2006) Significant increase of listeriosis in Germany – Epidemiological patterns 2001–2005. *Euro Surveillance* **11**(6), 85–88.

Kraehenbuhl, J.P. & Neutra, M.R. (2000) Epithelial M cells: differentiation and function. *Annual Review of Cell Developmental Biology* **16**, 301–332.

Lamont, J.T. (1992) Mucus: the front line of intestinal mucosal defense. *Annals of the New York Academy of Sciences* **664**, 190–201.

Larson, E. & Nirenberg, A. (2004) Evidence-based nursing practice to prevent infection in hospitalized neutropenic patients with cancer. *Oncology Nursing Forum* **31**, 717–725.

Ledner, W.M., Lemon, S.M., Kirkpatrick, J.W. *et al.* (1985) Frequency of illness associated with epidemic hepatitis A virus infections in adults. *American Journal of Epidemiology* **122**, 226–233.

Leon, F., Roldan, E., Sanchez, L. *et al.* (2003) Human small-intestinal epithelium contains functional natural killer lymphocytes. *Gastroenterology* **125**, 345–356.

Levine, W.C., Smart, J.F., Archer, D.L. *et al.* (1991) Foodborne disease outbreaks in nursing homes, 1975 through 1987. *Journal of the American Medical Association* **266**, 2105–2109.

Lew, J.F., Glass, R., Gangarosa, R.E. *et al.* (1991) Diarrheal deaths in the United States, 1979 through 1987. *Journal of the American Medical Association* **265**, 3280–3284.

Lopman, B.A., Adak, G.K., Reacher, M.H., & Brown, D.W.G. (2003) Two epidemiologic patterns of norovirus outbreaks: surveillance in England and Wales, 1992–2000. *Emerging Infectious Diseases* **9**, 71–77.

Lorber, B. (1997) Listeriosis. *Clinical Infectious Diseases* **24**, 1–11.

Mattner, F., Sohr, D., Heim, A. *et al.* (2006) Risk groups for clinical complications of norovirus infections: an outbreak investigation. *Clinical Microbiology and Infection* **12**, 69–74.

Maury, J., Nicoletti, C., Guzzo-Chambraud, L. & Maroux, S. (1995) The filamentous brush border glycocalyx, a mucin-like marker of enterocyte hyper-polarization. *European Journal of Biochemistry* **228**, 323–331.

Mayer, L. (2003) Mucosal immunity. *Pediatrics* **111**(Suppl S), 1595–1600.

McGuirk, P. & Mills, K.H. (2002) Pathogen-specific regulatory T cells provoke a shift in the Th1/Th2 paradigm in immunity to infectious diseases. *Trends in Immunology* **23**, 450–455.

Meyers, B.R. (1989) Infectious diseases in the elderly: an overview. *Geriatrics* **44**, 4–6.

Modlin, J.F. & Kinney, J.S. (1987) Prenatal enterovirus infections. *Advances in Pediatric Infectious Diseases* **2**, 57–78.

Mohle-Boetani, J.C., Farrar, J.A., Werner, S.B. *et al.* (2001) *Escherichia coli* O157: H7 and *Salmonella* infections associated with sprouts in California, 1996–1998. *Annals of Internal Medicine* **135**, 239–247.

Mowat, A.M. (2003) Anatomical basis of tolerance and immunity to intestinal antigens. *Nature Reviews Immunology* **3**, 331–341.

Mylonakis, E., Paliou, M., Hohmann, E.L. *et al.* (2002) Listeriosis during pregnancy: a case series and review of 222 cases. *Medicine* **81**, 260–269.

National Center for Health Statistics (2003) *Health, United States, 2003*. U.S. Department of Health and Human Services, Centers for Disease Control and Prevention, Hyattsville, MD.

O'Boyle, C.J., MacFie, J., Mitchell, C.J. *et al.* (1998) Microbiology of bacterial translocation in humans. *Gut* **42**, 29–35.

Ogunmodede, F., Jones, J.L., Scheftel, J. *et al.* (2005) Listeriosis prevention knowledge among pregnant women in the USA. *Infectious Disease and Obstetric Gynecology* **13**, 11–15.

Oppenheim, J.J., Biragyn, A., Kwak, L.W. & Yang, D. (2003) Roles of antimicrobial peptides such as defensins in innate and adaptive immunity. *Annals of the Rheumatic Diseases* **62**(Suppl 2), ii 17–ii 21.

Oshiro, L.S., Halet, C.E., Roberto, R.R. *et al.* (1981) A 27-nm virus isolated during an outbreak of acute infectious nonbacterial gastroenteritis in a convalescent hospital: a possible new serotype. *Journal of Infectious Diseases* **143**, 791–795.

Paul, W.E. & Seder, R.A. (1994) Lymphocyte responses and cytokines. *Cell* **76**, 241–251.

Pether, J.V.S. & Caul, E.O. (1983) An outbreak of foodborne gastroenteritis in two hospitals associated with a Norwalk-like virus. *Journal of Hygiene (Cambridge)* **91**, 343–350.

Rajakulendran, S., Gadsby, K., Allen, D. *et al.* (2006) Neutropenia while receiving anti-tumour necrosis factor treatment for rheumatoid arthritis. *Annals of the Rheumatic Diseases* **65**, 1678–1679.

Reid, J.A., Breckon, D. & Hunter, P.R. (1990) Infection of staff during an outbreak of viral gastroenteritis in an elderly persons' home. *Journal of Hospital Infection* **16**, 81–85.

Robinson, J.K., Blanchard, T.G., Levine, A.D. *et al.* (2001) A mucosal IgA-mediated excretory immune system in vivo. *Journal of Immunology* **166**, 3688–3692.

Rodriguez-Guillen, L., Vizzi, E., Alcala, A.C. *et al.* (2005) Calicivirus infection in human immunodeficiency virus seropositive children and adults. *Journal of Clinical Virology* **33**, 104–109.

Rolfe, R.D. (1997) Colonization resistance. In: Mackie, R.I. & White, B.A. (eds). *Gastrointestinal Microbiology*. Chapman & Hall, New York, pp. 501–536.

Ryan, C.A., Hargrett-Bean, N.T. & Blake, P.A. (1989) *Salmonella typhi* infections in the United States 1975–1984: increasing role of foreign travel. *Reviews of Infectious Diseases* **11**, 1–8.

Scialli, A.R. & Rarick, T.L. (1992) Salmonella sepsis and second trimester pregnancy loss. *Obstetrics and Gynecology* **79**, 820–821.

Shiloh, M.U. & Nathan, C.F. (2000) Reactive nitrogen intermediates and the pathogenesis of Salmonella and mycobacteria. *Current Opinion in Microbiology* **3**, 35–42.

Skirrow, M.B., Jones, D.M., Sutcliffe, E. & Benjamin, J. (1993) Campylobacter bacteraemia in England and Wales, 1981–1991. *Epidemiology and Infection* **110**, 567–573.

Smith, J.L. (1997) Long-term consequences of foodborne toxoplasmosis: effects on the unborn, the immunocompromised, the elderly and the immunocompetent. *Journal of Food Protection* **60**, 1595–1611.

Smith, J.L. (1998) Foodborne illness in the elderly. *Journal of Food Protection* **61**, 1229–1239.

Smith, J.L. (1999) Foodborne infections during pregnancy. *Journal of Food Protection* **62**, 818–829.

Smith, P.D. & Janoff, E.N. (2002) Gastrointestinal infections in HIV-1 disease. In: Blaser, M.J., Smith, P.D. Ravdin, J.I. *et al.* (eds). *Infections of the Gastrointestinal Tract*, 2nd edn. Lippincott Williams & Wilkins, Philadelphia, pp. 415–443.

Sugiyama, H. & Mills, D.C. (1978) Intraintestinal toxin in infant mice challenged intragastrically with *Clostridium botulinum* spores. *Infection and Immunity* **21**, 59–63.

Szekeres-Bartho, J. (1993) Endocrine regulation of the immune system during pregnancy. In: Chaouat, G. (ed). *Immunology of Pregnancy*. CRC Press, Boca Raton, FL, pp. 151–172.

Tamura, A., Soga, H., Yaguchi, K. *et al.* (2003) Distribution of two types of lymphocytes (intraepithelial and lamina-propria-associated) in the murine small intestine. *Cell and Tissue Research* **313**, 47–53.

Tannock, G.W. (1997) Modification of the normal Microbiota by diet, stress, antimicrobial agents, and probiotics. In: Mackie, R.I. & White, B.A. (eds). *Gastrointestinal Microbiology*. Chapman & Hall, New York, pp. 434–465.

Taupin, D. & Podolsky, D.K. (2003) Trefoil factors: initiators of mucosal healing. *Nature Reviews: Molecular Cell Biology* **4**, 721–732.

Todd, E.C.D. (1989) Preliminary estimates of costs of foodborne disease in the United States. *Journal of Food Protection* **52**, 595–601.

Townsend, T.R., Yolken, R.H., Bishop, C.A. *et al.* (1982) Outbreak of coxsackie Al gastroenteritis, a complication of bone-marrow transplantation. *Lancet* **1**, 820–823.

University of California, San Francisco Children's Hospital (UCSF) (2006) *Bone Marrow Transplant. Dietary Concerns During Bone Barrow Transplant.* Available from http://www.ucsfhealth.org/childrens/medical_services/cancer/bmt/diet.html. Accessed 8 March 2007.

U.S. Gateway to Government Food Safety Information (2007) Available from http://www.foodsafety.gov/. Accessed 2 March 2007.

Weinberger, M., Andorn, N., Agmon, V. *et al.* (2004) Blood invasiveness of*Salmonella* enterica as a function of age and serotype.*Epidemiology and Infection* **132**, 1023–1028.

Williams, N. (1998) T cells on the mucosal frontline. *Science* **280**, 198–200.

World Health Organization (WHO) (2004) *Technical Report: Risk Assessment of Listeria Monocytogenes in Ready-to-Eat Foods.* Available from ftp://ftp.fao.org/docrep/fao/007/y5394e/y5394e00.pdf. Accessed 2 March 2007.

Xia, Y., Hu, H.Z., Liu, S. *et al.* (2000) *Clostridium difficile* toxin A excites enteric neurones and suppresses sympathetic neurotransmission in the guinea pig. *Gut* **46**, 481–486.

Yolken, R.H., Bishop, C.A., Townsend, T.R. *et al.* (1982) Infectious gastroenteritis in bone marrow recipients. *New England Journal of Medicine* **306**, 1009–1012.

Ziegler, H.K., Skeen, M.J. & Pearce, K.M. (1994) Role of alpha/beta T and gamma/delta T cells in innate and acquired immunity. *Annals of the New York Academy of Sciences* **730**, 53–70.

6 Provision of Food in Healthcare Settings

Dinah Barrie

6.1 Introduction

The provision of food for healthcare facilities is completely different from that for other large organizations. Some patients have no appetite or suffer from nausea, some are debilitated due to chronic illness, malnutrition, malignancy or by treatment with powerful drugs, and others have difficulty with swallowing. Many patients are susceptible to infections, including food poisoning, because of their illness or as a result of treatment such as chemotherapy. Some patients and staff need special diets, for example, those who have coeliac disease or who cannot tolerate, or are allergic to, particular foods for example milk. The dietary requirements of different religious and ethnic groups have to be considered; these include eating meat according to Kosher or Halal regulations or meat only from certain animals. Some groups and individuals are vegetarians.

The healthcare management has to provide all the meals for their patients and to ensure that they have an adequate choice, that the food is nutritious, palatable and attractive and, above all, that it is microbiologically safe. Complicating factors are that patients may be admitted or discharged without warning and some may not be in their ward when food is being served, due to treatment or an investigation in another part of the hospital – the subsequent delay between the food leaving the kitchen and being delivered to the patient has serious implications for food safety. Many staff have some or most of their meals in the hospital. Visitors' and patients' relatives should be considered.

Food may be prepared on-site or in off-site production units from which it is sent to patients and staff. Microbiological safety involves using a Hazard Analysis Critical Control Point (HACCP) system, complying with food law, having written procedures for all food handing procedures, and training and supervising the catering staff and also all other members of staff who handle food, from the time that raw or cooked food enters the premises until it is served to the consumer (Chapter 8).

6.2 Food law in the UK

Providers of food are responsible for ensuring that their activities comply with food safety law. The most important laws for healthcare premises in Great Britain are:

- Food Safety Act 1990
- Food Hygiene (England) Regulations 2006, and similar regulations in Scotland and Wales

Food Safety Act 1990: The main provisions of this act came into force in January 1991 and provide the framework for all food legislation in Great Britain. The Act applies to anyone working in the production, processing, storage, distribution and sale of food, no matter how small or large their business. The Act aims to ensure that all food produced for sale is safe. The definition of 'food' includes anything used as an ingredient including additives such as colours, sweeteners and preservatives and also includes water used in food production or drawn from a tap in the course of a food business.

The Food Hygiene (England) Regulations 2006: These regulations (and similar regulations in Scotland and Wales) give effect to EC Regulations, including those relating to general hygiene for all food business operators, laid down in EC Regulation 852/2004 (EC, 2004a), and specific requirements for food businesses dealing with foods of animal origin, laid down in EC Regulation 853/2004 (EC, 2004b). They include a requirement that food businesses put in place, implement and maintain procedures based on the principles of HACCP. EC Regulation 2073/2005 (EC, 2005) covers implementation of measures on microbiological criteria for foodstuffs.

The regulations set out the basic hygiene requirements for all aspects of food preparation, processing and transport, premises and facilities and the personal hygiene of the staff. These are summarized in Table 6.1.

The regulations include temperature control requirements, which specify that any food (raw, ingredients, intermediate or finished products) likely to support the growth of pathogenic microorganisms or the formation of toxins must not be kept at temperatures which would result in a risk to health. The requirements specify that chill holding of any food that is likely to support the growth of pathogenic microorganisms or the formation of toxins must be at 8°C or lower. A holding temperature between 8°C and ambient temperature is allowed while food is being served but only if at those temperatures for less than 4 hours or during transfer (for a limited period which is consistent with food safety) to another site where it will be kept at or below 8°C.

Table 6.1 Summary of the structural and basic hygiene requirements of the regulation on the hygiene of foodstuffs.

Premises	Design, layout, construction siting and size	Adequate space that is easily cleaned and prevents the entry of pests
	Lavatories	Adequate number of flush lavatories connected to an effective drainage system. Not to open directly into rooms where food is handled
	Washbasins	Adequate number of washbasins with hot and cold running water, materials for cleaning hands and for hygienic drying. Where necessary, separate facilities for washing hands and washing food
	Ventilation	Suitable and sufficient natural or mechanical ventilation. Mechanical airflow from contaminated to clean area to be avoided. Filters and other parts of ventilation system should be readily accessible for cleaning or replacement
	Lighting	Adequate natural and/or artificial
	Drainage	Adequate and designed and constructed to avoid risk of contamination
	Changing facilities	Adequate for number of staff
Rooms where foods are prepared, treated or processed	Floors, walls, ceilings, windows, surfaces	Sound condition, easy to clean and where necessary to disinfect. Windows fitted, if necessary, with insect-proof screens
	Tools, utensils, equipment	Adequate facilities for cleaning, including hot and cold water supplies, and for disinfection and storage
	Washing food	Adequate facilities, sinks provided with hot and/or cold potable water
Transport	Conveyances and/or containers	Clean, in good repair, protect foods from contamination; where necessary easily cleaned and disinfected. Where several foods or products are transported, ensure separation
	Receptacles in vehicles and/or containers	Dedicated for food and in good condition, easily cleaned and/or disinfected
	Bulk foodstuffs	To be transported in receptacles and/or containers/tankers reserved for the purpose and marked clearly to show they are for foodstuffs only
	Vehicles	If used for different foodstuffs or for other products, ensure effective cleaning between loads
		Where necessary, can transport food at correct temperatures which can be monitored
		Foodstuffs and/or containers to be placed and protected to minimize contamination
Equipment	Articles, utensils	Of sound condition. In good repair. Easily cleaned
Waste	Food	Disposed of as soon as possible
	Containers	To be deposited in closable containers, sound, easy to clean and, where necessary, to disinfect
	Storage and removal	Provision to store food until removed Stores kept clean and pest proof. Waste should not be a source of contamination

Table 6.1 *(Continued).*

Water supply	Potable (drinking)	Adequate supply
	Clean	Clean water may be used for whole fishery products and clean seawater for other seafood including live, bivalve molluscs
	Recycled water	Not to present a risk of contamination
	Ice	To be made from potable water or, when used to chill whole fishery products, clean water. To be protected from contamination
Personal hygiene	Cleanliness and clothing	Staff working in a food handling area maintain a high standard of personal cleanliness and wear appropriate, clean, and where necessary protective, clothing
	Disease	No person suffering from, or a carrier of foodborne disease, or suffering from infected wounds, skin infections, sores or diarrhoea is to be permitted to handle food or enter any food handling area in any capacity if there is a likelihood of direct or indirect contamination
Foodstuffs	Raw	Items should not be accepted if they are liable to be contaminated, and even after normal sorting and/or preparatory or processing procedures the final product would be unfit for human consumption
	Storage	Conditions prevent deterioration and contamination
	Protection against contamination	Handling, storage, display, transport conditions protect against contamination
Wrapping and packaging	Materials used	Should not be contaminated, should not contaminate products, if re-used, should be easy to clean and, where necessary, disinfect
Heat treatment	Food marketed in hermetically sealed containers	Specific requirements regarding control of heating
Training	Food handlers	Should be supervised, and instructed and/or trained in food hygiene commensurate with their work
	Personnel responsible for development and maintenance of HACCP system or operation of guides	Should have received training in application of HACCP principles
	Persons working in certain food sectors	Comply with any national requirements for training

Data from EC, 2004a. The original document should be consulted for full details.

Hot holding requirements are that food that has been cooked or reheated, or is for service or on display for sale and is likely to support the growth of pathogenic microorganisms or the formation of toxins, must be kept at or above 63°C. If the temperature falls below 63°C, it would not be a contravention of the regulation if the period were less than 2 hours.

Food must be cooled as quickly as possible after cooking or after the final preparation if it is then to be kept at less than ambient temperatures.

The Food Safety (Sampling and Qualifications) Regulations 1990. These set out the procedures to be followed by enforcement officers when taking samples for analysis or microbiological examination. They also set out the qualification requirements for public analysts and food examiners.

6.3 Better hospital food plan and "Eating well in hospital" programme

In the UK, a plan for improvement of the National Health Service (DH, 2000) outlined the following main actions relating to better hospital food:

- A 24-hour NHS catering service with a new NHS menu, designed by leading chefs. It would cover continental breakfast, cold drinks and snacks on at least two occasions per day, light lunchtime meals and an improved two-course evening dinner.
- A national franchise for NHS catering would be examined to ensure that hospital food is provided by organizations with a national reputation for high-quality customer satisfaction.
- Hospitals would have new 'ward housekeepers' to ensure that the quality, presentation and quantity of meals meets patients' needs: that patients, particularly elderly patients, are able to eat the meals on offer, and that the service patients receive is genuinely available around the clock.
- Dieticians would advise and check on nutritional values in hospital food. Patients' views would be measured as part of the performance assessment framework.

The plan also provided for:

- Unannounced inspections of hospital food premises.
- Taking into account regional differences in eating habits, the varying needs of patients from paediatric to geriatric individuals, differences across ethnic groups, therapeutic diets and provision of food according to the clinical treatment that patients may be receiving.
- Special arrangements for hospitals with patients who are elderly, mentally ill or with learning disabilities.

As part of this plan the Better Hospital Food Programme was launched in 2001. Individuals within and outside the NHS, including patients' representatives, would provide advice and identify new standards for hospital food services. The Better Hospital Food Programme is due to be replaced by a new programme, 'Eating well in hospital' (NAO, 2006).

6.4 Catering systems

These include prime cooking (conventional cooking), cook-chill cooking and cook-freeze cooking.

In prime cooking food is prepared and cooked in the on-site kitchen, preferably as close to the time of consumption as possible. It should be served and consumed within 15 minutes of removal from hot or from cold trolleys.

Cook-chill and cook-freeze cooking are alternative methods that can give greater flexibility than conventional cooking in the preparation and serving of meals. Food is prepared throughout the working day, the portions are distributed, as required, from a central store to a hospital where the food is regenerated before serving; these methods enable a wider choice of palatable and nutritious food than is possible with conventional catering.

In *cook-chill catering* the food is fully cooked; according to UK guidelines the temperature at the centre of the food must be maintained at a temperature of at least 70°C for 2 minutes during cooking (DH, 1989; NHS, 1996). Chilling must begin as soon as possible but within 30 minutes of leaving the cooker, and the temperature should fall to between 0 and 3°C within 90 minutes. Typical methods of chilling are given (DH, 1989). If the food is to be portioned into smaller quantities this must be completed within 30 minutes of cooking. Wherever possible this handling after cooking must be carried out in a controlled environment with a maximum temperature of 10°C.

During storage and transport to the satellite (hospital) kitchen the chilled food must be kept below 3°C. Stores for holding pre-cooked chilled foods in quantity should be specially designed for the purpose and should have a temperature-recording system with an alarm to indicate when the air temperature in the store has risen to an unacceptable level (DH, 1989). If the temperature rises to 5°C but remains below 10°C for very short periods of time the food must be consumed as soon as possible and in any case within 12 hours. If the temperature rises higher than 10°C the food must be destroyed. To regenerate, reheating should commence as soon as possible, but always within 30 minutes of removal of the food from chill. The maximum life for cook-chill food is 5 days including the day of cooking and the day of consumption.

In the *cook-freeze system* the food is fully cooked, and according to UK guidelines the centre must reach a temperature of at least 70°C and be maintained at that temperature for 2 minutes. Portioning before freezing must be carried out under the same conditions as the portioning of cook-chilled food.

Freezing should commence as soon as possible, and in any event within 30 minutes, after cooking (DH, 1989). The temperature at the centre should be at least minus 5°C within 90 minutes of entering the freezer and subsequently reach a storage temperature of minus 18°C.

Air temperature measurement should be recorded and alarm devices should be as specified for chilled food.

During storage and transport to the hospital kitchen and also from there to the point of consumption, that is to the wards, staff dining rooms, meetings etc., the food must be kept at or below minus 18°C until regeneration. The shelf life of cook-freeze foods varies according to the type of food but in general it is up to 8 weeks. To regenerate, the frozen food is thawed and then held at or below 3°C and never above 10°C until reheated to a temperature of at least 70°C for 2 minutes or it is regenerated, without prior thawing, to reach a temperature of at least 70°C for 2 minutes. Food thawed in rapid thaw cabinets should be consumed with 24 hours of thawing. Detailed recommendations regarding distribution, reheating and service are provided (DH, 1989).

Considerable caution must be used if reheating is carried out using microwave ovens. Only ovens designed for commercial use must be used and these only according to the manufacturer's instructions (NHS, 1996), and it is essential that a temperature of at least 70°C for 2 minutes should be achieved in all parts of the food. A commercial microwave oven must have a suitable defrost programme if it is to be used for thawing frozen food.

The NHS guidance on Hospital Catering (NHS, 1996) advised that particular care *must* be paid to the cooking of 'at risk' foods, which must be well cooked, in order to achieve a core temperature of 75°C and then held at safe temperatures, if required for service, at 63°C or above. Foods considered at risk included in general:

- all foods for vulnerable care groups (e.g. the elderly and acutely ill patients)
- all proprietary foods used for enteral and tube feeds

specifically:

- poultry and pork
- minced meat products (e.g. for burgers and sausages)
- dried pulses (e.g. kidney beans)
- rice
- foods to be served cold
- foods being reheated (e.g. minced and pureed meats, poultry for soft diets) and dishes requiring reheating as part of their cooking process
- meat stocks and sauces

The guidelines included the advice that food temperatures and times must be controlled carefully and temperature probes used to record food temperatures (together with the proper use of alcohol wipes).

The particular temperatures that must be achieved and maintained during the stages of cooking, chilling/freezing, storage, transport and regeneration, and the maximum times allowed before portioning, chilling, freezing and from regeneration to serving are based on the effect on the growth and destruction of microbial pathogens (Table 6.2).

Table 6.2 The relationship between time/temperature and the growth and destruction of microbes.

Stages	Effect on growth of microbes
Use of good quality ingredients	Less likelihood of contamination
Cooking temperature, at 70°C for 2 min	Destroys most non-sporing pathogenic bacteria
Chilling to commence within 30 min and completed within 90 min	Minimizes growth of bacteria that have survived cooking
Freezing to commence and completed within 30 min of cooking with a centre temperature of minus 5°C within 90 min	Minimizes growth of bacteria that have survived cooking
Portioning of cooked food to be completed as soon as possible but always within 30 min	Minimizes growth of bacteria that have survived cooking
Storage of chilled food at 0–3°C	Bacteria multiply very slowly at these temperatures
Storage of frozen food at minus 18°C	Bacteria do not grow at this temperature
Regeneration at 70°C for 2 min Serve within 15 min of cooking	Any remaining vegetative bacteria are destroyed Reduces bacterial growth of any contaminating bacteria

Another major consideration in the use of these systems is their effect on nutrients and flavour. Loss of nutrients and of flavour occur if food is overcooked and if there is a delay between reheating and consumption. Rapid chilling causes less loss than slow cooling. Very slow losses occur in frozen cooked food.

The decision to introduce cook-chill/cook-freeze catering is taken according to local catering needs. Whether it is decided to use conventional, cook-chill, cook-freeze or a combination of the systems the Environmental Health Officer (EHO) can be an important member of the planning group.

The NHS Management Executive (NHS, 1992a) stated that the local Chief EHO, or his representative, and the NHS EHO can be invited when changes are being planned to kitchens or other food handling areas.

6.5 Contracts

Many hospitals have contracts with outside caterers for the supply of hospital meals. The NHS has set out contract standards, including a requirement that there must be a written Food Safety control system based on principles of HACCP (NHS, 1996) (Chapter 8). The standards also state that the Hospital Trust should nominate persons to advise the catering management team on all aspects of food hygiene and to help with training procedures to be followed, e.g. Consultant in Communicable Disease Control (CCDC); Control of Infection Officer, local EHOs. Under the NHS Supply Chain management (NHS Supply Chain, 2007), about 300 hospital food contracts are spread across the different catering systems, i.e. fresh, chilled, frozen and ambient groceries. Until September 2006, the NHS Logistics Authority, since 1 October 2006 NHS Supply Chain, has been utilized as the supply channel for the majority of ambient groceries, whilst suppliers deliver 'direct' to customers their fresh, chilled and frozen food requirements. Hospital Trusts are autonomous and can arrange their own contracts. NHS Supply Chain has the responsibility for negotiating national framework contracts with suppliers, which NHS Trusts may then use to buy a range of food items, and auditing the food safety of these suppliers (NAO, 2006); thus any Trust using one of their nominated suppliers will be covered by a due diligence defence. Hospitals may consider changing to a ready-prepared meals option when there is an inability to secure suitable staff, where space in hospital is deemed more suitable for medical purposes rather than catering facilities, where kitchens are in need of refurbishment and taking into account the improving quality of ready-prepared meals.

NHS Supply Chain liaises with a number of hospital representatives including the catering manager, nutritionist, supplies personnel and procurement leads/board leads etc.

With the introduction of the Better Hospital Food Plan, NHS Purchasing and Supply Agency (PASA) worked to source appropriate suppliers/products to meet the requirements of the plan. The continuation of this work has now migrated to the Food team within NHS Supply Chain.

Some larger suppliers are able to offer both traditional meals and authentic ethnic meals but, in general, the majority of authentic ethnic meals are sourced from specialist suppliers. Agreements have been developed for authentic ethnic meals and special diets, and NHS

Supply Chain works with religious/cultural leaders and the British Dietetic Association in order to ensure the products available are suitable for their particular consumers.

Under the NHS Supply Chain Food Hygiene contract, food safety and quality control are assured by the following measures (NHS, 2001). The successful supplier must be approved by the NHS Supply Chain appointed Hygiene Auditors in compliance with the NHS Code of Practice (NHS, 2001) or with the technical standard of the British Retail Consortium (BRC, 2005) prior to supplying food to a Health Authority. To meet the requirements of the Food Safety Act 1990, NHS Supply Chain requires all their food suppliers to be audited by announced and unannounced inspections by qualified food inspectors appointed by the NHS Supply Chain contractor responsible for due diligence defence. Currently (February 2007), this contractor is Support Training Services Ltd (STS). This is a prerequisite for new suppliers, while existing suppliers are audited in accordance with a pre-arranged timetable. The costs of the announced audits are borne by the contractor. Occasionally, unannounced audits and product testing are carried out at the expense of NHS Supply Chain (although this procedure will be subject to review). Product testing (including microbiological/nutritional content and ingredient compliance against specification) is undertaken.

NHS Supply Chain must be informed of all recipe changes a minimum of 3 months in advance of implementation and may re-evaluate the product to decide whether or not it is acceptable. The contractor must ensure that whilst working on health authority premises its staff are properly trained and instructed with regard to the tasks to be performed, standards of dress and all other relevant rules and procedures (e.g. health and safety at work, fire risks, fire precautions), etc.

6.6 Quality control

Everyone involved in the chain of food production (obtaining the raw materials, during manufacture, storage, preparation, cooking, transport and distribution) has a responsibility to ensure that food is safe from contamination and deterioration. Systems based on HACCP principles are designed to prevent the occurrence of problems that threaten food safety. HACCP is discussed in detail in Chapter 8. It relies on a food business taking responsibility for the safety of the food it produces by identifying potential hazards and the measures needed for their control. In a food safety context, hazards can be biological (bacteria, toxins, parasites and viruses); physical (stones, glass, metal fragments and packaging materials); and chemical (cleaning compounds and insecticides).

6.7 Inspections

6.7.1 *Inspections by EHOs*

EHOs carry out inspections of food processors and caterers in general, and of all food handling areas in healthcare settings. Inevitably most of the inspection in hospital centres on the main kitchens, but it includes all ward kitchens and the staff dining areas. Inspections are undertaken at a frequency between 6 months and 5 years and are unannounced where possible.

All aspects from the delivery, whether of raw food, cook-chill or cook-freeze food, the preparation, cooking, through to the point of consumption are considered. The primary aims of the inspection are to determine whether there are contraventions of the Food Safety Act 1990 and the Food Safety Regulations, and to advise on how these can be remedied. Inspection aims also to identify potential risks, indicate action that is needed or, when the risks are serious, to issue an enforcement notice. The observations of the inspector are based on the requirements of the Food Safety Act 1990 and the Food Safety Regulations as summarized in Table 6.1.

In addition, the inspector will be particularly concerned about temperature control, noting:

- if food is left at room temperature
- that there are enough refrigerators and deep freezes and that they are at the correct temperature
- that defrosting is carried out at the right temperature
- that food is cooked at the correct temperature and probe thermometers are used
- that food trolleys are pre-heated and the core (centre) temperature of the food (whether served from a conventional or regeneration trolley) is measured

The inspector clearly differentiates between recommendations of good hygiene practices and matters that are contraventions of food hygiene legislation. The assessment of the management and control systems, especially in relation to food hygiene and safety and use of the correct temperatures at all times, are important considerations.

6.7.2 *Inspections by hospital staff*

It has been advised that a representative of the hospital management, CCDC, Infection Control Doctor (ICD) and a member of the works department should together inspect all food handling areas twice a year (NHS, 1986) and record their findings. Recently, the author conducted a telephone survey of 20 of my microbiology colleagues; only three inspected the kitchens annually; several questioned whether it was necessary to do so. However, this author is fortunate to have taken part in several inspections with an experienced EHO and some of his findings and the reasons for his concerns are shown in Table 6.3 (Barrie, 1996).

6.8 Microbiological analysis

Microbiological assessment of food samples is performed when a cook-chill or cook-freeze catering system is being commissioned or altered during investigations of suspected food poisoning.

When the cook-chill/cook-freeze systems were being established microbiological sampling ensured that production standards were maintained, and was used selectively for high-risk items during the introductory phase (Wilkinson, 1988). The value of this surveillance was shown in a survey (Sandys & Wilkinson, 1988) when 3000 items were cultured during 18 months. Unsatisfactory results were investigated and faulty processes modified with an improvement in the subsequent results.

Table 6.3 Examples of observations made by an Environmental Health Officer, while examining food handling areas.

Observation	Reason for concern
Noting from the planned menus when the food being prepared is to be eaten	Food prepared too far in advance
Temperature readings of refrigerators, deep freezers, stores, ovens (during use)	Incorrect food storage, inadequate temperature control
Refrigerators and deep freezers in need of defrosting	Poor maintenance, inadequate temperature control
Food in bain-maries	Prolonged reheating, inadequate temperature control
Worn numbers on oven dials illegible	Inadequate temperature control
Items past 'use by' date, blown tins	Poor stock rotation Potential unfit/unsound food
Rubbish under large equipment	Cleaning schedules/standards
Cleanliness of environment/equipment	Cleaning schedules/standards
Guard on meat slicers taken apart safely and cleaned in the dishwasher?	Cleaning schedules/standards
Condition of the dishwasher?	Cleaning schedules/standards
Multipurpose cloths	Cross contamination
Separate areas/equipment/utensils for raw and cooked food	Cross contamination
Waste disposal	Pest control
Do cleaning staff know where insecticide has been sprayed?	Pest control
Cockroach traps	Pest control
Cockroaches in wall cavity of food trolleys	Pest control
Food brought in for patients	Safety of food preparation/storage at home

Standard methods for microbiological examination of food are provided by the Health Protection Agency (HPA, 2006a). The HPA also organizes a quality control system (HPA, 2006b) that involves sending, to laboratories that undertake microbiological analysis of food, two samples every 2 months and listing a requirement of the tests. Examinations covered include aerobic plate counts, *Bacillus cereus*, *Campylobacter* spp., *Clostridium perfringens*, *Escherichia coli* 0157, *Listeria monocytogenes*, *Salmonella* spp., *Staphylococcus aureus*, coliforms, Enterobacteriaceae and *Escherichia coli.*

6.9 Ward kitchens

6.9.1 *General requirements*

Observing the rules of food hygiene, keeping the food at the right temperature whether it be hot or cold, from the moment it reaches the ward until it is eaten, having clean and ordered equipment is just as important as in the main kitchens. Ward staff must understand this because failure can put patients at risk. The kitchens should be the responsibility of the ward manger who must provide written and readily available policies with a named person for each of them. Inevitably, this means having a series of 'rules'. An annual audit of ward kitchens can help in maintaining standards (Millward *et al.*, 1993).

In summary, the 'rules' are that:

- Staff must wash their hands before handling food and must not work if they have symptoms of possible gastrointestinal infection. Staff with lesions on exposed skin (hands, face, neck or scalp) that are actively weeping or discharging must be excluded from work until the lesions have heated (DH, 1995, 1996). They must not consume food or drink intended for patients or keep their own food in the cupboards or refrigerator. They must not allow patients or visitors to enter the ward kitchen.
- Food must be distributed from the trolley with the minimum of delay and in the case of cook-chill/frozen food within 15 minutes of regeneration. If the patient is not available hot food must be kept hot and discarded as waste after $1\frac{1}{2}$ hours, cold food must be refrigerated and discarded at the end of the day. Discarded or left-over food must be returned in the food trolley to the main kitchen or otherwise kept in a sealed container and discarded within 12 hours.
- Friends and relatives should be discouraged from bringing potentially hazardous foods into the hospital. Where food is brought in by relatives it must be placed in a container which is then labeled with patient's name and date, and discarded after 24 hours. Dry foods must be stored in sealed containers and out-of-date items discarded.
- Equipment, surfaces and utensils must be cleaned after use. The tap of the milk-dispensing machine must be dismantled daily and washed. The cleaning materials must be colour-coded and not used in patient areas. Mops should be washed, rinsed, squeezed out and inverted to dry.
- Cleaning schedules giving the method and frequency should be displayed.
- Sightings of pests or signs of infestation should be reported to the authorized officer immediately.

6.9.2 *Ward refrigerators*

These are solely for patients' food and never for medicines, blood, pathology samples or staff food (NHS, 1996). Daily temperature checks should be made to ensure that a satisfactory temperature of not higher than 5°C is maintained, temperatures should be recorded and the record kept with the refrigerator. Thermometers fitted inside the refrigerator facilitate this task. Some ward refrigerators have an external, digital temperature display, allowing continuous monitoring. Daily checks of the refrigerator should include an audit of cleanliness and of whether inappropriate food is being kept.

6.9.3 *Dishwashers*

Crockery and cutlery used for beverages and light snacks are usually washed in the ward kitchen. Washing up can be done manually or in a dishwasher. Washing up in a dishwasher is preferable but failing that the two-sink method, with very hot rinsing water, should be used, leaving articles to air dry. In addition, washing up by machine ensures a consistent standard of cleanliness and items are more likely to be subject to temperatures high enough to achieve disinfection. Also, staff are released for other tasks. One study showed that some machines may achieve acceptable cleaning results but do not disinfect, others disinfect but the items are not acceptably clean while, others can achieve both tasks (Cowan *et al.*, 1995). The desired machine should clean and disinfect, and its purchase can be arranged through NHS Supply Chain, the Office of Government Commerce or directly from a supplier.

Patients should not be allowed to wash dishes in the ward kitchens. Water sufficiently hot to wash dishes adequately could scald patients. Furthermore, patients may be the source of infectious agents that could contaminate kitchen items.

When inspecting ward kitchens this author has seen ward dishwashers that had upper and lower trays but the upper one was never used since it did not have its own water jets. On another occasion the dishwasher had completed a washing programme but 20 saucers had been stacked upside down on top of each other and the ones at the top of the heap were still dirty. A modicum of commonsense it always needed!

A cleaning schedule is needed and equally important, it must be firmly established who is responsible for this.

6.9.4 *Microwave ovens*

Where microwave ovens are used in the hospital setting they must be ones designed for commercial use and used only for reheating food prepared as part of the hospital catering system. Microwave ovens should not be used to reheat meals for patients absent from the ward when meals are being distributed or to reheat home-cooked food brought in by relatives. This is because it is impossible to guarantee that such food is safe. Because of the difficulty of preventing well- meaning ward staff from using microwave ovens for these purposes, such ovens should not be allowed in areas where there is access open to general hospital staff.

An additional problem with the use of microwave ovens by general hospital staff is the mistaken belief that microwaving renders food suitable for immunocompromised patients. The destruction of bacteria during microwave cooking is probably due solely to the heating effect of microwaves. Heating in microwave ovens is liable to be uneven, because of the uneven composition of foods, and harmful microorganisms may survive (USDA/FSIS, 2006); in the setting of a ward kitchen it is unlikely that the operator can carry out the temperature measurements needed to ensure adequate heating throughout. Microwave ovens must be maintained effectively and have a cleaning schedule.

6.9.5 *Ice making machines*

Machines for making ice and iced water are used in the care of patients, providing cool drinks, for mouth care and the relief of painful mouth conditions. These will be discussed in Chapter 7.

6.9.6 *Vending machines*

Many hospitals have vending machines, usually for the benefit of staff but they are often accessible to visitors who may give items from the machines to patients. Vending machines can be divided into three categories: (1) wrapped food products that do not require refrigeration, (2) chilled foods that require refrigeration and (3) hot and cold drinks (EVA, 2006). Drinks vending machines will be discussed in Chapter 7.

The hygiene of vending machines is controlled under Regulation EC No. 852/2004 (EC, 2004a) (Sprenger, 2005) and under the Food Hygiene (England) Regulations 2006. The European Vending Association (EVA), which develops standards in close co-operation with the North American Vending Association, has developed a Food Safety Management System for vending to assist operators to base their food safety management on HACCP principles (EVA, 2006).

Machines should be kept in good condition to avoid contamination of food. The machines must not be sited in direct sunlight or near to sources of heat that could affect the temperature of the food, and should be located in a position that facilitates thorough cleaning of the adjacent floor and wall surfaces (Sprenger, 2005).

6.10 Pest control

Cockroaches, Pharoh's ants, fleas, birds, rodents and cats are unwelcome visitors to healthcare premises. Warmth and the availability of food and harbourage sites ensure the presence and persistence of pests, a wide variety of which may transmit infection and can spoil food (Burgess & Chetwyn, 1979; Fotedar *et al.*, 1991; Pai *et al.*, 2004; Sprenger, 2005).

The presence of pests is completely incompatible with healthcare premises, and wherever possible, they should be eradicated completely from the site. Food handling areas must be inspected regularly for signs of infestation and for structural faults that may cause infestation, any sightings should be reported, recorded and appropriate action taken (NHS, 1996). Electrified flying insect killers should be sited in all food handling areas and cleaned and maintained regularly. A pest control specialist should monitor regularly food handling and associated areas such as waste disposal areas for signs of pest activity.

The NHS Management Executive (NHS, 1992b) recommended that General Managers and Chief Executives have procedures to ensure pest-free conditions in their premises.

It is recommended that each hospital has an authorized person with responsibility for all aspects of pest control. The authorized person should be trained in the recognition of pests; the means of controlling them, keeping a record of sightings, investigating reports, ensuring that action is taken and managing the contract. In most cases pest control is only part of the job and it is advisable that the person attends the 'Pest Control Management for the Health Service' course, (NHS, 1992b), which gives the nominated officers the skills to manage their pest control contractor and generally co-ordinate pest control activities on their site.

The majority of NHS Trusts have a contract with a pest control company. At the beginning of contract, the company carries out a programme to rid the site of specified pests, and subsequently visits the premises at regular intervals to monitor, inspect and possibly apply treatments in order to maintain the pest-free status. Locations visited will include the main food handling areas, and any areas of on-going or potential activity, or where pest sightings have been reported to the person monitoring the contract. If necessary, emergency visits may be needed. Special arrangements may need to be made to deal with non-specified pests. The authorized person should accompany the company on periodic inspections. Night inspections, when the activity of pests is at their greatest, may be necessary.

The author had an opportunity to take part in a night visit with an EHO to the main kitchens in a large hospital in 1982. Many cockroaches were seen; it was the last straw as the EHO had reported the infestation on numerous occasions; the next day he applied for closure of the kitchen.

For the pest control arrangement to work properly staff have an essential role:

- The authorized officer is responsible for maintaining pest awareness on the site, and ensuring that, through induction programmes, trust newsletters, briefings etc., that all staff know the importance of pest prevention, of reporting pest sightings, and how to report sightings.
- The night staff are more likely to realize that there is an infestation – cockroaches in particular are nocturnal.
- Catering staff and ward nurses should store food in lidded containers, not leave food out, and discard leftover food and waste promptly. Food must not be left out for the birds, cats, foxes or other pests.
- Catering and food handling staff should also have the appropriate food hygiene qualifications (these courses are run by a variety of organizations, including the Chartered Institute of Environmental Health, the Royal Society for Public Health etc.).
- The authorized person identifies structural defects which provide access or harbourage for pests, and liaises with maintenance staff over carrying out the repairs. The author remembers only too well that, many years ago, an outbreak of salmonella food poisoning among elderly-care patients persuaded estates staff to install netting on the main kitchen window, after a thrush was seen feeding on bowls of jelly deserts.

6.11 Conclusion

Food is an important part of medical care, and as hospital patients are at increased risk of illness if exposed to potential foodborne pathogens, it must be safe as well as nutritious. Catering managers are responsible for contracts with outside companies who supply food to hospitals. Catering staff are the main, but not the only, food handlers in healthcare setting as many other members of the hospital staff are involved, particularly the nursing staff who distribute food to the patients and auxiliary staff who may assist patients with their meals. Every opportunity should be taken to explain food hygiene practice and procedures, which may not be part of their usual training. Similarly, other staff should be considered.

Acknowledgements

The author is grateful to David Rowley, Environmental Health Services Manager, Epsom & Ewell Borough Council, for explaining and showing me how environmental health officers inspect food handling areas in hospitals; Nigel Watson, Business Manager, Food and Catering Facilities, NHS Supply Chain, for helpful information about NHS contracts for hospital food; Clive Boase, Principal Consultant, Pest Management Consultancy, for giving me up-to-date advice about pest control; Tony Grillo, Business Development Manager, Clenaware Systems, for information about Hospital Dishwashers; Kathy Brindle, Medical Secretary, Mayday University Hospital, for her patience and secretarial skills.

References

Barrie, D. (1996) The provision of food and catering services in hospital. *Journal of Hospital Infection* **33**, 13–33.

British Retail Consortium (BRC) (2005) *BRC Global Standard, Food*. Issue 4. Available from http://www.brc.org.uk/standards/downloads/food_std_background.pdf. Accessed 2 March 2007.

Burgess, N.R.H. & Chetwyn, K.N. (1979) Cockroaches in the hospital environment. *Nursing Times* **6**, 5–7.

Cowan, M.E., Allen, J. & Pilkington, F. (1995) Small dishwashers for hospital ward kitchens. *Journal of Hospital Infection* **29**, 227–231.

Department of Health UK (DH) (1989) *Chilled and Frozen. Guidelines on Cook-Chill and Cook-Freeze Catering Systems*. HMSO, London.

Department of Health, UK (DH)(1995) *Food Handlers. Fitness to Work. Guidelines for Food Businesses, Enforcement Officers and Health Professionals*. Department of Health, London.

Department of Health, UK (DH) (1996) *Food Handlers. Fitness to Work. Guidelines for Food Business Managers*. Department of Health, London.

Department of Health UK (DH) (2000) *The NHS Plan*. CM 4818-I. Available from http://www.dh.gov.uk/assetRoot/04/05/57/83/04055783.pdf. Accessed 2 March 2007.

EC (2004a) Regulation (EC) No. 852/2004 of the European Parliament and of the Council on the hygiene of foodstuffs. *Official Journal of the European Union* **L226/3**, 25 June 2004. Available from http://www.food.gov.uk/multimedia/pdfs/hiojregulation.pdf. Accessed 2 March 2007.

EC (2004b) Regulation (EC) No. 853/2004 of the European Parliament and of the Council of 29 April 2004 laying down specific hygiene rules for food of animal origin. *Official Journal of the European Union* **L226/22**, 25 June 2004. Available from http://www.food.gov.uk/multimedia/pdfs/h2ojregulation.pdf. Accessed 2 March 2007.

EC (2005) Commission Regulation (EC) No. 2073/2005 of 15 November 2005 on microbiological criteria for foodstuffs. *Official Journal of the European Union* **L338**, 22 December 2005. Available from http://www.food.gov.uk/multimedia/pdfs/microbiolcriteria.pdf. Accessed 2 March 2007.

European Vending Association (EVA) (2006) *EVA Food Safety Management System for Vending* (CD-based). Available from http://www.eva.be. Accessed 2 March 2007.

Food Hygiene (England) Regulations (2006) *Statuary Instrument 2006 No. 14*. Available from http://www.opsi.gov.uk/si/si2006/uksi_20060014_en.pdf. Accessed 2 March 2007.

Food Safety Act 1990 (c.16) The Stationery Office Ltd.

Food Safety (Sampling and Qualifications) Regulations (1990) *Statutory Instrument 1990 No. 2463*.

Fotedar, R., Shriniwas, U.B. & Verma, A. (1991) Cockroaches (*Blattella germanica*) as carriers of microorganisms of medical importance in hospitals. *Epidemiology and Infection* **107**, 181–187.

Health Protection Agency (HPA) (2006a) *National Standard Methods – Food*. Available from http://www.hpa-standardmethods.org.uk/pdf_sops.asp. Accessed 2 March 2007.

Health Protection Agency (HPA) (2006b) *Food External Quality Assessment Schemes. Standard Scheme and Extended Scheme*. Available from http://www.hpa.org.uk/cfi/quality/eqa/foodeqa.htm. Accessed 2 March 2007.

Millward, S., Barnett, J. & Tomlinson, D. (1993) A clinical infection control audit programme: evaluation of an audit tool used by infection control nurses to monitor standards and assess effective staff training. *Journal of Infection Control* **24**, 219–232.

National Audit Office (NAO) (2006) *Smarter Food Procurement in the Public Sector.* HC 963-1. The Stationery Office, London. Available from http://www.nao.org.uk. Accessed 2 March 2007.

National Health Service (NHS) (1986) *Health Service Catering: Hygiene*. HMSO, London.

National Health Service (NHS) (1992a) National Health Service (NHS) Management Executive. *Management of Food Services and Food Hygiene in the National Health Service.* Health Service Guideline (92) 34.

National Health Service (NHS) (1992b) National Health Service (NHS) Executive. *Pest Control Management for the Health Service*. Health Service Guideline (92) 35.

National Health Service (NHS) Executive (1996) *Hospital Catering. Delivering a Quality Service*. Department of Health, London, pp. 1–113.

National Health Service (NHS) (2001) *Code of Practice for the Manufacture, Distribution and Supply of Food Ingredients and Food-Related Products*. Available from http://www.pasa.doh.gov.uk/food/docs/code_of_practice_2001.pdf. Accessed 8 March 2007.

National Health Service (NHS) Supply Chain (2007) Available from http://www.supplychain.nhs.uk. Accessed 2 March 2007.

Pai, H.H., Chen, W.C. & Peng, C.F. (2004) Cockroaches as potential vectors of nosocomial infections. *Infection Control and Hospital Epidemiology* **25**, 979–984.

Sandys, G.H. & Wilkinson, P.J. (1988) Microbiological evaluation of a hospital delivered meal service using pre-cooked chilled foods. *Journal of Hospital Infection* **11**, 209–219.

Sprenger, R.A. (2005) *Hygiene for Management*, 12th edn. Highfield.co.uk Limited.

United States Department of Agriculture/Food Safety Inspection Service (USDA/FSIS) (2006) *Fact Sheets. Appliances and Thermometers. Cooking Safely in the Microwave Oven.* Available from http://www.fsis.usda.gov/Fact_Sheets/Cooking_Safely_in_the_Microwave/index.asp. Accessed 2 March 2007.

Wilkinson, P.J. (1988) Food Hygiene in hospitals. *Journal of Hospital Infection* **11**(Suppl 1), 77–81.

7 Provision of Water in Healthcare Settings

Paul R. Hunter

7.1 Introduction

Hospitals have a huge demand for water. Water is used for drinking in its natural state, in the preparation of food and drinks, for personal and environmental hygiene and for a wide range of medical purposes.

There is an increasing recognition of the health benefits of adequate hydration especially in the elderly and unwell. Inadequate hydration has been implicated as a risk factor in the development of pressure sores (Casimiro *et al.*, 2002), constipation (Klauser *et al.*, 1990), urinary tract infections (Eckford *et al.*, 1995) and falls (American Geriatrics Society, British Geriatrics Society and American Academy of Orthopaedic Surgeons Panel on Falls Prevention, 2001). There is also evidence that many elderly individuals do not drink enough water for adequate hydration, in part because of an impaired thirst response (Kenney & Chiu, 2001). A recent survey of water consumption in care homes found that many elderly residents consumed only about a half of their recommended daily amounts (Royal Society for the Promotion of Health, in cooperation with the Water for Health Alliance, 2003). Nursing and other healthcare staff have an important role in encouraging an adequate intake of fluids.

Clearly, problems with drinking water in community water systems, such as drinking waterborne outbreaks, will also affect hospital patients and staff. However, certain immune compromised hospital patients are also prone to a range of opportunistic pathogens that generally do not affect people in the community. In this chapter, we shall first discuss enteric pathogens and then go onto discuss the issues around the opportunistic pathogens and finally consider a range of issues with water, drinks and ice.

7.2 Enteric pathogens and potable water supplies

7.2.1 *Definition*

For the purposes of this chapter, the term enteric pathogen will cover those infections that are associated with gastroenteritis and will often affect otherwise healthy people. The nature of the enteric waterborne pathogens and the diseases that they cause will not be reviewed here as these have already been discussed in Chapter 2. Most of the pathogens capable of causing foodborne infections can also cause waterborne disease, though the likelihood that a particular foodborne pathogen will also spread by the waterborne route varies from one to another.

7.2.2 *Outbreaks of disease*

Table 7.1 lists the pathogens reported to have caused outbreaks of drinking water-associated illness in England and Wales or the US. In England and Wales, the predominant pathogens were *Cryptosporidium*, *Campylobacter* or *Escherichia coli* O157:H7. *Cryptosporidium* was mainly associated with public drinking water supplies whilst the two bacterial pathogens were mainly associated with small private supplies. The range of pathogens detected in the US is greater reflecting the larger number of small, inadequately treated water supplies and the inclusion of several US-managed territories in tropical countries. A substantially greater proportion of US outbreaks do not have a causative organism identified, in part reflecting the fact that such stool examination, especially for viruses, is less likely to be done in the

Table 7.1 Pathogens identified in outbreaks of waterborne disease in the England and Wales (1991–2004) or US (1991–2002).

	England and Wales		US	
Pathogen	Number of outbreaks	Number of affected individuals	Number of outbreaks	Number of affected individuals
Cryptosporidium sp.	34	3353	13	408,274
Giardia intestinalis	1	31	25	2293
Campylobacter	16	619	8	1146
Escherichia coli O157:H7	3	34	10	1040
Salmonella sp.	0	0	3	782
Shigella sp.	0	0	8	605
Vibrio cholerae	0	0	2	114
Plesiomonas shigelloides	0	0	1	60
Yersinia enterocolitica	0	0	1	12
Noroviruses	0	0	13	4531
Hepatitis A	0	0	2	56
Acute gastroenteritis of unknown aetiology	1	56	67	15,362

Adapted from Hunter *et al.*, 2003.

US than in the UK. Of those that do report a pathogen *Giardia* caused most outbreaks, but *Cryptosporidium* affected most people. Norovirus and *Campylobacter* are also common.

During outbreaks of waterborne disease associated with public water supplies hospitals may be affected, either because patients and staff become ill from drinking contaminated water or because of the imposition of restrictions on the use of water imposed by the public health team responsible for the outbreak management. Nevertheless, there are relatively few reports of hospital outbreaks of enteric illness due to contaminated water supplies reported in the world literature.

Two well-described outbreaks have been reported, the first from Finland and the second from the US. Neither of these outbreaks was due to contamination of the public supply but to problems within the hospital.

The Finnish outbreak involved *Campylobacter jejuni* in a hospital for rheumatic diseases during November and December 1986 (Rautelin *et al.*, 1990). A total 32 patients and 62 members of the staff developed gastroenteritis from whom 77 samples yielded *C. jejuni*. The hospital had a private borehole and the water was filtered and chlorinated. The same serotype of *Campylobacter* was isolated from water samples as from patients. The source of the problem was an aging water system that allowed ingress of contaminated water.

The American outbreak occurred in a Chicago hospital during July 1994 and affected medical residents and hospital administrative staff (Huang *et al.*, 1995). The causative organism in this case was a *Cyclospora* species, a protozoal pathogen. The failure with this system seems to have been that the water storage tanks were badly maintained and subject to contamination by bird faeces (Fox, 2003). There was a failure in a water pump such that levels within the storage tank fell and stagnant water from the bottom of the tank (where bird faeces may have accumulated) was then distributed to taps for consumption.

Another outbreak from a developed country was reported in 1999. This was due to norovirus and affected patients on a ward in a hospital in France (Schvoerer *et al.*, 1999). The outbreak occurred on a single ward but the investigators suspected a water source as virus was detected in four of seven tap water samples. As the authors pointed out, they could not exclude person-to-person transmission as the main route, with the water samples becoming positive due to soiling of the taps. It is this author's view that person-to-person transmission with contamination of the tap was the most likely explanation for the outbreak.

The message from the two well-described outbreaks is quite clear. Both these outbreaks resulted from a failure to adequately maintain the hospital plumbing infrastructure. The hospital in Finland was supplied from a private well and the failure was in the water treatment and distribution. In the Chicago outbreak, the water was from a good quality public supply but was subject to contamination due to failings in the management of the storage and distribution systems within the hospital. Even good quality water can become a risk to patients if the hospital engineers do not maintain the system adequately.

7.2.3 *Boil water notices*

During outbreaks of waterborne disease in the community, water utilities and health agencies can sometimes issue 'boil water notices' or 'boil water advisories'. These basically advise the public to boil any water intended for consumption. The notices are issued when the appropriate authority believes there is a high risk that the supply may be contaminated with organisms that are harmful to health. Boiling water is a very effective treatment for waterborne pathogens. All such pathogens will be killed by the time the water has come to a rolling boil. However, for such advisories to be effective all water that may be consumed needs to be boiled. This includes water intended for brushing teeth and oral hygiene. Alternately, a non-tap water source such as bottled water needs to be used.

Clearly, boil water notices can have very serious impacts on those health services in affected areas. Patients and staff can no longer drink unboiled tap water or even use it for brushing teeth. Tap water can no longer be used for oral hygiene or for dental work. Such restrictions could limit the provision of health care and especially dental care unless adequate alternatives are provided.

7.2.4 *Advice to boil water for severely immune compromised patients*

Most western countries have issued advice to certain immune compromised patients to boil all drinking water (CMO, 1999a; CDC, 2002). This advice was issued to reduce the risk of Cryptosporidiosis, which can be very severe in certain groups (Hunter & Nichols, 2002). In the UK, the initial advice for 'immune compromised patients' to boil water was clarified (CMO, 1999b). Current UK advice is that people whose T-cell function is compromised (this includes people with HIV infection who are immunosuppressed, children with severe combined immunodeficiency (SCID) and those with specific T-cell deficiencies, such as CD40 ligand deficiency, also known as Hyper IgM syndrome) should be advised to boil and cool their drinking water from whatever source.

7.3 Opportunistic pathogens and potable water supplies

So far in this chapter, we have concerned ourselves with those pathogens that pose a risk to communities generally. Attention will now be given to opportunistic pathogens. Biology-on-line defines an opportunistic pathogen as a pathogenic organism that is often normally a commensal, but which gives rise to infection in immunocompromised hosts. In the context of water, *Pseudomonas aeruginosa* is the most important such opportunist.

Hospitals are often large, old buildings with old and poorly designed plumbing networks. Thus, bacteria can grow after water leaves the mains often to quite high concentrations (Hunter & Burge, 1988). These bacteria, commonly known as heterotrophic plate count bacteria (HPC) were the subject of a recent World Health Organization review that concluded that they did not pose a risk to the health of the general population (Bartram *et al.*, 2003). However, this review did not address the issue of immune compromised patients. As has been discussed elsewhere in this book, hospitalized patients are often prone to infections with opportunistic pathogens that would not generally cause illness in otherwise healthy people (Chapter 5). Nosocomial opportunistic pathogens are those opportunistic pathogens that cause disease in hospitalized patients.

There are a large number of such opportunistic pathogens, but this review will restrict itself to those pathogens that have been identified as being associated with hospital water supplies. The most frequently described pathogen responsible for outbreaks linked

Table 7.2 Range of pathogens associated with nosocomial outbreaks linked to hospital water supplies.

Pathogen	Number of outbreaks	Sites of infection (number of outbreaks reporting infections at this site if >1)
Pseudomonas aeruginosa	11	Blood (7), lung (4), urine (4), trachea (2), CSF, catheter, sinuses, skin, wounds
Burkholderia cepacia	4	Blood (4)
Mixed *Ralstonia pickettii* and *Burkholderia cepacia*	1	Blood
Ralstonia pickettii	1	Blood and respiratory
Stenotrophomonas maltophilia	4	Blood (2), respiratory tract (2), stools (2), throat, urine, skin
Serratia marcescens	1	Eye, stools
Acinetobacter baumannii	2	Skin, wound, respiratory CVP
Aeromonas hydrophila	1	Blood
Sphingomonas paucimobilis	1	Blood
Chryseobacterium sp.	1	Blood
Mycobacterium avium	1	Disseminated
Mycobacterium fortuitum	3	Disseminated, wound, respiratory tract, sputum
Mycobacterium xenopi	2	Various, spine
Mycobacterium kansasii	1	Abscess, blood, bone, sputum, stomach, urine
Mycobacterium simae	1	Respiratory
Mycobacterium mucogenicum	1	Blood
Mixed *M. fortuitum* and *M. chelonae*	1	Sternal wound, prosthetic heart valves
Fusarium solani	1	Disseminated
Exophilia jeanselmei	2	Disseminated (1), blood (1)
Apergillus funigatus	1	Lungs

Adapted from Anaissie *et al.*, 2002.

to hospital water supplies is *Legionella*, though as the transmission route associated with Legionnaire's disease is by inhalation it falls outside the scope of this book.

Anaissie *et al.* (2002) produced a comprehensive review of reported outbreaks of nosocomial infection that had been linked to hospital water supplies. These authors conducted a Medline search of literature published between 1 January 1966 and 31 December 2001 using the keywords: water, hospital and infection. They identified 43 outbreaks of which 29 had strong supporting evidence of molecular relatedness between strains from patients and from the water supply. The range of pathogens in these 29 outbreaks is listed in Table 7.2. Using this same search strategy, a further 12 outbreaks were found that were published between 2002 and 2005 (Bukholm *et al.*, 2002; Nucci *et al.*, 2002; Perola *et al.*, 2002; Magalhaes *et al.*, 2003; Wang *et al.*, 2003; Conger *et al.*, 2004; Kendirli *et al.*, 2004; Kline *et al.*, 2004; Nasser *et al.*, 2004; Souza *et al.*, 2004; Bellazzi & Ciniselli, 2005; Moreira *et al.*, 2005).

The epidemiological evidence of the association between water and the outbreak across these reports is variable. Many of these outbreaks were deemed to be waterborne based on the finding of the same organism in water at the tap as in clinical samples. In such cases, it is not always obvious whether the patients were infected from the tap or the tap contaminated from people washing their hands after caring for the patients. Amongst the best evidence in favour of waterborne transmission is those outbreaks where the outbreak was terminated only by improving tap water hygiene (e.g. Bukholm *et al.*, 2002). Even when tap water was clearly implicated it is often not clear whether transmission was by consumption of the water or by use of the water for washing and other purposes. In general, most nosocomial outbreaks linked to water will be by the latter transmission pathway.

In addition to these outbreaks, there have been several pseudo-outbreaks due to waterborne organisms reported in the literature (Hunter, 1997a). Some of these were due to colonization of the patient, others were due to contamination of microbiological specimens during transport or analysis. Although the organisms associated with these pseudo-outbreaks did not directly cause disease, many patients received unnecessary therapy or hospitalization. The incidents also led to significant costs for the healthcare institution incurred in the investigation.

The great majority of the real outbreaks were not caused by water intended for consumption, but rather for water used in humidifiers or for washing patients or the environment and as such are beyond the scope of this review. However, particular attention needs to be given to *Pseudomonas aeruginosa*.

7.3.1 *Pseudomonas aeruginosa*

P. aeruginosa is ubiquitous in the hospital environment being found in most moist environments and especially in the drains from sinks and baths (Levin *et al.*, 1984; Doring *et al.*, 1991, 1993). A recent flurry of publications have addressed the issue of electronic taps, which appear to be more susceptible to contamination than manual taps (Chaberny & Gastmeier, 2004; Merrer *et al.*, 2005; van der Mee-Marquet *et al.*, 2005).

The role of tap water as a reservoir or transmission vehicle for *P. aeruginosa* is still strongly debated with some workers suggesting that tap water is the single most important preventable cause of nosocomial infection (Anaissie *et al.*, 2002; Exner *et al.*, 2005). Anaissie *et al.* (2002) estimated that in the US each year some 1400 people die from a waterborne nosocomial infection with *P. aeruginosa*. There is no doubt that several outbreaks of nosocomial infection due to *P. aeruginosa* probably from tap water have been reported in the literature (Table 7.2). What remains uncertain is the more general contribution that tap water makes to such infections. Not all waterborne outbreaks of *P. aeruginosa* will be identified as outbreaks or their association with tap water shown. Even when a waterborne outbreak has been successfully identified many will not be reported in the literature. It is reasonable to assume that the reported outbreaks remain very much the tip of the iceberg. Nevertheless, nosocomial infections with *P. aeruginosa* are responsible for a substantial burden of disease globally and the proportion due to water compared to cross infection unclear.

The non-outbreak evidence of waterborne pseudomonas infection comes largely from studies typing strains from human and environmental sources reviewed by Trautmann *et al.* (2005). However, as with the issue of outbreaks, the finding of similar or even identical strains in tap water and clinical infections does not prove cause and effect. Environmental strains do appear to be less virulent than human isolates (Fenner *et al.*, 2006). As yet no well-designed epidemiological studies have been published in the peer-reviewed literature in support of the hypothesis. This is clearly an area where randomized controlled trials of interventions aimed at removing *Pseudomonas* from tap water are needed.

General advice on infection control aspects of hospital water has been given by the US Centers for Disease Control and Prevention (2003). Much of this advice has been directed at controlling *Legionella* infections, by for example keep hot water systems above 51°C and cold water below 20°C. Specific advice of relevance to this chapter is the importance of hand washing and keeping sinks clean and disinfected on a regular basis. Some authorities recommend point of use filters in taps and shower heads to reduce the risk of waterborne infections, especially from opportunistic pathogens (Exner *et al.*, 2005). In this author's view, it is too early to recommend the widespread use of these filters other than in particularly at risk settings. Nevertheless, this is an area that requires further research.

In the absence of any pre-existing advice, we would make the following recommendations when filling jugs with water for patients to drink:

- Wash hands before filling jugs or other containers with water intended to be drunk by patients.
- Always wash jugs between use in a dish washer that achieves an adequate disinfection temperature.
- Only fill jugs with water from a sink designated for drinking water and where hand-washing and other washing is not allowed.
- Drinking water should not be left to stand at room temperature for more than 6 hours before changing, other than water left overnight.
- In intensive care units and on wards with severely immune compromised patients consider point-of-use-tap filters.
- If point-of-use filters are installed always follow the manufacturers' instructions on replacement times.

7.4 Ice and ice-making machines

Ice in hospitals and other healthcare settings deserves a special mention. Although relatively infrequent, there have been some reports of infections associated with the use of ice in drinks for medical purposes in hospitals (Newsom, 1968; Anon, 1993; Graman *et al.*, 1997). In one such outbreak, eight infections in haematology patients with *Stenotrophomonas maltophilia* could be traced to an ice-making machine on the ward where counts of the organism were $>10^5$ colony forming units (CFU)/mL of ice, which was used for making cold drinks (Anon, 1993). Other outbreaks include one due to *Flavobacterium* sp. on an intensive care unit (Stamm *et al.*, 1975), cryptosporidial infections in patients with Aids (Ravn *et al.*, 1991) and nosocomial legionellosis (Bangsborg *et al.*, 1995; Graman *et al.*, 1997). Community outbreaks of norovirus (Cannon *et al.*, 1991) and *Shigella sonnei* (Anon, 1996) linked to commercially available ice have also been described. In the case of the norovirus outbreak this was due to commercially produced ice sold on through retail outlets in several US states.

As well as causing real outbreaks, ice machines have been implicated in pseudo-outbreaks of *Mycobacterium fortuitum* and *Mycobacterium gordonae* (Panwalker & Fuhse, 1986; Gebo *et al.*, 2002; Labombardi *et al.*, 2002). Pseudo-outbreaks occur when an organism contaminates specimens or colonizes patients so that the investigating microbiologist believes the patient is infected. Such pseudo-outbreaks can cause significant harm to the patient through unnecessary treatment and investigations.

These outbreaks have led various workers to investigate the microbiological quality of ice in hospitals or in the community or to audit the maintenance of ice makers in their hospital.

Wilson and co-workers (1997) undertook a survey of the microbiological quality of ice from 27 ice makers in their hospital in Northern Ireland. A wide variety of potentially pathogenic microorganisms were detected in these samples, though all in relatively low numbers (<10 CFU/mL). The most commonly detected organisms were *Bacillus* spp. in 93% of samples, Coagulase-negative staphylococci (85%), *Pseudomonas* spp. (78%) and *Acinetobacter* spp. (33%). No infections related to these machines were identified in patients during the time of the study. These results were consistent with an earlier study by Burnett *et al.* (1994). In a community study of ice from various sources in the UK, Nichols and colleagues (2000) surveyed 4346 ice samples from retail and catering premises, mostly intended for cooling drinks. Of the samples of ice intended to cool drinks some 9% yielded coliforms, 1% *E. coli*, and 1% enterococci, all in numbers greater than 10^2 CFU/100 mL, and from 11% the aerobic plate count at 37°C was greater than 10^3 CFU/mL. Community surveys of ice in developing countries have shown even greater contamination including the detection of frank enteric pathogens (Kruy *et al.*, 2001; Falcao *et al.*, 2004).

An audit of the use of five types of ice-making machine on three sites in a hospital in Birmingham, UK, showed that there was no control over, or designated responsibility for, the purchase and subsequent maintenance of ice-making machines (King, 2001). As a result, the machines were poorly maintained with no record or evidence of the machines being routinely cleaned or defrosted. Following this audit several recommendations were made including:

- the introduction of a policy for the purchase, use and maintenance of ice machines as part of the Infection Control policy manual;
- the requirement to consult with the Infection Control Team for advice prior to purchasing ice machines;
- ice made in automated ice machines should not be given to immunocompromised patients.

The infection control team should advise that the only ice-making machines that are purchased are those that make small cubes which cannot be touched by casual users.

Advice on the care and maintenance of ice storage chests and ice making machines has been published by CDC (Manangan *et al.*, 1998). Hygienic precautions during handling of ice are outlined. The monthly to quarterly schedule for cleaning ice-making machines and weekly to monthly schedule in the case of ice-storage chests – involves disconnecting the machine, discarding all the ice, disassembling any removable parts of the machine that come into contact with the water used to make the ice, cleaning with detergent, rinsing and drying. Then circulating ice-making machines with 50–100 ppm chlorine solution, which is left for 4 hours, drained, flushed with tap water and allowed to dry. Ice-storage chests are rinsed with 10–50 ppm of chlorine solution and then allowed to dry.

7.5 Bottled water

Sales of bottled water in most developed countries have increased substantially over the past few decades and it is certain that bottled water will be seen in the healthcare environment, being brought in by staff, visitors and patients. Bottled waters are usually classified as bottled natural mineral waters or simply bottled waters (Barrell *et al.*, 2000).

In Europe, bottled natural mineral water is governed by various Directives that are enacted into law in the member states (Barrell *et al.*, 2000). Bottled natural mineral water should be free from: parasites and pathogenic organisms; *Escherichia coli*, coliforms, faecal streptococci and *Pseudomonas aeruginosa* <1 CFU/250 mL; and sulphite-reducing clostridia <1 CFU/50 mL. However, there is no standard for heterotrophic plate count other than in the first 12 hours after bottling. The standards for other bottled waters are similar but *Escherichia coli*, coliforms, faecal streptococci a need to be <1 CFU/100 mL and sulphite-reducing clostridia <1 CFU/20 mL. *Pseudomonas aeruginosa* is not regulated in these waters. Many authors have noted high heterotrophic plate counts ($>10^3$ CFU/mL) in a varying proportion of non-carbonated, bottled waters (Hunter, 1993). Carbonated waters have much lower heterotroph counts.

There have been a few outbreaks associated with bottled water world wide (Hunter, 1997b), the most notable being an outbreak of cholera in Portugal during 1974. Each of these outbreaks was associated with water that would not have passed current EU standards. More recently, there has been an outbreak in a healthcare setting where bottled water was implicated in an outbreak of norovirus in a German veterans centre (Petersen *et al.*, 2000). However, insufficient data has been presented to determine the strength of evidence behind this association.

This author sees no reason to exclude bottled waters from the hospital and healthcare environment. However, bottled waters, unless otherwise sterilized, should not be given to patients on intensive therapy units, on haematology wards where severely neutropaenic patients are nursed, or to other patients on low microbial diets.

Another form of bottled water is the water dispenser, which consists of a large bottle or other container inverted over a unit, often containing some cooling mechanism, and a tap (Hunter & Barrell, 1999). In one study we found that coliforms were present in 12.5% of samples and at concentrations of >100 CFU/100 mL in 1.8%. *Pseudomonas aeruginosa* was present in 3.6% samples at counts of >100 CFU/100 mL. Heterotrophic plate counts were $>10^4$ CFU/mL in 26.6% of samples after incubation at 22°C and in 8.9% of samples after incubation at 37°C. The worst results were from machines where the container had been refilled from the mains water. In our view such machines should not be situated in clinical areas where patients for whom low microbial diets are indicated have access. If such dispensers are situated in clinical settings the bottle should be refilled professionally and not from a tap within the hospital.

7.6 Drinks vending machines

Vending machines dispensing both hot and cold drinks have become ubiquitous in hospital environments over the past 30 years. Several types of drinks vending machines are available. Premix machines mix all the ingredients of the drink, including water before loading. Postmix machines store all ingredients separately and water is added when the final drink is dispensed. Fresh-brew machines brew the final drink by adding hot water to tea leaves or coffee grounds as the drink is dispensed (Hunter, 1992).

This author's first foray into water microbiology was finding high coliform counts (32, 108 and >180 CFU /100 mL) in three samples of drinking water from a drinks dispenser constructed of a clear plastic reservoirs above a chiller unit (Hunter, 1985). Such machines were not designed to dispense only water, rather orange drink which is sufficiently acidic to prevent growth of coliform organisms. A further study of coin-operated vending machines in the hospital and elsewhere found that 44% of 25 water samples yielded coliforms and from 2 (8%) the count was >100 coliforms/100 mL (Hunter & Burge, 1986). Robertson (1985) reported finding *E. coli* in 23% of swabs from vending machine nozzles. Robertson also noted that the hygienic quality of machines depended on their maintenance by the operators.

Design of drinks vending machines has improved considerably since the early days (Chaidez *et al.*, 1999; Schillinger & Du Vall Knorr, 2004). These recent studies from the US have seen fewer machines colonized by coliforms. However, total coliforms were found in up to 20% of samples and *P. aeruginosa* in 23% at counts of up to 79 CFU/500 mL. In addition, heterotrophic plate counts were >500 CFU/mL on about 50% of occasions.

The other findings from these studies have been the reinforcement of evidence that poor microbiological quality of water from vending machines is associated with poor maintenance of the machine; in particular, long intervals between servicing and inadequate cleaning (Hunter, 1992; Schillinger & Du Vall Knorr, 2004).

Despite the findings discussed above, this author is only aware of one outbreak of disease linked to such machines reported in the literature, due to consumption of drinking chocolate contaminated with *Bacillus cereus* (Nelms *et al.*, 1997).

Provided drinks vending machines are maintained properly this author sees no reason why they should not be placed within hospitals. When placed within a hospital or other healthcare setting, drinks vending machines must have an appropriate contract for their cleaning and maintenance. The reports of isolations of *P. aeruginosa* from some machines mean that they should not be placed in intensive care units, or on haematology wards where severely neutropaenic patients are nursed.

References

American Geriatrics Society, British Geriatrics Society and American Academy of Orthopaedic Surgeons Panel on Falls Prevention (2001) Guidelines for the prevention of falls in older persons. *Journal of the American Geriatrics Society* **49**, 664–672.

Anaissie, E.J., Penzak, S.R. & Dignani, C. (2002) The hospital water supply as a source of nosocomial infections. *Archives of Internal Medicine* **162**, 1482–1492.

Anon (1993) Ice as a source of infection acquired in hospital. *Communicable Disease Report. CDR Weekly* **3**(53), 241.

Anon (1996) *Shigella sonnei* outbreak associated with contaminated drinking water – Island Park, Idaho, August 1995. *Journal of the American Medical Association* **275**, 1071.

Bangsborg, J.M., Uldum, S., Jensen, J.S. & Bruun, B.G. (1995) Nosocomial legionellosis in three heart–lung transplant patients: case reports and environmental observations. *European Journal of Clinical Microbiology and Infectious Disease* **14**, 99–104.

Bartram, J., Cotruvo, J., Exner, M. *et al.* (eds) (2003) *Heterotrophic Plate Counts and Drinking-Water Safety, The Significance of HPCs for Water Quality and Human Health.* IWA publishing on behalf of the World Health Organization, London.

Barrell, R.A.E., Hunter, P.R. & Nichols, G. (2000) Microbiological standards for water and their relationship to health risk. *Communicable Disease and Public Health* **3**, 8–13.

Bellazzi, R. & Ciniselli, F. (2005) From the Mailing List SIN: expected and unexpected professional risks for the nephrologists – reflections from an outbreak of *Burkholderia cepacia* bacteremia in a hemodialysis unit. *Giornale Italiano di Nefrologia* **22**, 376–380.

Bukholm, G., Tannaes, T., Kjelsberg, A.B.B. & Smith-Erichsen, N. (2002) An outbreak of multidrug-resistant *Pseudomonas aeruginosa* associated with increased risk of patient death in an intensive care unit. *Infection Control & Hospital Epidemiology* **23**, 441–446.

Burnett, I.A., Weeks, G.R. & Harris, D.M. (1994) A hospital study of ice-making machines: their bacteriology, design, usage and upkeep. *Journal of Hospital Infection* **28**, 305–313.

Cannon, R.O., Poliner, J.R., Hirschhorn, R.B. *et al.* (1991) A multistate outbreak of Norwalk virus gastroenteritis associated with consumption of commercial ice. *Journal of Infectious Diseases* **164**, 860–863.

Casimiro, C., Garcia-de-Lorenzo, A. & Usan, L. (2002) Prevalence of decubitus ulcer and associated risk factors in an institutionalized Spanish elderly population. *Nutrition* **18**, 408–414.

Centers for Disease Control and Prevention (CDC) (2002) Guidelines for preventing opportunistic infections among HIV-infected persons – 2002. *Morbidity and Mortality Weekly Reports. Recommendations and Reports* **51**(RR08), 1–46.

Centers for Disease Control and Prevention (CDC) (2003) Guidelines for environmental infection control in health-care facilities: recommendations of CDC and the Healthcare Infection Control Practices Advisory Committee (HICPAC). *Morbidity and Mortality Weekly Report* **52**(RR10), 1–42.

Chaberny, I.F. & Gastmeier, P. (2004) Should electronic faucets be recommended in hospitals? *Infection Control & Hospital Epidemiology* **25**, 997–1000.

Chaidez, C., Rusin, P., Naranjo, J. & Gerba, C.P. (1999) Microbiological quality of water vending machines. *International Journal of Environmental Health Research* **9**, 197–206.

Chief Medical Officer (1999a) *Cryptosporidium* in water: advice to the immunocompromised. *CMO's Update* **21**, 2. Available from http://www.dh.gov.uk/assetRoot/04/01/36/47/04013647.pdf. Accessed 2 March 2007.

Chief Medical Officer (1999b) *Cryptosporidium* in water: clarification of the advice to the immunocompromised. *CMO's Update* **23**, 4. Available from http://www.dh.gov.uk/assetRoot/04/01/35/68/04013568.pdf. Accessed 2 March 2007.

Conger, N.G., O'Connell, R.J., Laurel, V.L. *et al.* (2004) *Mycobacterium simae* outbreak associated with a hospital water supply. *Infection Control & Hospital Epidemiology* **25**, 1050–1055.

Doring, G., Horz, M., Ortelt, J. *et al.* (1993) Molecular epidemiology of *Pseudomonas aeruginosa* in an intensive care unit. *Epidemiology and Infection* **110**, 427–436.

Doring, G., Ulrich, M., Muller, W. *et al.* (1991) Generation of *Pseudomonas aeruginosa* aerosols during handwashing from contaminated sink drains, transmission to hands of hospital personnel, and its prevention by use of a new heating device. *Zentralblatt fur Hygiene und Umweltmedizin* **191**, 494–505.

Eckford, S.D., Keane, D.P., Lamond, E. *et al.* (1995) Hydration monitoring in the prevention of idiopathic urinary tract infections in premenopausal women. *British Journal of Urology* **76**, 90–93.

Exner, M., Kramer, A., Lajoie, L. *et al.* (2005) Prevention and control of health care-associated waterborne infections in health care facilities. *American Journal of Infection Control* **33**(5, Suppl 1), S26–S40.

Falcao, J.P., Falcao, D.P. & Gomes, T.A.T. (2004) Ice as a vehicle for diarrheagenic *Escherichia coli*. *International Journal of Food Microbiology* **91**, 99–103.

Fenner, L., Richet, H., Raoult, D. *et al.* (2006) Are clinical isolates of *Pseudomonas aeruginosa* more virulent than hospital environmental isolates in amebal co-culture test? *Critical Care Medicine* **34**, 823–828.

Fox, K.R. (2003) Engineering considerations in the investigation of waterborne outbreaks. In: Hunter, P.R., Waite, M. & Ronchi, E. (eds). *Drinking Water and Infectious Disease: Establishing the Links*. CRC Press, Boca Raton, FL, pp. 97–103.

Gebo, K.A., Srinivasan, A., Perl, T.M. *et al.* (2002) Pseudo-outbreak of *Mycobacterium fortuitum* on a Human Immunodeficiency Virus Ward: transient respiratory tract colonization from a contaminated ice machine. *Clinical Infectious Diseases* **35**, 32–38.

Graman, P.S., Quinlan, G.A. & Rank, J.A. (1997) Nosocomial legionellosis traced to a contaminated ice machine. *Infection Control and Hospital Epidemiology* **18**, 637–640.

Huang, P., Weber, J.T., Sosin, D.M. *et al.* (1995) The first reported outbreak of diarrheal illness associated with *Cyclospora* in the United States. *Annals of Internal Medicine* **123**, 409–414.

Hunter, P.R. (1985) Misuse of chilled drinks dispensers. *Journal of Hospital Infection* **6**, 434.

Hunter, P.R. (1992) Bacteriological, hygienic and public-health aspects of food and drink from vending machines. *Critical Reviews in Environmental Control* **22**, 151–167.

Hunter, P.R. (1993) A review: the microbiology of bottled natural mineral waters. *Journal of Applied Bacteriology* **74**, 345–353.

Hunter, P.R. (1997a) *Waterborne Disease: Epidemiology and Ecology*. Wiley, Chichester.

Hunter, P.R. (1997b) Outbreaks of disease linked to bottled waters and their relevance to risk assessment. In: Caroli, G., Levré, E., Baggiani, A. & Pollicino, G. (eds). *Proceedings of the 1st International Congress on Mineral Waters and Soft Drinks, Legislation, Quality Control and Production*. Pacini Editore, Pisa, pp. 275–282.

Hunter, P.R., Andersson, Y., Von Bonsdorff, C.H. *et al.* (2003) Surveillance and investigation of contamination incidents and waterborne outbreaks. In: Dufour, A., Snozzi, M. & Koster, W. (eds). Assessing Microbial Safety of Drinking Water: Improving Approaches and Methods. OECD/WHO, Paris, pp. 205–236.

Hunter, P.R. & Barrell, R. (1999) Microbial quality of drinking water from office water dispensers. *Communicable Disease and Public Health* **2**, 67–68.

Hunter, P.R. & Burge, S.H. (1986) The microbiological quality of drinks from vending machines. *Journal of Hygiene* **97**, 497–500.

Hunter, P.R. & Burge, S.H. (1988) Monitoring the bacteriological quality of potable waters in hospital. *Journal of Hospital Infection* **12**, 289–294.

Hunter, P.R. & Nichols, G. (2002) The epidemiology and clinical features of cryptosporidium infection in immune-compromised patients. *Clinical Microbiology Reviews* **15**, 145–154.

Kendirli, T., Ciftci, E., Ince, E. *et al.* (2004) *Ralstonia pickettii* outbreak associated with contaminated distilled water used for respiratory care in a paediatric intensive care unit. *Journal of Hospital Infection* **56**, 77–78.

Kenney, W.L. & Chiu, P. (2001) Influence of age on thirst and fluid intake. *Medicine and Science in Sports and Exercise* **33**, 1524–1532.

King, D. (2001) Ice machines – an audit of their use in clinical practice. *Communicable Disease and Public Health* **4**, 49–52.

Klauser, A.G., Beck, A., Schindlbeck, N.E. & Muller-Lissner, S.A. (1990) Low fluid intake lowers stool output in healthy male volunteers. *Zeitschrift fur Gastroenterologie* **28**, 606–609.

Kline, S., Cameron, S., Streifel, A. *et al.* (2004) An outbreak of bacteremias associated with *Mycobacterium mucogenicum* in a hospital water supply. *Infection Control & Hospital Epidemiology* **25**, 1042–1049.

Kruy, S.L., Soares, J.L., Ping, S. & Sainte-Marie, F.F. (2001) Microbiological quality of "ice, ice cream, sorbet" sold on the streets of Phnom Penh; April 1996 – April 1997. *Bulletin de la Societe de Pathologie Exotique* **94**, 411–414.

Labombardi, V.J., O'Brien, A.M. & Kislak, J.W. (2002) Pseudo-outbreak of *Mycobacterium fortuitum* due to contaminated ice machines. *American Journal of Infection Control* **30**, 184–186.

Levin, M.H., Olson, B., Nathan, C. *et al.* (1984) *Pseudomonas* in the sinks in an intensive care unit: relation to patients. *Journal of Clinical Pathology* **37**, 424–427.

Magalhaes, M., Doherty, C., Govan, J.R.W. & Vandamme, P. (2003) Polyclonal outbreak of *Burkholderia cepacia* complex bacteraemia in haemodialysis patients. *Journal of Hospital Infection* **54**, 120–123.

Manangan, L.P., Anderson, R.L., Arduino, M.J. & Bond, W. (1998) Sanitary care and maintenance of ice-storage chests and ice-making machines in health care facilities. *American Journal of Infection Control* **26**, 111–112.

Merrer, J., Girou, E., Ducellier, D. *et al.* (2005) Should electronic faucets be used in intensive care and hematology units? *Intensive Care Medicine* **31**, 1715–1718.

Moreira, B.M., Leobons, M.B.G.P., Pellegrino, F.L.P.C. *et al.* (2005) *Ralstonia pickettii* and *Burkholderia cepacia* complex bloodstream infections related to infusion of contaminated water for injection. *Journal of Hospital Infection* **60**, 51–55.

Nasser, R.M., Rahi, A.C., Haddad, M.F. *et al.* (2004) Outbreak of *Burkholderia cepacia* bacteremia traced to contaminated hospital water used for dilution of an alcohol skin antiseptic. *Infection Control and Hospital Epidemiology* **25**, 231–239.

Nelms, P.K., Larson, O. & Barnes-Josiah, D. (1997) Time to *B. cereus* about hot chocolate. *Public Health Report* **112**, 240–244.

Newsom, S.W. (1968) Hospital infection from contaminated ice. *Lancet* **ii**, 620–622.

Nichols, G., Gillespie, I. & de Louvois, J. (2000) The microbiological quality of ice used to cool drinks and ready-to-eat food from retail and catering premises in the United Kingdom. *Journal of Food Protection* **63**, 78–82.

Nucci, M., Akiti, T., Barreiros, G. *et al.* (2002) Nosocomial outbreak of *Exophiala jeanselmei* fungemia associated with contamination of hospital water. *Clinical Infectious Diseases* **34**, 1475–1480.

Panwalker, A.P. & Fuhse, E. (1986) Nosocomial *Mycobacterium gordonae* pseudoinfection from contaminated ice machines. *Infection Control* **7**, 67–70.

Perola, O., Nousiainen, T., Suomalainen, S. *et al.* (2002) Recurrent *Sphingomonas paucimobilis* – bacteraemia associated with a multi-bacterial water-borne epidemic among neutropenic patients. *Journal of Hospital Infection* **50**, 196–201.

Petersen, L.R., Ammon, A., Hamouda, O. *et al.* (2000) Developing national epidemiologic capacity to meet the challenges of emerging infections in Germany. *Emerging Infectious Diseases* **6**, 576–584.

Rautelin, H., Koota, K., von Essen, R. *et al.* (1990) Waterborne *Campylobacter jejuni* epidemic in a Finnish hospital for rheumatic diseases. *Scandinavian Journal of Infectious Diseases* **22**, 321–326.

Ravn, P., Lundgren, J.D., Kjaeldaard, P. *et al.* (1991) Nosocomial outbreak of cryptosporidiosis in AIDS patients. *British Medical Journal* **302**, 277–280.

Robertson, P. (1985) The modern drinks vending machine – a link in the food poisoning chain? *Environmental Health* **94**, 281–285.

Royal Society for the Promotion of Health, in cooperation with the Water for Health Alliance (2003) *Water Provision in Care Homes For the Elderly.* Available from http://www.rsph.org/water/survey.asp. Accessed 2 March 2007.

Schillinger, J. & Du Vall Knorr, S. (2004) Drinking-water quality and issues associated with water vending machines in the city of Los Angeles. *Journal of Environmental Health* **66**, 25–31.

Schvoerer, E., Bonnet, F., Dubois, V. *et al.* (1999) A hospital outbreak of gastroenteritis possibly related to the contamination of tap water by a small round structured virus. *Journal of Hospital Infection* **43**, 149–154.

Souza, A.V., Moreira, C.R., Pasternak, J. *et al.* (2004) Characterizing uncommon *Burkholderia cepacia* complex isolates from an outbreak in a haemodialysis unit. *Journal of Medical Microbiology* **53**, 999–1005.

Stamm, W.E., Colella, J.J., Anderson, R.L. & Dixon, R.E. (1975) Indwelling arterial catheters as a source of nosocomial bacteremia. An outbreak caused by *Flavobacterium* species. *New England Journal of Medicine* **292**, 1099–1102.

Trautmann, M., Lepper, P.M. & Haller, M. (2005) *Ecology of Pseudomonas aeruginosa* in the intensive care unit and the evolving role of water outlets as a reservoir of the organism. *American Journal of Infection Control* **33**(5, Suppl 1), S41–S49.

van der Mee-Marquet, N., Bloc, D., Briand, L. *et al.* (2005) Non-touch fittings in hospitals: a procedure to eradicate *Pseudomonas aeruginosa* contamination. *Journal of Hospital Infection* **60**, 235–239.

Wang, S.H., Sheng, W.H., Chang, Y.Y. *et al.* (2003) Healthcare-associated outbreak due to pan-drug resistant *Acinetobacter baumannii* in a surgical intensive care unit. *Journal of Hospital Infection* **53**, 97–102.

Wilson, I.G., Hogg, G.M. & Barr, J.G. (1997) Microbiological quality of ice in hospital and community. *Journal of Hospital Infection* **36**, 171–180.

8 Practical Implementation of Food Safety Management Systems in Healthcare Settings

Jon-Mikel Woody and Dianne L. Benjamin

8.1 Active managerial control

8.1.1 *What is the goal of food service operations in healthcare settings and what is presently being done to achieve this goal?*

The goal of food service establishments operating in healthcare settings is to produce safe, quality food for consumers. Since the onset of regulatory oversight of food service operations, regulatory inspections have emphasized the recognition and correction of food safety violations that exist at the time of the inspection. Recurring violations have traditionally been handled through re-inspections or enforcement activities such as fines, suspension of permits or closures. Operators of food service establishments routinely respond to inspection findings by correcting violations, but often do not implement proactive systems of control to prevent violations from recurring. While this type of inspection and enforcement system has done a great deal to improve basic sanitation and to upgrade facilities, it has emphasized reactive rather than preventive measures to food safety. Additional measures must be taken on the part of operators and regulators to better prevent or reduce foodborne illness.

8.1.2 *Who has the ultimate responsibility for providing safe food to the consumer?*

The responsibility of providing safe food to the consumer is shared by many people in every stage in the production of food, including consumers, themselves. Food for patients in healthcare settings may be prepared within the hospital or institution or may be obtained from external suppliers. The responsibility for providing safe food is shared by staff responsible for contracts with suppliers, the suppliers themselves, as well as catering staff and other healthcare professionals, including nursing and voluntary staff.

8.1.3 *How can foodborne illness be reduced?*

The United States Centers for Disease Control and Prevention (CDC) Surveillance Report for 1993–1997 (CDC, 2000) identifies the most significant contributing factors to foodborne illness. Five of these broad categories of contributing factors are termed by the US Food and Drug Administration (FDA) as 'foodborne illness risk factors'. These five broad categories are:

- food from unsafe sources
- inadequate cooking
- improper holding temperatures
- contaminated equipment
- poor personal hygiene

In 1998, FDA initiated a project designed to determine the incidence of foodborne illness risk factors in retail and food service establishments. Inspections focusing on the occurrence of foodborne illness risk factors were conducted in establishments throughout the US (FDA, 2000). The data collection project was repeated in 2003 (FDA, 2004). An additional data collection project is planned for 2008.

These reports support the concept that operators of retail and food service establishments must be proactive and implement food safety management systems that will prevent, eliminate or reduce the occurrence of foodborne illness risk factors. By reducing the occurrence of foodborne illness risk factors, foodborne illness can also be reduced.

8.1.4 *How can the occurrence of foodborne illness risk factors be reduced?*

To effectively reduce the occurrence of foodborne illness risk factors, operators of food service establishments in healthcare settings must focus their efforts on achieving active managerial control. The term 'active managerial control' is used to describe the responsibility for developing and implementing food safety management systems to prevent, eliminate or reduce the occurrence of foodborne illness risk factors.

Active managerial control means the purposeful incorporation of specific actions or procedures by management into the operation of their business to attain control over foodborne illness risk factors (FDA, 2006). It embodies a preventive rather than reactive approach to food safety through a continuous system of monitoring and verification.

Many tools can be used by industry to provide active managerial control of foodborne illness risk factors. Regulatory inspections and follow-up activities must also be proactive by using an inspection process designed to assess the degree of active managerial control that food service operators in healthcare settings have over the foodborne illness risk factors.

In addition, regulators must assist operators in developing and implementing strategies to strengthen existing systems to prevent the occurrence of foodborne illness risk factors. Elements of an effective food safety management system may include the following (FDA, 2006):

- certified food protection managers who have shown a proficiency in required information by passing a test that is part of an accredited programme
- standard operating procedures (SOPs) for performing critical operational steps in a food preparation process, such as cooking or cooling
- standardized recipe cards that contain the specific steps for preparing a food item and the food safety critical limits, such as final cooking temperatures, that need to be monitored and verified
- purchase specifications
- equipment and facility design and maintenance
- monitoring procedures
- record keeping
- employee health policy for restricting or excluding ill employees
- manager and employee training
- on-going quality control and assurance
- specific goal-oriented plans, like risk control plans (RCPs), that outline procedures for controlling foodborne illness risk factors.

A food safety management system based on Hazard Analysis and Critical Control Point (HACCP) principles contains many of these elements and provides a comprehensive framework by which an operator can effectively control the occurrence of foodborne illness risk factors.

8.2 Introduction to HACCP

8.2.1 *What is HACCP and how can it be used by operators of food service establishments in healthcare settings?*

The HACCP system was developed originally in the US to provide microbiologically safe food for astronauts on space missions. HACCP principles provide a systematic approach to identifying, evaluating and controlling food safety hazards. The importance of HACCP principles in contributing to food safety has been recognized by food safety professionals since the 1960s (Tompkin, 1994; ILSI Europe, 1997; Stevenson & Bernard, 1999; Jouve, 2000). Food safety hazards are biological, chemical or physical agents that may cause a food to be unsafe for human consumption. Because a HACCP programme is designed to ensure that hazards are prevented, eliminated or reduced to an acceptable level before a food reaches the consumer, it embodies the preventive nature of 'active managerial control'.

Active managerial control through the use of HACCP principles is achieved by identifying the food safety hazards attributed to products, determining the necessary steps that will control the identified hazards, and implementing on-going practices or procedures to ensure safe food.

Like many other quality assurance programmes, HACCP provides a common-sense approach to identifying and controlling problems that are likely to exist in an operation. Consequently, many food safety management systems that exist in food service operations in healthcare settings already incorporate some, if not all, of the principles of HACCP. Combined with good basic sanitation, a solid employee training programme, and other pre-requisite programmes, a food safety management system based on HACCP principles will prevent, eliminate or reduce the occurrence of foodborne illness risk factors that lead to out-of-control hazards.

While the operator is responsible for developing and implementing a system of controls to prevent foodborne illness risk factors, the role of the regulator is to assess whether the system the operator has in place is achieving control of foodborne illness risk factors. Using HACCP principles during inspections will enhance the effectiveness of routine inspections by incorporating a risk-based approach. This helps inspectors focus their inspection on evaluating the effectiveness of food safety management systems to control foodborne illness risk factors.

8.2.2 *What are the seven HACCP principles?*

The HACCP approach was presented first at a food safety conference in 1971, and since then its use has been endorsed by governmental and trade bodies worldwide (Jouve, 2000). The European Union regulations require that food businesses put in place, implement and maintain procedures based on HACCP principles (Chapters 1 and 6). The seven basic principles were described by the Codex Alimentarius Commission and endorsed by the US National Advisory Committee on Microbiological Criteria for Foods (NACMCF) in 1992 and 1997 (NACMCF, 1992, 1997) and are as follows:

Principle 1: Conduct a hazard analysis
Principle 2: Determine the critical control points (CCPs)
Principle 3: Establish critical limits at CCPs
Principle 4: Establish monitoring procedures
Principle 5: Establish corrective actions
Principle 6: Establish verification procedures
Principle 7: Establish record-keeping and documentation procedures

This chapter will provide a brief overview of each of the seven principles of HACCP. A more comprehensive discussion of these principles was published by the NACMCF (1997). Following the overview, a practical scheme for applying and implementing the HACCP principles in healthcare settings is presented, based on the FDA Manual for the voluntary use of HACCP principles for operators of food service and retail establishments (FDA, 2006).

8.2.3 *What are prerequisite programmes?*

In order for a HACCP system to be effective, a strong foundation of procedures that address the basic operational and sanitation conditions within an operation must first be developed and implemented. These procedures are collectively termed 'prerequisite programmes'. When prerequisite programmes are in place, more attention can be given to controlling hazards associated with the food and its preparation. Prerequisite programmes may include such things as:

- buyer specifications
- vendor certification programmes
- premises and equipment that are well designed, equipped and maintained
- training programmes
- personal hygiene programmes
- allergen management
- recipe/process specifications
- first-in-first-out (FIFO) procedures (stock rotation)
- emergency preparedness and response
- other SOPs
- cleaning programmes
- pest control

Basic prerequisite programmes should be in place to:

- protect products from contamination by biological, chemical and physical food safety hazards
- control bacterial growth that can result from temperature abuse
- maintain equipment

Additional information about prerequisite programmes and the types of activities usually included in them can be found in the FDA's Retail HACCP manuals (FDA, 2006) or in NACMCF (1997).

8.3 The HACCP principles

8.3.1 *Principle 1: conduct a hazard analysis*

8.3.1.1 *What is a food safety hazard?*

A hazard is a biological, chemical or physical agent property that may cause a food to be unsafe for human consumption.

8.3.1.2 *What are biological hazards?*

Biological hazards include bacterial, viral and parasitic microorganisms (Chapter 2). Table 8.1 shows a listing of selected biological hazards. Bacterial pathogens are a major

Table 8.1 Selected biological hazards, associated foods and control measures.

Hazard	Associated foods	Control measures
Bacteria		
Bacillus cereus (intoxication caused by heat stable, preformed emetic toxin and infection by heat labile, diarrhoeal toxin)	Meat, poultry, starchy foods (rice, potatoes), puddings, soups, cooked vegetables	Cooking, cooling, cold holding, hot holding
Campylobacter jejuni	Poultry, raw milk	Cooking, handwashing, prevention of cross-contamination
Clostridium botulinum	Vacuum-packed foods, reduced oxygen packaged foods, under-processed canned foods, garlic-in-oil mixtures, time/temperature abused baked potatoes/sautéed onions	Thermal processing (time + pressure), cooling, cold holding, hot holding, acidification and drying, etc.
Clostridium perfringens	Cooked meat and poultry, cooked meat and poultry products including casseroles, gravies	Cooling, cold holding, reheating, hot holding
E. coli O157:H7 (other shiga toxin-producing *E. coli*)	Raw ground beef, raw seed sprouts, raw milk, unpasteurized juice, foods contaminated by infected food workers via faecal–oral route	Cooking, no bare hand contact with RTE foods, employee health policy, handwashing, prevention of cross-contamination, pasteurization or treatment of juice
Listeria monocytogenes	Raw meat and poultry, fresh soft cheese, paté, smoked seafood, deli meats, deli salads	Cooking, date marking, cold holding, handwashing, prevention of cross-contamination
Salmonella spp.	Meat and poultry, seafood, eggs, raw seed sprouts, raw vegetables, raw milk, unpasteurized juice	Cooking, use of pasteurized eggs, employee health policy, no bare hand contact with RTE foods, handwashing, pasteurization or treatment of juice
Shigella spp.	Raw vegetables and herbs, other foods contaminated by infected workers via faecal–oral route	Cooking, no bare hand contact with RTE foods, employee health policy, handwashing
Staphylococcus aureus (preformed heat stable toxin)	RTE PHF foods touched by bare hands after cooking and further time/temperature abused	Cooling, cold holding, hot holding, no bare hand contact with RTE food, handwashing
Vibrio spp.	Seafood, shellfish	Cooking, approved source, prevention of cross-contamination, cold holding
Parasites		
Anisakis simplex	Various fish (cod, haddock, fluke, pacific salmon, herring, flounder, monkfish)	Cooking, freezing
Taenia spp.	Beef and pork	Cooking
Trichinella spiralis	Pork, bear and seal meat	Cooking
Viruses		
Hepatitis A and E	Shellfish, any food contaminated by infected worker via faecal–oral route	Approved source, no bare hand contact with RTE food, minimizing bare hand contact with foods not RTE, employee health policy, handwashing
Other viruses (rotavirus, norovirus, reovirus)	Any food contaminated by infected worker via faecal–oral route	No bare hand contact with RTE food, minimizing bare hand contact with foods not RTE, employee health policy, handwashing

RTE, ready-to-eat; PHF, potentially hazardous food (time/temperature control for safety food).

cause of confirmed foodborne disease outbreaks and cases. Although thorough cooking destroys the vegetative cells of foodborne bacteria, spores of bacteria such as *Bacillus cereus*, *Clostridium botulinum* and *Clostridium perfringens* survive cooking and may germinate and grow if food is not properly cooled or held after cooking. The toxins produced by the vegetative cells of *Bacillus cereus*, *Clostridium botulinum* and *Staphylococcus aureus* may not be destroyed to safe levels by reheating. Post-cook recontamination with vegetative cells of bacteria such as *Salmonella* and *Campylobacter jejuni* is also a major concern for operators providing food in healthcare settings.

Viruses such as norovirus and hepatitis A, and rotavirus are directly related to contamination from human faeces, and norovirus may be transmitted via droplets in the air. In limited cases, foodborne viruses may occur in raw commodities contaminated by human faeces (e.g. shellfish harvested from unapproved, polluted waters and fruit and vegetables contaminated in the field or during harvesting [Chapter 2, Section 14]). Contamination of food by viruses may also result from contamination by food handlers or from unclean equipment and utensils. Unlike bacteria, a virus cannot multiply outside of a living cell. The relatively mild heat treatment used in cooking shellfish may not inactivate hepatitis A virus and norovirus (Chapter 2, Section 14). Obtaining food from approved sources, practising no bare-hand contact with ready-to-eat (RTE) food as well as proper handwashing, and implementing an employee health policy to restrict or exclude ill employees are important control measures for viruses.

The protozoa *Cryptosporidium hominis*, *C. parvum* and *Giardia duodenalis* have human and animal hosts, whereas present evidence indicates that humans may be the only source of *Cyclospora cayetanensis* (Chapter 2). Foodborne infection has been associated with consumption of raw fruits and vegetables contaminated with human or animal faecal material, and with contamination of foods by infected food handlers. Important control measures are obtaining foods from sources where contamination is prevented, implementing an employee health policy, avoiding bare-hand contact with RTE foods and handwashing.

The parasites *Anisakis simplex*, *Taenia* spp. and *Trichinella spiralis* are most often animal host-specific, but can include humans in their life cycles. Parasitic infections are commonly associated with undercooking meat products or cross-contamination of RTE food with raw animal foods, untreated water or contaminated equipment or utensils. Like viruses, parasites do not grow in food, so control is focused on destroying the parasites and/or preventing their introduction. Adequate cooking destroys parasites. In addition, parasites in fish to be consumed raw or undercooked can also be destroyed by effective freezing techniques. Parasitic contamination by ill employees can be prevented by proper handwashing, no bare-hand contact with RTE food and implementation of an employee health policy to restrict or exclude ill employees.

8.3.1.3 *What are chemical hazards?*

Chemical hazards may be naturally occurring or may be added during the processing of food. High levels of toxic chemicals may cause acute cases of foodborne illness, while chronic illness may result from low levels. Table 8.2 provides examples of chemical hazards.

Table 8.2 Common chemical hazards, associated foods and control measures.

Chemical hazards	Associated foods	Control measures
Naturally occurring:		
Scombrotoxin	Primarily associated with tuna fish, mahi-mahi, blue fish, anchovies bonito, mackerel; also found in cheese	Check temperatures at receiving; store at proper cold holding temperatures; buyer specifications: obtain verification from supplier that product has not been temperature abused prior to arrival in facility
Ciguatoxin	Reef fin fish from extreme SE US, Hawaii and tropical areas; barracuda, jacks, king mackerel, large groupers and snappers	Ensure fin fish have not been caught: • purchase fish from approved sources • fish should not be harvested from an area that is subject to an adverse advisory
Tetrodoxin	fish (Fugu; Blowfish)	Do not consume these fish
Mycotoxins		
Aflatoxin	Corn and corn products, peanuts and peanut products, cottonseed, milk and tree nuts such as Brazil nuts, pecans, pistachio nuts and walnuts. Other grains and nuts are susceptible but less prone to contamination	Check condition at receiving; do not use moldy or decomposed food
Patulin	Apple juice products	Buyer specification: obtain verification from supplier or avoid the use of rotten apples in juice manufacturing
Toxic mushroom species	Numerous varieties of wild mushrooms	Do not eat unknown varieties or mushrooms from unapproved source
Shellfish toxins		Ensure molluscan shellfish are:
Paralytic shellfish poisoning (PSP)	Molluscan shellfish from NE and NW coastal regions of North America (NA); mackerel, viscera of lobsters and Dungeness, tanner and red rock crabs	• from an approved source; and properly tagged and labelled
Diarrhoetic shellfish poisoning (DSP)	Molluscan shellfish in Japan, western Europe, Chile, NZ, eastern Canada	
Neurotoxin shellfish poisoning (NSP)	Molluscan shellfish from Gulf of Mexico	
Amnesic shellfish poisoning (ASP)	Molluscan shellfish from NE and NW coasts of NA; viscera of Dungeness, tanner, red rock crabs and anchovies	
Pyrrolizidine alkaloids	Plant food containing these alkaloids. Most commonly found in members of the Borginaceae, Compositae and Leguminosae families	Do not consume of food or medicinals contaminated with these alkaloids
Phytohaemmagglutinin	Raw red kidney beans (undercooked beans may be more toxic than raw beans)	Soak in water for at least 5 h. Pour away the water. Boil briskly in fresh water, with occasional stirring, for at least 10 min
Added chemicals:		
Environmental contaminants: pesticides, fungicides, fertilizers, insecticides, antibiotics, growth hormones	Any food may become contaminated	Follow label instructions for use of environmental chemicals. Soil or water analysis may be used to verify safety

Table 8.2 (*Continued*)

Chemical hazards	Associated foods	Control measures
PCBs	Fish	Comply with fish advisories
Prohibited substances (21 CFR 189)	Numerous substances are prohibited from use in human food; no substance may be used in human food unless it meets all applicable requirements of the FDC Act	Do not use chemical substances that are not approved for use in human food
Toxic elements/compounds Mercury	Fish exposed to organic mercury: shark, tilefish, king mackerel and swordfish	Pregnant women/women of childbearing age/nursing mothers, and young children should not eat shark, swordfish, king mackerel or tilefish because they contain high levels of mercury
	Grains treated with mercury-based fungicides	Do not use mercury containing fungicides on grains or animals
Copper	High acid foods and beverages	Do not store high acid foods in copper utensils; use backflow prevention device on beverage vending machines
Lead	High acid food and beverages	Do not use vessels containing lead
Preservatives and food additives: Sulphiting agents (sulphur dioxide, sodium and potassium bisulphite, sodium and potassium metabisulphite)	Fresh fruits and vegetables Shrimp Lobster Wine	Sulphiting agents added to a product in a processing plant must be declared on labelling Do not use on raw produce in food establishments
Nitrites/nitrates Niacin	Cured meats, fish, any food exposed to accidental contamination, spinach Meat and other foods to which sodium nicotinate is added	Do not use more than the prescribed amount of curing compound according to labelling instructions. Sodium nicotinate (niacin) is not currently approved for use in meat or poultry with or without nitrates or nitrates
Flavour enhancers Monosodium glutamate (MSG)	Asian or Latin American food	Avoid using excessive amounts
Chemicals used in retail establishments (e.g. lubricants, cleaners, sanitizers, cleaning compounds and paints)	Any food could become contaminated	Address through SOPs for proper labelling, storage, handling and use of chemicals; retain Material Safety Data Sheets for all chemicals
Allergens	Foods containing or contacted by: Milk Egg Fish Crustacean shellfish Tree nuts Wheat Peanuts Soybeans	Use a rigorous sanitation regime to prevent cross contact between allergenic and non-allergenic ingredients

8.3.1.4 *Food allergens as food safety hazards*

Recent studies indicate that over 11 million Americans suffer from one or more food allergies. A food allergy is caused by a naturally occurring protein, which is referred to as an 'allergen', in a food or a food ingredient.

The following foods can cause a serious allergic reaction in sensitive individuals; these foods account for 90% or more of all food allergies:

- milk
- egg
- fish (such as bass, flounder or cod)
- crustacean shellfish (such as crab, lobster or shrimp)
- tree nuts (such as almonds, pecans or walnuts)
- wheat
- peanuts
- soybeans

8.3.1.5 *What are physical hazards?*

Illness and injury can result from foreign objects in food. These physical hazards can result from contamination or poor procedures at many points in the food chain from harvest to consumer, including those within the food establishment.

8.3.1.6 *What is the purpose of the hazard analysis principle?*

The purpose of hazard analysis is to develop a list of food safety hazards that are reasonably likely to cause illness or injury if not effectively controlled.

8.3.1.7 *How is the hazard analysis conducted?*

The process of conducting a hazard analysis involves two stages:

1. hazard identification
2. hazard evaluation

Hazard identification focuses on identifying the food safety hazards that might be present in the food given the source of the food, the food preparation process used, the handling of the food, the facility and general characteristics of the food itself. During this stage, a review is made of the ingredients used in the product, the activities conducted at each step in the process, the equipment used, the final product, and its method of storage and distribution, as well as the intended use and consumers of the product. Based on this review, a list of potential biological, chemical or physical hazards is made at each stage in the food preparation process.

In stage 2, the hazard evaluation, each potential hazard is evaluated based on the severity of the potential hazard and its likely occurrence. The purpose of this stage is to determine which of the potential hazards listed in stage 1 of the hazard analysis warrant control in the HACCP plan. Severity is the seriousness of the consequences of exposure to the hazard. Considerations made when determining the severity of a hazard include understanding the impact on vulnerable groups in healthcare settings. Consideration of the likely occurrence is usually based upon a combination of experience, epidemiological data and information in the technical literature. Hazards that are not reasonably likely to occur are not considered in a HACCP plan. During the evaluation of each potential hazard, the food, its method of preparation, transportation, storage, and persons likely to consume the product should be considered, to determine how each of these factors may influence the likely occurrence and severity of the hazard being controlled.

Upon completion of the hazard analysis, a list of significant hazards that must be considered in the HACCP plan is made, along with any measure(s) that can be used to control the hazards. These measures, called control measures, are actions or activities that can be used to prevent, eliminate or reduce a hazard. Some control measures are not essential to food safety, while others are. Control measures essential to food safety, like proper cooking, cooling, and refrigeration of RTE, potentially hazardous foods (PHFs), are applied at CCPs in the HACCP plan (discussed later). The term control measure is used because not all hazards can be prevented, but virtually all can be controlled. More than one control measure may be required for a specific hazard. Likewise, more than one hazard may be addressed by a specific control measure (e.g. proper cooking).

The term 'potentially hazardous food' (time/temperature control for safety food, TCS food) means a food that requires time/temperature control to limit growth of pathogenic microorganisms or toxin production (FDA/CFSAN, 2005).

8.3.2 *Principle 2: determine critical control points*

8.3.2.1 *What is the critical control point?*

A CCP means a point or procedure in a specific food system where loss of control may result in an unacceptable health risk. Control must be applied at this point and is *essential* to prevent or eliminate a food safety hazard or reduce it to an acceptable level. Each CCP will have one or more control measures associated with it. Common examples of CCPs include cooking, cooling, hot holding and cold holding of RTE PHFs (i.e. foods that require time/temperature control for safety). Because pathogenic microorganisms may be associated with raw animal and plant foods, proper execution of control measures at each of these operational steps is essential to prevent or eliminate food safety hazards or reduce them to acceptable levels.

8.3.2.2 *Are quality issues considered when determining CCPs?*

CCPs are only used to address issues with product safety. Actions taken on the part of the establishment, such as FIFO, or in relation to foods that are unlikely to transmit pathogenic microorganisms or toxins, are to ensure food quality rather than food safety and therefore should not be considered as CCPs unless they serve a dual-purpose of ensuring food safety.

8.3.2.3 *Are the CCPs the same for everyone?*

Different facilities preparing similar food items may identify different hazards and the CCPs. This can be due to differences in each facility's layout, equipment, selection of ingredients and processes employed. In mandatory HACCP systems, there may be rigid regulatory requirements regarding what must be designated a CCP. In voluntary HACCP systems, hazard control may be accomplished at CCPs or through prerequisite programmes.

8.3.3 *Principle 3: establish critical limits*

8.3.3.1 *What is a critical limit and what is its purpose?*

A critical limit is a prescribed parameter (e.g. minimum and/or maximum value) that must be met to ensure that food safety hazards are controlled at each CCP. A critical limit is used to distinguish between safe and unsafe operating conditions at a CCP. Each control measure at a CCP has one or more associated critical limits. Critical limits may be based upon factors like temperature, time, moisture level, water activity (a_w) or pH. They must be scientifically-based and measurable.

8.3.3.2 *What are examples of critical limits?*

Examples of critical limits are the time/temperature parameters for cooking chicken (in the US, heating at an internal temperature of at least 165°F [74°C] for 15 s). In this case, the critical limit designates the *minimum* heat treatment required to eliminate food safety hazards or reduce them to an acceptable level. The critical limit for the acidification of sushi rice, a pH of $\leq$4.6, sets the *maximum* limit for pH necessary to control the growth of spore- and toxin-forming bacteria. Critical limits may be derived from regulatory standards such as the FDA Food Code (FDA/CFSAN, 2005), other applicable guidelines, performance standards or experimental results.

8.3.4 *Principle 4: establish monitoring procedures*

8.3.4.1 *What is the purpose of monitoring?*

Monitoring is the act of observing and making measurements to help determine if critical limits are being met and maintained. It is used to determine whether the critical limits that have been established for each CCP are being met.

8.3.4.2 *What are examples of monitoring activities?*

Examples of monitoring activities include visual observations and measurements of time, temperature, pH and water activity. If cooking chicken is determined to be a CCP in an operation, then monitoring the internal temperature of a select number of chicken pieces immediately following the cook step would be an example of a monitoring activity. The temperature of an oven or fryer and the time required to reach an internal temperature in the chicken of at least 165°F (74°C) at every position in the cooker could also be monitored.

8.3.4.3 *How is monitoring conducted?*

Typically, monitoring activities fall under two broad categories:

- measurements
- observations

Measurements usually involve time and temperature but also include other parameters such as pH. If an operation identifies the acidification of sushi rice as a CCP and the critical limit as the final pH of the product being $\leq$4.6, then the pH of the product would be measured to ensure that the critical limit is met.

Observations involve visual inspections to monitor the presence or absence of a food safety activity. If date marking is identified as a CCP for controlling *Listeria monocytogenes* in RTE deli meats, then the monitoring activity could involve making visual inspections of the date marking system to monitor the sell, consume or discard dates.

8.3.4.4 *How often is monitoring conducted?*

Monitoring can be performed on a continuous or intermittent basis. Continuous monitoring is always preferred when feasible as it provides the most complete information regarding the history of a product at a CCP. For example, the temperature and time for an institutional cook-chill operation can be recorded continuously on temperature recording charts.

If intermittent monitoring is used, the frequency of monitoring should be conducted often enough to make sure that the critical limits are being met.

8.3.4.5 *Who conducts monitoring?*

Individuals directly associated with the operation (e.g. the person in charge of the establishment, chefs and departmental supervisors) are often selected to monitor CCPs. They are usually in the best position to detect deviations and take corrective actions when necessary. These employees should be properly trained in the specific monitoring techniques and procedures used.

8.3.5 *Principle 5: establish corrective actions*

8.3.5.1 *What are corrective actions?*

Corrective actions are activities that are taken by a person whenever a critical limit is not met. Discarding food that may pose an unacceptable food safety risk to consumers is a corrective action. However, other corrective actions such as further cooking or reheating a product can be used provided food safety is not compromised. For example, a restaurant should continue cooking hamburgers that have not reached an internal temperature of 155°F (68.3°C) for 15 seconds until the proper temperature is met. Clear instructions should be developed detailing who is responsible for performing the corrective actions, the procedures to be followed, and when.

8.3.6 *Principle 6: establish verification procedures*

8.3.6.1 *What is verification?*

Verification includes those activities, other than monitoring, that determine the validity of the HACCP plan and show that the system is operating according to the plan. Clear instructions should be developed detailing who is responsible for conducting verification, the frequency of verification and the procedures used.

8.3.6.2 *What is the frequency of verification activities? What are some examples of verification activities?*

Verification activities are conducted frequently, such as daily, weekly, monthly and include the following:

- observing the person doing the monitoring and determining whether monitoring is being done as planned
- reviewing the monitoring records to determine if they are completed accurately and consistently
- determining whether the records show that the frequency of monitoring stated in the plan is being followed
- ensuring that corrective action was taken when the person monitoring found and recorded that the critical limit was not met
- confirming that all equipment, including equipment used for monitoring, is operated, maintained and calibrated properly

8.3.7 *Principle 7: establish record keeping procedures*

8.3.7.1 *Why are records important?*

Maintaining documentation of the activities in a food safety management system can be vital to its success. Records provide documentation that appropriate corrective actions were taken when critical limits were not met. In the event that an establishment is implicated in a foodborne illness, documentation of activities related to monitoring and corrective actions can provide proof that reasonable care was exercised in the operation of the establishment. Documenting activities provides a mechanism for verifying that the activities in the HACCP plan were properly completed. In many cases, records can serve a dual purpose of ensuring quality and food safety.

8.3.7.2 *What types of records are maintained as part of a food safety management system?*

There are at least five types of records that could be maintained to support a food safety management system:

- records documenting the activities related to the prerequisite programmes

- monitoring records
- corrective action records
- verification and validation records
- calibration records

8.3.8 *Conduct periodic validation*

Once the food safety management system is established it should be reviewed periodically to ensure that food safety hazards are controlled when the system is implemented properly. Changes in suppliers, products or preparation methods may require a re-validation of the food safety management system, a small change in supplier, product or preparation method may have a significant effect.

8.4 The process approach – a practical application of HACCP in healthcare settings to achieve active managerial control

8.4.1 *Why focus on HACCP principles in healthcare settings?*

There are important differences between using HACCP principles in a food safety management system developed for food manufacturing plants and applying these same principles in a food safety management system developed for use in foodservice establishments, including those in healthcare settings.

Since the 1980s, operators and regulators have been exploring the use of the HACCP principles in restaurants, grocery stores, institutional care facilities and other retail food establishments. During this time, much has been learned about how these principles can be used in these varied operations. Most of this exploration has centred around the focal question of how to stay true to HACCP and still make the principles useful to an industry that encompasses the broadest range of conditions.

Unlike industries such as canning, other food processing, and dairy plants, food service establishments in healthcare settings are not easily defined by specific commodities or conditions. Consider the following characteristics that food service establishments share that set them apart from most food processors:

1. Employee and management turnover is exceptionally high in food establishments, especially for entry level positions. This means the many employees or managers have little experience and food safety training must be provided continuously.
2. There is an almost endless number of production techniques, products, menu items and ingredients used which are not easily adapted to a simple, standardized approach. Changes occur frequently and little preparation time is available.

There is a tremendous amount of diversity in food service establishments and their varying in-house resources to implement HACCP. That recognition is combined with an understanding that the success of such implementation is dependent upon establishing realistic and useful food safety strategies that are customized to the operation.

8.4.2 *What is the process approach?*

When conducting the hazard analysis, food manufacturers are usually handling a limited number of products at a time and follow the flow of each product. By contrast, in food service operations, foods of all types are worked together to produce the final product. This makes a different approach to the hazard analysis necessary. Conducting the hazard analysis by using the food preparation processes common to a specific operation is often more efficient and useful for food service operators. This is called the 'process approach' to HACCP.

The process approach can best be described as dividing the many food flows in an establishment into broad categories based on activities or stages in the preparation of the food, then analysing the hazards and placing managerial controls on each grouping.

8.4.3 *What are the three food preparation processes most often used by food service establishments in healthcare settings and how are they determined?*

The flow of food in a food service establishment is the path that food follows from receiving through service or sale to the consumer. Several activities or stages make up the flow of food and are called operational steps. Examples of operational steps include receiving, storing, preparing, cooking, cooling, reheating, holding, assembling, packaging, serving and selling. The terminology used for operational steps may differ between food service and retail food store operations.

Most food items produced in a food service establishment can be categorized into one of three preparation processes based on the number of times the food passes through the temperature danger zone between 41°F (5°C) and 135°F (57.2°C):

- Process 1: food preparation with no cook step

  Example flow: Receive → Store → Prepare → Hold → Serve

  (other food flows are included in this process, but there is no cook step to destroy pathogens)
- Process 2: preparation for same day service

  Example flow: Receive → Store → Prepare → Cook → Hold → Serve

  (other food flows are included in this process, but there is only one trip through the temperature danger zone)
- Process 3: complex food preparation

  Example flow: Receive → Store → Prepare → Cook → Cool → Reheat → Hot Hold → Serve

  (other food flows are included in this process, but there are always two or more complete trips through the temperature danger zone)

A summary of the three food preparation processes in terms of number of times through the temperature danger zone can be depicted in a danger zone diagram. Although foods produced using process 1 may *enter* the danger zone, they do not pass all the way through it. Foods that go through the danger zone only once are classified as same day service, while foods that go through more than once are classified as complex food preparation.

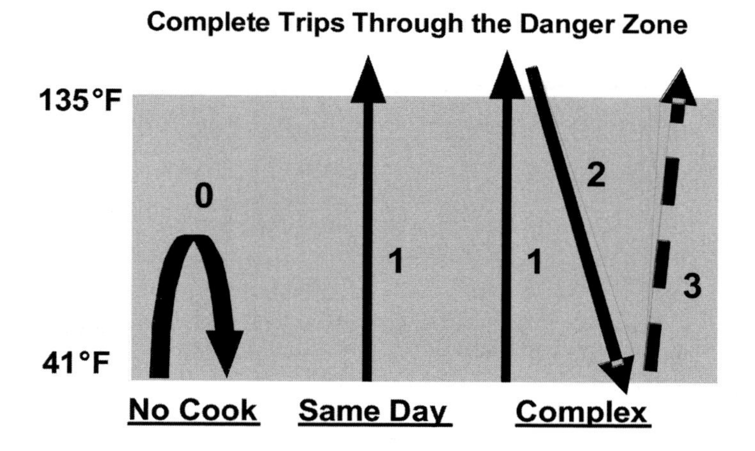

8.4.4 *How is a hazard analysis conducted in process HACCP?*

In the process approach to HACCP, conducting a hazard analysis on individual food items is time and labour intensive and is generally unnecessary. Identifying and controlling the hazards in each food preparation process achieves the same control of risk factors as preparing an HACCP plan for each individual product.

For example, an establishment has dozens of food items (including baked chicken and baked meatloaf) in the 'Preparation for Same Day Service' category. Each of the food items may have unique hazards, but regardless of the individual hazards, control via proper cooking and holding will generally ensure the safety of *all* of the foods in this category. An illustration of this concept follows:

- Even though they have unique hazards, baked chicken and meatloaf are items frequently grouped in the 'Same Day Service' category (process 2).
- *Salmonella* spp. and *Campylobacter*, as well as spore-formers, such as *Bacillus cereus* and *Clostridium perfringens*, are significant biological hazards in chicken.
- Significant biological hazards in meatloaf include *Salmonella* spp., *E. coli* O157:H7, *Bacillus cereus* and *Clostridium perfringens.*
- Despite their different hazards, the control measure used to kill vegetative pathogens in both these products is cooking to the proper temperature for the required time.
- Additionally, if the products are held after cooking, then proper hot holding or (time control) limitation of the holding time is also required to prevent the outgrowth of spore-formers that are not destroyed by cooking.

As with product-specific HACCP, critical limits for cooking remain specific to each food item in the process. In the scenario described above, the FDA specifies that the cooking step

Table 8.3 Examples of hazards and control measures for same day service items.

Process 2: preparation for same day service		
Example products	Baked meatloaf	Baked chicken
Example biological hazards	*Salmonella* spp. *E. coli* O157:H7 *Clostridium perfringens* *Bacillus cereus* Various faecal–oral route pathogens	*Salmonella* spp. *Campylobacter* *Clostridium perfringens* *Bacillus cereus* Various faecal–oral route pathogens
Example control measures	Refrigeration at 41°F (5°C) or below	Refrigeration at 41°F (5°C) or below
	Cooking at a food internal temperature of 155°F (68°C) for 15 s	Cooking at a food internal temperature of 165°F (74°C) for 15 s
	Hot holding at 135°F (57°C) or above or time control	Hot holding at 135°F (57°C) or above or time control
	Good personal hygiene (no bare hand contact with RTE food, proper handwashing, exclusion/restriction of ill employees)	Good personal hygiene (no bare hand contact with RTE food, proper handwashing, exclusion/restriction of ill employees)

RTE, ready-to-eat food.

for chicken requires a final internal temperature of 165°F (74°C) for 15 seconds to control the pathogen load for *Salmonella* spp. (FDA/CFSAN, 2005). Meatloaf, on the other hand, is a ground beef product and requires a final internal temperature of 155°F (68°C) for 15 seconds to control the pathogen load for both *Salmonella* spp. and *E. coli* O157:H7. Some operational steps such as refrigerated storage or hot holding have critical limits that apply to all foods.

Table 8.3 further illustrates this concept. Note that the only unique control measure applies to the critical limit of the cooking step for each of the products. Other food safety hazards and control measures may exist that are not depicted here.

Tables 8.4–8.6, at the end of the chapter, show sample HACCP tables for processes 1–3.

8.4.5 *How is the process approach helpful in determining the measures that must be implemented to manage the foodborne illness risk factors?*

Even though variations in foods and in the three food preparation process flows used to prepare them are common, the control measures will generally be the same based on the number of times the food goes through the temperature danger zone. Several of the most common control measures associated with each food preparation process are discussed in this chapter. Retail or food service establishments should use these simple control measures as the core of their food safety management systems; however, there may be other risk factors unique to an operation or process that are not listed here. Each operation should be evaluated independently.

In developing a food safety management system, active managerial control of risk factors common to each process can be achieved by implementing control measures at CCPs or by implementing prerequisite programmes. This is explained in more detail in the Operator's Manual (FDA, 2006).

8.4.5.1 *Facility-wide considerations*

In order to have active managerial control over personal hygiene and cross-contamination, certain control measures must be implemented in all phases of the operation. All of the following control measures should be implemented regardless of the food preparation process used:

- *No bare-hand contact with RTE foods (or use of a pre-approved, alternative procedure)* to help prevent the transfer of viruses, bacteria or parasites from hands to food.
- *Proper handwashing* to help prevent the transfer of viruses, bacteria or parasites from hands to food.
- *Restriction or exclusion of ill employees* to help prevent the transfer of viruses, bacteria or parasites from hands to food. Does the facility have a management policy that does not allow ill food workers to prepare food? Conditions for restriction or exclusion of workers are given in subparagraph 2-201.12 of the FDA Food Code 2005. The requirements for employees working in an establishment serving a highly susceptible population (meaning immunocompromised people, preschool children, older adults and people in facilities such as hospitals, nursing homes and senior centres) are more stringent than for employees not working in such establishments.
- *Prevention of cross-contamination* of RTE food or clean and sanitized food-contact surfaces, with soiled cutting boards, utensils, aprons, etc., or raw animal foods.

8.4.5.2 *Food preparation process 1 – food preparation with no cook step*

Example flow : Receive → Store → Prepare → Hold → Serve

Several food flows are represented by this particular process. Many of these food flows are common to both retail food stores and food service facilities, while others only apply to retail operations. Raw, RTE food like sashimi, raw oysters and salads are grouped in this category. Components of these foods are received raw and will not be cooked before consumption. It should be noted that raw seafood including sashimi and oysters should not be used in settings that serve a highly susceptible population (FDA/CFSAN, 2005).

Foods cooked at the processing level but that undergo no further cooking before being consumed are also represented in this category. Examples of these kinds of foods are deli meats, cheeses and other pasteurized dairy products (such as yogurt).

All the foods in this category lack a cook step *while at the food service facility*; thus, there are no complete trips through the danger zone. It is particularly important that such foods are obtained from an approved supplier. Purchase specifications should be required by the healthcare establishment to ensure that foods received are as safe as possible. Without a kill step to destroy pathogens, preventing further contamination by ensuring that employees follow good hygienic practices is an important control measure.

Cross-contamination must be prevented by properly storing RTE food away from raw animal foods and soiled equipment and utensils. Foodborne illness may result from RTE food being held at unsafe temperatures for long periods of time, allowing growth of bacteria.

In addition to the facility-wide considerations, a food safety management system involving this food preparation process should focus on ensuring active managerial control over the following:

- *Cold holding or using time alone* to control bacterial growth and toxin production.
- *Food source* (e.g. for shellfish due to concerns with viruses, natural toxins and *Vibrio* and for certain marine finfish intended for raw consumption due to concerns with ciguatera toxin, for eggs, salad vegetables, berry fruits).
- *Receiving temperatures* (e.g. certain species of marine finfish due to concerns with scombrotoxin).
- *Date marking* of RTE PHF held for more than 24 hours, to control the growth of psychrophiles such as *Listeria monocytogenes*.
- *Freezing* certain species of fish intended for raw consumption, due to parasite concerns; note that the Food Code (FDA/CFSAN, 2005) states that raw fish may not be served in a food establishment that serves a highly susceptible population.
- *Cooling* from ambient temperature to prevent the growth of foodborne pathogenic bacteria.

8.4.5.3 *Food preparation process 2 – preparation for same day service*

Example flow: Receive →Store →Prepare →Cook →Hold →Serve

In this food preparation process, food passes through the danger zone only once in the retail or food service facility before it is served or sold to the consumer. Food is usually cooked and held hot until served, e.g. fried chicken, but can also be cooked and served immediately. In addition to the facility-wide considerations, a food safety management system involving this food preparation process should focus on ensuring active managerial control over the following:

- *Cooking* to destroy non-sporing bacteria and parasites.
- *Hot holding or (using time alone) limiting holding time* to prevent the outgrowth of spore-forming bacteria.

Approved food source, proper receiving temperatures and proper cold holding before cooking would also be important, particularly if dealing with certain marine finfish because of concerns with ciguatera toxin and scombrotoxin.

8.4.5.4 *Food preparation process 3 – complex food preparation*

Example flow: Receive →Store →Prepare →Cook →Cool →Reheat →Hot Hold →Serve

Foods prepared in large volumes or in advance for next day service usually follow an extended process flow. These foods pass through the temperature danger zone more than one time; thus, the potential for the growth of spore-forming or toxigenic bacteria is greater in this process. Failure to adequately control food product temperatures is one of the most frequently encountered risk factors contributing to foodborne illness. Food handlers should minimize the time foods are at unsafe temperatures.

In addition to the facility-wide considerations, a food safety management system involving this food preparation process should focus on ensuring active managerial control over the following:

- *Cooking* to destroy non-sporing bacteria, viruses and parasites.
- *Cooling* to prevent the outgrowth of spore-forming or toxin-forming bacteria (Chapter 2, Section 5).
- *Hot and cold holding or using time and temperature* to control bacterial growth and toxin formation.
- *Date marking* of RTE PHF (TCS food) held for more than 24 hours, to control the growth of psychrophiles such as *Listeria monocytogenes*.
- *Reheating* for hot holding, if applicable.
- *Reduced Oxygen Packaging* Food Processing Criteria for Reduced Oxygen Packaging are given in Annexe 6 of the FDA Food Code 2005.

Approved food source, proper receiving temperatures and proper cold holding before cooking are also be important, particularly if dealing with certain marine finfish due to concerns with ciguatera toxin and scombrotoxin.

8.5 Advantages of using the principles of HACCP

8.5.1 *What advantages does using HACCP principles offer operators providing food in healthcare settings?*

Rather than relying solely on periodic feedback from inspections by regulatory agencies, an establishment operator who implements a food safety management system based on HACCP principles emphasizes continuous problem solving and prevention. Additionally, HACCP enhances and encourages communication between food providers and regulators.

A food safety management system based on HACCP principles offers many other advantages to food providers. One advantage is that such a system may provide a method for achieving active managerial control of multiple risk factors associated with an entire operation. Other advantages include:

- reduction in product loss
- increase in product quality
- better inventory control
- consistency in product preparation
- increase in cost-effectiveness
- increased employee awareness and participation in food safety

8.5.2 *What advantages does using HACCP principles offer regulators of food service establishments?*

Traditional inspections are relatively resource-intensive, inefficient and reactive rather than preventive in nature. Using traditional inspection techniques allows for a satisfactory 'snapshot' assessment of the requirements of the code at the time of the inspection. Unfortunately, unless an inspector asks questions and enquires about the activities and procedures being utilized by the establishment even at times when the inspector is not there, there is no way to know if an operator is achieving *active* managerial control.

With the limited time often available for conducting inspections, regulators must focus their attention on those areas that clearly have the greatest impact on food safety – foodborne illness risk factors. By knowing that there are only a few control measures that are essential to food safety and focusing on these during the inspection, an inspector can assess the operator's active managerial control of the foodborne illness risk factors.

Regulators can provide invaluable feedback to an operator through their routine inspections. This is especially useful when utilizing a risk-based approach. By incorporating HACCP principles into routine inspections, an inspector can provide an operator with the constructive input needed to establish the control system necessary to bring the foodborne illness risk factors back under continuous control.

8.6 Summary

In order to prevent foodborne illness, those persons responsible for the provision of food in healthcare settings must achieve active managerial control of the risk factors contributing to foodborne illness. Combined with basic sanitation, employee training and other prerequisite programmes, the principles of HACCP provide an effective system for achieving this objective.

The goal in applying HACCP principles in healthcare settings is to have staff at all levels take purposeful actions to ensure safe food. The process approach simplifies HACCP principles for use in retail and food service. This practical and effective method of hazard control embodies the concept of active managerial control by providing an on-going system of simple control measures that will reduce the occurrence of risk factors that lead to out-of-control hazards.

The role of regulatory professionals is to conduct risk-based inspections using HACCP principles to assess the degree of control food providers have over the foodborne illness risk factors. Regulators can assist food providers in achieving active managerial control of risk factors by using a risk-based inspection approach to identify strengths and weaknesses and suggesting possible solutions and improvements.

Acknowledgements

The authors wish to thank the many federal, state and local regulatory officials, as well as industry colleagues, who were instrumental in the development of the concepts contained in this chapter.

Sample HACCP Tables

Table 8.4a Process #1 – Food Preparation with No Cook Step

FOOD/MENU ITEMS:						
HAZARD(S)	CRITICAL CONTROL POINTS (List Only the Operational Steps that are CCPs)	CRITICAL LIMITS	MONITORING	CORRECTIVE ACTIONS	VERIFICATION	RECORDS
PREREQUISITE PROGRAMS						

Table 8.4b Process #1 – Food Preparation with No Cook Step

MENU ITEMS/PRODUCTS:							
PROCESSSTEP	**HAZARD(S)**	**CCP(Y/N)**	**CRITICAL LIMITS**	**MONITORING**	**CORRECTIVE ACTIONS**	**VERIFICATION**	**RECORDS**
RECEIVE							
STORE							
PREPARE							
HOLD							
SERVE							
Prerequisite Programs							

Table 8.5a Process #2 – Preparation for Same Day Service

FOOD/MENU ITEMS:						
HAZARD(S)	**CRITICAL CONTROL POINTS (List Only the Operational Steps that are CCPs)**	**CRITICAL LIMITS**	**MONITORING**	**CORRECTIVE ACTIONS**	**VERIFICATION**	**RECORDS**
PREREQUISITE PROGRAMS						

Table 8.5b Process #2 – Preparation for Same Day Service

MENU ITEMS/PRODUCTS:							
PROCESSSTEP	HAZARD(S)	CCP(Y/N)	CRITICAL LIMITS	MONITORING	CORRECTIVE ACTIONS	VERIFICATION	RECORDS
RECEIVE							
STORE							
PREPARE							
COOK							
HOLD							
SERVE							
Prerequisite Programs							

Table 8.6a Process #3 – Complex Food Preparation

FOOD/MENU ITEMS:						
HAZARD(S)	CRITICAL CONTROL POINTS (List Only the Operational Steps that are CCPs)	CRITICAL LIMITS	MONITORING	CORRECTIVE ACTIONS	VERIFICATION	RECORDS
PREREQUISITE PROGRAMS						

Table 8.6b Process #3 – Complex Food Preparation

MENU ITEMS/PRODUCTS:							
PROCESSSTEP	**HAZARD(S)**	**CCP(Y/N)**	**CRITICAL LIMITS**	**MONITORING**	**CORRECTIVE ACTIONS**	**VERIFICATION**	**RECORDS**
RECEIVE							
STORE							
PREPARE							
COOK							
COOL							
REHEAT							
HOLD							
SERVE							
Prerequisite Programs							

References

Centers for Disease Control and Prevention (CDC) (2000) Surveillance for foodborne disease outbreaks – United States, 1993–1997. *Morbidity and Mortality Weekly Report* **49**(No. SS-1), 1–51.

FDA (2000) *Report of the FDA Retail Food Program Database of Foodborne Illness Risk Factors*. FDA Baseline Report. Available from http://www.cfsan.fda.gov/~dms/retrsk.html. Accessed 2 March 2007.

FDA (2004)*Report on the Occurrence of Foodborne Illness Risk Factors in Selected Institutional Foodservice, Restaurant, and Retail Food Store Facility Types*. Available from http://www.cfsan.fda.gov/~dms/retrsk2.html. Accessed 2 March 2007.

FDA (2006) *Managing Food Safety: A Manual for the Voluntary Use of HACCP Principles for Operators of Food Service and Retail Establishments*. Available from http://www.cfsan.fda.gov/~acrobat/hret2.pdf. Accessed 2 March 2007.

FDA/CFSAN (2005) *Food Code*. Available from http://www.cfsan.fda.gov. Accessed 2 March 2007.

ILSI Europe (1997) *A Simple Guide to Understanding and Applying the Hazard Analysis Critical Control Point Concept*, 2nd edn. ILSI Europe Concise Monograph Series. ILSI Europe, Brussels.

Jouve, J.-L. (2000) Good manufacturing practice, HACCP, and quality systems. In: Lund, B.M., Baird-Parker, T.C. & Gould, G.W. (eds). *The Microbiological Safety and Quality of Food*. Aspen Publishers, Gaithersburg, MD, pp. 1627–1655.

National Advisory Committee on Microbiological Criteria for Food (NACMCF) (1992) Hazard Analysis and Critical Control Point System. *International Journal of Food Microbiology* **16**, 1–23.

National Advisory Committee on Microbiological Criteria for foods (NACMCF) (1997) *Hazard Analysis and Critical Control Point Principles and Application Guidelines*. Available from http://www.cfsan.fda.gov/~comm/nacmcfp.html. Accessed 2 March 2007.

Stevenson, K.E. & Bernard, D.T. (1999) *HACCP: A Systematic Approach to Food Safety*, 3rd edn. Food Products Association, Washington, DC.

Tompkin, R.B. (1994) HACCP in the meat and poultry industry. *Food Control* **5**, 153–161.

Index